S0-BLI-380

PROCESSES IN PHOTOREACTIVE POLYMERS

PROCESSES IN PHOTOREACTIVE POLYMERS

EDITED BY

V.V. KRONGAUZ

DSM Desotech, Inc.

Elgin, IL

A.D. TRIFUNAC

Argonne National Laboratory

Argonne, IL

CHAPMAN & HALL

I(T)P An International Thomson Publishing Company

New York • Albany • Bonn • Boston • Cincinnati • Detroit • London • Madrid • Melbourne • Mexico City • Pacific Grove • Paris • San Francisco • Singapore • Tokyo • Toronto • Washington

Cover design: Saeed Sayrafiezadeh, emDASH inc.
Illustration—*Point de vue du Gras*—used with permission of The University of Texas at Austin, the Gernsheim Collection, Harry Ransom Humanities Research Center, Austin, Texas.

Printed in the United States of America

For more information, contact:

Chapman & Hall
One Penn Plaza
New York, NY 10119

International Thomson Publishing
Berkshire House 168-173
High Holborn
London WC1V 7AA
England

Thomas Nelson Australia
102 Dodds Street
South Merlbourne, 3205
Victoria, Australia

Nelson Canada
1120 Birchmount Road
Scarborough, Ontario
Canada, M1K 5G4

Chapman & Hall
2-6 Boundary Row
London SE1 8HN

International Thomson Editores
Campos Eliseos 385, Piso 7
Col. Polanco
11560 Mexico D.F. Mexico

International Thomson Publishing Gmbh
Königwinterer Strasse 418
53228 Bonn
Germany

International Thomson Publishing Asia
221 Henderson Road
#05-10 Henderson Building
Singapore 0315

International Thomson Publishing-Japan
Hirakawacho-cho Kyowa Building, 3F
1-2-1 Hirakawacho-cho
Chiyoda-ku, 102 Tokyo
Japan

1 2 3 4 5 6 7 8 9 10 XXX 01 00 99 97 96 95

Library of Congress Cataloging-in-Publication Data

Processes in photoreactive polymers / editors, V.V. Krongauz and A.D. Trifunac
p. cm.
Includes bibliographical references.
ISBN 0-412-98401-6
1. Photopolymers. I. Krongauz, V.V. (Vadim V.), 1953- II. Trifunac, A.D. (Alexander D.), 1944-
QD382.P45P76 1994
661.8'08--dc20 94-3136
CIP

Please send your order for this or any Chapman & Hall book to **Chapman & Hall, 29 West 35th Street, New York, NY 10001, Attn: Customer Service Department.** You may also call our Order Department at 1-212-244-3336 or fax your purchase order to 1-800-248-4724.

For a complete listing of Chapman & Hall's titles, send your requests to **Chapman & Hall, Dept. BC, One Penn Plaza, New York, NY 10119.**

Contents

Preface vii
Contributors ix

Section I. ***Photoreactive Polymers: History and State of the Art***

1. Asphalt as the World's First Photopolymer—Revisiting the Invention of Photography 3
Jean-Louis Marignier
2. Recent Developments in Radiation Curing Chemistry 34
Christian Decker

Section II. ***Processes in Photoreactive Polymers***

1. Time Resolved Laser Spectroscopy of Excited State Processes in Photoimaging Systems 59
Jean-Pierre Fouassier
2. Hexaarylbiimidazolyl: The Choice Initiator for Photopolymers. History of Synthesis and Investigation of Photochemical Properties 90
Koko Maeda
3. Visible Light Photoinitiation Systems Based on Electron and Energy Transfer Processes 111
Tsuguo Yamaoka and Kazuhiko Naitoh
4. Investigation of Electron Transfer Between Hexaarylbiimidazole and a Visible Sensitizer 170
Yi Lin, Andong Liu, Alexander D. Trifunac, and Vadim V. Krongauz
5. Diffusion in Polymer Matrix and Anisotropic Photopolymerization 185
Vadim V. Krongauz

6. Benzoin Ether Photoinitiators Bound to Acrylated Prepolymers 260
Kwang-Duk Ahn
7. Photosensitive Liquid Crystalline Polymers 278
Carolyn Bowry-Devereaux

Section III. *Photoreactive Polymers in Advanced Applications*

1. Holographic Recording Materials 307
Roger A. Lessard, Rupak Changkakoti, and Gurusamy Manivannan
2. Photoresists and Their Development 368
Nigel P. Hacker

Index 405

Preface

The development of photosensitive materials in general and photoreactive polymers in particular is responsible for major advances in the information, imaging, and electronic industries. Computer parts manufacturing, information storage, and book and magazine publishing all depend on photoreactive polymer systems. The photo- and radiation-induced processes in polymers are also active areas of research. New information on the preparation and properties of commercially available photosensitive systems is constantly being acquired. The recent demand for environmentally safe solvent-free and water-soluble materials also motivated changes in the composition of photopolymers and photoresists. The interest in holographic recording media for head-up displays, light scanners, and data recording stimulated development of reconfigurable and visible light sensitive materials. Photoconductive polymerizable coatings are being tested in electrostatic proofing and color printing. The list of available initiators, polymeric binders, and other coating ingredients is continually evolving to respond to the requirements of low component loss (low diffusivity) and the high rate of photochemical reactions.

Owing to the fast expansion of the field, many aspects of photopolymer applications and behavior remain without a proper commentary in the older monographs and publications describing photopolymeric systems. Outdated information and erroneous assumptions about the photopolymer properties, on one hand, proliferate unrealistic expectations about the rates and yields of polymerization that are achievable and, on the other hand, result in the underestimation of the versatility of photopolymer systems. The goal of this book is to illustrate the complexity of photopolymer chemistry and the diversity of applications of these materials.

The reader will see that research and development opportunities abound in the fields of photochemistry and polymer chemistry. We hope to emphasize that polypolymer chemistry is still far from being "mature" and future discoveries may still yield, as in the past, spectacular returns on the investment in the research.

The mechanistic description of processes in photosensitive and photoreactive polymeric systems is presented in this monograph. Visible and infrared sensitive polymeric materials, diffusion control of the image quality, liquid crystal polymeric imaging materials, polymer backbone initiators, high activity initiators, and new holographic polymers are described. The mechanism of photoinitiation by electron transfer and photopolymerization mechanisms in films are discussed in some detail. The history of the field and the state of the art in the field of the photo- and radiation-curable polymers are presented. The discussion of science and technology of the resists and holographic materials will help the reader to understand the directions in the development of the electronic and imaging systems. The concepts that have not yet found wide acceptance received particular attention.

We compiled this book as a reference material for scientists who are working in fields associated with photoreactive polymer chemistry, and for students in the fields of chemistry, chemical engineering, imaging, and materials science. This book should be useful to those working on the development of new photosensitive materials, and the extensive references will help the readers to further their knowledge of photosensitive polymers.

We are grateful to all the contributing authors who devoted their time and energy to writing the reviews for this monograph. We would like to thank Ms. Laura Bowers for typing and for assistance in correcting and assembling the manuscripts. We would also like to express our appreciation to our families, friends, and co-workers for their support and encouragement.

V.V.K. and A.D.T.
September 1994

Contributors

Kwang-Duk Ahn, Korea Institute of Science and Technology, Polymer Chemistry Laboratory, P.O. Box 131, Cheongryang, Seoul 130-650, KOREA

Carolyn Bowry-Devereaux, GEC-Marconi Ltd., Hirst Research Centre, East Lane, Wembley, Middlesex HA9 7PP, UNITED KINGDOM

Rupak Changkakoti, Centre d'Optique, Photonique et Laser (COPL), Faculte des Sciences et Genie, Pavillon Vachon, Universite Laval, Quebec, CANADA G1K 7P4

Christian Decker, Universite de Haute-Alsace, Ecole National Supérieure de Chimie de Mulhouse, Laboratoire de Photochimie Générale, Unité de Recherche Associée au CNRS #431, 3, rue Alfred Werner 68093 Mulhouse, FRANCE

Jean-Pierre Fouassier, Universite de Haute-Alsace, Ecole National Supérieure de Chimie de Mulhouse, Laboratoire de Photochimie Générale, Unité de Recherche Associée au CNRS #431, 3, rue Alfred Werner 68093 Mulhouse, FRANCE

Nigel P. Hacker, IBM Research Division, Almaden Research Center, 650 Harry Road, San Jose, CA 95120-6099, U.S.A.

Vadim V. Krongauz, DSM Desotech, Inc., Fiber Optic Materials, 1122 St. Charles Street, Elgin, IL 60120, U.S.A.

Roger A. Lessard, Centre d'Optique, Photonique et Laser (COPL), Faculte des Sciences et Genie, Pavillon Vachon, Universite Laval, Quebec, CANADA G1K 7P4

Yi Lin, Chemistry Division, Argonne National Laboratory, Argonne, IL 60439, U.S.A.

Andong Liu, Chemistry Division, Argonne National Laboratory, Argonne, IL 60439, U.S.A.

Koko Maeda, Ochanomizu University, Department of Chemistry, Faculty of Science, Bunkyoku, Tokyo 112, JAPAN

Gurusamy Manivannan, Centre d'Optique, Photonique et Laser (COPL), Faculte des Sciences et Genie, Pavillon Vachon, Universite Laval, Quebec, CANADA G1K 7P4

Jean-Louis Marignier, Laboratorie de Physico-Chimie des Rayonnements, Batiment 350, 91405 Orsay, Cedex, FRANCE

Kazuhiko Naitoh, Chiba University, Department of Image Science, Faculty of Engineering, Yayoi-Cho 1-33, Inage-ku Chiba, 263, JAPAN

Alexander D. Trifunac, Chemistry Division, Argonne National Laboratory, Argonne, IL 60439, U.S.A.

Tsuguo Yamaoka, Chiba University, Department of Image Science, Faculty of Engineering, Yayoi-Cho 1-33, Inage-ku Chiba, 263, JAPAN

Processes in Photoreactive Polymers

I

Photoreactive Polymers: History and State of the Art

1

Asphalt as the World's First Photopolymer—Revisiting the Invention of Photography

Jean-Louis Marignier

The announcement of the invention of the daguerreotype was proclaimed by François Dominique Arago on August 19, 1839 at the Academy of Sciences in Paris,[1] but the true invention of photography occurred somewhat earlier.

In fact, 10 years before, in 1829, at Chalon-sur-Saône, in the French province of Burgundy, a Frenchman named Joseph Nicéphore Niépce wrote a text entitled *Notice about Heliography*, which started with these words: *"The discovery I made, that I call Heliography, consists in the spontaneous reproduction, by the action of light, with the graduations of hue from black to white, of the images received in the camera obscura."* [2]

He then explained: *"Light in its state of composition and of decomposition, acts chemically on substances. It is absorbed, it combines with them and confers new properties on them. Thus, it increases the natural consistency of some of them; it even solidifies them and makes them more or less insoluble, according to the duration or the intensity of its action. This is in short the principle of my discovery."* [2]

In the course of his communication in 1839, Arago had rightly enough mentioned the work of Niépce, but the description he gave was very erroneous and was followed by a list of unjustified criticisms. The audience did not pay any attention to Niépce's process, which was immediately forgotten. So it is now the habit to date the birth of photography as 1839 and to attribute the authorship to Daguerre. Moreover, it is now known that the first photographic process was not a silver process, but was based on a photopolymer principle.

It was in 1824 that Niépce succeeded, for the first time in the world, in fixing on a support an image received in a camera obscura. He was then 59 years old and it was the result of nearly 20 years of research. He then improved his process, which attained its most perfect form in 1828. Only one image made by Niépce, a photograph, has come to us until now. It is a view taken from the window of his house at St. Loup de Varennes, near Chalon-sur-Saône. This image is now in the possession of the University of Texas at Austin.[3] The know-how of Niépce's process had never been transmitted, so that it was completely unknown until the work we did in 1989.[4,5] For revisiting the process of Niépce's work more closely we referred exclusively to the original manuscripts; i.e., the *Notice* and the letters written by the inventor and his correspondents.[6–9]

1. The Prehistory of Photography

It seems that the idea of photography had never been proposed before the end of the eighteenth century. Curiously, at a moment when the question of the nature of light was a most burning one and when some scientists started to be interested in studying the effect of light on different compounds, no one had expressed the idea of photography.

The first experiments directly related to photography were due to an Englishman, Thomas Wedgwood, the son of the founder of the famous factory of glazed earthenware. At the end of the eighteenth century, he used white paper or white leather moistened with a solution of silver nitrate, a compound that darkened under the light. He then placed flat objects such as leaves, flowers or paintings on glass, in contact with the paper, and exposed the paper to daylight. It darkened except where it had been protected from the light. When the objects were removed from the paper, their silhouettes remained on it. But these images were not stable because the silver nitrate that had not been transformed during the exposure continued to darken when people looked at the image under light. Wedgwood also tried to put his paper in a camera obscura but did not succeed in producing any image. He declared: *"The images formed by means of a camera obscura, have been found to be too faint to produce, in any moderate time, an effect upon the nitrate of silver."*[10]

2. Nicéphore Niépce

Nicéphore Niépce was born in 1765 in Chalon-sur-Saône. After studying physics and chemistry with the Oratorian friars, he joined the army during the French revolution and left it in 1794. He then returned to his home town where he was occupied until the end of his life in managing the family inheritance constituted of numerous estates. With Claude, his older brother by two years, he started some research on a new principle for an engine, and they invented the first internal combustion engine for which in 1807 they received a patent for 10 years. The explosion was achieved first with lycopodium powder, then with a mixture of coal and resins, and finally with oil of white petroleum; i.e., some kind of fuel. This invention never interested anyone—at that time only water vapor was considered a modern energy principle. In 1816, one year before the completion date of the patent rights, Claude left Chalon-sur-Saône for Paris and then for England to sell the invention of their engine. From this time on the two brothers communicated through an abundant and regular mail, and it is because of these numerous letters that the work presented here has been possible.

3. Niépce's First Research Toward the Invention of Photography

3.1. The First Negative Photograph

In 1816, Niépce, 51 years old, decided to start new research, to try to fix the images projected in the camera obscura. He first used silver salts deposited on paper. Unaware of the work of Wedgwood, he employed muriate of silver (now called silver chloride) instead of nitrate and obtained the first negative image of a landscape taken from one window of his house. Like Wedgwood, he did not find any method of fixing the image so that the silver chloride that was not transformed by light continue to blacken when observed under light.

3.2. The Steps Toward Success

Because of the instability of the negative proof, Niépce did not succeed when he tried to obtain a positive proof by contact printing of the negative. In order to directly obtain a positive image, he turned his interest toward photosensitive compounds that discolored under light instead of blackening as silver chloride did. He tried to used some iron oxide and "black oxide of manganese." Even though he obtained some good results with these compounds, he always came up against the problem of fixing the image; i.e., of eliminating the compound that had not been transformed by light.

To make very stable images, he tried to find a process that would give images engraved in a support.[11] For this purpose he studied the effect of light on acids. In his mind he imagined that if there existed a transformation of the acid with light, it would lead to a variation of the acidity. Consequently he imagined that when projecting an image on an acid spread on a support (copper or limestone), the variation of acidity due to the variation of light intensity would engrave the support to a varying degree according to the hues of the original image.

Unfortunately, acids are not decomposed by light, and after some experiments, Niépce abandoned this idea. But after these studies he looked at the problem in a different way. Through his experiments he understood that it was not necessary to use a compound in which transformation under light was directly visible; but rather that the change in a property, even if it was not visible, could lead to the production of an image by undergoing reaction with the support or another chemical compound. After this, Niépce turned his interest to every compound that interacts with light.

He then read in a chemical treatise that the resin extracted from a tree, the Gaïac, is yellow but turns green under light. More interesting to Niépce was the fact that after light irradiation this resin is less soluble in alcohol, which must be very pure to dissolve it. He immediately understood that this variation of solubility could be used to fix the image, because of the possibility of eliminating the resin that had not been transformed by light. Unfortunately, although he observed some color changes under the direct action of sunlight, he could not obtain the variation of solubility. But this idea stayed in his mind.

In his chemical readings, he learned of the properties of phosphorus. White phosphorus becomes red when exposed to light. This is a transformation between two allotropic forms. He explained: "*This fluid* [i.e., light] *turns it rapidly from white to yellow and from yellow to dark red which finally turns blackish. Lampadius alchohol*[12] *which easily dissolves white phosphorus does not react with red phosphorus and to melt the latter a much greater heat is required than to melt the former. Exposed to air red phosphorus does not become deliquescent as white phosphorus which, after absorption of oxygen transforms into phosphorous acid. This acid has the same consistency as oil and corrodes the stone as the mineral acids do. I have checked that all this is true.*"[8]

When he performed the experiments directly under sunlight Niépce obtained results in agreement with what he read, but when he tried to make an image in the camera obscura, no transformation occurred. This was probably due to the absence of ultraviolet rays inside the camera obscura, contrary to the direct sunlight.

Niépce wanted to use phosphorus spread as a varnish on limestone. Because of the ability of phosphorus to inflame spontaneously in air, he built what we now call a glovebox, a box flushed with "*nitrous gas.*" But this did not exclude air sufficiently, and Niépce burnt his hand during the preparation of his stone and never used phosphorus again.

Niépce's idea of making images with phosphorus was not really bad because around 1860 Draper succeeded in obtaining some reproductions of the ultraviolet part of the sun spectra by making a varnish of phosphorus under nitrogen between two glasses. He obtained a red image.[13]

4. The Invention of Photography

Niépce then experimented with a mineral resin, "*asphalt known as bitumen of Judea.*" This compound corresponds to the residue obtained from the distillation of petroleum after elimination of all volatile and liquid products. It looks like coal and from black becomes brown when powdered. From antiquity, the Dead Sea was a natural source of bitumen, which came up regularly as solid blocks from the bottom to the surface. The Egyptians used it to embalm mummies. It was also employed to make boats watertight or to cover the pavements in Babylonia.

Niépce discovered that, like resin of gaïac, bitumen becomes less soluble in its normal solvents when it is exposed to light. But with bitumen Niépce succeeded in using this property to fix the images. He dissolved the powder of bitumen in lavender oil to make a concentrated solution. This was a viscous black-brown liquid. The dissolution was very slow; it took approximately one day. Next, he spread this solution on a support (at the beginning, glass and limestone) and dried it by heating. This led to the production of a bright golden brown varnish. Under the action of light this varnish became insoluble in the lavender oil. After exposure, Niépce immersed the varnish with its support in a bath of lavender oil diluted white petroleum.[14] Parts of the images that had not received light were still soluble in lavender oil, while those illuminated were not dissolved and stayed on the support. The variation of solubility made it possible to reproduce the difference between light and shadow. As the bitumen was brown and stayed in the area that was illuminated, whereas it was dissolved and removed from the area in the dark, the image was negative. For Niépce, this was not a good representation of the original but, because of the stability of this negative, he continued to work on this compound in order to finally obtain a positive.

4.1. The First Photopolymer and the First Photograph

Remembering his idea of obtaining engraved images, Niépce tried immersing the support in acid covered with the bitumen image in order to etch the support in the place where it was no longer protected by the bitumen varnish. He used this method from 1824 to 1826 with different supports such as stone, copper, or tin plate.

This was a successful method for the reproduction of line drawings. As early as 1822 Niépce reproduced a drawing of Pope Pius VII on a glass plate. He explained in 1823 how he proceeded. He took a drawing on paper and covered it with a varnish, which made it translucent. He then put this paper in contact with the support covered with bitumen and exposed all of it to the direct light of the sun. In the lavender oil solution the bitumen was dissolved at the exact place of the lines of the drawing where the bare support appeared. In 1822 he used glass as a support, and the after-treatment with acid appeared only in 1823, when he started using limestone with bitumen.

In 1824 Niépce succeeded in reproducing an image from nature recorded only with a camera obscura as he wrote to his brother: ''*I have the satisfaction to let you know at last that thanks to the improvement of my procedures, I have finally succeeded in obtaining a viewpoint such as I desired. . . . The image of the objects is represented with an astonishing sharpness and fidelity, down to the minutest details, and with their most delicate hues.*'' Niépce stated that bitumen reproduced, in negative, all the variations of light of the recorded image.

He then tried to etch this view and asked a printer from Dijon to print the image on paper. The man was so surprised to look at Niépce's stone that he did not do anything. That is why Niépce then asked for the services of a specialist of etching in Paris to print his images on paper. In 1825 he changed the support and worked on copper plate and on pure tin and pewter in 1826.

He rapidly obtained encouraging results in the case of the reproduction of line drawings, but not for the etching of views with a continuous gradation of tones. For these there arose a new problem concerning the engraving of the hues of the images that were presented on the bitumen varnish. In actuality, the difficulties were due to the very good property of bitumen to be absolutely waterproof. For this reason, acid only reacts with the support when it is absolutely unprotected by the varnish and does not penetrate through the slightest film of bitumen like the one that is barely visible in places which correspond to dark hues in the original image. Hence, acid etched only what corresponded to the black of the image.

In 1827 he obtained images on a very thin film of bitumen, probably underexposed in his box camera, and stated that in this case it was possible to see the image as positive when observed under a high light (the sun) with a dark reflection on the metal. An example of this kind of image realized by Niépce himself on a tin plate is kept at the Harry Ransom Center of the University of Texas at Austin. It represents the view from one window of Niépce's house and is at present the most ancient photograph in the world (Photograph 1-1).

In 1828, Niépce improved his process by using silver plates as support. After the production of the image on the bitumen he put the plate in a box containing some crystals of iodine allowed to evaporate spontaneously. In fact, the box was saturated with iodine vapors, which are very corrosive. These vapors oxidized the silver plate where it was not protected by the varnish. A thin layer of silver iodide resulted, which transformed into black silver particles under the action

Photograph 1-1. *The* Point de vue du Gras. *This heliograph was made by Niépce using a camera obscura in 1827. It represents the view from the window of the room where he used to work. Among the numerous heliographs he made with a camera obscura, it is the only one that has not been lost. It is now at the University of Texas at Austin. This photographic reproduction was made in 1952 with a special treatment that enhances the contrast. Finally it was touched up to give an idea of what it looks like when observed visually. It gives an incorrect idea of the real image (see text). (© Gernsheim collection, Harry Ransom Humanity Research Center, Austin, Texas.)*

of light. Consequently the light areas of the bitumen image became black. The dark brown ones protected the silver plate from attack by iodine vapors. In the intermediate case of a remaining thin film of bitumen, the iodine vapors penetrated the varnish and partially oxidized the silver, leading to a very thin layer of iodide that became grey after the action of daylight. The thicker the film of bitumen, the weaker the effect of the iodine vapors; i.e., the darker the brown color of the bitumen, the lighter the grey on the silver plate. After all these operations, Niépce removed the bitumen varnish with alcohol. This method led to images with a broad range of gradations of grey as in modern-day black and white photographs.

At present about nine plates of tin or pewter etched by Niépce himself are in the possession of museums in France and in Great Britain. Only one view taken with a camera obscura has survived until now, viz., the one in Austin made on tin plate.

In Paris in 1827, Niépce had met Louis Jacques Mandé Daguerre, a painter who was a specialist in making stage settings with the aid of the camera obscura. In 1829, Niépce needed a better camera than the one he had and asked Daguerre to cooperate with him in order to jointly improve his invention. In December 1829 the two men signed a contract in which Niépce committed himself to divulging the secret of his invention to Daguerre, and Daguerre had to reveal all he knew about improvements of the camera obscura in the field of application of Niépce's invention.

5. Heliography

On December 14, 1829 Niépce gave Daguerre a text entitled: *Notice sur l'Héliographie.*[15] This was the name that Niépce had given to his process (from the Greek *Helios*: the sun and *Graphein*: to write—the writing by the sun).

As we shall see, the process described by Niépce is similar to what we now call photolithography. Let us summarize the different steps of Niépce's process. They show that he imagined all the possibilities of the photographic process. To obtain the bitumen image it is necessary to perform the following steps:

- Dissolving powdered bitumen in lavender oil
- Spreading the solution on the support
- Drying the support covered with the solution of bitumen
- Exposing the support to light
- Dissolving the unirradiated areas in white petroleum mixed with oil of lavender
- Rinsing with water and drying.

At this stage the image obtained is a negative because the brown varnish is eliminated in the dark portions of the images. To obtain a positive image, Niépce used this bitumen image in the following three ways:

1. *To obtain an image engraved in the support (copper or tin).* He employed the etching method used by artists as famous as Rembrandt to engrave the metal with acids. This technique was totally successful for the reproduction and printing of drawings. Thus Niépce is the inventor of photoengraving,

whose principle is still the same in today's printing processes. For many years after Niépce, until around 1930, bitumen was employed in printing, to make the photoengraved plates used for printing photographs. The main improvement was the interposition of a transparent checkered screen between the image and the varnish, which allowed the reproduction of the halftones as very small squares side by side.

2. *Without further treatment, in cases in which the varnish is very thin and underexposed.* Such images can be seen negatively or positively, depending on the orientation of the light and on the reflection of the metal when observed.
3. *To obtain a positive image on silver.* He put the silver plate[16] with the bitumen image in a box containing iodine crystals allowed to evaporate spontaneously. Oxidation of the area of the plate insufficiently protected by the bitumen led to the formation of a thin layer of photosensitive silver iodide that became black under the action of light. After removing the varnish, this method led to black and white photographs with a resolution and a gradation of tones similar to those of modern day photographs.

Let us now analyze each of these steps in detail.

6. Formation of the Images with Bitumen

6.1. Preparation of the Bitumen Solution in Niépce's Process

Niépce described his preparation as follows: "*I half fill a glass with this powdered bitumen. I pour on it, dropwise, essential oil of lavender until the bitumen no longer absorbs it but is only well soaked in it. I then add enough of this essential oil to reach a level about three lines*[17] *above the mixture which must be covered and left to a mild heat until the extra oil is saturated by the coloring matter of the bitumen. If this varnish does not have the necessary consistency, it is left to evaporate, in the open air, in a small dish, protected from moisture.*"

Our initial work was performed with solutions of bitumen prepared according to Niépce's writings. Such a preparation rapidly exhibited an aging process during which the behavior of the bitumen varnish changed progressively. This led us to investigate the properties of bitumen in order to understand this evolution of solutions.

6.2. What is Bitumen?

The answer to this question is not trivial. There are different kinds of bitumen: (1) the viscous liquid bitumen and (2) the solid bitumen. Bitumen is a compound that originates from petroleum. In 1837 Boussingault[18] made a classification of different products extracted from petroleum. He gave the name of *petrolene* to the liquid separated from liquid bitumen by distillation at high temperature and he called the corresponding residue *asphaltene*, which is similar to solid bitumen. In the case of heliography we are concerned only with asphaltene described at present as a dark brown to black solid that looks like coal.[19] The description given by Niépce was: ''*This substance, which is less friable than the usual resin, is blackish and very opaque. Its fractured edge is sharp and shiny like coal, it becomes brownish when it is reduced to a very fine dust, and takes fire as easily as resin.*'' [8] At present, asphaltene is defined as the fraction that is not soluble in *n*-pentane.[20]

The bitumen we employed was purchased in powder form from Prolabo. It is insoluble up to 75% in *n*-pentane, indicating that it is mostly asphaltene. Experiments carried out with the crude PROLABO bitumen, as compared with the fraction insoluble in *n*-pentane, yielded different results in terms of photosensitivity and image quality. Moreover, some attempts made with the soluble fraction permitted us to obtain unstable images of high quality, thus indicating that these soluble hydrocarbons have a much lower sensitivity, but a good ability to reproduce a wide range of hues.

It is evident from Niépce's description that he proceeded with a saturated solution of bitumen. We have performed the same kind of preparation and used it to obtain images. When the solution is used one day after the dissolution, it gives a varnish that is not very photosensitive and that requires a long exposure time (for example, 5 hours under direct sunlight). At that moment the images obtained are contrasted. After one week, the varnish is more sensitive and leads to images of good quality for etching and for iodine treatment. The exposure time under direct sunlight is about 3 hours. After one month the same solution becomes more viscous and give a varnish much more sensitive to light (1.5 hours under sunlight) but the images are without any contrast. They are also unsuitable for etching as it is impossible to eliminate the varnish completely at any point of the plate, as if it was already reticulated before the exposure to light.

6.3. Studies of Different Fractions Extracted from Bitumen

By mixing bitumen with an excess (greater than 60×) volume of a series of simple alkanes such as isopentane, *n*-pentane, hexane, heptane, and octane, it was possible to separate different fractions of bitumen. After evaporation of alcane, the dry soluble fraction called asphaltene was dissolved in oil of lavender and spread on a plate to make a varnish. For the soluble fraction in isopentane, no matter how long the exposure to light, it was impossible to reach a variation of solubility and hence to obtain an image. By mixing bitumen with an excess volume of *n*-heptane, only 66% was insoluble. The soluble fraction obtained after evaporation of *n*-heptane and then dissolved in lavender oil gave a varnish very weakly sensitive to light. During immersion in the bath of white petroleum mixed with oil of lavender, the image appeared, but insolubilization was not sufficient and the image rapidly vanished after complete dissolution. On the contrary, the insoluble fraction obtained both in *n*-pentane and in *n*-heptane was much more sensitive to light, but gave images with no contrast and that were sometimes difficult to dissolve after exposure to light.

These observations led us to understand the aging process observed in Niépce's preparation. A few days after the preparation in oil of lavender only the soluble products observed in the dissolution in simple alkanes are dissolved. This explains the low sensitivity of fresh dissolution of bitumen in lavender oil. Day after day what we call asphaltenes, corresponding to the products insoluble in simple alkanes, are slowly dissolved in lavender oil and the photosensitivity is enhanced, with a good quality to the images. After one month all the asphaltenes are dissolved and the images obtained are without contrast, while the photosensitivity is better than for solutions with asphaltenes only in lavender oil.

It is a disadvantage of Niépce's preparation that a variation in the composition of the varnish is achieved progressively. To avoid this, we pursued our investigations with solutions of bitumen in oil of lavender without any excess of powder which could be carried out with the liquid during the spreading operations and cause some damage to the final varnish (dots, holes, and inhomogeneity in the varnish). To ensure that the dissolution was complete, the solutions were stirred with a rotating magnet for 24 hours. Typically our solutions were made of 150 g of bitumen (Prolabo) for 1l of lavender oil. The effects of concentration and the stirring method (rotation or ultrasonication) of this solution have been studied from the point of view of the quality of the final image and indirectly of the light process.

It appears that asphaltenes are the compounds that play the major role in the transformations of the varnish exposed to light; thus, we will describe their properties only. Numerous investigations have been carried out in different laboratories to analyze and to determine the structure of asphaltenes. Actually, it is impossible to give a perfect chemical and molecular description of asphaltenes because of the very complex mixture of different compounds that enter in their composition. Numerous analyses have been performed in order to tentatively assign hypothetical structures of asphaltene. After the separation of different components by differential solubility in hydrocarbons such as *n*-pentane, *n*-heptane, and decane, different products have been isolated with various molecular weights. Analysis using physical methods (such as X rays, nuclear magnetic resonance, mass spectrometry, infrared spectroscopy, vapor pressure spectrometry, differential thermal analysis, densimetric methods, electron microscopy, and small angle scattering) and chemical methods (such as oxidation, alkylation, and halogenation) have been performed.[21]

The main constituents are carbon, hydrogen, nitrogen, oxygen, and sulfur. A constancy in the H/C atomic ratio is observed: 1.15±0.05 (82±3 wt % carbon; 8.1±0.7 wt % hydrogen), which favors a definite composition of asphaltene. Oxygen contents vary from 0.3% to 4.9%, the O/C ratio varying from 0.003 to 0.045, and sulfur contents vary from 0.3% to 10.3%, so that the S/C ratio ranges from 0.001 to 0.049. A smaller variation is observed in the nitrogen content, which varies from 0.6% to 3.3%, giving a N/C ratio of about 0.015±0.008.[22]

In spite of all these data, it remains difficult to give an exact structure of asphaltene. However, it is commonly accepted that these compounds are condensed polynuclear aromatic ring systems bearing alkyl side chains. The number of rings varies from six to twenty in the most important systems. Hypothetical structures of asphaltene have been proposed.[19,23] An example is given in Fig. 1-1.

Investigations of the macrostructure of asphaltene by X-ray diffraction allowed the determination of an organized structure[24–27] such as the one represented in Fig. 1-2.

It appears that asphaltene is naturally organized and that the molecules are already in some definite position that might favor crosslinking reactions induced by light. It must be remarked that in all these studies no attention has been paid to the effect of light on asphaltene and that the small samples observed are never protected from light at any stage of their analysis.

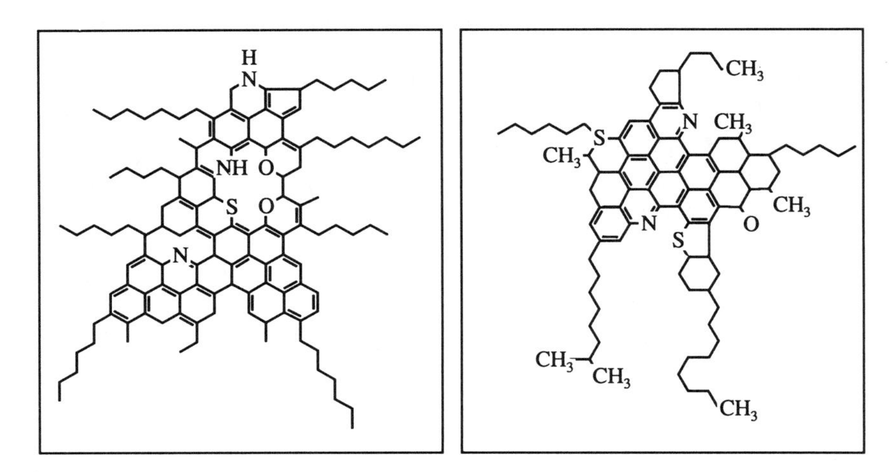

Figure 1-1. *Hypothetical structures of a petroleum asphaltene. (left: Ref. 36, right: Ref. 19.)*

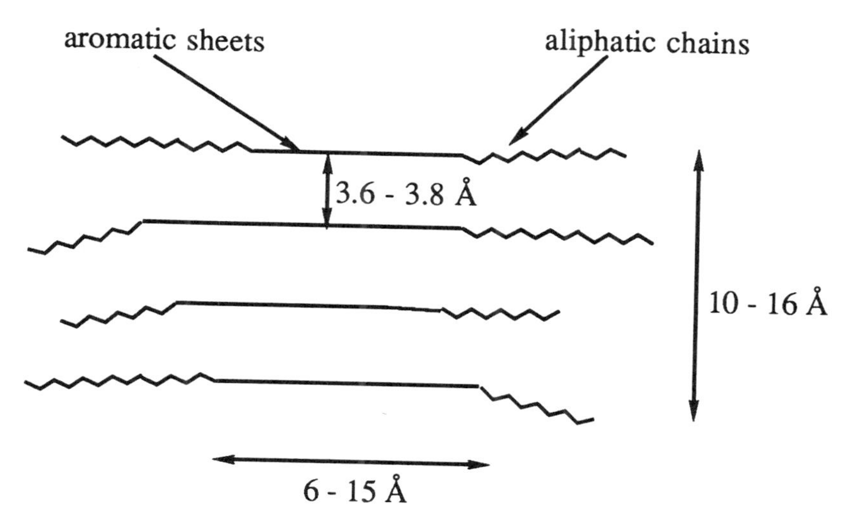

Figure 1-2. *Representation of an asphaltene from X-ray analysis. (Ref. 19.)*

6.4. Principle of the Formation of Images with Bitumen

Under the influence of light, bitumen undergoes a reticulation process.[28] The mechanism of this process is unknown, but some studies do give scarce informations. Thus it is known that the process does not take place in the absence of oxygen.[29] It has been found that the photosensitivity of bitumen is enhanced when sulfur atoms are added to the structure of asphaltene. Photo-oxidation of asphaltene has been studied from the point of view of the absorption of oxygen induced by UV light.[30] It was shown that light markedly enhanced the natural property of asphaltene to absorb oxygen. Fig. 1-3 indicates the principle of the formation of the bitumen image for a given intensity of light.

For the varying intensities of light that occur in the projection of an image, our experiments have evidenced a proportional effect of light on the reticulation process, as shown in Fig. 1-4 where the density of the film of bitumen deposited on a glass support is plotted as a function of the variation of intensity of incident light.

This shows that asphaltene is able to reproduce the various hues of an image as is easily observable at first glance of a negative bitumen image. This effect results in different thicknesses of the varnish on the plate. Such an ability to reproduce tones is one of the properties that are required to provide an image, but another important property is the ability to precisely ascertain the geometry of the different shapes projected on the plate.

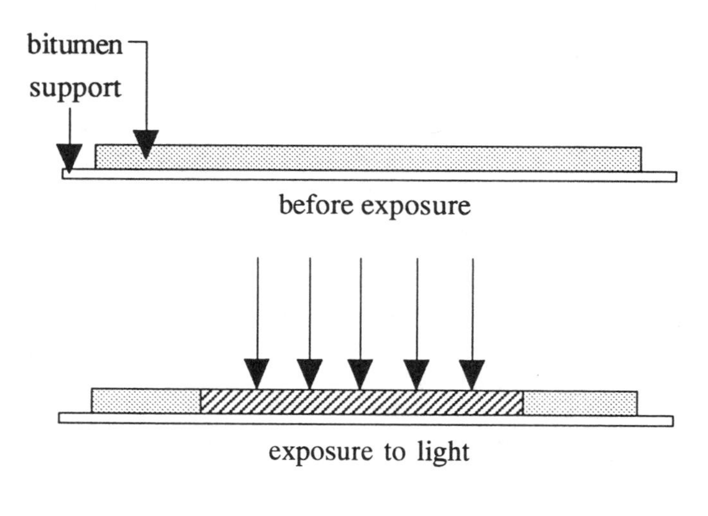

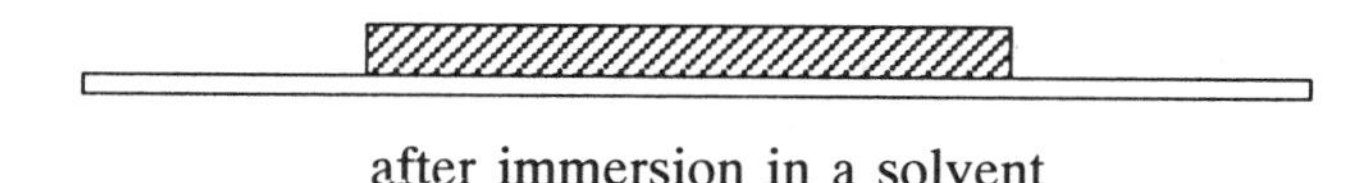

Figure 1-3. *Principle of image formation with bitumen of Judea.*

This ability is measured as the resolution. Our measurements have shown that a film of bitumen is able to reproduce details as precise as 80 lines/millimeter (mm). It can be deduced from this result that the effect of light is perfectly reproduced and that the photochemical process does not involve a chain mechanism. This means that each photon gives only very few reactions of reticulation and explains why the process needs a large amount of light and consequently such a long exposure time.

6.5. Spectral Sensitivity of Bitumen

Fig. 1-5 shows the absorption spectrum of a film of bitumen prepared on a silica plate following Niépce's method.

The absorption spectrum shows that this compound absorbs only UV and blue light. In heliographic applications the light is always filtered by the lens of the camera or by a glass and transparent paper in the case of the reproduction of drawings. Thus, only wavelengths above 350 nm are transmitted to the bitumen

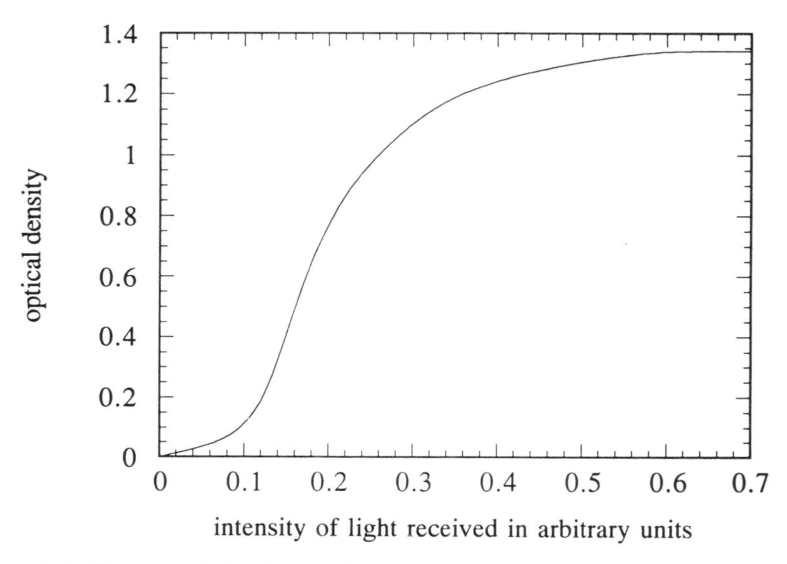

Figure 1-4. *Variation of the density of the varnish with the intensity of light after dissolution in diluted lavender oil.*

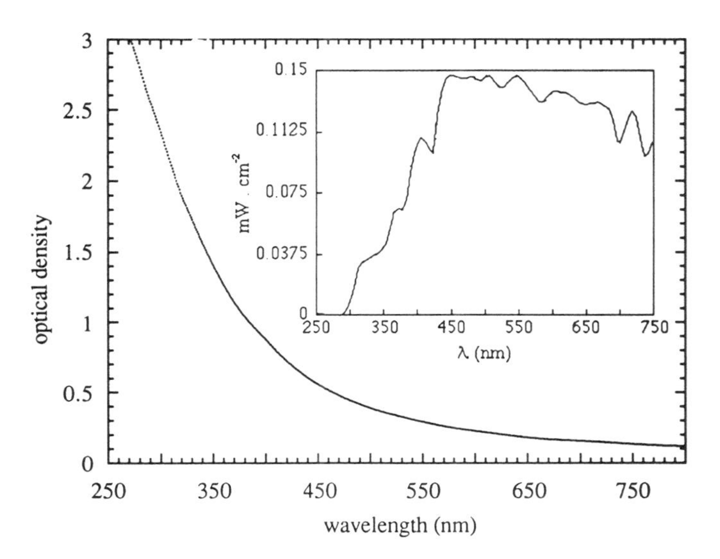

Figure 1-5. *Absorption spectrum of a film of bitumen prepared by Niépce's method on a silica plate. (Insert: Solar spectrum.)*

varnish. The insert of Fig. 1-5 shows the emission spectra of the sun as received on earth. In Niépce's time the sun was the only source of light available.

To determine maximum efficiency with regard to wavelength, we irradiated the bitumen through color filters. It turned out that light above 470 nm is inefficient and that the efficiency is almost constant in the range 300–450 nm. It must be mentioned that this small domain of sensitivity is the same for silver halides used in photography. These halides are sensitive only below 400 nm and must be sensitized by the addition of different dyes to be chromatized so as to have the same photosensitivity at every wavelength.

6.6. Achievement of the Photosensitive Varnish

Niépce has indicated that the varnish must be the thinnest possible and have a high consistency. He has also mentioned that in his process the support plays a nonnegligible role and that the clearer the support the shorter is the time of exposure: "*My experience enabled me to know that the effect is greatest when the support on which the image draws itself is whitest.*"[7] We are going to show that these facts established by Niépce are true.

As Niépce did, we spread the bitumen solution with a paintbrush on supports such as glass, limestone, copper, tin, and silver. Niépce indicated that he dried his solution by heating on a hot metallic plate, but did not give the temperature or the approximate duration of this operation. Throughout our work it appeared that this step is important in determining the quality of the image. Trying different temperatures and durations of heating, we were able to show that there are optimal temperature-duration conditions. We were led to carry out this operation at 90° C for 20 minutes.

At this temperature, 20 minutes is not the shortest time required to obtain the drying of the varnish; the drying is complete in 10 minutes. When the varnish is dried below 90° C, the sensitivity of the bitumen decreases and a longer exposure time is required, but more important is the fact that it adheres poorly to the support (from which it spontaneously removed during the revealing dissolution in the solution of lavender oil). If the drying time is longer than 20 minutes at 90° C or if the temperature is higher than 90° C, an important thermal reticulation process takes place and the varnish becomes insoluble, even in the absence of exposure to light. In other words, it can be said that this post-coating thermal treatment has the effect, first, of inducing a prereticulation process that acts as a pre-exposure of the varnish and allows a decrease of the exposure time, and, second, of conferring a strong adherence of the film on the support.

In order to determine the thickness of the prepared varnish, we first recorded the absorption spectra of a solution of bitumen in lavender oil. This led us to determine the value of the extinction coefficient at 400 nm:

$$\epsilon_{400\ \text{nm}} = 6400\ \text{g}^{-1}\ \text{liter cm}^{-1}$$

Assuming that this coefficient is the same for films of bitumen, it is then possible to measure the thickness of the film. In a normal preparation of a film such as the one used to record the spectra of Fig. 1-5, the thickness is 1.5 micron. Fig. 1-6 shows that all the light at 400 nm is absorbed in 3 microns in such a film.

For good efficiency, it is necessary to absorb all the light, but the 3-micron film that corresponds to this situation does not give the best configuration because only a weak intensity of light interacts with the varnish at the surface of the support. After irradiation, this part is not sufficiently reticulated, while the surface of the varnish can be totally insoluble. During the revealing step, the varnish in contact with the support dissolves and the well-formed image is stripped off the support. Consequently, it is better to prepare a thinner varnish to ensure that the part in contact with the support becomes sufficiently reticulated. This thickness depends on the concentration of the bitumen solution. If the concentration is high, the varnish must be thinner. This supports Niépce's conclusion that an important property of the bitumen solution is that the film of varnish must be the thinnest possible.

The effect of the support indicated by Niépce can easily be understood in

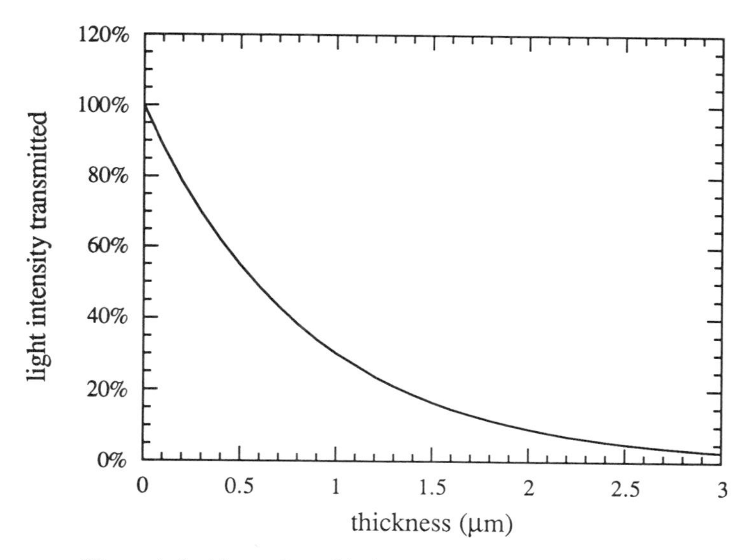

Figure 1-6. *Absorption of light vs. thickness of the varnish.*

terms of reflection of light. Let us look at what happens in the varnish in the case of a reflecting support such as a mirror for example. Because all the light is not absorbed in the varnish, as we have just explained, the fraction that is not absorbed is reflected by the support and re-enters the varnish to generate additional reactions of reticulation. The absorption follows the behavior indicated in Fig. 1-7.

With a reflecting support:

- The varnish receives more light, which allows a decrease in exposure time. This agrees fully with the observations of Niépce (letter of October 18, 1824).

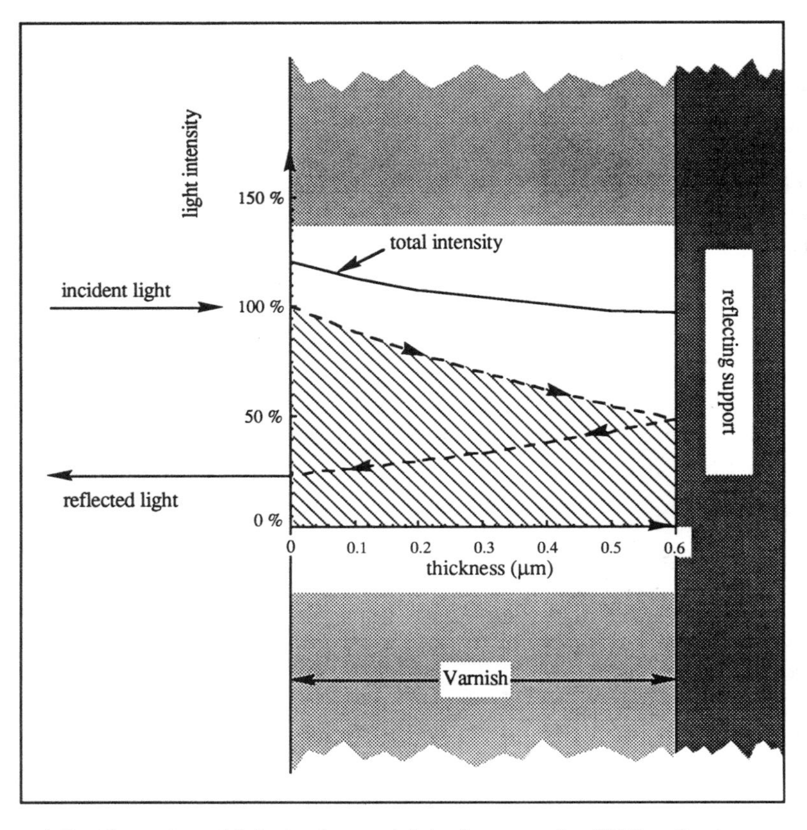

Figure 1-7. *Absorption of light in the varnish in the case of a 100% reflecting support.* ------: *Intensity of the incident and reflected light,* ——: *Added intensity of the incident and reflected light.*

- The intensity of light received is more homogeneous in the whole thickness of the varnish with an inevitable small decrease in the depth. The reticulation is better and gives a better quality of the varnish during the revealing treatment (no risks of stripping off).

All these measurements only show with greater precision what Niépce had stated, observed, and deduced from experiments. These results are useful for understanding, optimizing, and achieving a good reproducibility of the method. We can admire Niépce's abilities as an experimenter—very good capacities of observation together with a fertile imagination—which led him in 1829 to find the best conditions for the preparation of the varnish.

6.7. Sensitivity to Light-Exposure Time

Niépce used his photosensitive plates in two ways: in the camera obscura and in contact with a drawing on a paper made translucent by coating it with a varnish.

We have obtained reproductions of drawings by the latter method of Niépce, which is now currently called contact printing. The exposure time needed to obtain an image, using sunlight through an oiled paper, is about 4 hours. This time is in agreement with the one given by Daguerre, the associate of Niépce, who mentioned in his booklet that the time required for such experiments was 3–4 hours.[31] Also, J. B. Biot, one of the members of the commission named by Arago to make an enquiry about the inventions of Niépce and Daguerre, has written that 4–5 hours were necessary to reproduce a drawing by Niépce's method.[32] This agreement between our results and the data given by collaborators or contemporaries of Niépce confirm that we operated under the same condition as Niépce and that the bitumen we used had approximately the same sensitivity as the one used by Niépce.

Concerning the experiments in the camera obscura, we have no precise indications of the exposure time, either from Niépce or from Daguerre, who were the only persons who practiced the process. The same J. B. Biot who got his information from Daguerre wrote that an exposure time as long as two or three days was necessary to obtain an image by means of the camera obscura under the brightest light.[32]

From the letters of Niépce we know the characteristics of the lens that he put on the hole of his camera obscura. At the beginning of his research (before 1828), it was a simple biconvex lens. Among all the lenses he bought, the one that possessed the most luminous properties was 3 inches in diameter with a focal length of 12 inches. (This corresponds to 81 mm for the diameter and a 324.8-mm focal length.)[33] Thus, it is possible to know what is called in photography the aperture of Niépce's lens. This is given by the ratio between the focal length and the diameter of the lens. For Niépce's lens, the aperture was 4 (f:4). Every lens or group of lenses that has the same ratio aperture transmits the same quantity of light into the camera.

The camera we used was equipped with an f:4 aperture lens. First we measured with a luxmeter the intensity of light inside the camera exposed in front of a landscape under the sunshine. It appeared that very little light came through the lens. For example, the direct sunlight was found to be around 100,000 lux, while inside the camera only one-twentieth of this same light reflected by the landscape was transmitted (5000 lux). We can immediately understand that the exposure time with the camera must be multiplied by the same factor of 20, leading to an exposure time of about 40 to 60 hours. This means exposing the plate in the camera during four or five days, depending on the date, but around the longest daylight days in summer.

This result has been confirmed by experiments. To obtain an image in the camera obscura in the days of June, it was necessary to expose the bitumen plate for a long time. Some attempts to obtain an image in two days were unsuccessful, but with three days' of exposure, it was possible to obtain an underexposed image, with particular properties that we will discuss later. In terms of ISO, the modern-day unit of sensitivity of photographic films, the bitumen corresponds to 10^{-6} ISO, whereas today's films are usually 100 ISO. This enormous difference is due to the fact that in the bitumen process all the transformations must be achieved by the light, while in the later photographic processes—from the daguerreotype onward—light only begins the chemical transformation of silver halides, which is then completed in a second step performed in the laboratory, by using a chemical product: the developer. This step, called the development, consists of an amplification of the effect of light by a factor of about 10^9, which allows exposure of the photosensitive system to light for only a very short time, actually a fraction of a second.

After revisiting Niépce's process it has become possible to understand many paragraphs of his letters, and it appears clearly that the time required to obtain an image was five days, as for example in his letter of September 16, 1824, written on a Thursday, where Niépce explained that an image on glass just started to be exposed in the camera would be ready on Monday the 20th.[7]

7. Different Types of Images Obtained with Bitumen

7.1. Images Made with Bitumen

As we have already explained, Niépce had found different ways to use the negative bitumen image. In his first successful experiments on stone in 1824, he explained the effect that made it possible to see the negative image as a positive: "*As this counterproof is almost uncolored, one can appreciate the effect only when looking at the stone obliquely; only then it becomes visible to the eye, thanks to the shadows and of the reflections of light, and this effect, I may say, my dear friend, has really something magic.*"[7]

Our experiments on limestone showed this effect clearly. Even if it is perfectly polished, the stone is never as bright as a mirror. With a correctly exposed varnish, the inversion of the negative image occurred because the very shiny varnish reflecting the light appeared very bright while the matter stone always appeared darker. Photographs 1-2 and 1-3 show the two visions of an engraving reproduced in the bitumen spread on a limestone.

At that time, Niépce did not pay any attention to this phenomenon; he took some interest in it only during the summer of 1827. The unique photograph made by Niépce in a camera obscura that has survived until now was achieved during the summer of 1827 and is an image of precisely this type as we concluded from the observations we made on Niépce's plate at Austin in 1990.[3] The support is a pure tin plate that is still as bright as a mirror after more than 160 years. The image on this support is the bitumen negative obtained by Niépce in his camera. Niépce did not pursue treatment after obtaining this image. Under

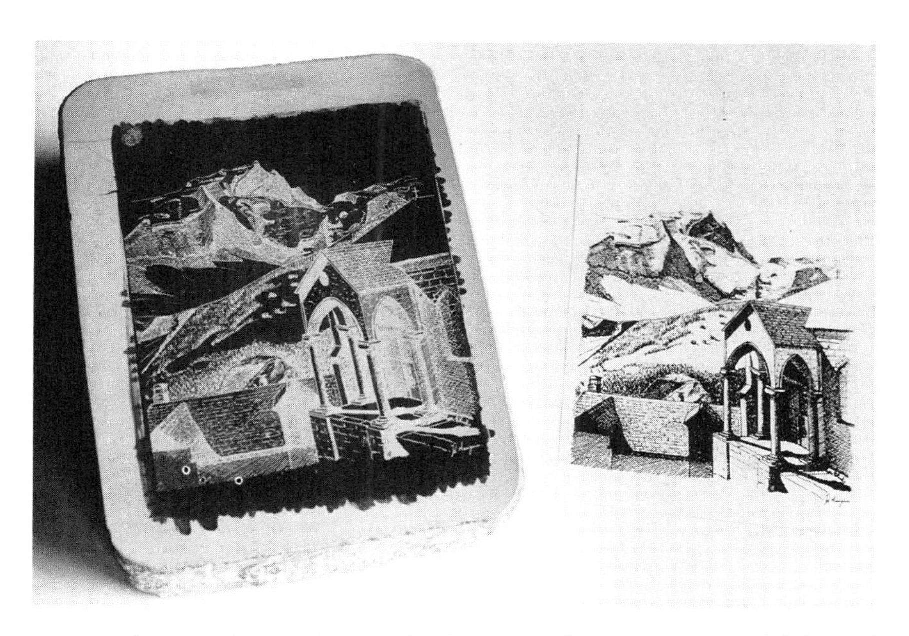

Photograph 1-2. *Left: reproduction of a drawing on limestone. In normal lighting the image is negative. Right: the original drawing for comparison.*

Photograph 1-3. *The limestone of Photograph 1-2 photographed in front of a window. The reflection of bright light makes the image appear as positive.*

normal lighting, we see the negative. But when observed in a dark place and near an intense source of light oriented obliquely, the reflection by the metal appears dark, while the varnish appears lighter. This leads to a visually positive image.

Niépce brought this plate to England during a journey he made from December 1827 to February 1828 when he tried to present his invention to the Royal Society in London. For this purpose he wrote a *Note* in which he described this kind of plate: "*My framed drawings made on tin, will undoubtedly be found too weak in tone. This defect arises principally from the fact that the highlights do not contrast sufficiently with the shades resulting from the metallic reflection.*" [7]

We have made a reproduction of this plate in order to show its real appearance.[3] Photograph 1-4 shows the image as it is seen naturally and Photograph 1-5 shows the same plate under lighting that makes it appear positive.

Photograph 1-4. *Reproduction of the* Point de vue du Gras. *In normal lighting the image is negative.*

All the white areas on this black and white photograph are in reality of the pale golden brown color of the bitumen, while the black areas correspond to the reflection by the bare metal of the dark environment.

Our reproduction has been made by using the famous black and white touched-up photograph shown at the beginning of this article (Photograph 1-1), projected on a tin plate coated with bitumen solution and underexposed. After dissolution in lavender oil, the remaining varnish was very thin and mat. For this reason, the varnish behaves like alignments of microcrystals reflecting the ambient light, whatever the orientation of the plate may be. When the metal presents a dark reflection, the varnish reflects the light and is clearer than the bare metal areas. Hence, the image is positive. The reproduction we show cannot have the same quality as the original by Niépce because we used the touched-up photograph, which is a bad reproduction.

7.2. Images Engraved in the Support

As we have already explained, this method, similar to the etching process, is successful only in the case of the reproduction of line drawings, in which tones

Photograph 1-5. *The plate of Photograph 1-4 as it is seen when placed in dark surroundings, with light coming in obliquely.*

are represented by a series of lines. Niépce had understood this problem and worked to improve this method for the contact printing reproduction of drawings such as the Cardinal d'Amboise shown in Photograph 1-6.

Plates such as the one of the Cardinal d'Amboise are *incunabula* of the history of the printing process because they are the first photomechanical reproductions in the world and prove that Niépce was also the inventor of photoengraving. While it is impossible to engrave half-tones of a bitumen image, it is easy to etch the finest details of engravings when drawings are made of lines only, thanks to the high resolution of this resin and its great watertightness. By using an appropriate acid it is possible to engrave exactly the image reproduced. That is why Niépce's technique can also be considered as the precursor of what is now called photolithography, used to make microcircuits in electronics. By the same method as Niépce's, electronic circuits are now made by projecting their image on photoresists that are similar to the bitumen, but that react with UV light or X rays within a few seconds. The image obtained in the resin coated on a support such as silicon is then revealed by a solvent, just as in Niépce's method. The microcircuit is finally obtained by submitting the support and the resin to an acid (wet etching), as in Niépce's

Photograph 1-6. *Photograph of a heliograph made by Niépce in 1826. It is a reproduction of a drawing* Le cardinal d'Amboise *etched on a tin plate. This image is now in the Niépce Museum in Chalon-sur-Saône. (© Musée Nicéphore Niépce, Chalon-sur-Saône.)*

method or to a gas (dry etching), as in a method also discovered by Niépce, which we will describe next.

7.3. Positive Half-tone Images on Silver

In 1828 Niépce first alluded to another form of his process in his *Note* written to the Royal Society of London. After commenting on the weak contrast of the image on tin that he had brought to England, he wrote: *"It should be easy to remedy this by giving more whiteness and brightness to those parts which represent the effects of light; and by receiving the impressions of this fluid* [the light] *on well polished and burnished plated silver; then the contrast between white and black would be more marked; and the latter colour, intensified by means of some chemical agent, would lose this bright reflection which hinders the vision and even produces a sort of clash."*[7]

The principle imaged by Niépce is based on the idea that it is necessary to blacken the bare metallic areas that appeared bright in the negative image. After that it becomes possible to remove the bitumen varnish, allowing emergence of the areas corresponding to the highlights as the bright metal. Plated silver was chosen for two reasons. The first was its ability to be polished as a mirror brighter than other metals. The second reason, which had attracted the interest of Niépce, was the ability of this metal to turn black when brought in contact with iodine vapors. Note that iodine had just been discovered in 1811 by Courtois, and its properties were only known from the work of Gay-Lussac in 1813. This shows that Niépce was aware of recent discoveries in chemistry.

Niépce achieved bitumen images on silver and then put the plate covered with the image into a box that contained crystals of iodine. The latter evaporate spontaneously at room temperature, so the box covered with glass was completely saturated with iodine vapors. These vapors are very corrosive and oxidize silver in a few minutes, forming a thin layer of silver iodide microcrystals that are photosensitive. Under UV or blue light, electrons are ejected from iodide and scavenged by silver ions to give silver atoms. Progressively the ionic crystals become metallic crystals of silver that appear black to the eye. The bare bright metal areas of the negative image have then turned black. After removing the varnish, the image is now positive in black and white. The diagrams of Fig. 1-8 show the different steps of this procedure.

It is only by revisiting the practice of this process that it has been possible to show the performance and progress attained, in comparison with the previous one. Before carrying out practical tests, we knew that iodine was very efficient to oxidize and to blacken silver but one could imagine that, as in the case of etching, only the unprotected areas would react with this vapor to give highly contrasted images. But that is not so, and this method leads to images of high quality, similar to our modern-day photographs.

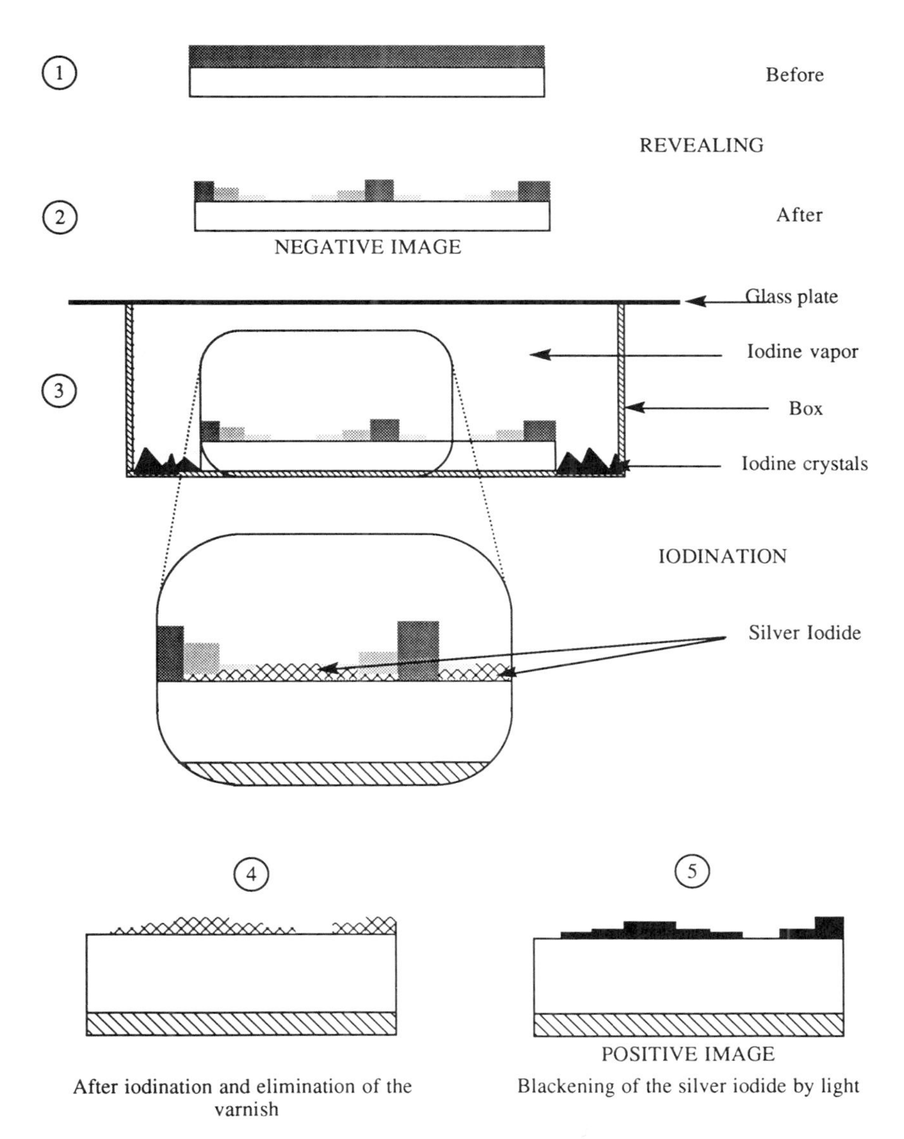

Figure 1-8. *Schematic representation of iodine vapor treatment of bitumen images on silver.*

The advantage of iodine is its ability to diffuse through the varnish, unlike the acid. This diffusion is governed by the thickness and the density of the varnish and is a slow process that can be controlled. When the varnish is thick and dense, iodine reaches the silver support after time t. If the plate is submitted to iodine vapors during a time shorter than t, silver is not corroded by iodine. The varnish acts as a filter through which the iodine vapors pass. The value of time t refers to the areas that present the highest density of varnish, for they correspond to the highlights that must be reproduced by the bare metal. During this time, the other less dense areas have been exposed to the same quantity of iodine, and only a fraction of this vapor has reached the metal. This fraction is inversely proportional to the thickness and the density of the varnish. Consequently, the quantity of silver iodide produced at the silver surface is also inversely proportional to the thickness of the varnish. After the transformation of silver iodide into black metallic particles of silver, the result is the exact negative image of the original image made in bitumen. Because bitumen gives negative images, the final image is a perfect positive.

The advantage of this technique compared to the etching process is the ability to reproduce half-tones. Moreover, all the details of the bitumen negative image are reproduced. Photograph 1-7 shows a negative bitumen image obtained by means of a camera obscura equipped with a lens of aperture f/3.5 and a focal length of 100 mm. The dimensions of the plate are 80×90 mm. Such a lens gives an image that does not cover all the plate, and a round halo can be observed on the negative. The exposure time was five days in bright sunlight. Photograph 1-8 shows the image obtained after treatment with iodine vapor of a second image identical to the one of Photograph 1-7. The very high quality of the definition and of the reproduction of the half-tones must be pointed out. These two properties are two of the three basic criteria required to make a photograph. The third one is the exposure time, which was much too long to permit this process to be widely used.

All the experiments mentioned above were described by Niépce before December 1829. After this date he collaborated with Daguerre in his research, but the two men never succeeded in making any progress on bitumen. They progressively changed the photosensitive compound and worked on plant resins such as rosin, turpentine of various origins, and also the residue obtained from the distillation of lavender oil.[34]

8. Conclusion

Niépce died suddenly on July 5, 1833. At that time, his process remained secret, and Niépce never had the glory of being known as the inventor of photography.

Bitumen of Judea appears to have very low sensitivity. This explains why the

Photograph 1-7. *The negative bitumen image of a heliograph made in a camera obscura in June 1990, under the same conditions as Niépce's.*

different steps of the whole process, e.g., the preparation of the varnish and its dissolution after exposure, can be carried out under room light without damage. It seems that the determination of the exposure time was the major obstacle preventing the reproduction of Niépce's process. Some historians have decided on an exposure time of 8 hours for a bright, sun-illuminated landscape. Despite the absence of any reference to Niépce's writings, this hypothesis has been extended so far as to become a genuine quotation. The exposure time of Niépce's process has been discussed in a few of our publications.[34,35]

Concerning photography, it is clear that the bitumen process could not survive because of the very long exposure time. But the discovery of Niépce opened the way for his successors. Let us emphasize that Niépce is not only the inventor of photography, but also of the photoengraving processes. Moreover, he can be regarded as the "father" of the microlithography now used to produce electronic microcircuits with a photoresist, a compound similar to bitumen of Judea, although it exhibits a higher sensitivity.

Photograph 1-8. *An image identical to the one of Photograph 1-7, after treatment with iodine vapors followed by elimination of the varnish.*

Acknowledgement

The author would like to thank Mrs. Michèle Minana for her technical assistance during this work.

References

1. F. D. Arago, Compte Rendus Acad. Sci., TIX, No. 8, p. 250 (1839).
2. L. J. M. Daguerre, *Historique et description des procédés du Daguerréotype* (Paris, 1839).
3. J. L. Marignier, Le Photographe, No. 1480, p. 50 (1990).
4. J. L. Marignier, *Nature* 346, 115 (1990).
5. J. L. Marignier, *J. Chim. Phys.* 88, 865 (1991).
6. V. Fouque, *La vérité sur l'invention de la photographie. Nicéphore Niépce sa vie, ses essais, ses travaux* (Paris, 1867).
7. T. P. Kravetz, *Documentii po istorii izobretenia Fotografii* (Académie des Sciences d'URSS, Léningrad-Moscou, 1949).
8. J. N. Niépce, *Lettres 1816–1817* (Rouen, 1973).
9. J. N. Niépce, *Correspondances 1825–1829* (Rouen, 1974).

10. Th. Wedgwood and Sir H. Davy, *J. Royal Inst. of G. Brit.* 170–174 (1802).
11. J. N. Niépce, Letters of 2 June 1816.
12. Liquid CS_2.
13. E. Becquerel, *La lumière ses causes et ses effets* (T2, Paris, 1867), p. 81.
14. Like petroleum for lamps.
15. Notice about Heliography.
16. In fact it was a copper plate covered with a thin foil of silver.
17. The Paris line is an old unit of measurement of the eighteenth century. It corresponds to 2.25 millimeters.
18. M. Boussingault, *Ann. Chim. Phys.* 64, 141 (1837).
19. J. G. Speight and S. E. Moschopedis, *Adv. Chem. Ser.* 195, 1 (1981).
20. R. B. Long, *Adv. Chem. Ser.* 195, 17 (1981).
21. T. F. Yen, *Adv. Chem. Ser.* 195, 39 (1981) and references therein.
22. J. G. Erdman, W. E. Hanson, and T. F. Yen, *J. Chem. Eng. Data* 49, 2363 (1977).
23. It is necessary to keep in mind the words written by J. G. Speight: *"A key feature in the current concept of asphaltene structure is believed to be the occurrence of condensed polynuclear aromatic clusters, which may contain as many as twenty individual rings in account for approximately 50% of the asphaltene carbon. However, it would be naive to presume that precise (or meaningful 'average') molecular structures can be deduced by means of any spectroscopic technique when too many assumptions (incorporating several unknown factors) are required to derive the structural formulae."* (*Adv. Chem. Ser.* 195, 7 [1981]).
24. T. F. Yen, J. G. Erdman, and S. S. Pollack, *Anal. Chem.* 33, 1587 (1961).
25. J. P. Dickie, M. N. Haller, and T. F. Yen, *J. Colloid Int. Sci.* 29, 475 (1969).
26. S. S. Pollack and T. F. Yen, *Anal. Chem.* 42, 623 (1970).
27. J. PH. Pfeiffer and R. N. J. Saal, *J. Phys. Chem.* 44, 139 (1940).
28. J. Kosar, *Light Sensitive Systems*, (Wiley and Sons, New York, 1965).
29. E. Chevreul, Compte Rendus Acad. Sci., Session of August 28th, 1854.
30. R. R. Thurston and E. C. Knowles, *Ind. Eng. Chem.* 33, 322 (1941).
31. L. J. M. Daguerre, *Historique et description des procédés du Daguerreotype* (Paris, 1839), p. 42.
32. J. B. Biot, Le journal des Savants, Mars 1839.
33. At the beginning of the nineteenth century in France, this inch was the Paris inch, corresponding to 27.07 mm.
34. J. L. Marignier, Le Photographe, No. 1499, p. 26 (Nov. 1992).
35. J. L. Marignier, *Photo Antiquaria*, Hürth (Germany) p. 17, April 1991.
36. J. G. Speight, *Adv. Chem. Ser.* 217, 201 (1988).

33

elopments in uring Chemistry

Christian Decker

1. Summary

Radiation curing has found a growing number of applications in recent years, because of the unique characteristics of this advanced technology—mainly its great speed, selective cure, and 100% solid formulation. The performances of some newly developed photoinitiators (PI) are briefly described in this survey on the chemistry of UV-curable systems. With these compounds, substantial progress has been achieved with respect to both the cure speed and the cure penetration for clear and pigmented coatings exposed to UV radiation or visible light. Differential through cure profiles and PI decay curves, recorded by means of real-time IR and UV spectroscopy, are presented. This method of kinetic analysis proved of much value in assessing the efficiency of new photoinitiators and in determining the importance of the internal filter effect due to the presence of light stabilizers, PIs, or pigments. The monomer used to lower the formulation viscosity plays a key role, for it acts both on the cure speed and on the polymerization extent, as well as on the final properties of the cured material. The performances of different types of monomers in radiation-curable resins are reported, in particular acrylates containing a cyclic structure or a silicone group in the monomer unit, epoxy silicones, and vinyl ethers. Some of these new monomers were found to exhibit an outstanding reactivity, with formation of crosslinked polymers that show remarkable mechanical properties. Such high-performing compounds are expected to find their major openings in areas in which fast extensive cure, as well as product quality, are of prime importance.

2. Introduction

Radiation curing has now become a well-accepted technology that has found its main applications in various industrial sectors in which ultrafast cure and high quality product are required.[1–6] Ultraviolet (UV) or electron beam (EB) curing is indeed unique in that it achieves, selectively in the exposed areas, a quasi-instantaneous transformation of a liquid resin into a solid polymer. The ability to tailor polymer properties, together with very high cure rates, has resulted in the technique being adopted worldwide in an ever-increasing number of end uses. The radcure market, which has long been restricted to the coating industry for producing fast-drying varnishes and printing inks, has progressively invaded new sectors of applications, mainly in microelectronics, photoimaging, adhesives, and composites. Today, the UV curing technology has gained a predominant position in some specific industrial sectors, such as in the graphic arts, the coating of optical fibers, and laser-stereolithography.

During the last decade, great progress has been achieved in the chemistry of radiation-curable systems, with the development of ever-more-efficient photoinitiators (for a comprehensive, up-to-date review, see Ref. 7) and highly reactive new monomers[4,8–11] and oligomers[5,12] for both clear and pigmented formulations.[13] There is currently a large choice of commercially available products, thus allowing one to select the proper formulation components in order to meet the cure speed and product properties requirements for the particular application considered. This is the main reason why research and development in radiation curing has shifted over the years from basic chemistry toward the application sector, taking full advantage of the remarkable performance of radiation-curable systems. This tendency is best illustrated by the fact that the proportion of "chemistry" papers presented at RadTech conferences has dropped from 50% to 25% over the past 10 years, with a concomitant increase of the "applications" papers. As a result of the nice work accomplished by chemists, today's trend in radiation curing appears to be more toward the development of new end-uses, rather than new products.

In this paper, we will first outline some of the important requirements that are now imposed on radiation curing chemistry, before reporting the progress made in these areas with respect to both the photoinitiators and the monomers. Special attention has been directed toward the chemistry of light-induced curing, as this sector still covers more than 90% of the radiation curing market. However, because the main difference between UV and EB curing lies in the initiation step, any progress made in the development of new monomers and oligomers is likely to be directly applicable to EB formulation.

3. Today's Requirements in Radiation Curing Chemistry

Despite the great progress achieved in UV and EB formulation,[13] several pending problems still need to be addressed in order to ensure further expansion, and overcome the manifold barriers to using this new technology.[14]

3.1. Cure Speed

The development of highly efficient photoinitiators and of very reactive monomers and functionalized oligomers has led to such an increase of the photosensitivity of today's UV-curable resins that the speed of cure is no longer a top priority, except for some specific applications such as optical fiber coating, stereolithography, and direct imaging. Most of the research efforts devoted to improving the formulation reactivity are focused in two main areas where more effective photoinitiating systems are required in order to meet the end user demands:

- The UV curing of pigmented coatings, i.e., paints and lacquers, where it is still difficult to achieve a fast deep-through cure of thick films, because of the screen effect of the pigment particles
- The curing with visible light required for some specific applications, e.g., dental materials, printing plates, or laser imaging, where cure rates are significantly lower than with UV radiation.

Another area in which faster cure would be highly desirable is that of photoinitiated cationic polymerization, which has experienced a slower growth than the widely used radical polymerization of acrylates, partly because of a lower reactivity.

3.2. Cure Extent

A well-known feature in the synthesis of tridimensional networks by polymerization of multifunctional monomers is that chain propagation usually stops before the reaction is completed. This is mainly due to the gelation of the medium undergoing crosslinking polymerization and related mobility restrictions of the reactive sites in the cured material. Other factors responsible for the cure limitation include the formulation viscosity, the stiffness of the polymer formed, the consumption of the photoinitiator, and the inhibiting effect of oxygen. In most radiation curing applications, it is necessary to reach degrees of conversion as high as possible, because the amount of unreactive monomer is known to adversely affect the long-term properties of the cured material. This can be achieved by using monofunctional diluents and by selecting the proper resin-to-monomer ratio, in order to maintain fluidity and molecular mobility up to high conversions. Unfortunately, switching from multi- to monofunctional monomers usually causes a severe reduction in the cure speed. It has thus become a real challenge to find low viscosity reactive diluents that would afford both fast and complete polymerization.

An additional difficulty arises in coatings cured by UV light, as the radiation filter effect of the photoinitiator limits the penetration of light and leads to the development of a through-cure differential.[15] Poor adhesion often results from the surface-to-depth cure gradient and the resulting insufficient polymerization at the coating-substrate interface. The usual way to solve this problem consists of reducing the light absorbance of the sample by decreasing the PI concentration, but at the expense of the cure speed. Another solution is to use a photoinitiator that undergoes an effective bleaching reaction upon photolysis, thus allowing UV radiation to penetrate deeper into the sample, as curing proceeds.

High conversion and deep-through cure are especially difficult to achieve with UV light in thick pigmented coatings, whereas homogeneous cure is readily obtained by means of electron beam irradiation, thus ensuring a good adhesion. Indeed, one of the important requirements in the coating industry is to achieve a good adhesion by matching the characteristics of the resin formulation with that of the substrate surface. Reducing the shrinkage and internal stress and improving the resin wettability are effective ways to help improve interfacial adhesion.

3.3. Properties of Cured Polymers

The properties of radiation-cured polymers depend primarily on the chemical structure, functionality, and concentration of the various formulation components, as well as on the cure extent and on the irradiation conditions (temperature, presence of air). The type of monomers and oligomers selected, together with their relative proportion, will directly affect the mechanical properties of the cured product. Monofunctional monomers, associated with aliphatic polyurethane chains, lead to elastomeric films that are well suited to coat flexible supports. By contrast, multifunctional monomers, associated with epoxy derivatives, lead to stiff and hard materials that are more appropriate to protect rigid substrates.

In some surface treatment applications, one of the prime objectives is to produce hard but still flexible coatings that exhibit a great resistance to both shocks and scratching. Such conflicting requirements are often difficult to achieve with a single formulation, even by careful selection of the formulation constituents. They can usually be fulfilled only through the double layer concept: a soft, impact-resistant coating covered by a hard top coat. Other applications require formulations that show little shrinkage upon curing (stereolithography), or that produce polymer materials having a great resistance to heat, chemistry, moisture, weathering, and harsh environment conditions.

3.4. Safety

One of the great advantages of radiation curing technology is that these 100% solid formulations contain no solvent, thus preventing the emission of polluting vapors during the curing process. This advantage is partly offset by the fact that both free radical and cationic curing products may present a technological hazard and can be potent skin irritants or sensitizers. Some monomers commonly used as reactive diluents in UV and EB formulations, such as hexanediol diacrylate (HDDA), pentaerythritol triacrylate (PETA), or *n*-vinyl pyrrolidone (NVP), are being progressively replaced by less harmful products.

A considerable research effort was recently initiated by chemists to address this serious problem and to try to develop environmentally friendly diluents that would cause less eye and skin irritation, and have a lower Draize index. For instance, alkoxylated acrylate monomers have been found to be much less irritating than the conventional acrylates, while still exhibiting similar cure performances.[16] The potential of nonacrylate systems is also being increasingly explored. Provided that a standard industrial practice is maintained, the radcure technology is not presenting more or less of a handling hazard than any existing coating formulation.[14]

4. Recent Progress in UV Curing Chemistry

The most significant improvements in UV-curable systems have been obtained by acting on the nature of both the photoinitiator (PI) and the resin components, in particular the reactive diluent. Some of the progress made in this area over the past few years will be described here, with special emphasis being placed on the performance under usual working conditions, i.e., thin films exposed to intense radiation in the presence of air.

4.1. New Photointiators

The main research efforts on photoinitiation are currently concentrated in two important directions:

1. The development of photoinitiators best suited for the UV curing of pigmented coatings, e.g., printing inks, paints, or lacquers widely used in the graphic arts and the surface protection of materials
2. The development of highly sensitive photoinitiators absorbing in the visible range, mainly for imaging and recording applications.

Other related problems of concern for the end users have also been recently investigated, such as the differential through cure due to the PI or pigment screen effect, the fate of the photoinitiator, and its photolysis kinetics. In this respect, it should be mentioned that the negative effects due to the long-term migration of the residual PI and photoproducts can be successfully overcome by the use of copolymerizable photoinitiators.[17]

(a) *UV Photoinitiators*

Among the photoinitiators newly developed for UV-curing applications, special attention should be given to an α-aminoalkylphenone recently introduced by Ciba under the trade name Irgacure 369, which proved to be very efficient in both clear and pigmented acrylic systems.[18] Upon UV exposure, this compound undergoes a rapid photocleavage reaction that generates the highly reactive free radicals needed to initiate the polymerization process:

$$O(CH_2CH_2)_2N{-}C_6H_4{-}\underset{\underset{O}{\|}}{C}{-}\overset{C_2H_5}{\underset{CH_2C_6H_5}{C}}{-}N(CH_3)_2 \xrightarrow{h\nu} O(CH_2CH_2)_2N{-}C_6H_4{-}\underset{\underset{O}{\|}}{\dot{C}} + CH_3{-}CH_2{-}\underset{CH_2C_6H_5}{\dot{C}}{-}N(CH_3)_2$$

Irgacure 369

The marked increase in the cure speed provided by this photoinitiator is illustrated by the polymerization profiles shown in Fig. 2-1 for a clear epoxy-acrylate coating. It was shown to result primarily from a larger absorption of near-UV radiation (300–400 nm), rather than from a greater yield in radical production.[19] This is the reason why the performance improvement was more pronounced in pigmented systems than in clear varnishes.[18] Moreover, a fast photobleaching reaction was found to occur with Irgacure 369,[20] thus allowing UV radiation to penetrate progressively deeper into clear formulations and providing extensive through cure of samples a few centimeter thick.[21] The major drawback of this very efficient photoinitiator is that upon irradiation it generates slightly colored products.

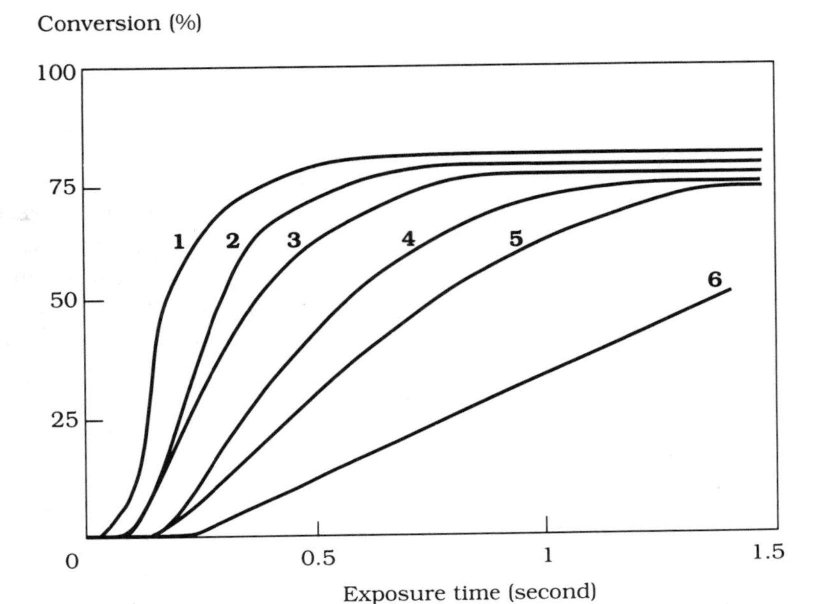

Figure 2-1. *Influence of the photoinitiator (1%) on the UV curing of an epoxy-acrylate resin. Light-intensity: 60 mW/cm². 1: Irgacure 369, 2: Lucirin TPO, 3: Quantacure CPTX, 4: Irgacure 651, 5: Darocur 1173, 6: Benzophenone + MDEA.*

Acylphosphine oxides belong to another class of α-cleavage photoinitiators that have also proved quite efficient for curing rapidly clear and pigmented coatings, especially in O_2 free conditions.[22]

$$H_3C\text{–}C_6H_2(CH_3)_2\text{–}C(=O)\text{–}P(=O)(C_6H_5)_2 \xrightarrow{h\nu} H_3C\text{–}C_6H_2(CH_3)_2\text{–}\dot{C}(=O) + \cdot P(=O)(C_6H_5)_2$$

Lucirin TPO (BASF)

Both the benzoyl and the phosphinoyl radicals were found to initiate the polymerization of acrylates, vinyl ethers, or styrene. One of the great advantages of acylphosphine oxides is that the bleaching reaction leads to the formation of low-colored photoproducts. TPO thus appears as one of the best photoinitiators to cure up to 100-μm-thick white lacquers by UV irradiation, especially when it is combined with α-hydroxyketones (Darocur 4263 and 4265).

A substantial improvement in the UV curing efficiency of pigmented systems has recently been achieved by using propoxy-substituted thioxanthones as photoinitiators or sensitizers.[23] The best overall performance was obtained with the 1-chloro-4-isopropoxy thioxanthone (trade name Quantacure CPTX, available at Octel Chemicals), with respect to cure speed, hardness, and depth cure. In the presence of hydrogen donors, such as tertiary amines, CPTX excited states produce the very active α-aminoalkyl radicals that will, in turn, initiate the polymerization of acrylated resins.

O Cl
$+ R_1-N(R_2)-CH_2-R_3 \xrightarrow{h\nu}$
OH Cl
$+ R_1-N(R_2)-\dot{C}H-R_3$
O
C_3H_7
O
C_3H_7

Quantacure CPTX

Initiation

Propoxy-substituted thioxanthones can also act as sensitizers, passing on their high triplet energy to a photoinitiator that then produces free radicals. In a blue offset litho ink based on epoxy-acrylates, maximum cure speed was thus obtained by combining Irgacure 369 with Quantacure CPTX.[23] This type of photoinitiator system proved particularly effective for curing rapidly and to a full depth up to 100-μm-thick colored pigmented coatings.

(b) Visible Photoinitiators

Photopolymerizable systems that can be cured rapidly by visible radiation are required for some specific applications, such as direct laser imaging, holography, or photopolymerization color printing. A large variety of dyes have been tested as initiators or sensitizers,[24] but their industrial use in radiation curing has been restricted so far by their relatively low initiation efficiency and their poor dark stability. High performance visible photoinitiators have been developed in the past few years to overcome these limitations[7], thus opening new perspectives for visible light curing applications.

A fluorinated diaryltitanocene (Irgacure 784 from Ciba) has been recently developed to initiate the polymerization of acrylates under visible light exposure.[21] Because of its great photosensitivity and its wide absorption range, up to 500 nm, this photoinitiator is ideally suited for use of the previously mentioned visible light curing applications, as well as for the curing of UV opaque formulations used in the microelectronic industry and in dental compositions.

Most of the other visible photoinitiator systems consist of an association of a photoreducible organic dye (such as eosin, methylene blue, cyanines, or ketocoumarins) with different types of coinitiators, such as tertiary amines, peroxides, organotin compounds, borate salts, and imidazoles.[7] Synergistic effects can be observed by combining some of these photoinitiator systems with a chloromethyl substituted triazine[25], as illustrated by Fig. 2-2 for the eosin + MDEA system. For a polyester acrylate resin containing 1% eosin, the rate of polymerization was found to increase from 11 to 80 $\times$ 10^{-2} mol/ls by the addition of as little as 0.5% of this triazine. The titanocene + triazine combi-

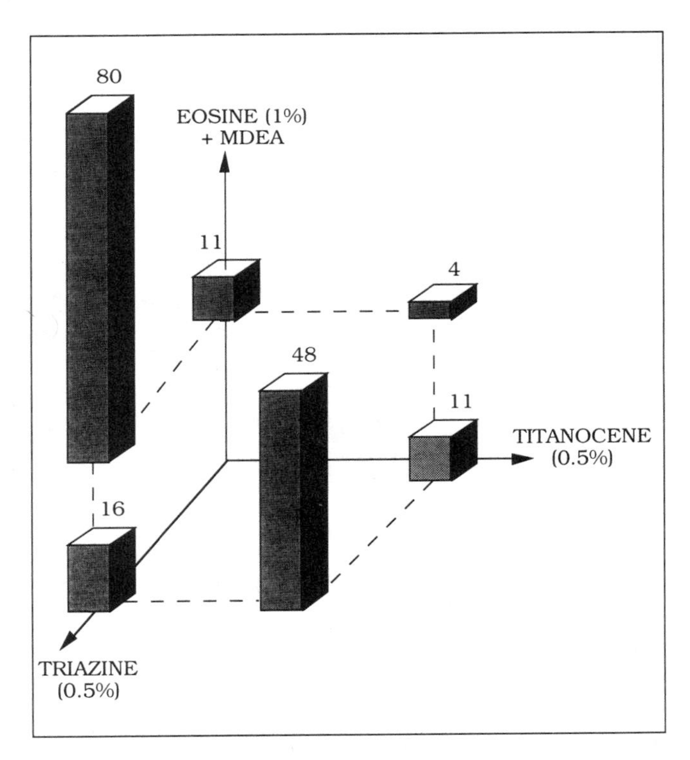

Figure 2-2. *Influence of the photoinitiator on the curing of an acrylate resin (Ebecryl 80 + Acticryl CL-959 in a 4 to 1 ratio) exposed to visible light in the presence of air. Film thickness: 40 μm; light-intensity: 16 mW/cm². Numbers refer to the polymerization rate (in 10^{-2}* mol/liter·s).

nation proved also to be highly efficient, whereas the eosin + MEDA + titanocene combination performed poorly, because of some antagonist effect of the tertiary amine with the titanocene. For a given acrylate resin irradiated in the presence of air, the best performing system was only two times less sensitive in the visible range than the Irgacure 651-based formulation in the UV range. Another type of photoinitiator system has recently been shown to be particularly effective for achieving a fast deep-through cure of acrylic paints.[26]

(c) Differential Through Cure

By contrast to the EB curing that develops uniformly throughout the irradiated coating, photoinitiated curing follows a surface-to-depth gradient, because of the limited penetration of light in those systems. In a clear formulation, UV radiation is absorbed mainly by the photoinitiator, so that the cure depth is directly controlled by the PI concentration. For each specific application, the best compromise must be found between cure speed and cure depth, the two extremes being either a uniform but slow deep-through cure for low absorbing coatings, or a rapid but differential through cure for highly absorbing coatings (at an optical density of 0.5, the bottom layer receives three times less light than the top layer).

It should be mentioned that if the PI photoproducts absorb at shorter wavelengths than the photoinitiator, UV radiation can progressively penetrate deeper into clear coats, thus rendering the cure process more uniform. Some photoinitiators (such as phosphine oxides or morpholino ketones) are particularly prone to such a photobleaching process (see next section). This phenomenon is illustrated by Fig. 2-3, which shows the cure depth profiles of a 40-μm-thick polyurethane-acrylate coating after various exposure times. It can be seen that the polymerization gradient progressively smooths down and that a quasi-uniform through cure is achieved after one second of irradiation.

The situation is quite different with pigmented coatings or with clear coats containing UV absorber type photostabilizers. Here the incident light is rapidly attenuated as it passes into the film, thus leading to a sharp cure gradient (Fig. 2-4). Prolonged UV exposure (>3 seconds) is then required in order to obtain sufficient curing at the coating/substrate interface and thus ensure a good adhesion.

(d) Photolysis of the Initiator

Although extensive research efforts have been devoted to the synthesis of photoinitiators and their mechanism of photolysis, there is scant information about the actual PI disappearance rate during UV curing operations. This quantity is still of prime importance, for it governs the rate of production of initiating species, which will in turn determine the rate of polymerization.

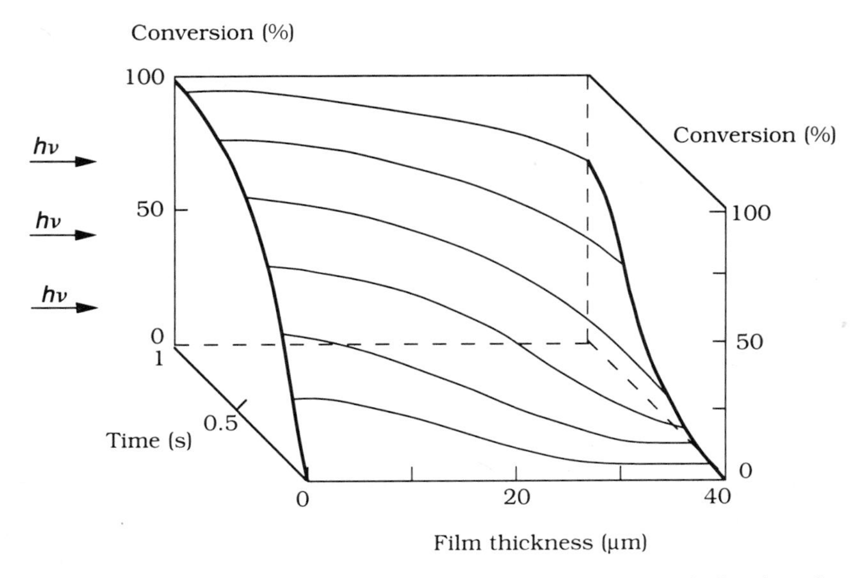

Figure 2-3. *Variation of the cure distribution profiles of a 40-μm-thick polyurethane-acrylate coating upon UV exposure.*

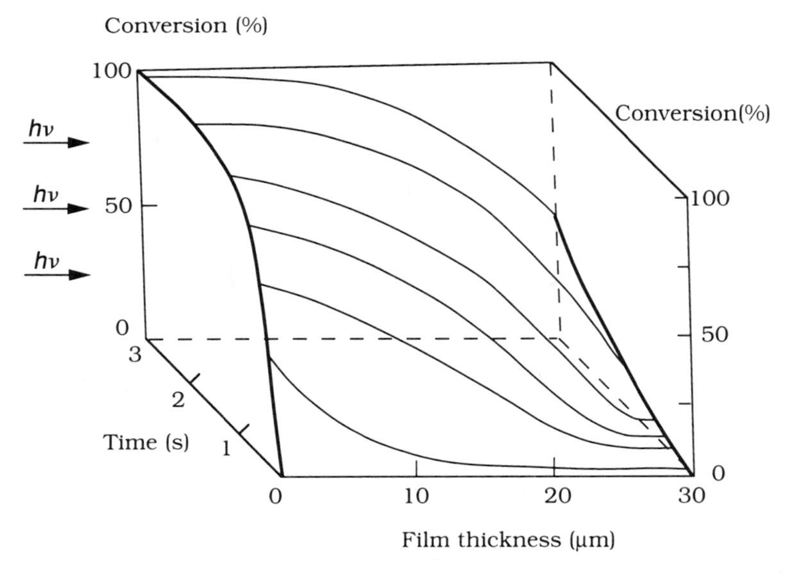

Figure 2-4. *Influence of a UV absorber (1.5% Tinuvin 900) on the cure distribution profiles of a 30-μm-thick polyurethane-acrylate coating exposed to UV radiation.*

Real-time UV (RTUV) spectroscopy was recently used to record the loss profiles of radical and cationic photoinitiators, under conditions similar to those used in UV curing applications.[20] With the morpholino-ketone Irgacure 369, photobleaching was shown to occur according to an exponential law (Fig. 2-5), the photolysis rate being at any time proportional to the PI concentration:

$$\frac{\log\,[\mathrm{PI}]_t}{[\mathrm{PI}]_0} = -kt$$

As expected for a direct photolysis process, the rate constant k was found to increase linearly with the light-intensity (I_0), the PI lifetime dropping to values below one second at high I_0 values (>200 mW/cm^2). One of the distinct advantages of RTUV spectroscopy is to permit a fast and accurate evaluation of the photolysis quantum yield (ϕ_{-PI}) for a given photoinitiator, simply by measuring the decay rate constant and using the equation:

$$\phi_{-\mathrm{PI}} = \frac{k\ (s^{-1})}{10^3 \times \epsilon\ (L\ \mathrm{mol}^{-1}\ \mathrm{cm}^{-1} \times I_0\ (\mathrm{einstein\ s}^{-1}\ \mathrm{cm}^{-2})}$$

where ϵ is the PI molar extinction coefficient. Its value was found to be inde-

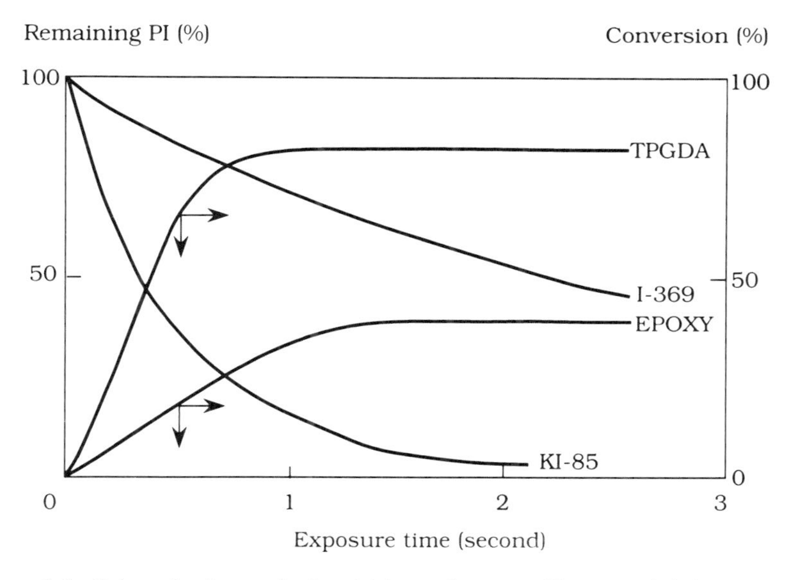

Figure 2-5. *Polymerization and photoinitiator decay profiles recorded by RTIR and RTUV spectroscopy for a TPGDA film containing 0.3% Irgacure 369 and a cycloaliphatic diepoxy film containing 5% Degacure KI-85, upon UV exposure at 55 mW/cm². Film thickness: 25 μm.*

pendent of the light-intensity, the PI concentration, the film thickness, and the presence of oxygen.[20]

It can be seen in Fig. 2-5 that the polymerization of the acrylated resin develops substantially faster than the PI photolysis, so that the cured polymer contains a large amount of unreacted photoinitiator. A quite different behavior was observed in the photoinitiated cationic polymerization of a cycloaliphatic diepoxide (Cyracure UVR 6110). The triarylsulfonium salt (Degacure KI-85) was rapidly photolyzed, thus causing an early stop of the polymerization (Fig. 2-5). Despite the fast PI photolysis and the related high initiation rate, the cationic polymerization of the epoxy monomer develops at a slower pace than in acrylic systems, because of much shorter kinetic chain lengths.[20] Further progress towards greater cure speed is thus expected to result from a substantial increase in the monomer reactivity, rather than in the initiation efficiency.

Real-time UV spectroscopy is a most valuable method to monitor quantitatively fast photolysis reactions and thus make an accurate evaluation of the photosensitivity of new initiators. When employed in conjunction with real-time IR spectroscopy, it allows both the amount of residual photoinitiator and the amount of unreacted monomer to be determined, at any stage of the polymerization. These new techniques should therefore be of great interest in the radcure industry to evaluate these two quantities, which are known to affect the long-term properties of UV-cured polymers.

4.2. New Monomers

Both the monomer and the functionalized prepolymer employed in radiation-curable formulations will determine the basic characteristics of the cured product. The chemical structure and functionality of the oligomer were shown to greatly affect the thermal stability and the chemical and weathering resistance, as well as the mechanical properties, such as the resistance to abrasion, scratching, and shocks. For example, aliphatic polyurethane chains give highly flexible and soft materials, whereas the stiff aromatic polyether or polyester chains yield hard and scratch resistant coatings.

The monomer used to lower the formulation viscosity also plays a key role, for it acts both on the cure speed and on the polymerization extent, as well as on the properties of the crosslinked polymer formed. An increase in the functionality of the reactive diluent was shown to accelerate the curing process, but at the expense of the overall conversion, thus leading to a polymer that contained a substantial amount of residual unsaturation.[27] The increased crosslink density also makes the cured polymer harder but less flexible and more brittle. Some newly developed monomers undergoing either radical or cationic polymerization will now be described at some length, with special emphasis being laid on their performance in UV-curable formulations.

(a) Acrylate Monomers

The chemical structure of the monomer has a strong influence on the cure kinetics and on the characteristics of the polymer formed.

Acrylates with cyclic structures. Introducing an oxazolidone function in the structural unit of a monoacrylate was found to greatly increase the cure speed.[27,28] When this monomer (Acticryl CL 959 from SNPE) was used as the reactive diluent in a polyurethane-acrylate resin, nearly 100% conversion was reached within milliseconds of exposure, with the formation of a soft, highly flexible, and impact resistant coating. By contrast, an epoxy-acrylate based resin generated a tough and hard, but still flexible polymer film after UV curing. An additional advantage of the oxazolidone-acrylate is that it is miscible in water and can be polymerized in a formulation containing up to 10% water.

Even better results were recently obtained with monoacrylate monomers containing a cyclic carbonate function (Acticryl CL 1042 and CL 1058 from SNPE), which showed outstanding reactivity.[11,29] This permits one to substantially reduce the photoinitiator concentration and thus cure with UV radiation sections a few centimeter thick. The light-induced homopolymerization was found to develop both faster and more extensively than with the usual di- or triacrylate monomers (HDDA, TMPTA, PETA), as shown by the conversion vs. exposure time curves of Fig. 2-6. A most surprising feature observed with these cyclic monomers containing only one acrylate function was that the UV-cured homopolymer remained totally insoluble in the organic solvents, just as with di- or triacrylates. This result strongly suggests that an efficient crosslinking process is taking place, probably through a chain transfer reaction.[30]

A distinct advantage of such nontoxic and low-irritating monomers is to impart both hardness and flexibility to UV-cured polyurethane-acrylate coatings (Table 2-1), which also exhibit good resistance to organic solvents, chemicals, and weathering. These new acrylate monomers should find their main uses in areas in which a fast and complete cure is required to produce polymer materials with tailor-made properties.

Silicone-acrylates. Silicone-based polymers have experienced a fast growth during the last decade because of their unique properties, and are used in an ever-growing number of applications.[31] Radiation curing has proved to be a particularly efficient technique to produce rapidly crosslinked silicone polymers by photopolymerization of multifunctional monomers and oligomers containing silicone groups in their structural unit. (For a comprehensive review on this subject, see Ref. 32.)

The rate of polymerization of phenyl-silicone-based di- and triacrylates was shown to be much greater than that of standard diluents such as hexanediol

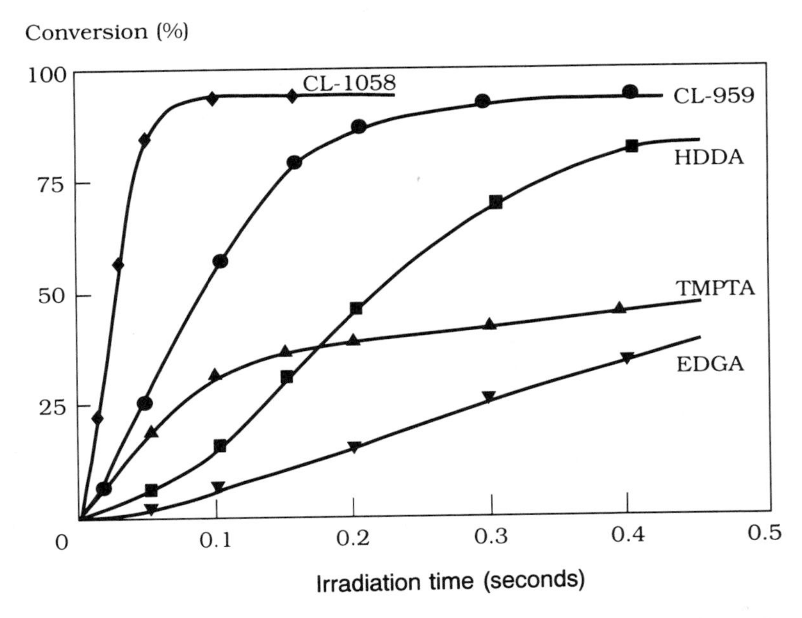

Figure 2-6. *Photopolymerization of acrylic monomers in the presence of air. [Irgacure 651]: 5%; film thickness: 25 μm; light-intensity: 500 mW/cm^2.*

diacrylate (HDDA), tripropylene glycol diacrylate (TPGDA), and trimethylolpropane triacrylate (TMPTA).[33] This effect is most probably due to the presence of the aromatic group, rather than to the presence of the silicone, as purely aliphatic silicone-triacrylates were found to cure significantly slower than TMPTA. An interesting and important feature of the films formed from the trifunctional silicone acrylates is the lack of visible shrinkage upon curing.[33]

Most of today's UV-curable acrylated silicones consist of either acrylate end-capped linear polydimethylsiloxanes or of silicone chains containing pendant acrylate groups.[34–37] The versatility of silicone-acrylates in regard to curing, coatability, compatibility, and release control make them ideally suited for a variety of applications, in particular for coatings having a desired release level.[38] Some newly developed silicone-acrylates show an outstanding reactivity, with reported line speeds of 600 m/min for both UV and EB curing in an inert atmosphere. As expected, atmospheric oxygen has a strong inhibition effect on such radical-induced polymerization,[34] especially in those resins that are highly permeable to oxygen.

Silicone-based polymers are known for their softness, thermal stability, insulating qualities, biocompatibility, high substrate wettability, and great permeability to gases. Such remarkable properties account for the increased use of

Table 2-1. Performance Analysis of Various Acrylate Monomers in UV Curing of a Polyurethane-Acrylate Coating in the Presence of Air. Actilane 20: 48%; Monomer: 48%; Irgacure 651: 4%. Light-intensity: 500 mW/cm^2; film thickness: 25 μm.

Monomer	Reactivity[a] (s^{-1})	Residual[b] Unsaturation (%)	Persoz[b] Hardness (s^{-1})	Mandrel[b] Flexibility (mm)
Monoacrylate				
- ether :EDGA	9	2	30	0
- oxazolidone:Acticryl CL 959	130	3	80	0
- carbonate :Acticryl CL 1058	125	4	270	0
:Acticryl CL 1042	165	6	100	0
Diacrylate				
- ether : HDDA	23	16	150	2
- carbonate : TPGDA	27	10	130	1
: Acticryl CL 993	200	14	250	2
Triacrylate				
TMPTA	110	36	270	5

[a]Reactivity = maximum rate of polymerization/initial acrylate concentration.
[b]Tack-free UV-cured coating.

UV-curable silicone acrylates in various sectors of applications,[39] mainly in the coating industry for the surface treatment of optical fibers, papers, and textiles. They also show good performance as pressure-sensitive adhesives, electronic circuit encapsulents, separation membranes, and release coatings.[32]

(b) Epoxy-Silicone Monomers

Epoxy-silicones ranging from low molecular weight siloxane monomers to high molecular weight polymers have been developed for a wide range of radiation-curable applications.[40–43] Among the most successful are those containing cycloaliphatic epoxy-functional silicones, which undergo a fast ring-opening cationic polymerization upon UV irradiation in the presence of iodonium or sulfonium salts.

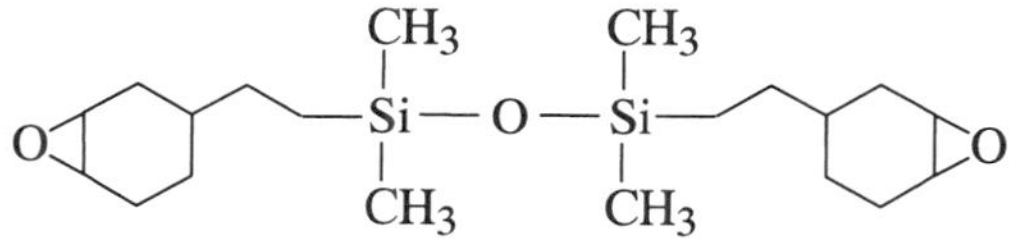

Difunctional cycloaliphatic epoxy-silicone

Copolymerization with alcohols or polyols was shown to improve the physical properties of the final product, with no loss of cure speed.[40] Novel expoxy-fluorosilicones hve recently been synthesized and found to combine exceptionally fast UV cure with good solvent resistance, making them potential candidates for conformal coatings or other applications in the automotive and aerospace industries.[44]

Epoxy-functional silicones are also well suited for EB curing applications.[45] The monomers containing cyclohexylepoxy groups show outstandingly high cure rates that compare favorably with those of the widely used multifunctional acrylates. In the presence of iodonium or sulfonium salts, an EB dose as low as 2 Mrad proved to be sufficient to achieve an extensive cationic-type polymerization. Epoxy monomers containing cyclic siloxane rings are also available, as well as multifunctional epoxy monomers with star and branched structures.[42] One of the additional benefits of epoxy-silicones is their ability to increase the cure rate of conventional organic diepoxides when incorporated as a blend.[44]

The excellent physical and chemical properties of the radiation-cured film (hardness, heat and solvent resistance, flexibility) make these epoxy-containing silicone monomers highly attractive for potential applications as coatings, adhesives, inks, and composites. Most noteworthy are the very high glass transition temperatures that were obtained for some of the cured polymers, up to 200° C after 5 seconds of UV irradiation at room temperature and the related remarkable thermal stability of such crosslinked materials.[42]

(c) Vinyl Ether Monomers

Today's growing concern for environmental issues of the radiation curing industry has led to the development of nonacrylate formulations that should be more friendly for both the manufacturer and the end user. In this respect, vinyl ethers appear as an attractive alternative. These monomers combine low toxicity with great reactivity, while also being excellent viscosity reducers. Several reports have appeared in the past few years on radiation-curable vinyl ether-based resins containing either divinyl ether diluents or vinyl ether-functionalized esters, urethanes, ethers, or siloxanes[46–49] (for a comprehensive review see Ref. 50).

Vinyl ether monomers are readily polymerized by a cationic mechanism in the presence of photoacid initiators. Monomers such as triethylene glycol divinyl ether (Rapi-Cure DVE-3 from ISP) have been used as reactive diluents for UV- and EB-curable epoxy resins to increase their reactivity.

$$CH_2{=}CH{-}O{-}(CH_2{-}CH_2{-}O)_3{-}CH{=}CH_2$$

Rapicure DVE-3

The UV-induced polymerization of DVE-3 was found to develop initially as fast as for triacrylates, but the reaction continued up to 100% conversion instead of leveling at 50% conversion[51] (Fig. 2-7). This is most probably due to the much lower viscosity of DVE-3 and the elastomeric character of the polymer formed, which gives a greater mobility to the reactive sites.

The combination of vinyl ether monomers and vinyl ether urethane oligomers, along with onium salt catalysts, provides versatile new UV- or EB-curable systems.[47] The curing was reported to proceed at high rates in the presence of oxygen and to produce coatings with desirable physical properties. Highly crosslinked polymers, showing a low degree of swelling, have been recently obtained by photoinitiated cationic polymerization of vinyl ether-styrene copolymers, which develops with long kinetic chains.[52] Similarly, the divinyl ether of bisphenol A was shown to cure rapidly under UV irradiation, to give a tough film having a low moisture absorption.[49] It is an excellent candidate for adhesive and UV coating applications, in particular on epoxy fiberglass.

One of the problems encountered in photoinitiated cationic polymerization is the poor solubility of onium salts in the usual monomers. A new reactive diluent has been especially designed for solubilizing cationic photoinitiators, in particular aryl sulfonium salts: the propenyl ether of propylene carbonate (Rapi-Cure PEPC from International Specialty Products).

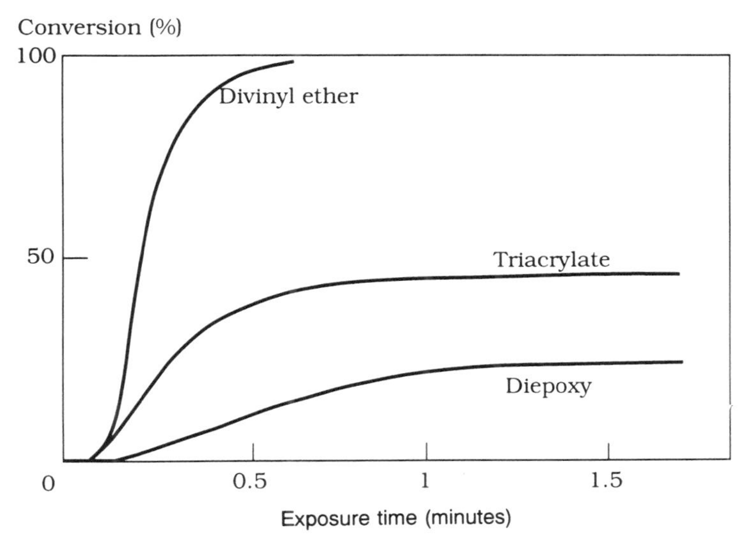

Figure 2-7. *Cure profiles obtained by differential scanning calorimetry analysis for vinyl ether, triacrylate, and diepoxy monomers exposed to UV radiation in an inert atmosphere (from Ref. 51).*

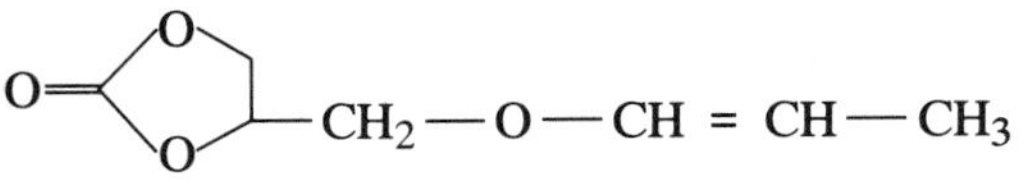

$$O=C\langle\text{cyclic carbonate}\rangle-CH_2-O-CH=CH-CH_3$$

PEPC

This low viscosity monomer proved particularly effective in the UV curing of vinyl ethers, cycloaliphatic epoxy resins, and epoxidized natural oils.[53] Besides its great solubilization capacity, PEPC's low toxicity, odor, and color make it easy to handle and use.

It should be mentioned that vinyl ethers can also polymerize by a free radical mechanism, in the presence of maleate or fumarate functionalized oligomers.[54,55] By using α-cleavage photoinitiators (such as Darocur 1173), the fastest copolymerization was observed for stoichiometric amounts of vinyl ether and maleate groups, as shown by Fig. 2-8. Because neither the vinyl ether nor the maleate monomer undergo significant polymerization when exposed to UV or EB radiation alone, a charge transfer complex must be formed in order to account for the great reactivity of the vinyl ether + maleate mixture. The cure speed of the low viscosity and low odor formulation is still significantly less than that of typical acrylate systems, for both UV and EB irradiation. Vinyl ether monomers have also been shown to polymerize readily with acrylates, by using either free radical[56] or hybrid cationic/free radical[57,58,59] photoinitiation, to form interpenetrating polymer networks. Unique film properties were achieved by replacing NVP or HDDA by such monomers. Vinyl ether monomers appear thus to offer a safe alternative to the reactive diluents currently being used in the radcure industry.[56]

5. Conclusion

Significant progress has been achieved in radiation curing chemistry during the past few years, with the development of new highly reactive photoinitiators and monomers. Future research efforts should be directed to areas where improvements in some of the performances of both the formulation and the cured product would be highly desirable in order to overcome some of today's limitations of this technology. For example, various methods have been proposed to reduce the strong inhibition effect of oxygen in radiation-curable systems, but so far none has proven really satisfactory. Another key issue in ensuring a steady growth of the radiation curing technology lies in the development of environment-friendly products that would still meet the performance of the compounds currently used in the radcure market.

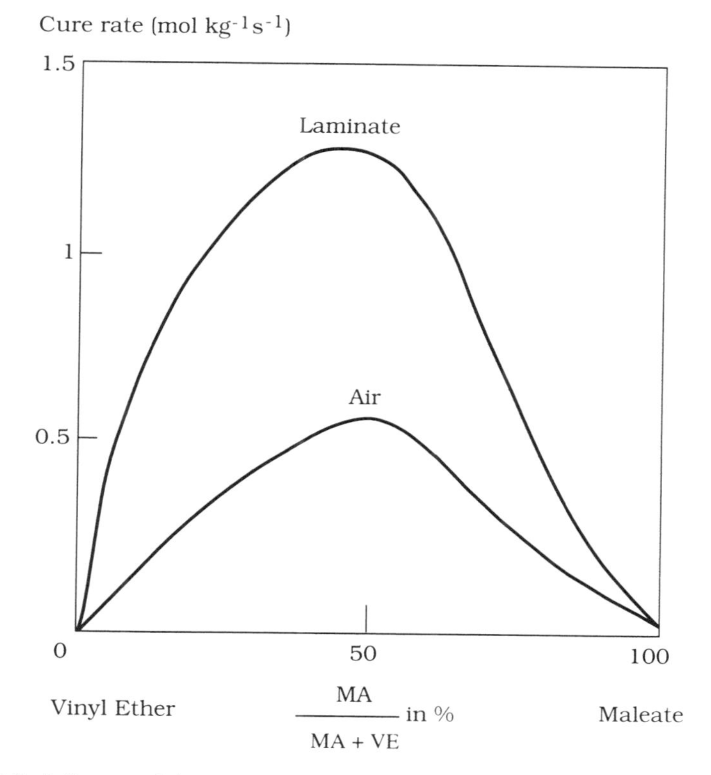

Figure 2-8. *Influence of the composition on the cure rate of vinyl ether (VE)-maleate (MA) resins exposed to UV light. [Darocur 1173]: 2%; Rapicure DVE-3 (GAF); Desolite R (Desotech).*

The distinct advantages of radiation curing processing—great speed, selective cure, and 100% solid formulation—have contributed to the continuous expansion of this technology, which has found a growing number of applications in recent years, mainly in the adhesives, composites, and coatings industry, as well as in microlithography, direct imaging, and stereolithography. Further applications of specially designed products are likely to appear in some new areas, such as:

- The manufacture of safety glasses by a cheaper and faster process than the now-used thermal treatment under very high pressure
- The long-term protection of organic materials by weather-resistant coatings
- The ultrafast drying of paints or lacquers

- The laser patterning of high-resolution optical guides for optoelectronic applications.

The future of radiation curing lies above all in the hands of chemists, with their ability to create new materials with tailor-made properties, and also in the skill of the engineer to fully exploit the performance and great potential of this advanced technology.

References

1. S. P. Pappas, ed., *UV-curing: Science and Technology*, Vol. 2 (Technology Marketing Corporation, Norwalk, 1985).
2. C. Decker, *J. Coat. Technol.* **59** (751), 97 (1987).
3. C. Decker, *Handbook of Polymer Science and Technology*, Vol. 3, N. P. Cheremisinoff, ed. (M. Dekker, New York, 1989), p. 541.
4. C. E. Hoyle and J. F. Kinstle, eds., *Radiation Curing of Polymeric Materials* (ACS Symp. Ser. 417), American Chemical Society, Washington, D.C., 1990.
5. P. K. T. Oldring, ed., *Chemistry and Technology of UV and EB Formulation for Coatings, Inks and Paints*, Vols. 1 and 2 (SITA Techn. Ltd., London, 1991).
6. S. P. Pappas, ed., *Radiation Curing. Science and Technology* (Plenum Press, New York, 1992).
7. K. Dietliker, *Chemistry and Technology of UV and EB Formulation for Coatings, Inks and Paints*, Vol. 3 (SITA Technol. Ltd., London, 1991).
8. R. S. Tu in Ref. 1, p. 143.
9. W. R. Watt in Ref. 1, p. 247.
10. C. Decker and K. Moussa, *Europ. Polym. J.* **27**, 403 (1991).
11. C. Decker and K. Moussa, *Makromol. Chem.* **192**, 507 (1991).
12. B. Martin in Ref. 1, p. 107.
13. P. K. T. Oldring, ed., *Chemistry and Technology of UV and EB Formulation for Coatings, Inks and Paints*, Vol. 4 (SITA Technol. Ltd., London, 1991).
14. R. Holman, *PRA Radnews*, Autumn 1992, p. 3.
15. L. R. Gatechair and A. M. Tiefenthaler in Ref. 4, p. 27.
16. Y. J. Wildi, *Proc. RadTech Conf.*, Florence 1989.
17. P. Battista, P. Giaroni, G. Li Bassi, C. Angiolini, C. Carlini, and N. Lelli, *Proc. RadTech 91 Conf.*, Edinburg 1991, p. 655.
18. V. Desobry, K. Dietliker, R. Hüsler, L. Misev, M. Rembold, G. Rist, and W. Rutsch in Ref. 4, p. 92.
19. C. Decker and K. Moussa, *Makromol. Chem.* **191**, 963 (1990).
20. C. Decker, *J. Polym. Sci. Polym. Chem. Ed.* **30**, 913 (1992).
21. W. Rutsch, H. Angerer, V. Desobry, K. Dietliker, R. Hüsler, *Proc. Intern. Conf. Organic Coatings. Science and Technology*. Athens 1990, p. 423.
22. M. Jacobi, A. Henne, and A. Boettcher, *Polym. Paint Colour J.* **175**, 636 (1985).
23. W. A. Green, A. W. Timms, and P. N. Green, *J. Radiation Curing* **19**(4), 11 (1992).

24. D. F. Eaton, *Adv. Photochem.* **13**, 427 (1986).
25. G. Buhr, R. Dammel, and C. R. Lindley, *Polym. Mater. Sci. Eng.* **61**, 269 (1989).
26. J. P. Fouassier and L. Catilaz, *Proc. RadTech Europe 93*, Mediterraneo 1993.
27. C. Decker and K. Moussa in Ref. 4, p. 439.
28. C. Decker and K. Moussa, *Europ. Polym. J.* **27**, 403 and 881 (1991).
29. C. Decker and K. Moussa, *Makromol. Chem. Rapid Commun.* **11**, 159 (1990).
30. C. Decker and K. Moussa, *Proc. RadTech Conf.*, Boston 1992, p. 260.
31. J. M. Zeigler and F. W. G. Fearon, ed., *Silicon-Based Polymer Science. A Comprehensive Resource* (Amer. Chem. Soc., Washington D.C., 1990).
32. A. F. Jacobine and S. T. Nakos in Ref. 6, p 181.
33. R. S. Davidson, R. Ellis, S. Tudor, and S. A. Wilkinson, *Polymer* **33**, 3031 (1992).
34. U. Müller, B. Strehmel, and J. Neuenfeld, *Makromol. Chem. Rapid Commun.* **10**, 539 (1989).
35. U. Müller, H. J. Timpe, and J. Neuenfeld, *Europ. Polym. J.* 27, 621 (1991).
36. U. Müller, S. Jockusch, and H. J. Timpe, *J. Polym. Sci. Polym. Chem. Ed.* **30**, 2755 (1992).
37. J. E. Thompson and J. Cavezzan, *Proc. RadTech 92 Conf.*, Boston, 1992, p. 212.
38. T. E. Hohenwarter, Jr., *Proc. RadTech 92*, Boston 1992, p. 108.
39. D. Wewers, *Proc. RadTech Europ. Conf.*, Edinburgh 1991, p. 1.
40. R. P. Eckberg, K. D. Riding, and D. E. Farley, *Proc. RadTech 90 Conf.*, Vol. 1, Chicago, 1990, p. 358.
41. R. P. Eckberg and K. D. Riding in Ref. 4, p. 382.
42. J. V. Crivello and J. L. Lee in Ref. 4, p. 398.
43. K. D. Riding, *Proc. RadTech 92*, Boston, 1992, p. 112.
44. R. P. Eckberg and E. R. Evans, *Proc. RadTech 92 Conf.*, Boston, 1992, p. 541.
45. J. V. Crivello, M. Fan, and D. Bi, *Proc. RadTech 92 Conf.*, Boston 1992, p. 535.
46. R. J. Brautigam, S. C. Lapin, and J. R. Snyder, *Proc. RadTech 90 Conf.*, Vol. 1, Chicago, 1990, p. 99.
47. S. C. Lapin in Ref. 4, p. 363.
48. S. C. Lapin and J. R. Snyder, *Proc. RadTech 90 Conf.*, Vol. 1, Chicago, 1990, p. 410.
49. Y. Okamoto, P. Klemarczyk, and S. Levandoski, *Proc. RadTech 92 Conf.*, Boston, 1992, p. 559.
50. S. C. Lapin in Ref. 6, p. 241.
51. S. R. Sauerbrunn, D. C. Armbruster, and P. D. Shiekel, *Proc. RadTech 90 Conf.*, Vol. 1, Chicago, 1990, p. 303.
52. M. P. Lin, T. Ikeda, and T. Endo, *J. Polym. Sci. Polym. Chem. Ed.* **30**, 2569 (1992).
53. J. S. Plotkin, J. A. Dougherty, M. Miller, K. S. Narayanan, and F. J. Vara, *Proc. RadTech 92 Conf.*, Boston, 1992, p. 703.
54. G. K. Noren, A. J. Tortorello, and J. T. Vandeberg, *Proc. RadTech 90 Conf.*, Vol. 2, Chicago, 1990, p. 201.
55. J. J. Shouten, G. K. Noren, and S. C. Lapin, *Proc. RadTech 92 Conf.*, Boston 1992, p. 167.
56. J. R. Snyder and G. D. Green, *Proc. RadTech 92 Conf.*, Boston, 1992, p. 703.
57. F. J. Vara and J. A. Dougherty, *Proc. RadTech 89 Conf.*, Florence, 1989, p. 523.
58. C. Decker and D. Decker, *Proc. RadTech 94 Conf.*, Orlando, 1994, p. 68.
59. C. Decker and D. Decker, *Intern. IUPAC Symposium on Macromolecules*, Akron, 1994.

II

Processes in Photoreactive Polymers

1

Time Resolved Laser Spectroscopy of Excited State Processes in Photoimaging Systems

Jean-Pierre Fouassier

The applications of lasers in the polymer photochemistry area spring from the meeting point of two fascinating research fields: Photopolymers and the Development of Laser Techniques. Lasers can be used as irradiation sources or in probing devices. In this chapter, we will primarily discuss the investigation of the main processes involved in the excited states of photosensitive organic molecules—photoinitiators, photosensitizers, photoinitiator combinations—that can be used in photoimaging systems for laser-induced polymerization, laser writing, or holographic recording.

Ultraviolet[1] curing has gained general acceptance since the 1960s. Its major applications are found in fields such as coatings, adhesives, inks, microcircuits, and printing plates. In typical applications (e.g., in reprography and holography, for direct writing, information storage, computer-driven pattern formation, 3D machining, manufacture of holographic optical elements) the possibility of using visible light sources, especially lasers, can be considered.[2] For this purpose, sensitizer absorption and wavelength of light emission have to be matched. Photopolymerization of organic multifunctional monomers under visible light can be induced through two different processes, photoinitiation or photosensitization.

The choice of photoinitiators is partly governed by the requirement of a high curing speed. Many reactive monomers and efficient photoinitiators have been synthesized and are commercially available. They make it possible to formulate a broad range of appealing photocurable coatings. Still, the fact remains that improved photoinitiator systems exhibiting ever-faster curing speeds and/or increased photosensitivity are required. The intrinsic reactivity in the excited states determines the interest of a given photoinitiator with reference to its efficiency. However, other factors, e.g., high molecular absorption coefficients and a broad range of spectral absorption (limiting the inner filter reabsorption in pigmented media), synergistic effects, yellowing, extractability, and the effect on the long-term stability of the material, must be considered as important items. Very attractive fields of research and development involve the roles played by decisive parameters affecting the reactivity of a given structure, such as chemical substitution in well-chosen positions, the character of water solubility, acid releasing ability, and copolymerization availability. Although the reactivity of photoinitiators can be successfully investigated through steady-state experiments, laser spectroscopy has proved a most convenient tool for the real-time investigation of the excited state dynamics and the processes involved.[3]

This chapter, which is concerned with a brief review of recent papers and patents relevant to this field of imaging, is focused on a discussion of the excited state processes (in solution or in bulk) in selected examples of efficient photosensitive systems for laser-induced polymerization reactions.

1. The Laser-Induced Polymerization Reaction

The basic features bear a strong resemblance to those of light-induced reactions under conventional lamps or flash lamps. The chemical and photochemical processes correspond to photopolymerization, photocrosslinking, or photomodification. Only the first process will be considered in this chapter. A photoinduced polymerization reaction is initiated through the decomposition of a photoinitiator PI under light exposure:

$$\text{PI} \xrightarrow{h\nu_1} \text{R}^{\bullet}$$

$$\text{R}^{\bullet} + \text{monomer} \rightsquigarrow \text{polymer}$$

$$\text{or PI} \xrightarrow{h\nu_2} \text{cationic species C}^{+}$$

$$\text{C}^{+} + \text{monomer} \longrightarrow \text{polymer}$$

The matching between the absorption spectra of the photosensitive organic compounds and the emission wavelengths of the light source (especially the laser lines) is obviously of prime importance. Photosensitizers (PS) have been frequently incorporated in the monomer/oligomer matrix:

$$PS \xrightarrow{h\nu_2} PS^*$$

$$PS^* + PI \rightsquigarrow \text{reactive species}$$

$$\longrightarrow PI^* \longrightarrow R^{\bullet} \text{ (or } C^+\text{)}$$

in order to achieve light absorption in a wavelength range where PI is transparent and/or to enhance the decomposition of PI (Fig. 1-1).

2. Examples of Photosensitive Systems for Laser-Induced Polymerization Reactions

A suitable substitution on the thioxanthone[4] backbone leads to molecules exhibiting higher wavelengths of maximum absorption. These compounds undergo

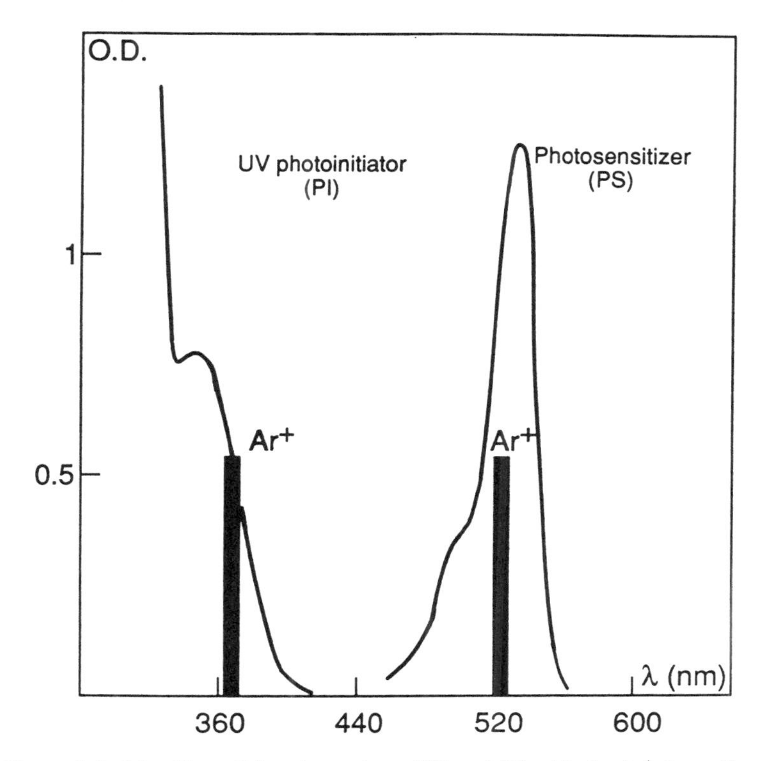

Figure 1-1. *Matching of the absorption of Pl and PS with the Ar^+ laser lines.*

the well-known electron transfer process from an amine, followed by a proton transfer reaction. No significant difference in the excited state processes has been observed.[5]

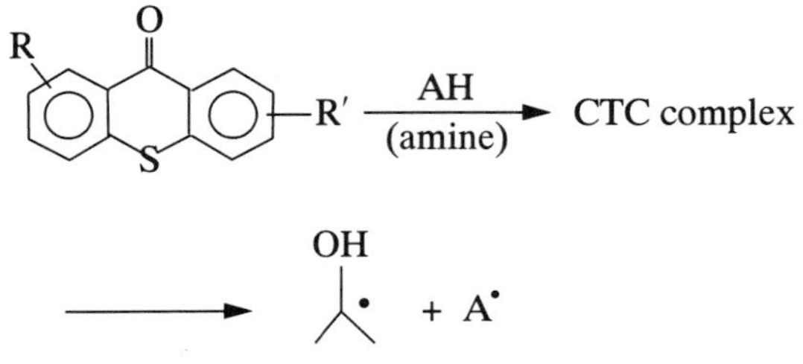

R, O, S, R′ —AH (amine)→ CTC complex

→ OH • + A•

Unsaturated ketone sensitizers (based on a chalcone structure) absorb radiation in the broad visible wavelength range. They can be used as photoinitiators or in conjunction with imidazolyl dimers and a free radical producing hydrogen or electron donor[6] or a thiopyrylium dye.[7] In the presence of dyes, imidazole derivatives are sensitive up to 400 nm. Upon irradiation, this molecule yields the corresponding triarylimidazolyl radical by cleavage of the single covalent bond. The addition of mercaptobenzoxazole as a hydrogen donor and dye sensitizer extends the efficiency.[8] Efficient holographic recording has been achieved with such a system.

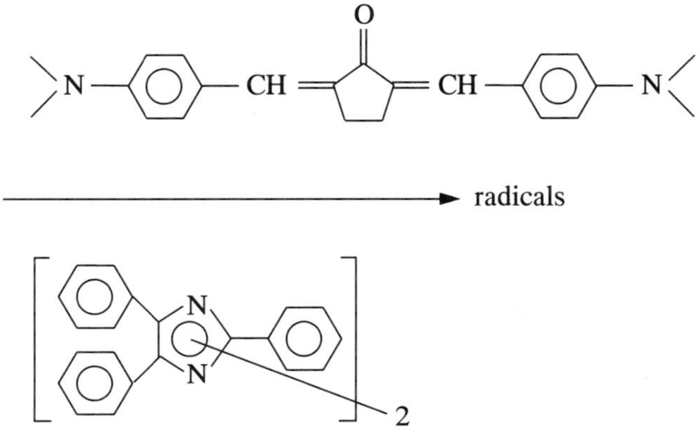

N — CH = O = CH — N

→ radicals

[N N]$_2$

Substituted ketocoumarins (whose light sensitivity ranges from 350–550 nm) can be chosen to match the imposed irradiation wavelength of a large variety of light sources and have been shown to be efficient triplet sensitizers for photocrosslinkable polymers.[9] In addition, if combined with amines, phenoxyacetic acid, or alkoxy-pyridinium salts, they are able to induce a free radical polymerization of acrylic monomers, whose initiation step was explained in terms of charge transfer or electron transfer between the additive and the ketocoumarin.[10]

The main advantage of metal salts as photoinitiators is their spectral sensitivity to visible light, because they are generally colored. The limitation is their water solubility, which governs the nature of the polymerization medium. In the past, various salts, complexes or chelates, have been proposed.[11] In the same way, iron arene complexes are suitable photoinitiators for cationic polymerization.[12] Their direct photosensitivity extends up to 550 nm. Photosensitive systems for microlithography, based on organometallic photoinitiators (e.g., titanocene derivatives[12]), have been proposed.

Perester structures are effective in vinyl polymerization, yielding alkoxy and aryloxy free radical pairs through homolytic decomposition. Depending on the nature of the chromophore moiety, strong absorption can be achieved over a wide spectral range.[13]

Organoborates were recently shown to be very effective because of an electron transfer in the intra-ion pair between the borate anion and the photosensitize cation (e.g., a cyanine dye moiety).[14]

Quinones, benzoquinone, camphorquinone, and anthraquinonic dyes have been recognized as polymerization initiators when irradiated with an argon ion laser or polychromatic visible light. Quinone derivatives are also able to sensitize the decomposition of cobalt hexamine complexes. Applications are found in the field of microelectronics and in curable sealants. Dye-initiated photopolymerization processes are usually considered.[15] The reaction occurs primarily through photoreduction of the dye (e.g., in the presence of amines, toluene sulfonic acid, or diketones). Dyes make up a large class of molecules, extensively used to sensitize the polymerization of acrylic monomers (e.g., xanthene, thiazine, acridine, anthraquinone, cyanine, and merocyanine). The same holds true for porphyrins, thiopyrylium salts, and typical dyes. Improvement of the photoreactivity of the sensitive recipe has been achieved by adding a benzoyl oxime ester,[16] or a phenylacetophenone,[17] or an onium salt[18] derivative to a mixture of eosin and amine. The addition of amino ketones to eosin[19] is used for printing plates with high resolution, useful in coating technologies or photographic relief imaging. The development of multicomponent photoinitiating systems has been recently achieved for 3D stereolithography,[20] laser printing plates,[21] and holographic recording.[8,22]

Cationic photopolymerization is currently initiated by a large variety of organic salts,[23] such as diaryliodonium, triarylsulfonium, aryldiazonium, triarylpyrylium, benzylpyridinium thiocyanate, dialkylphenacylsulfonium, dialkylhydroxyphenylsulfonium, and phosphonium salts.

Photosensitization in the presence of dyes is generally possible when they fail to absorb in the visible range. Diazonium salts are photolyzed with N_2 release. Their decomposition is sensitized by pyrazolone dyes, metal oxalates, tetraphenylporphyrin, and radicals. The primary steps of the initiation mechanism in iodonium and sulfonium salts still involve an electron transfer and a subsequent intra-pair reaction to generate the Brönsted acid HX.[24] The photosensitivity has been extended to higher wavelengths through an electron transfer reaction in the presence of aromatic rings or dyes, or through radical reaction in the presence of ketones. Measurements of the rate constant of the electron transfer suggest that this process is efficient.

Trialkylstannanes are a class of polymerization initiators sensitized by excited photoreducible dyes.[11] Transition metal carbonyl derivatives behave as efficient initiators of photopolymerization and agents of photocrosslinking in the application area of photographic processes. Dyes (such as cyanines and xanthenes) are effective sensitizers.

The different systems thus briefly reviewed operate through direct or sensitized absorption of a photon of an appropriate energy that corresponds to the energy difference between the ground and the excited singlet states. In specific applications such as holography,[25] it may be worthwhile to find systems working under two simultaneous irradiations of different wavelengths. To meet this requirement, several systems involving two-photon photochemistry have been developed; e.g., biacetyl or tetrazine,[25] a complex mixture containing a dye and a singlet oxygen trap,[26] or a combination of a spiropyran and an iodonium salt.[27] The first two systems are particularly suitable for recording holograms in red or near infrared light and for applications in the manufacture of holographic optical elements (HOEs).

3. Examples of Excited State Processes in Visible Laser Light Photosensitive Systems

This section discusses the excited state processes encountered in several typical systems that are photosensitive in visible laser light.

3.1. One Component Systems

(a) Iron-Arene Salts

The most widely used organometallic derivative for cationic polymerization in visible light that has found technical acceptance is based on iron-arene complexes.[12] The initiation occurs through a ligand exchange with three monomer units. The reactive excited state is presumably short lived.

$Fe^{+}PF_6^-$

$\xrightarrow{h\nu}$ $Fe^{+}PF_6^-$ +

$\longrightarrow$ Fe $(\;)_3^+$ PF_6^-

(b) Peresters

Nanosecond and picosecond spectroscopy have recently shown that the O—O bond cleavage occurs on the picosecond time scale.[28,29]

Dye C(=O)—O—O—R

$\xrightarrow{h\nu}$ Dye C(=O)—O$^\bullet$ $^\bullet$O—R

$\longrightarrow$ Dye$^\bullet$ + CO_2 + $^\bullet$O—R

In UV-absorbing peresters, a sensitizer such as a thiopyrylium salt can be used, so that decomposition occurs through visible light excitation (see Section 3.2).

(c) Organic Borates

Initiation of the polymerization reaction occurs through an intramolecular electron transfer between the cation moiety (organic dye, pyrylium salt) and the borate anion:[14]

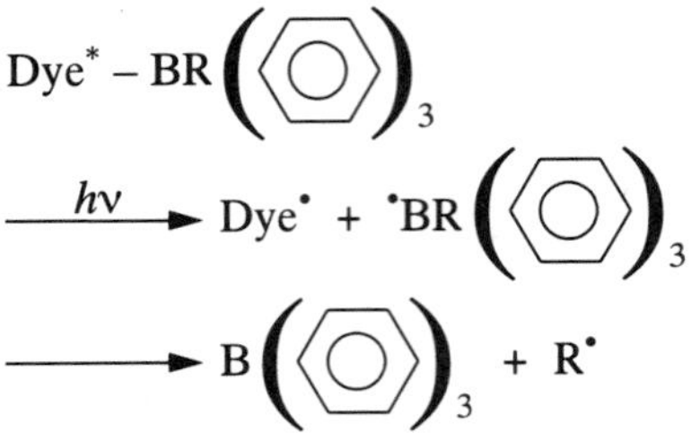

$$\text{Dye}^* - \text{BR}(\text{C}_6\text{H}_5)_3 \xrightarrow{h\nu} \text{Dye}^\bullet + {}^\bullet\text{BR}(\text{C}_6\text{H}_5)_3 \longrightarrow \text{B}(\text{C}_6\text{H}_5)_3 + \text{R}^\bullet$$

(d) Titanocenes

The primary step of the reaction is explained in terms of a fast ring slippage reaction yielding a highly electrophilic species that reacts with acrylates:[30]

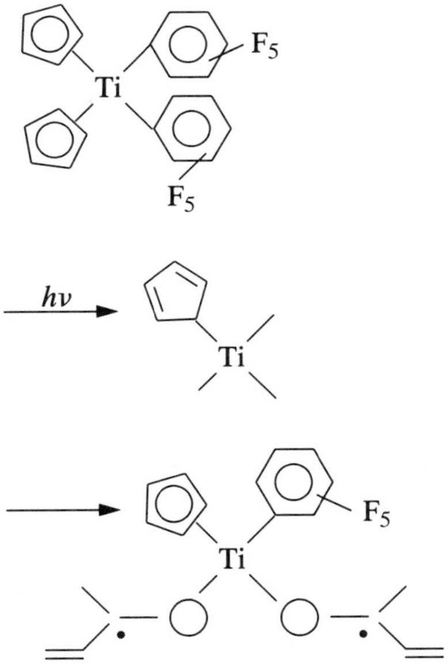

F_5 Ti Ti F_5 F_5 $h\nu$ Ti

3.2. Two Component Systems

Well-known examples of mechanisms are discussed as follows.

(a) Electron Transfer/Proton Transfer[11]

This process is illustrated and well documented with dyes, e.g.,

$$\text{Dye} \xrightarrow{h\nu} \text{Dye}^*$$

$$\xrightarrow[\text{AH}]{\text{amine}} \text{CT Complex}$$

$$\longrightarrow \text{Dye H}^\bullet + \text{A}^\bullet$$

(b) Energy Transfer

The energy transfer between PI and a suitable photosensitizers PS (where the relative triplet energy levels meet the requirements for an efficient process) has been recently exemplified by the combination thioxanthone derivative/substituted morpholino ketones (Fig. 1-2)[31], e.g.:

$$\text{TX (thioxanthone, COOR)} \xrightarrow{h\nu} {}^1\text{TX}^* \longrightarrow {}^3\text{TX}^*$$

$${}^3\text{TX}^* + CH_3S-C_6H_4-C(=O)-C(<)-N(\text{morpholino}) \longrightarrow {}^1\text{TX} + {}^3\left(CH_3S-C_6H_4-C(=O)-C(<)-N(\text{morpholino})\right)^*$$

$$\longrightarrow CH_3S-C_6H_4-\dot{C}(=O) \quad {}^\bullet C(<)-N(\text{morpholino})$$

The electron transfer has also been shown to occur according to the nature of PS, PI, and the medium:[31]

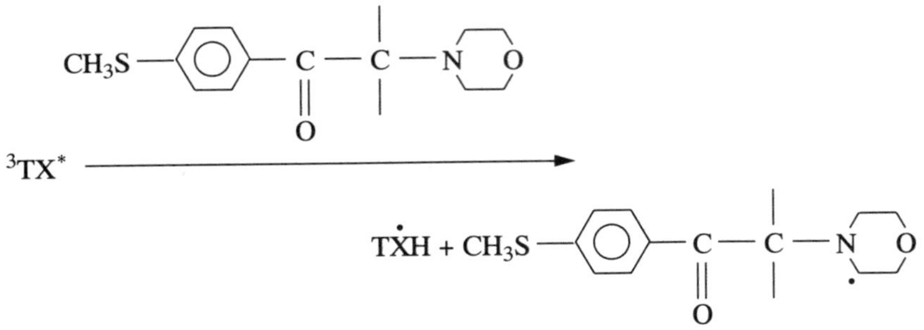

$$
{}^{3}TX^{*} \xrightarrow{CH_3S-C_6H_4-C(=O)-C(<)-N(CH_2CH_2)_2O} \dot{T}XH + CH_3S-C_6H_4-C(=O)-C(<)-\dot{N}(CH_2CH_2)_2O
$$

(c) Photoinduced Bond Cleavage Via Electron Transfer Reaction

Well-chosen examples[32] have been reported in past years regarding the sensitized decomposition of other morpholino ketones (where energy transfer cannot occur) in the presence of various PS (ruthenium bispyridine, dicyano (DCA) or tricyano (TCA) anthracene, and thioindigo dyes):

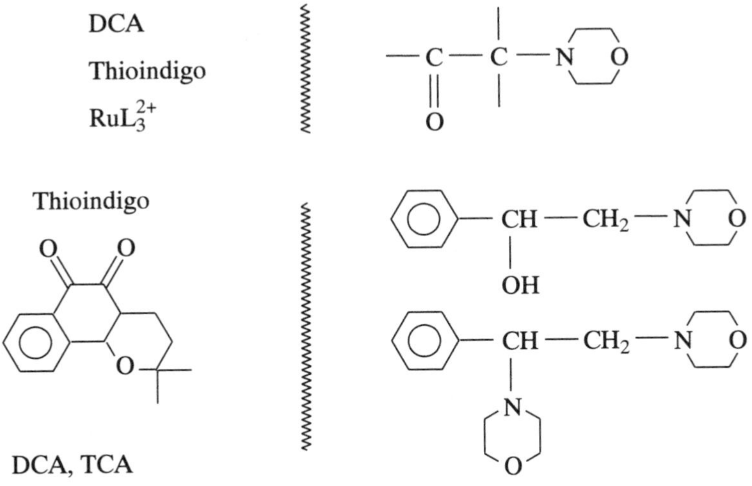

DCA
Thioindigo
RuL_3^{2+}

$-C(=O)-C(<)-N(CH_2CH_2)_2O$

Thioindigo

$C_6H_5-CH(OH)-CH_2-N(CH_2CH_2)_2O$

DCA, TCA

$C_6H_5-CH(N(CH_2CH_2)_2O)-CH_2-N(CH_2CH_2)_2O$

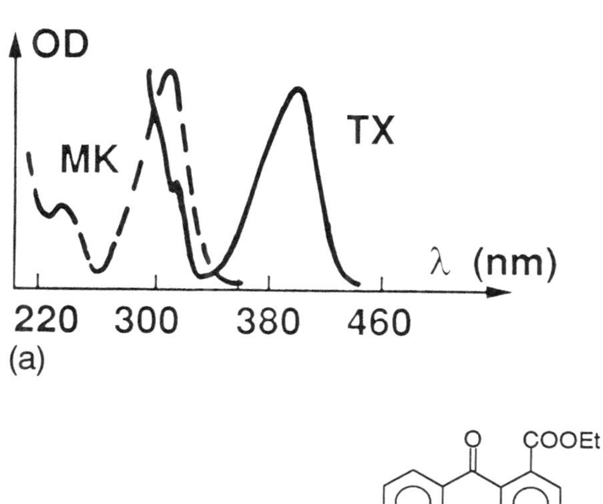

$E_T \sim 63$ Kcal / Mol

PI	R	E_T (kcal / mol)	
1a	-H	65	e^- transfer
1b	$-O-CH_3$	65	e^- transfer
1c	$-S-CH_3$	61	energy transer
1d	$-N(CH_3)_2$	63	e^- transfer (from $-N(CH_3)_2$

(b)

Figure 1-2. *Energy transfer vs. electron transfer in thioxanthones (TX)/morpholino ketones (MK). (a) Absorption spectra; (b) Triplet energy levels; (c) Stern-Volmer plots of the reciprocal value of the triplet state lifetime of thioxanthone as a function of the photoinitiator concentration. Solvent: Methanol; wavelength of detection: λ = 620 nm. (d) Rate constants of excitation transfer as a function of the substitution in thioxanthone.*

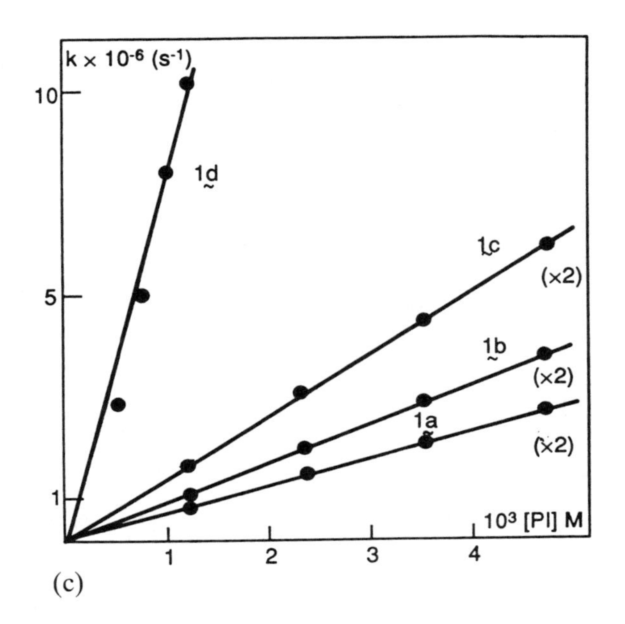

(c)

R1	R2	R3	R4	10-6 k_T ($M^{-1}s^{-1}$)	E_T (kcal M^{-1})	
H	COOR'	H	H	240	63	ET
H	H	CH(CH3)2	H	60	61.5	ET
CH3	H	H	COOR	12	58.5	eT

Solvent : Toluene

(d)

Figure 1-2. *Continued*

For RUL_3^{2+} as a photosensitizer, the following mechanism is proposed:

$$PS \xrightarrow{h\nu} PS^*$$

$$PS^* + C_6H_5{-}C({=}O){-}C(C_6H_5)({-}){-}N(CH_2CH_2)_2O$$

$$\longrightarrow PS^{\bullet -} + \left(C_6H_5{-}C({=}O){-}C({-})({-}){-}N(CH_2CH_2)_2O\right)^{\bullet +}$$

$$\longrightarrow C_6H_5{-}\dot{C}{=}O + ({-})C(C_6H_5){=}\overset{+}{N}(CH_2CH_2)_2O$$

$$\xrightarrow{PS^{\bullet -}} PS + {}^{\bullet}C({-})(C_6H_5){-}N(CH_2CH_2)_2O$$

(d) Electron Transfer

The sensitized decomposition of onium salts is observed in the presence of dyes, hydrocarbons, ketones, and radicals:[33]

$$>C{=}O \xrightarrow{h\nu} S_1 \longrightarrow T_1$$

$$T_1 + C_6H_5{-}\overset{+}{I}{-}C_6H_5 \longrightarrow >\overset{\bullet +}{C}{-}O + C_6H_5{-}\dot{I}{-}C_6H_5$$

$$\longrightarrow C_6H_5{-}I + C_6H_5^{\bullet}$$

e.g., $T_1 + (CH_3)_2CHOH \longrightarrow \;>C^{\bullet}—OH$

$$\xrightarrow{\;Ph-\overset{+}{I}-Ph\;} \;>\overset{+}{C}—OH + Ph-\overset{\bullet}{I}-Ph$$

$$\longrightarrow \;>C{=}O + H^+$$

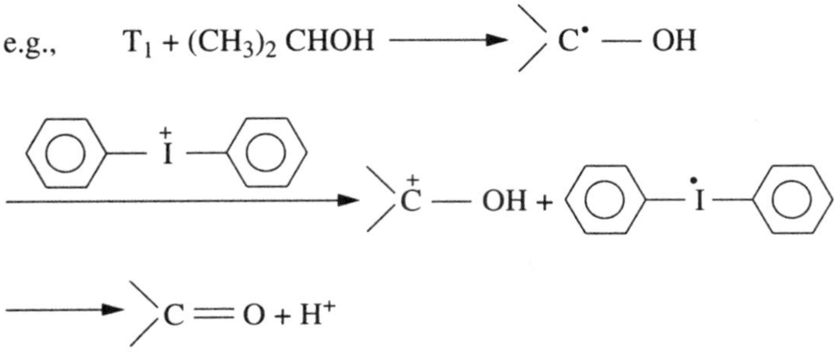

(e) Sensitized Decomposition of Peresters.

Thiopyrylium salt/perester As previously mentioned, if one proceeds under visible light exposure, PS can sensitive the decomposition of peresters such as benzophenone tetraperester (BTTB). When PS is a thiopyrylium salt derivative (TP), the excitation transfer occurs in the triplet state of TP, which is rather long lived (>10 μs in N_2 saturated acetonitrile), and leads to a radical. The mechanism is not clearly understood. BTTB itself can be cleaved, in UV light, within one nanosecond:[34(a)]

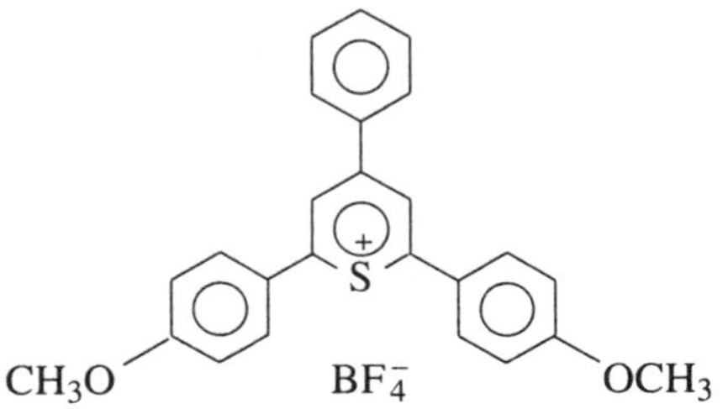

CH_3O BF_4^- OCH_3 $\overset{+}{S}$

TP

C O $C—O—OC(CH_3)_3$ O

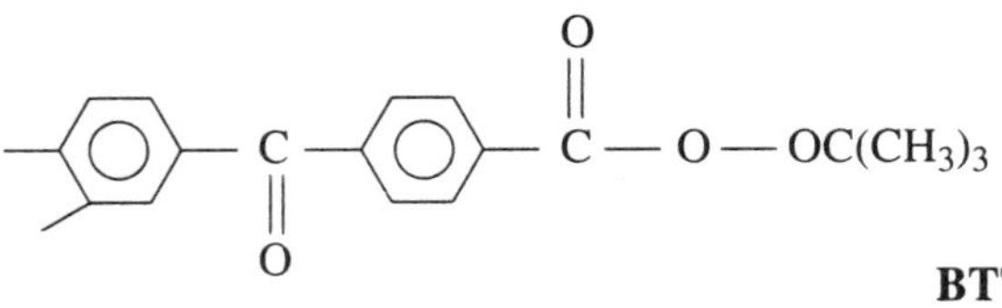

BTTB

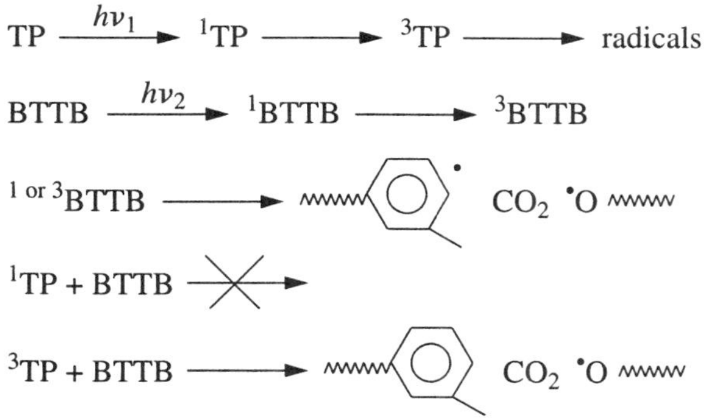

$$TP \xrightarrow{h\nu_1} {}^1TP \longrightarrow {}^3TP \longrightarrow \text{radicals}$$

$$BTTB \xrightarrow{h\nu_2} {}^1BTTB \longrightarrow {}^3BTTB$$

$${}^{1\text{ or }3}BTTB \longrightarrow \text{(aryl radical)} \quad CO_2 \quad {}^{\bullet}O$$

$${}^1TP + BTTB \not\longrightarrow$$

$${}^3TP + BTTB \longrightarrow \text{(aryl)} \quad CO_2 \quad {}^{\bullet}O$$

The mechanism of the triplet interaction between TP and BTTB is rather unclear.[34(b)] As shown in Fig. 1-3, the excitation transfer cannot be explained on the basis of either an energy transfer process from a triplet TP excited molecule to BTTB or an electron transfer process. The energy transfer from TP to a vibrationally hot O—O bond of the ground state, however, might be invoked.

Thioxanthene dye/perester. Another example relates to the system based on a thioxanthone dye and BTTB[35] where the following mechanism is proposed:

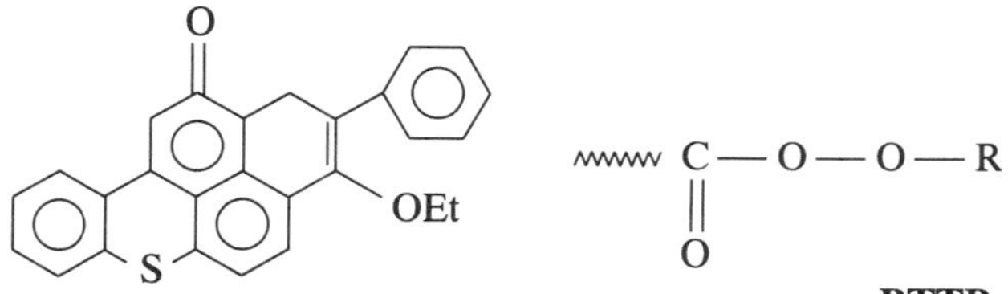

O OEt S C—O—O—R O **BTTB**

$$D \xrightarrow{h\nu} {}^1D^* \longrightarrow {}^3D^* + h\nu'$$

$${}^1D^* \xrightarrow{BTTB} BTTB^{\bullet -}$$

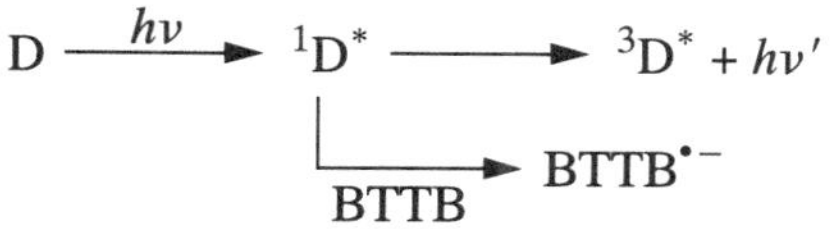

3.3. Three Component Systems

The basic idea of these systems is to try to enhance the photosensitivity by a judicious combination of several components. Several efficient systems have been recently reported, e.g.:

- Ketone/bromo derivative/amine[36]
- Ketone/amine/onium salt[18,37]
- Dye/amine/onium salt[38,39]

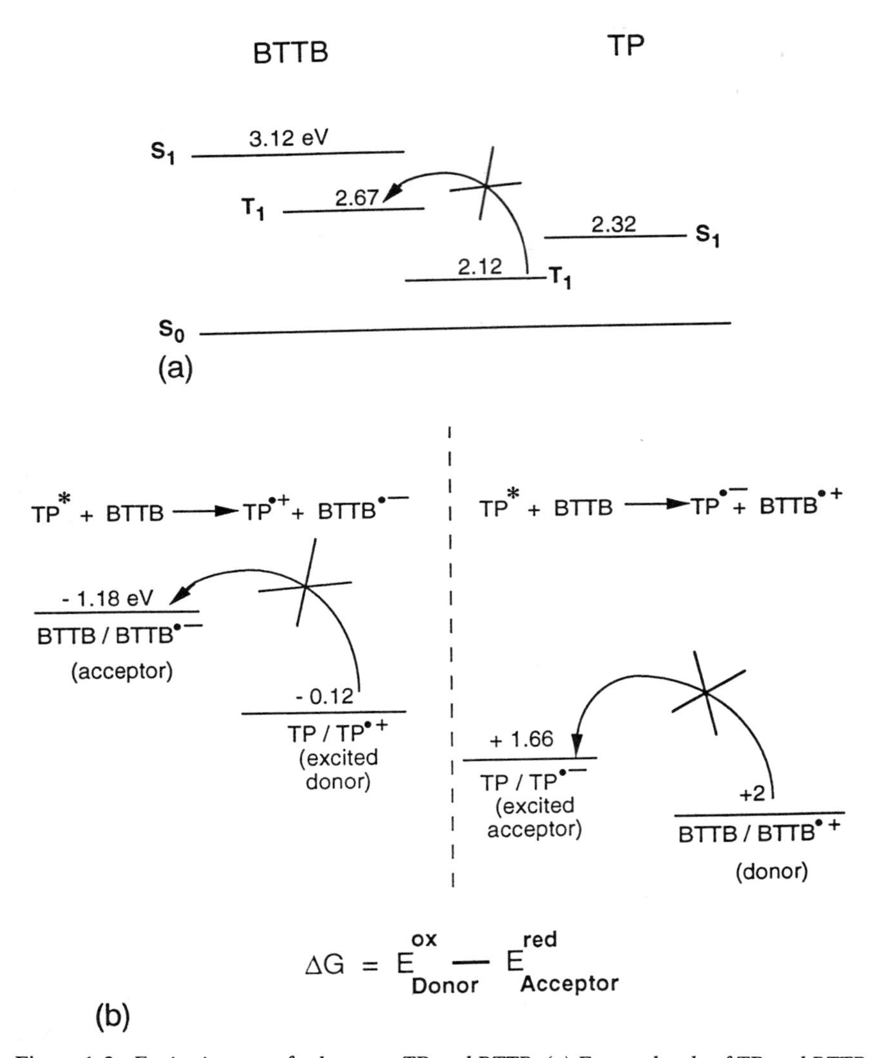

Figure 1-3. *Excitation transfer between TP and BTTB. (a) Energy levels of TP and BTTB. (b) Corrected values of the redox potentials of the excited state; ground state values: [From Ref. 34(b)]. As is known,* ΔG *should be* <0 *for having an electron transfer process occur. According to the reaction involved,* E^{ox} *and* E^{red} *refer to ground or excited state.*

- Dye/amine/ketone[40]
- Dye/triazine/electron donor[41]
- Dye/amine/bromo derivative[42] or dye/CBr_4.[22(e)]

Excited state processes have been investigated in three particular systems and will be discussed now.

(a) The Ketocoumarin/Amine/Onium Salt System

Ketocoumarin (KC)/amine is well known[9] as a two component system working through electron transfer and proton transfer (Fig. 1-4),[5(a)] e.g.,

$$\xrightarrow{h\nu} S_1 \longrightarrow T_1$$

$$T_1 + \text{amine (AH)} \rightsquigarrow A^{\bullet} + KCH^{\bullet}$$

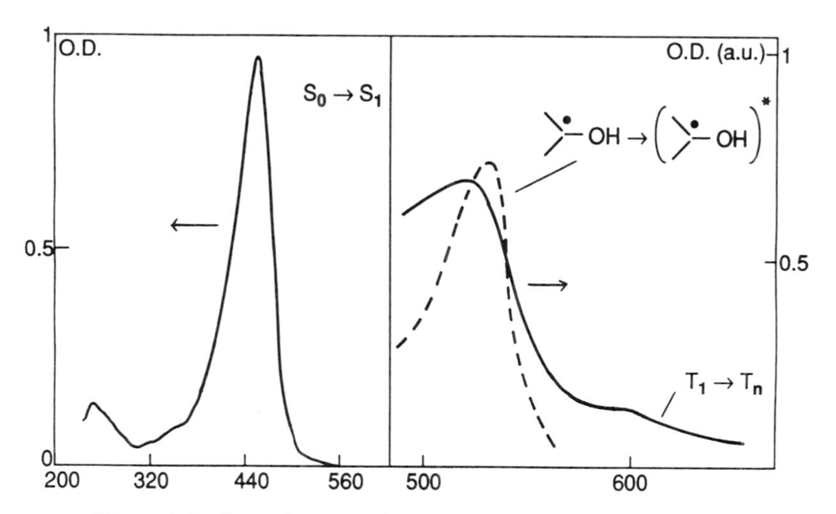

Figure 1-4. *Ground state and transient absorption spectra of KC.*

In the presence of onium salts, KC plays the role of an electron donor:

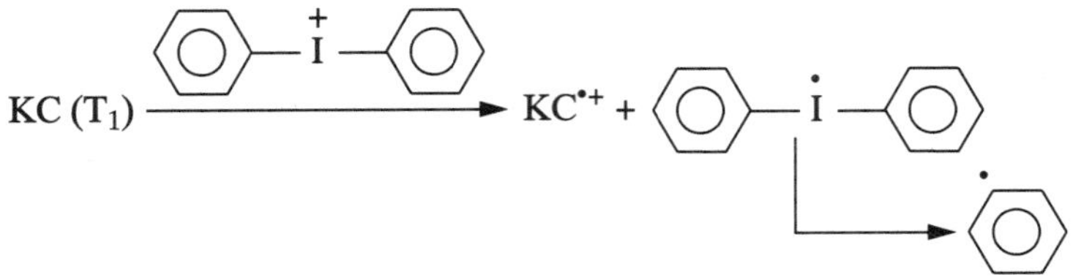

$$KC\,(T_1) \xrightarrow{\phi-\overset{+}{I}-\phi} KC^{\bullet +} + \phi-\overset{\bullet}{I}-\phi \longrightarrow \phi^{\bullet}$$

In the three component combination, both electron transfer processes (involving the amine and the onium salt, respectively) compete, and the balance is dependent on the nature of the ketocoumarin and the amine[5(a),43] which govern the rate constant of the reaction. Careful examination of the relative efficiency of KC/amine, KC/onium salt, and KC/amine/onium salt under laser exposure, shown below:

	Efficiency	
	λ = 368 nm	λ = 488 nm
PDO	100	
KC/ϕ_2I^+		25
$KC/\phi_2I^+/MDEA$		100
KC/MDEA		20

suggests[5(a)] that the higher reactivity of the three component systems cannot be explained only on the basis of a favorable interaction in the excited states. A plausible explanation involves the quenching of the ketyl type radicals $KCH^{\bullet}$ (by the onium salt) that are known as terminating agents of the growing macromolecular polymer chains. This process reduces the concentration of $KCH^{\bullet}$ and increases the concentration of phenyl initiating radicals:

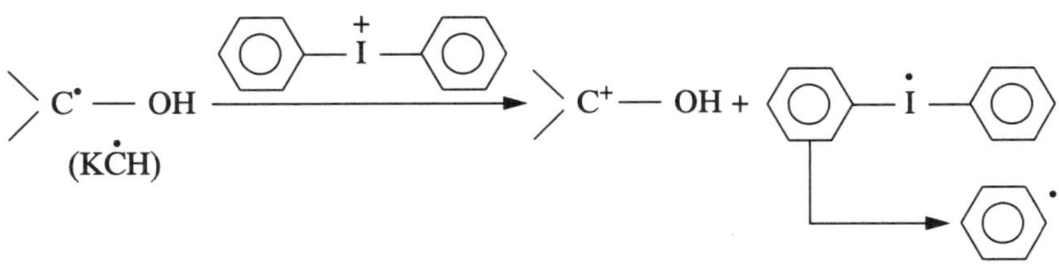

$$\underset{(K\dot{C}H)}{>C^{\bullet}-OH} \xrightarrow{\phi-\overset{+}{I}-\phi} >C^{+}-OH + \phi-\overset{\bullet}{I}-\phi \longrightarrow \phi^{\bullet}$$

The same behavior is observed in the presence of other ketones (Ket) such as benzophenone, benzyl, and thioxanthone derivatives[44], so that a very general mechanism can be proposed:

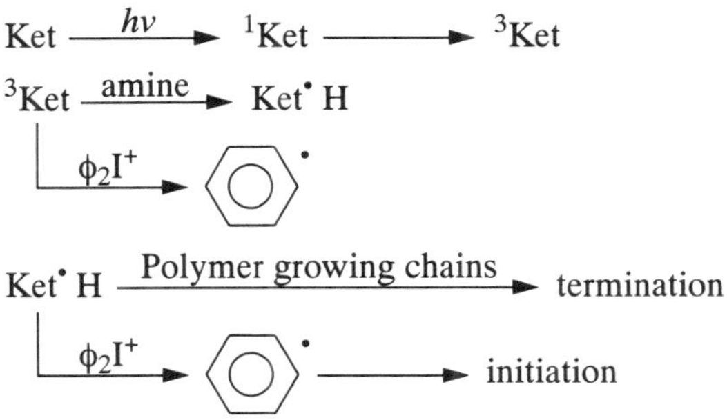

$$\text{Ket} \xrightarrow{h\nu} {}^1\text{Ket} \longrightarrow {}^3\text{Ket}$$

$${}^3\text{Ket} \xrightarrow{\text{amine}} \text{Ket}^{\bullet}\text{H}$$

$${}^3\text{Ket} \xrightarrow{\phi_2 I^+} C_6H_5^{\bullet}$$

$$\text{Ket}^{\bullet}\text{H} \xrightarrow{\text{Polymer growing chains}} \text{termination}$$

$$\text{Ket}^{\bullet}\text{H} \xrightarrow{\phi_2 I^+} C_6H_5^{\bullet} \longrightarrow \text{initiation}$$

(b) The Eosin/Amine/Ketone and Eosin/Amine/Onium Salt System

The combination of the compounds listed below has been extensively investigated, in solution, through time-resolved laser spectroscopy.[38,40]

D: (eosin: tetrabromo xanthene dye with O, OH, CO_2Na substituents)

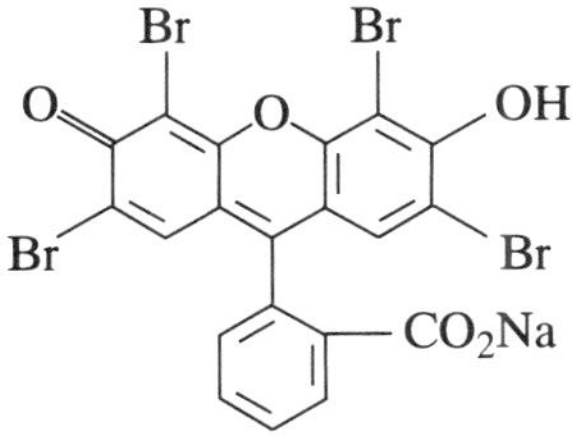

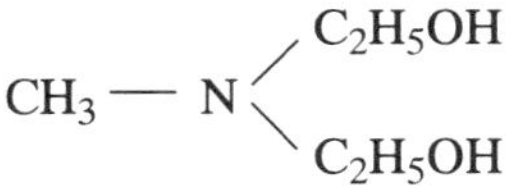

AH (MDEA): $CH_3-N(C_2H_5OH)_2$

PDO: $C_6H_5-C(=O)-C(CH_3)=N-OCOC_2H_5$

DMPA: $C_6H_5-C(=O)-C(OCH_3)_2-C_6H_5$

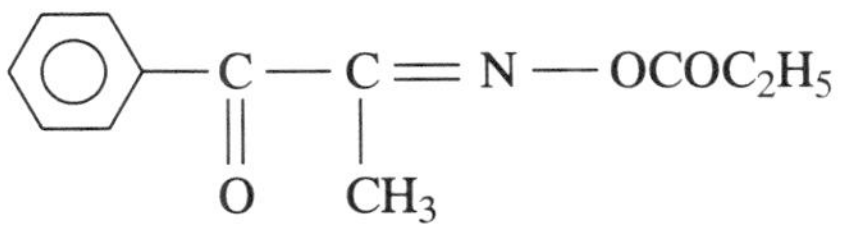

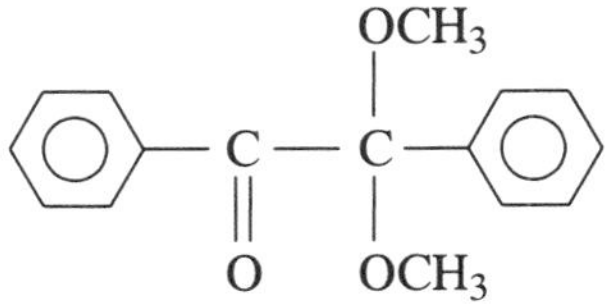

DPI: $C_6H_5-\overset{+}{I}-C_6H_5$ AsF_6^-

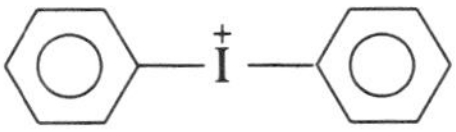

Under light exposure, eosin (D) leads to the generation of the reduced form $D^{\bullet -}$ and the oxidized form $D^{\bullet +}$ (Fig. 1-4 and Fig. 1-5).[38,40] The singlet and triplet state interact with MDEA, ϕ_2I^+:

$$\phi_2I^+: \quad D \xrightarrow{h\nu} {}^1D^* \longrightarrow {}^3D^*$$

$$\underset{}{{}^3D^*} \longrightarrow \underset{(red)}{D^{\bullet -}} + \underset{(ox)}{D^{\bullet +}}$$

$${}^1D^*, {}^3D^* \xrightarrow{MDEA} A^\bullet + DH^\bullet$$

$${}^1D^*, {}^3D^* \xrightarrow{\phi_2I^+} \phi_2I^\bullet \longrightarrow \phi^\bullet$$

$${}^3D^* \xrightarrow{AH} 4.3 \times 10^6\ M^{-1}\ s^{-1}$$

$${}^1D^* \xrightarrow{AH} 8 \times 10^8\ M^{-1}\ s^{-1}$$

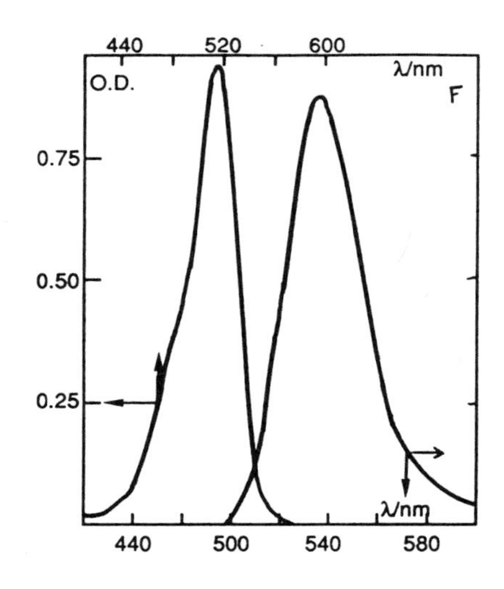

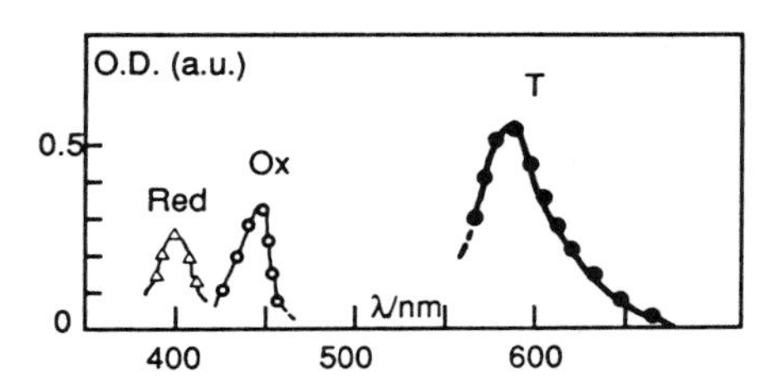

Figure 1-5. *Absorption spectra of D, $D^{\bullet -}$, $D^{\bullet +}$, T_1 and fluorescence spectrum of eosin.*

In the presence of the iodonium salt, a ground state complex is formed:

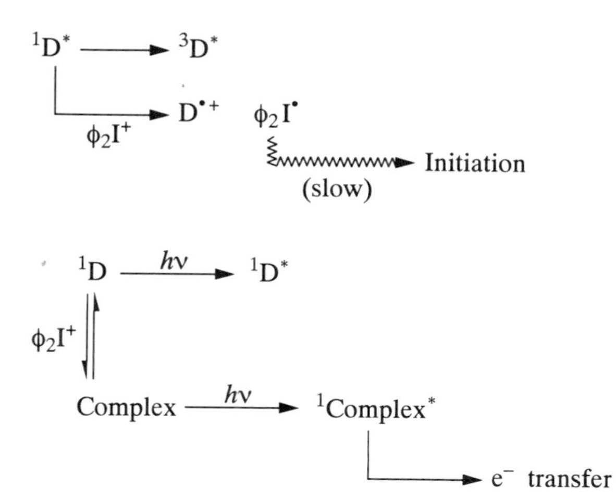

This complex is destroyed in the presence of the amino so that a strong quenching of the S_1 state is noted in eosin/amine/DPI:

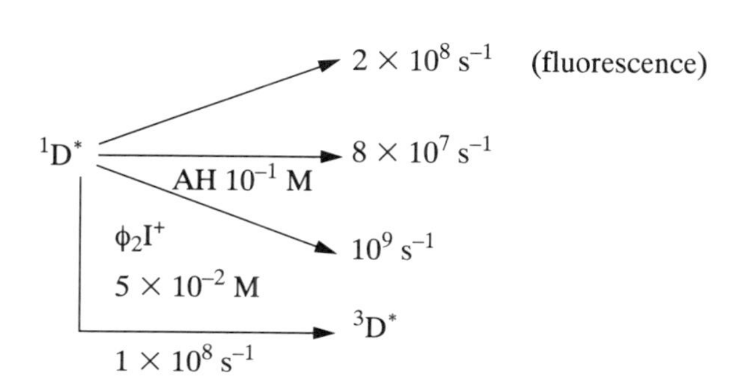

The interaction of the S_1 and T_1 states with DMPA and PDO is less efficient and can be disregarded. Monomer quenching (e.g., with acrylates) is negligible. The reduced and oxidized forms of eosin can react with DPI, PDO, and DMPA according to:

$$D^{\bullet -} + \phi_2 I^+ \qquad k = 10^{10}\ M^{-1}\ s^{-1}$$
$$D^{\bullet +} + PDO \qquad k = 2 \times 10^9\ M^{-1}\ s^{-1}$$
$$D^{\bullet -} + DMPA \qquad k = 10^8\ M^{-1}\ s^{-1}$$

The yields ϕ_x of the different processes in the different combinations are shown below for the singlet and the triplet states:

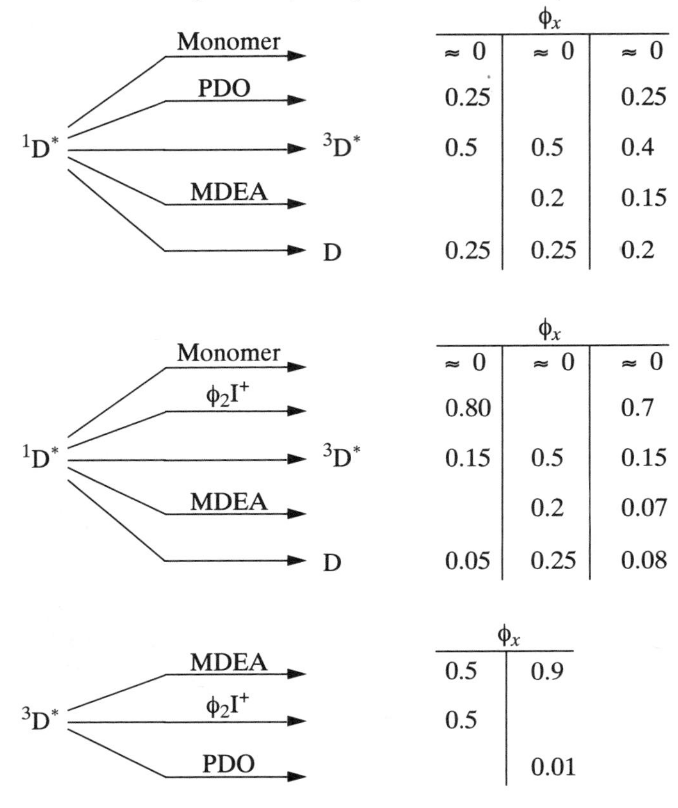

$^1D^*$ →	ϕ_x		
Monomer	≈ 0	≈ 0	≈ 0
PDO	0.25		0.25
$^3D^*$	0.5	0.5	0.4
MDEA		0.2	0.15
D	0.25	0.25	0.2

$^1D^*$ →	ϕ_x		
Monomer	≈ 0	≈ 0	≈ 0
ϕ_2I^+	0.80		0.7
$^3D^*$	0.15	0.5	0.15
MDEA		0.2	0.07
D	0.05	0.25	0.08

$^3D^*$ →	ϕ_x	
MDEA	0.5	0.9
ϕ_2I^+	0.5	
PDO		0.01

The initiation of the acrylate polymerization occurs through the generation of:

1. The $A^{\bullet}$ radical in eosin / amine
2. The radical in eosin / ϕ_2I^+

When PDO is present, the following reactions may occur:

$$^1D^*, {}^3D^* \xrightarrow{\text{MDEA}} A^{\bullet} + DH^{\bullet}$$

$$^1D^* \xrightarrow{\text{PDO}} PDO^{\bullet +} \xrightarrow[\text{(slow)}]{} \text{cleavage}$$

$$PDO^{\bullet +} \xrightarrow{\text{MDEA}} PDO + AH^{\bullet +}$$

$$AH^{\bullet +} \longrightarrow A^{\bullet}$$

Measurements of the rate and the degree of polymerization in acetonitrile

(AN) or THF allow the ratio between the rate constant of propagation k_p and the rate constant of termination k_t as well as the relative quantum yield of initiation ϕ_i to be calculated in the various systems considered:

Eosin	$\alpha = k_p^2/k_t$(rel)	ϕ_i(rel)
/MDEA/AN	1.2	38
/$\phi_i I^+$/MDEA/AN	7.1	23
/PDO/MDEA/AN	2.4	150
/MDEA/THF	3.8	85

It stands to reason that:

1. The polarity and the H donor character of the medium can affect ϕ_i and α.
2. PDO plays a role mostly in the photoinitiation step (ϕ_i).
3. $\phi_2 I^+$ has a strong influence on α (as well as THF), and results in a value close to that obtained in the polymerization of MMA in the presence of DMPA ($\alpha = 7.2$). This effect is explained by a quenching of the $DH^{\bullet}$ radicals (that act as chain-terminating agents) and the subsequent generation of new initiating radicals:

$$DH^{\bullet} \xrightarrow{\phi_2 I^+} D + H^+ + \phi_2 I^{\bullet}$$

$$DH^{\bullet} \xrightarrow{PDO^{\bullet +}} D + H^+ + PDO$$

$$DH^{\bullet} \xrightarrow{THF} DH_2 + THF^{\bullet}$$

The increase in iodobenzene concentration, as followed by gas chromatography mass spectrometry, is well supported by this mechanism:

$$\text{Eosin} / \phi_2 I^+ \longrightarrow \phi I$$

$$\text{Eosin} / \phi_2 I^+ / \text{MDEA} \longrightarrow \phi I \ (\uparrow\uparrow)$$

$$D \xrightarrow{\phi_2 I^+} \phi_2 I^{\bullet} \longrightarrow \boxed{\phi I}$$

$$D \xrightarrow{AH} D^{\bullet +} \longrightarrow DH^{\bullet}$$

$$DH^{\bullet} \xrightarrow{\phi_2 I^+} \phi_2 I^{\bullet} \longrightarrow \boxed{\phi I}$$

This study, conducted in a model system in solution, should be considered as a possible basis for the design of efficient three component systems A / B / C, schematically depicted as:

(1) $A \xrightarrow{h\nu} A^*$

$A^* \xrightarrow{B}$ } radicals → initiation ($R^\bullet$)

$A^* \xrightarrow{C}$ } radicals → scavenging ($S^\bullet_c$)

(2) $R^\bullet$ + monomer $\xrightarrow{\text{monomer}}$ polymer

(3) $S_c^\bullet + C \longrightarrow S_c^+ + C^{\bullet -}$

$S_c^\bullet$ + ~~~~~~• ⟶ polymer chain termination

(4) $C^{\bullet -}$ ⟶ radicals

Eosin/amine/PDO or DPI; ketocoumarin/amine/ϕ_2I^+ (as shown before) obey this very general mechanism and exhibit a high efficiency under laser light exposure.

	Efficiency	
	λ = 368 nm	λ = 488 nm
PDO	100	
PDO/MDEA	100	
Eosin/MDEA		3.3
Eosin/PDO	1	
Eosin/PDO/MDEA		5
Eosin/ϕ_2I^+		5
Eosin/ϕ_2I^+/MDEA		25

Other systems under investigation exhibit the same behavior; these include:

- Ketocoumarin/*N*-phenylglycine/ϕ_2I^+.[43]
- Benzophenone or benzyl or thioxanthone/MDEA/ϕ_2I^+.[44]

Presumably, the same holds true in the case of thioxanthene dye TXD/amine AH/ϕ_2I^+:[39]

TXD

$$\text{TXD} + AH \xrightarrow{h\nu} TXD^{\bullet -}\ AH^{\bullet +} \longrightarrow TXD^{\bullet}H + A^{\bullet}$$

$$\left.\begin{matrix} TXD^{\bullet -} \\ \text{or} \\ TXD^{\bullet}H \end{matrix}\right. \xrightarrow{\phi_2 I^+} \left.\begin{matrix} TXD \\ \text{or} \\ TXD^{\bullet +}H \end{matrix}\right. + \phi_2 I^{\bullet}$$

and ketone/amine/bromocompound:[36,45]

$$\underset{\text{Ket}}{>C{=}O} + AH \xrightarrow{h\nu} Ket^{\bullet -}\ AH^{\bullet +} \longrightarrow Ket^{\bullet}H + A^{\bullet}$$

$$Ket^{\bullet}H + CBr_4 \longrightarrow Ket^{\bullet}H^+ + CBr_3^{\bullet} + Br^- \longrightarrow Ket + H^+$$

4. Reactivity of Photoinitiating Systems in Bulk Monomer/Oligomer Media in Air

The following effects may be expected:

1. Deactivation of excited states by oxygen
2. Quenching of radicals by O_2
3. Detrimental interaction of the reactive excited state PI* or PS* with
   - Monomers
   - H donors
   - Additives, environment.

The relative importance of all these effects is dependent on the magnitude of the biomolecular quenching rate constant k_q of the above-mentioned processes and of the process leading to the generation of reactive species in multicomponent systems, e.g.:

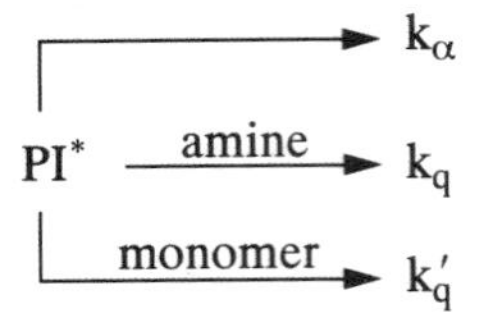

The k_q value is dependent on the diffusion rate constant that is connected with the viscosity of the medium. In molecules sensitive to "polarity" changes (e.g., thioxanthones) (Fig. 1-6),[46] k_q might also be dependent on the chemical nature of the monomer/oligomer matrix, even at a constant viscosity.

Recently reported data show the following trends:

1. Even in the presence of a high concentration of acrylate functions, the triplet state of thioxanthone derivatives remains long lived (e.g., 2500 ns) in HDDA

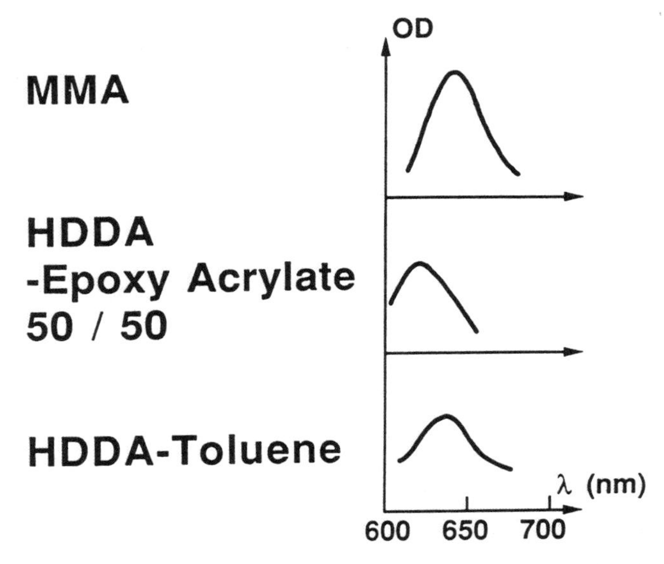

Figure 1-6. *Photochemistry in bulk media polarity effect as shown by the shift of the triplet-triplet absorption.*

(hexanediol diacrylate)/epoxy acrylate (50:50) in comparison with the value (~50 ns) in a low viscosity monomer medium (e.g., MMA).[47]

2. The interaction between triplet states of, e.g., ketones and O_2, is weak and can be disregarded. The lifetime is reduced from 2500 ns in N_2 (in the previous example) to 1850 ns in air.[48]
3. The interaction with amines competes very efficiently; in the presence of 2% w/w of MDEA, the lifetime drops to 280 ns.[48]
4. The same holds true for the interaction between a photosensitizer and a photoinitiator. In the example already considered, the triplet state lifetime of the thioxanthone derivative (PS) in the presence of a morpholino ketone (PI) is shortened to ~70 ns so that the calculated yield of this process is ~0.73 in N_2 and 0.67 in air.[46]

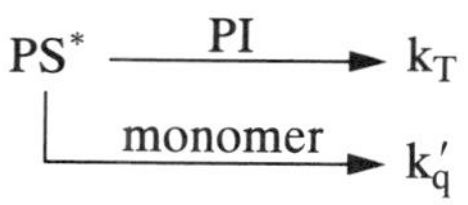

$$PS^* \xrightarrow{PI} k_T \qquad PS^* \xrightarrow{\text{monomer}} k'_q$$

The trends in the results that are currently recorded in a general investigation of PS/PI interaction in ketones in bulk might be used as a first approach to what happens in other PS/PI sensitive to visible light in laser imaging applications. To conclude, it can be stated that in currently employed efficient photoinitiators or photosensitizers, the expected oxygen quenching of the reactive excited states should be completely ruled out in the viscous film matrix. In a general way, the same should hold true for the process of monomer quenching, but the effect might be dependent on the relative values of k'_q and k_q (that are proportional to the diffusion rate constant $k_{diff.}$) compared to k_x. When going from an organic solvent (toluene or methanol) to a typical matrix (HDDA/epoxy acrylate) the viscosity increase results in a ~100-fold decrease of $k_{diff.}$:

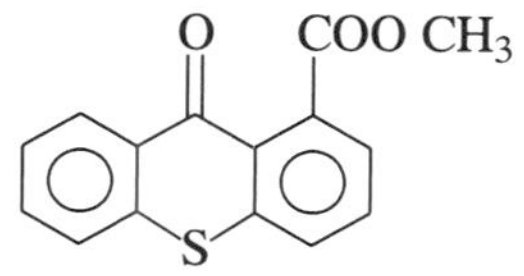

	η (cP)	Calculated $k_{diff.}$ 10^6 $M^{-1}s^{-1}$
TMPTA	40	167
TMPTA/EpAcr. (66/33)	185	36
HDDA	5.2	1240
HDDA/Ep. Acr. (50/50)	58	115
Toluene	0.53	12,400
Methanol	0.52	12,400

The effect of viscosity on k_T at a constant polarity is exemplified by the following data in the system containing a thioxanthone derivative and a morpholino ketone:

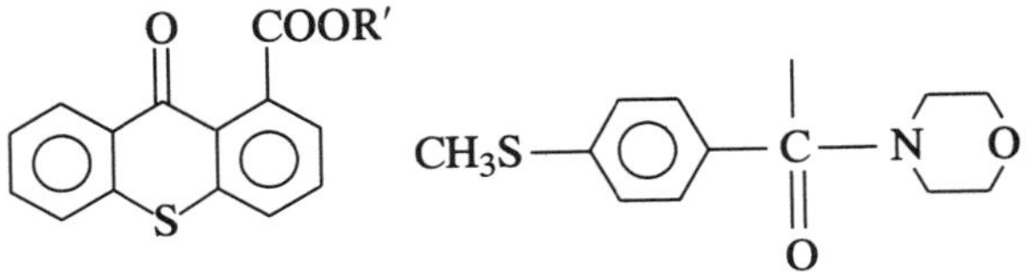

		η(cP)	10^{-6} k_T ($M^{-1}s^{-1}$)
MeOH	(5/0)	0.54	630
	(4.0)	1	575
	(1/4)	7.4	280
Et. glycol	(0/5)	15.6	180

When viscosity and polarity effects are present, nothing can really be predicted on their combined influence, except that k_T both decreases with increasing η and tends to increase with increasing polarity. The following results show that k_T is not substantially affected by a ~10-fold increase in viscosity:

	τ (ns)	$10^{-6}k_T$ ($M^{-1}s^{-1}$)	ϕ_T	η (cP)
TMPTA	650	80	0.72	40
HDDA	650	95	0.76	5.2
HDDA/EP. Acr.	2500	22	0.73	58

In such a medium (HDDA/Ep. Acr.), the triplet state T_1 of the thioxanthone mentioned above is expected to react with O_2, the monomer, and MDEA according to:

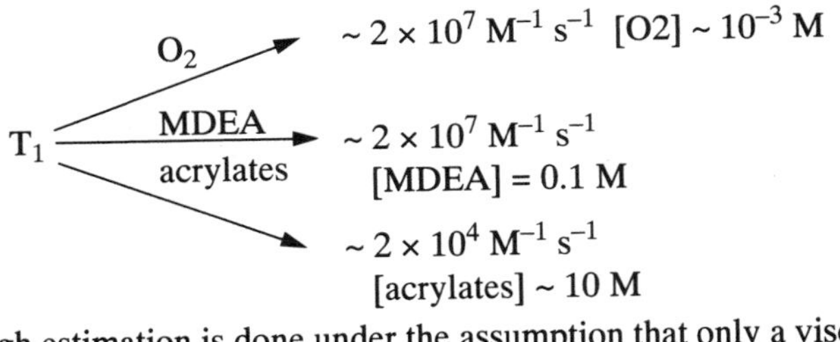

T_1 → (O_2) ~ 2×10^7 M^{-1} s^{-1} [O2] ~ 10^{-3} M

T_1 → (MDEA) ~ 2×10^7 M^{-1} s^{-1} [MDEA] = 0.1 M

T_1 → (acrylates) ~ 2×10^4 M^{-1} s^{-1} [acrylates] ~ 10 M

This rough estimation is done under the assumption that only a viscosity effect is present (which is not always correct, as explained previously); nevertheless, it correlates with the experimental results listed in (1) to (4) and the conclusions derived above. When the triplet state is short lived, monomer quenching in bulk cannot compete with the generation of radicals, because of a cleavage process. In the first example (thioxanthone/acrylates), the situation might change if acrylates are replaced by styrene moieties, because the quenching of ketone triplet states by this monomer is known to be more efficient.

In energy transfer experiments where, for example, the thioxanthone derivative is used as a photosensitizer, the picture is as follows (in HDDA/Ep. Acr.).

O COOR

S

$+ T_1$ —(monomer, bulk)→ $4 \times 10^5\ s^{-1}$

$+ T_1$ —(PI)→ $11 \times 10^5\ s^{-1}$

PI : CH_3 S — C(=O) — C — N O

In that system, the rate constant of quenching of T_1 (PS) by PI is 22×10^6 $M^{-1}s^{-1}$ compared to $630 \times 10^6\ M^{-1}s^{-1}$ is neat MeOH.

Extensive work is now focused on a general approach to the photochemical reactions in bulk media. Further investigations on excited state processes in visible laser light photosensitive systems by means of nanosecond and picosecond laser spectroscopy are in progress. They should promote the design of efficient photoinitiating combinations for laser imaging technologies.

References

1. (a) S. P. Pappas, ed., *UV-curing: Science and Technology* Tech. Mark. Corp. Stamford, 1986 (Plenum Press, New York, 1992). (b) C. G. Roffey, *Photopolymerization of Surface Coatings* J. Wiley and Sons, New York, 1982. (c) J. P. Fouassier and J. F. Rabek, eds., *Radiation Curing in Polymer Science and Technology* (Elsevier, London, 1993). (d) P. Olding, ed., *Chemistry and Technology of UV and EB for Coatings, Inks and Paints* (Sita Techn., London, 1991).
2. J. P. Fouassier and J. F. Rabek, eds., *Lasers in Polymer Science and Technology: Applications* (CRC Press, Boca Raton, FL, 1989).
3. (a) J. P. Fouassier in *Photopolymerization: Science and Technology*, Ed. N. S. Allen, ed. (Elsevier Appl. Sci. Publ., London, 1989). (b) J. P. Fouassier in *Photochemistry and Photophysics*, J. F. Rabek, ed. (CRC Press, Boca Raton, FL, **2**, 1, 1990). (c) J. P. Fouassier, *Progress in Org. Coatings* 18, 227 (1990). (d) W. Schnabel, "Application of Laser Flash Photolysis to the Study of Photopolymerization Reaction in Non Aqueous Systems," in *Lasers in Polymer Science and Technology: Applications*, J. P. Fouassier and J. F. Rabek, eds. (CRC Press, Boca Raton, FL 1989). (e) W. Schnabel in *New Trends in Photopolymerization*, N. S. Allen and J. F. Rabek, eds. (North-Holland Publ., Amsterdam, 1985).

4. (a) W. Fisher, V. Kuita, H. Zweifel, and L. Felder, Fr. Patent 8108981 (1981). (b) M. J. Davis, G. Gawne, P. N. Green, and W. A. Green, Chem. Spec., Manchester (1986).
5. (a) J. P. Fouassier and S. K. Wu, *J. Appl. Polym. Sci.* **44**, 1779 (1992). (b) J. V. Crivello, J. P. Fouassier, and D. Burr, to be published.
6. A. G. Anderson, Du Pont, Europ. Patent 811029792 (1981).
7. A. Hayashi, Y. Goto, and M. Kakayama, Jpn. Kokay Tokkyo Koho JP 01 29337, 1985.
8. W. K. Smothers, B. M. Monroe, A. W. Weber, and D. E. Keys, *SPIE OE/Lase Conf. Proc.*, Los Angeles, **1212** (1990).
9. W. G. Kerkstroetter and S. Farid, *J. Photochem.* **35**, 71 (1986).
10. S. K. Wu, J. P. Fouassier, J. K. Zang, and D. Burr, *Photogr. Sci. and Photochem.* **2**, 47 (1989).
11. D. Eaton, *Tapp. Cur.* **156**, 199 (1990).
12. A. Roloff, K. Meier, and M. Riediker, *Pure Appl. Chem.* **58(9)**, 1267 (1986).
13. (a) D. C. Neckers, I. I. Abu Abdoun, and L. Thys, *Macromol.* **17(3)**, 283 (1984). (b) D. Xu, A. Vanloon, S. M. Linden, and D. C. Neckers, *J. Photochem.* 38, 357 (1987).
14. G. B. Schuster, *Proc. XIII IUPAC Symp. Photochem.*, Warwick (1990).
15. D. F. Eaton, *Adv. Photochem.* **18**, 427 (1986).
16. Scott Bader, Europ Patent 793030784 (1980).
17. R. Patel, Europ. Patent 97012 (1983).
18. J. P. Fouassier, E. Chesneau, and M. Le Baccon, *Makromol. Chem.* **9**, 223 (1988).
19. R. Baumann, W. Singer, and K. Fritzsche, Ger. Patent 160084 (1983).
20. G. Sudesh Kamar and D. C. Neckers, *Macromolecules* **24**, 4322 (1991).
21. (a) T. Yamaoka, K. Koseki, and Y. Goto, Belg. Patent 897694 (1984). (b) M. Kawabata, M. Harada, and Y. Takimoto, U.S. Patent 4868092 (1989). (c) Y. Takimoto, *Proc. Lasers in Graphic Arts*, San Diego (1988).
22. (a) C. Carre and D. J. Lougnot, *J. Optics* **21**, 147 (1990). (b) C. Carre and D. J. Lougnot, *Opto 90* (ESI Publ., Paris, 1990). (c) J. P. Fouassier, C. Carre, and D. J. Lougnot, *SPIE Proc.* **1213**, 201 (1990). (d) J. P. Fouassier, *SPIE Proc.* **1559**, 76 (1991). (e) D. J. Lougnot and C. Turck, *Pure Appl. Opt.* 1, 251 (1992).
23. J. V. Crivello, *Adv. in Polym. Sci.* **62**, 3 (1984).
24. (a) J. L. Dektar and N. P. Hacker, "Photochemistry of Diaryliodonium Salts," *J. Org. Chem.* 55, 639 (1990). (b) J. L. Dektar and N. P. Hacker, "Photochemistry of Sulfonium Salts," *J. Amer. Chem. Soc.* **112(16)**, 6004 (1990).
25. C. Brauchle, U. P. Wild, D. M. Burland, G. C. Bjorklund, and D. C. Alvarez, *IBM J. Res. Devel.* **26**, 217 (1982).
26. (a) C. Carre, D. Ritzenthaler, D. J. Lougnot, and J. P. Fouassier, *Optics Lett.* 12, 646, (1987). (b) D. J. Lougnot, C. Carre, D. Ritzenthaler, and J. P. Fouassier, *J. Appl. Phys.* 63(10), 4841 (1988). (c) D. J. Lougnot, C. Carre, and J. P. Fouassier, *Makromol. Chem., Macromol. Symp.* **24**, 209 (1989).
27. (a) K. Ichimura and M. Sakuragi, *J. Polym. Sci., Polym. Lett.* **26**, 185 (1988). (b) M. J. Jeudy and J. J. Tobilard, *Opt. Comm.* **13**, 25 (1975).
28. D. E. Falvey and G. B. Schuster, "Picosecond Time Scale Dynamics of Perester Photodecomposition," *J. Amer. Chem. Soc.* **108**, 7419 (1986).

29. A. Morlino, M. D. Bohorquez, D. C. Neckers, and M. Rodgers, *J. Am. Chem. Soc.* **113**, 3599 (1991).

30. B. Klingert, A. Roloff, B. Urwyler, and J. Wirz, *Kelv. Chem. Acta* **71**, 1858 (1988).

31. (a) J. P. Fouassier and D. J. Lougnot, in ''Radiation Curing on Polymeric Materials,'' C. E. Hoyle and J. F. Kinstle, eds., *ACS Symp. Series 417* (Washington D.C., 1990). (b) G. Rist, V. Desobry, K. Detliker, J. P. Fouassier, and D. Rulhmann, *Macromolecules* **25**, 4182 (1992).

32. L. Y. C. Lee, X. Ci, C. Giannotti, and D. G. Whitten, *J. Amer. Chem. Soc.* **108**, 175 (1986).

33. (a) R. J. De Voe, M. R. V. Sahyun, E. Schmidt, N. Serpone, and D. K. Sharma, ''Electron Transfer Sensitized Photolysis of Onium Salts,'' *Can. J. Chem.* **66ℒ**, 319 (1988). (b) J. P. Fouassier, D. Burr, and J. V. Crivello, ''Time-Resolved Laser Spectroscopy of the Sensitized Photolysis of Iodonium Salts,'' *J. Photochem. Photobiol., A: Chem.* **49**, 318 (1989).

34. (a) F. Morlet-Savary, J. P. Fouassier, T. Matsumoto, and H. Inomata, to be published. (b) Y. Goto, E. Yamada, M. Nakayama, K. Tokumaru, and T. Arai, *J. Jap. Chem. Soc.* **6**, 1027 (1987).

35. T. Yamaoka, *Polymers for Adv. Techn.* **1**, (1991).

36. M. Harada, M. Kawabata, and Y. Takimoto, *J. of Photopolymer Sci. and Techn.* **2**, 199–204 (1989).

37. M. Harada, M. Kawabata, and Y. Takimoto, *Proc. Radiation Curing Asia* **20–22**, 454–460 (1986).

38. J. P. Fouassier and E. Chesneau, *Makromol. Chem.* **192**, 1307–1315 (1991).

39. M. Harada, Y. Takimoto, N. Noma, and Y. Shirota, *J. of Photopolymer Sci. and Tech.* **4**(1), 51–54 (1991).

40. J. P. Fouassier and E. Chesneau, *Makromol. Chem.* **192**, 245–260 (1991).

41. Hoechst Ag., Europ. Patent 364735 (1990).

42. M. Kawabata, K. Kimoto, and Y. Takimoto, Europ. Patent 211615 (1991).

43. J. P. Fouassier, D. Ruhlmann, Y. Takimoto, M. Harada, and M. Kawabata, *J. Imaging Sci. Tech.* **37**, 208 (1993).

44. J. P. Fouassier, D. Ruhlmann, Y. Takimoto, M. Harada, and M. Kawabata, *J. Polym. Sci., Polym. Chem. Ed.* **37**, 2245 (1993).

45. J. P. Fouassier, F. Morlet-Savary, A. Erddalane, Y. Takimoto, M. Harada, M. Kawabata, and Y. Sumiyoshi, Macromolecules (in press).

46. D. Ruhlmann and J. P. Fouassier, *Eur. Polym. J.* **29**, 505, 1993.

47. J. P. Fouassier and D. Ruhlmann, *J. Photochem.* **61**, 47 (1991).

48. J. P. Fouassier, D. Ruhlmann, and A. Erddalane, *Macromolecules* **26**, 727, 1993.

ylbiimidazolyl: The Cn... Initiator for Photopolymers. History of Synthesis and Investigation of Photochemical Properties

Koko Maeda

1. Discovery of a Photochromic Compound

In 1957, the late Prof. Hayashi and I began to investigate the chemiluminescence mechanism of lophine (2,4,5-triphenylimidazole), which was the first discrete chemiluminescent organic compound, described by Radziszewski[1] in 1877. Because the chemiluminescence intensity of lophine in KOH-ethanol solution with oxidants was not very strong, the chemiluminescence reaction was carried out in a dark room. When an aqueous solution of $K_3[Fe(CN)_6]$ was added dropwise to the alkaline solution of lophine as an oxidant, a circle of light appeared on the surface of the reaction mixture encircling the area in which the oxidant solution was added, and by stirring the reaction mixture a yellowish light was emitted from the solution. When the chemiluminescence reaction was carried out in daylight, no chemiluminescence light emission was observed, and a purple-colored circle appeared on the surface of the solution encircling the area where the oxidant solution was added. Further stirring of the solution caused the color to disappear and the solution became yellow. We thought that the purple substance should be an intermediate in the chemiluminescent reaction of lophine, and wanted to isolate the purple-colored substance. In further studies to determine the origin of the quickly fading purple color, a significant amount of aqueous solution of $K_3[Fe(CN)_6]$ was added to the reaction mixture. The color of the solution became purple, and a pale purple precipitate separated out. The precipitate was filtered out and repeatedly washed with water until the washing showed no reaction of $Fe(CN)_6^{4-}$ ion. The dried precipitate was an almost colorless powder. When this solid was pressed by a glass rod or spatula on the filter paper, a purple color appeared. (This precipitate turned out to be a piezochromic compound). Benzene was used to recrystallize this almost colorless solid. The solid showed large solubility in benzene and gave a deep purple-colored solution. Upon standing in a dark, cold place, the color of the solution gradually changed to pale yellow. After concentration of the benzene solution under reduced pressure followed by standing in the dark, a pale yellow crystalline substance was separated out. Recrystallization of this substance from ethanol gave fine, pale yellow prisms with a melting point (MP) of 199–201° C (decomposition). Although the elemental analysis (found: C, 85.2; H, 5.5; N, 9.4%, calculated for $C_{21}H_{15}N_2$: C, 85.38; H, 5.11; N, 9.49%) showed that the product had a composition nearly equal to lophine, its properties were quite different from those of lophine ($C_{21}H_{16}N_2$). When the pale yellow product was irradiated with sunlight or a high-pressure mercury lamp in benzene or other organic solvents and in the solid at room temperature, its pale yellow color turned rapidly into purple, both in air and in vacuo. Upon standing in the dark, the purple color faded to the original color in about one day in the case of the solid, and faded rapidly in benzene even at room temperature. This reversible phenomenon could be repeatedly observed, especially in a nitrogen atmosphere or in vacuo. We recognized the pale yellow compound as a new photochromic compound **1**.[2] From 2,4,5-tri(*p*-tolyl)imidazole, 2,4,5-tri(*p*-chlorophenyl)imidazole and 2-(*p*-chlorophenyl)-4,5-diphenylimidazole, we obtained the corresponding photochromic compounds **2**, **3**, and **4**, respectively, by a method similar to the preparation of photochromic compound **1**.

2,4,5-triaryl-imidazole $\xrightarrow[K_3Fe(CN)_6,\ O_2]{KOH - EtOH}$ piezochromic compound $\xrightarrow{C_6H_6}$ photochromic compound

2. Photochromic Behavior of the New Compounds

The pale yellow ethanol solution of the photochromic compound **1** showed an absorption spectrum with a maximum at 270 nm ($\epsilon = 1.6 \times 10^4$ mol dm^{-3} cm^{-1})

which obeyed Beer's law. Upon irradiation with sunlight or a UV lamp, the pale yellow ethanol solution quickly turned to purple and gave rise to two maxima at 348 and 554 nm in the optical spectrum with a decrease in the absorbance at 270 nm. Upon standing in the dark, the purple color gradually faded with a decrease in the absorbances at 348 and 554 nm and recovery at 270 nm showing an isosbestic point at 305 nm. Fig. 2-1[3] shows the photochromic color change in the absorption spectrum of **1** in benzene. The change in absorption intensity with time of a band due to the photocolored species (347 nm in benzene and 346 nm in hexane) is shown in Fig. 2-2. The maximum absorbance in the photostationary state depended on the intensity of the irradiation light, the concentration of **1**, the kind of solvent, and the temperature. The rate of fading in the dark was not accelerated by irradiation with visible light. The time duration of photochromic activity varied with the solvent and was sensitive to the presence of oxygen. For example, in a deaerated benzene solution, the photochromic activity was retained for one year or more. On the other hand, in ethanol or toluene, the duration of photochromic activity was very short, even under oxygen-free conditions, because of hydrogen abstraction from the solvent to give lophine. Furthermore, when nitrogen oxide gas was introduced into the purple-

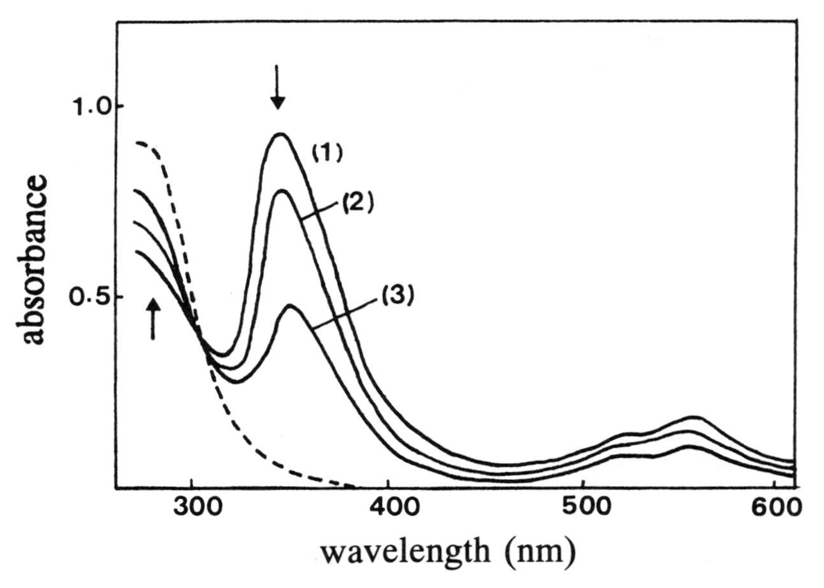

Figure 2-1. *Photochromic change in the absorption spectrum of photodimer* **1** *in benzene (Ref. 3).* ------: *before irradiation;* ——: *after irradiation; (1)* t = *0, (2)* t = *6 min, (3)* t = *24 min; temperature: 15° C, concentration of* **1** *is* 5.49×10^{-5} *mol* dm^{-3}.

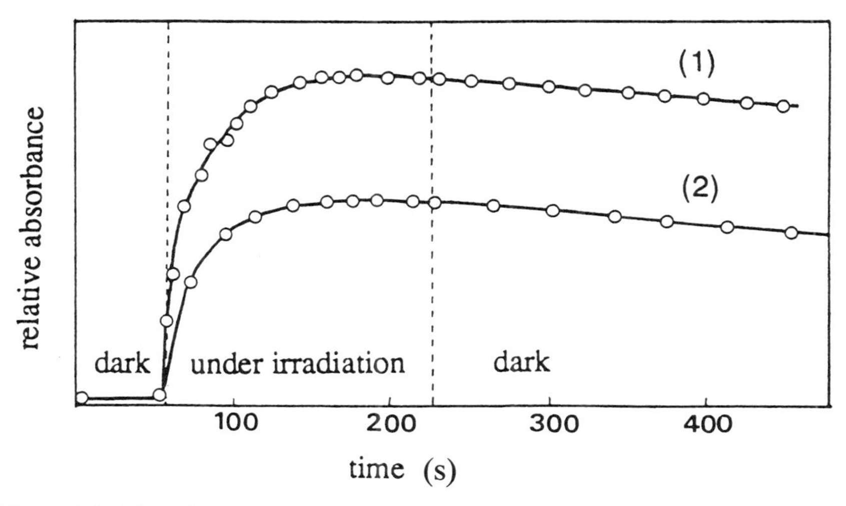

Figure 2-2. *Photochromic change in the absorbance of photocolored species of* **1** *(Ref. 3). (1) in benzene, (2) in* n-*hexane; temperature: 16.0° C, concentration of* **1** *(1) 5.28* $\times$ 10^{-5} *mol dm*$^{-3}$*, (2) 5.4* $\times$ 10^{-5} *mol dm*$^{-3}$.

colored solution, the color disappeared immediately. These observations suggested that the purple coloration was due to the formation of a free radical. ESR absorption[4] was detected (JEOL JEP-1 spectrometer) with the purple species in a deaerated benzene solution (g = 2.03) as well as in the irradiated solid state of **1**.

Before irradiation of **1** in the solid state, ESR absorption was not apparent; however, in a benzene solution of **1**, which was prepared under dim light, a weak ESR signal was detected at room temperature and was enhanced with an increase of the temperature while in the dark. Upon cooling the solution, the ESR signal intensity decreased with fading of the pale purple color . These observations indicate that the photochromic compound **1** exhibits thermochromism in solution. The solid sample also showed thermochromism at temperatures above 100° C up to its MP (190° C). The molecular weight of **1** was determined in a benzene solution to be ca. 600 by a cryoscopic method[5] (0.1127 g of **1**/ 20.13 g of C_6H_6) under the light of a tungsten lamp. This value corresponds to the molecular weight of 590 for the dimer of triphenylimidazolyl $C_{42}H_{30}N_4$, whereas the apparent molecular weight measured immediately after irradiation for five minutes with sunlight at about 6° C was 500 (freezing point depression Δt = 0.057° C). From the apparent molecular weight, the degree of photodissociation in benzene was estimated to be 0.18. Other photochromic compounds (**2**, **3**, and **4**) also showed photochromic change in the absorption spectrum similar to that of **1**, with the appearance of an ESR signal attributed to the formation of radicals. These data strongly suggested that all photochromic compounds **1**, **2**, **3**, and **4** are dimers of triarylimidazolyl radicals (Table 2-1) and that the photochromic phenomenon is due to the reversible radical dissociation of hexaarylbiimidazolyl into the triarylimidazolyl radical.

3. Kinetic Study of Photochromism in Solution

In order to confirm that the photochromism mechanism of hexaarylbiimidazolyls **1–4** is a reversible reaction, the rates of decrease of the radical concentration in the photochromic systems were measured by ESR methods[3,6] (JEOL JES 3B

Table 2-1. Photochromic Hexaarylbiimidazolyl

photodimer

	R^2	R^4	R^5
1	H	H	H
2	Me	Me	Me
3	Cl	Cl	Cl
4	Cl	H	H

Photodimer (molecular formula)	MP (°C) (decomp)	λ_{max}^{EtOH}/(nm) before irradiation	after irradiation (color)
1 $C_{42}H_{30}N_4$	191–192	270	270, 348, 554 (purple)
2 $C_{48}H_{42}N_4$	194–195	280	280, 365, 583 (bluish violet)
3 $C_{42}H_{24}N_4Cl_6$	213–214	274	274, 368, 585 (violet)
4 $C_{42}H_{28}N_4Cl_2$	219–220	268	268, 366, 575* (violet)

*Measured in benzene as the solubility of **4** is too low for measurement of the colored form.

spectrometer) and also by absorption spectroscopy[3,7] (Hitachi JPS2 spectrophotometer). The reaction rate of radical recombination to form the dimer is expected to obey the second-order rate law.

The single peak ESR signal shown by a purple oxygen-free benzene solution of **1** (1.48×10^{-4} mol dm^{-3}) gradually decreased with time in the dark. The rate of decrease of the relative signal intensity was measured by the signal height *h* (in cm), according to the following procedure. After completing the recording of the first ESR scan, the sweep direction of the magnetic field was reversed and the second ESR scan was immediately recorded (Fig. 2-3). This was repeated about 20 times in about 900 s at 23° C. The line width of the signals was 9.3 gauss and the signal height *h* was assumed to be proportional to the

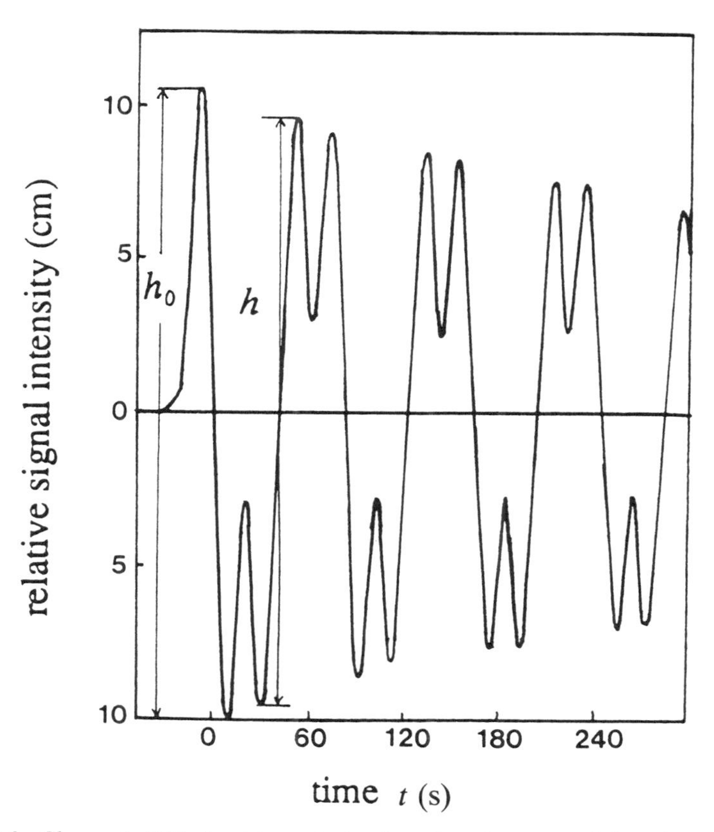

Figure 2-3. *Change in ESR signal intensity in photochromic system of photodimer* **1** *with time in the dark (Ref. 6). Temperature: 23.0° C, concentration of* **1** *is* 1.48×10^{-4} *mol* dm^{-3}.

radical concentration. The height h was used to represent the radical concentration in the reaction rate expression for the recombination of the radical. The differences between the reciprocals of h at time t and h_0 at $t = 0$ $(1/h - 1/h_0)$ were plotted vs. time t. The plots gave a straight line, as shown in Fig. 2-4, indicating that the rate of decrease in the ESR signal intensity obeyed the second-order rate law. A similar result was obtained for **2** as shown by the straight line (2) in Fig. 2-4.

The radical recombination reaction was also monitored by spectrophotometric measurements. In order to obtain the molar absorptivity of the triarylimidazolyl radical, the degree of photodissociation α of **1** in the photostationary state in benzene (conc. = 3.8×10^{-5} mol dm^{-3}) at 15.0° C was estimated to be about 0.30 from the change in absorbance at 280 nm ($\epsilon = 2.6 \times 10^4$ mol dm^{-3} cm^{-1}) in benzene measured before and after 180-s irradiation. The degree of photo-

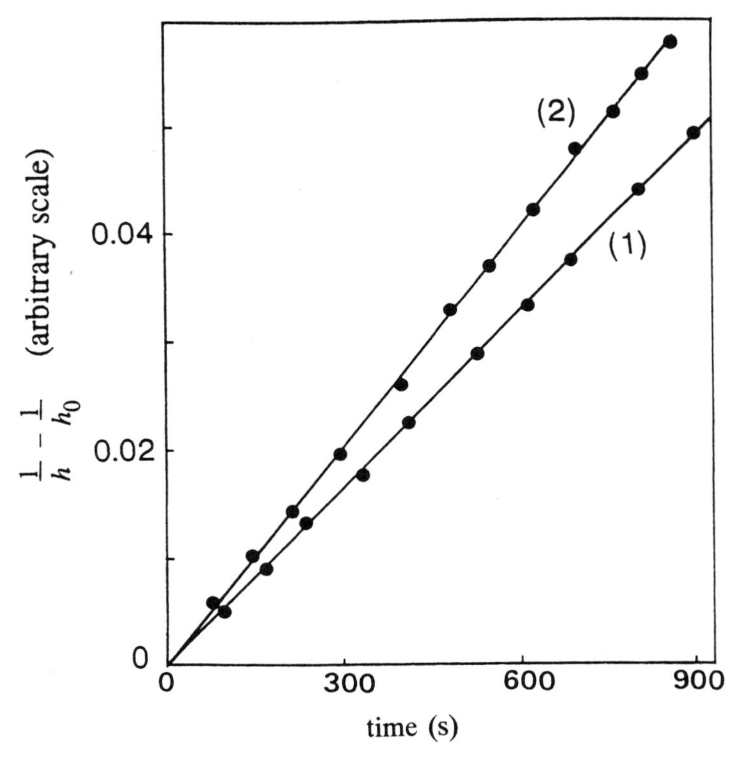

Figure 2-4. *Plots of* $(1/h - 1/h_0)$ *vs. time in the fading process of photochromism of the photodimers in benzene (Ref. 6).* h: *relative ESR signal intensity (cm); (1) photodimer* **1**, *(2) photodimer* **2**.

dissociation of the other photodimers in the photostationary state shown in Table 2-2 were also estimated in a similar way.

From the absorbances at 347 and 554 nm and the degree of photodissociation of α of **1** in benzene, the molar absorptivities of the triphenylimidazolyl radical at 347 nm and 554 nm were estimated to be about 4.8×10^4 and 8.4×10^3, respectively. The molar absorptivities of other radicals estimated by a similar method are shown in Table 2-3.

From the measurements of decrease in absorbance at the absorption maxima of the triarylimidazolyl radicals, the rate of recombination of the radical was obtained. Absorbance at the absorption maximum (at 354 nm), A_λ, is the sum of the absorbance A_p (photochromism) and A_t (thermochromism). A_p was obtained by subtraction of A_t obtained before irradiation from A_λ. The concentration of the triarylimidazolyl radicals C_r at time t was calculated from the relation $A_p = \epsilon C_r l$ using the molar absorptivities that were described in Table 2-3. The values of $(1/C_r = 1/C_{r0})$ of **1** and **2** increased linearly with time as shown in Figs. 2-5 and 2-6, respectively. From these data, the mechanism of the new photochromic system was ascertained to be the reversible radical dissociation of hexaarylbiimidazolyl to triarylimidazolyl shown in the following scheme. Activation energies for the recombination of the triarylimidazolyl radicals in the photochromic systems are shown in Table 2-4.[3]

hexaarylbiimidazolyl $\underset{\Delta}{\overset{h\nu}{\rightleftharpoons}}$ triarylimidazolyl radical

	R^1	R^2	R^3
1	H	H	H
2	Me	Me	Me
3	Cl	Cl	Cl
4	Cl	H	H

Furthermore, as previously described, the photodimers also showed a thermochromic color change in solution and in the solid state. The equilibrium constants K for the thermal dissociation of **1** in benzene were estimated at different temperatures; the values at 49.0, 54.0, and 60.5° C were 1.2×10^{-7}, 1.9×10^{-7}, and 4.9×10^{-7}, respectively, based on the absorbance values at 554 nm in the several concentrations of **1** (1.05×10^{-2} to 3.35×10^{-3} mol dm^{-3}). The plot of log K vs. the reciprocal of absolute temperature gave a

Table 2-2. Degree of Photodissociation of Hexaarylbiimidazolyl in the Photostationary State in Benzene at 15° C

Photodimer	Conc. × 10^5 (mol dm^{-3})	Degree of Dissociation (α)
1	3.38	0.30
2	1.60	0.35
3	1.61	0.25
4	1.45	0.25

straight line and from the slope of the straight line, ΔH for the dissociation of the photodimer **1** in benzene was estimated to be 109 kJ mol^{-1}.[3]

4. Chromotropism of Hexaarylbiimidazolyl with Isomerization

As previously mentioned, oxidation of lophine with ferricyanide produces purple triphenylimidazolyl radical, which rapidly precipitates when a decrease of the radical solubility occurs by addition of a large quantity of aqueous solution of the oxidant. The precipitated solid was found to exhibit piezochromism; that is, the almost colorless solid turned deep purple upon grinding in a mortar or upon pressing with a glass rod on filter paper at room temperature. The colored solid reverted very slowly to the original color after standing for several days in the dark at room temperature. Elevated temperature (around 50° C) caused faster recovery. We named the precipitated solid "piezo-dimer." A very intense ESR signal ($g = 2.003$) was induced upon rubbing of the piezo-dimer, and it decreased very slowly. The absorption spectrum of the piezochromic colored solid was

Table 2-3. Molar Absorptivity of Triarylimidazolyl Radical (R·) at Absorption Maximum in Benzene

Radical	$\lambda_{max}^{benzene}$/(nm)	($\epsilon \times 10^{-4}$/mol dm^{-3} cm^{-1})
1R·	348 (4.8),	554 (0.84)
2R·	366 (6.3),	587 (1.6)
3R·	367 (6.5),	588 (1.8)
4R·	366 (7.4),	575 (1.6)

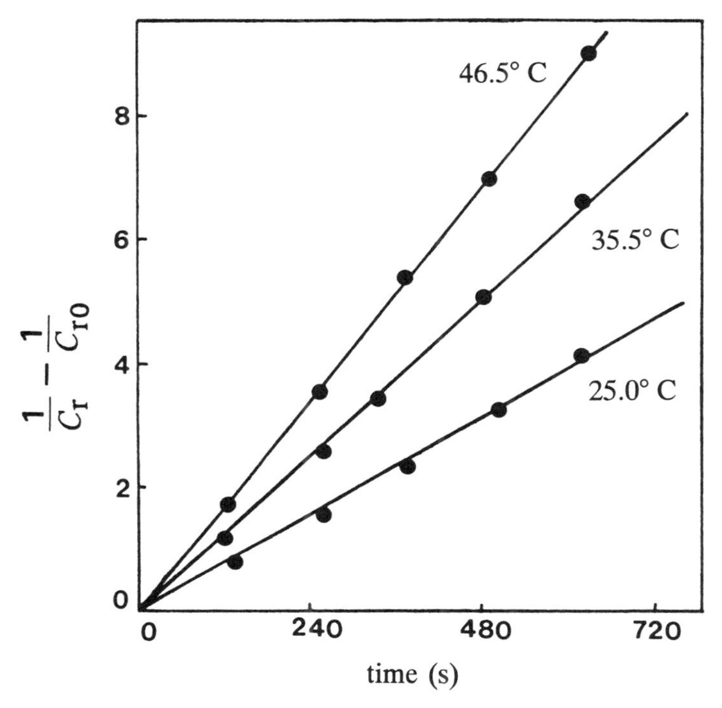

Figure 2-5. *Plot of* $(1/C_r - 1/C_{r0})$ *vs. time in the fading process of photochromism of photodimer* **1** *in benzene (Ref. 7). Concentration of* **1** *is* 1.44×10^{-5} *mol* dm^{-3}; C_r: *radical concentration at time* t, C_{r0}: *radical concentration at time* t = *0.*

measured by both the opal-glass and KBr pellet transmission methods to give a similar spectrum to that of the photochromic colored solid of **1**. The piezo-dimer, when dissolved in benzene, was found to be isomerized to the photodimer **1** via dissociation into the purple-colored triphenylimidazolyl radical. In 1972, Tanino et al.[8] reported that, in addition to the piezochromic dimer and photochromic dimer **1**, another photochromic dimer (dimer B)[8] and nonphotochromic dimer (dimer C)[8] were obtained by refluxing of a benzene solution containing the piezo-dimer in the dark and then separating **1**, dimer B, and dimer C by means of TLC on silica gel. From a comparison of the ^{1}H NMR and IR spectra of these three dimers with the corresponding dimers obtained from the mono-, di-, and tri-*p*-substituted triphenylimidazole, the structures of **1**, dimer B, and dimer C were assigned to be isomeric dimers in which two imidazolyl radicals were connected at the 1-2′ position, 1-4′ position, and 2-4′ position, respectively.

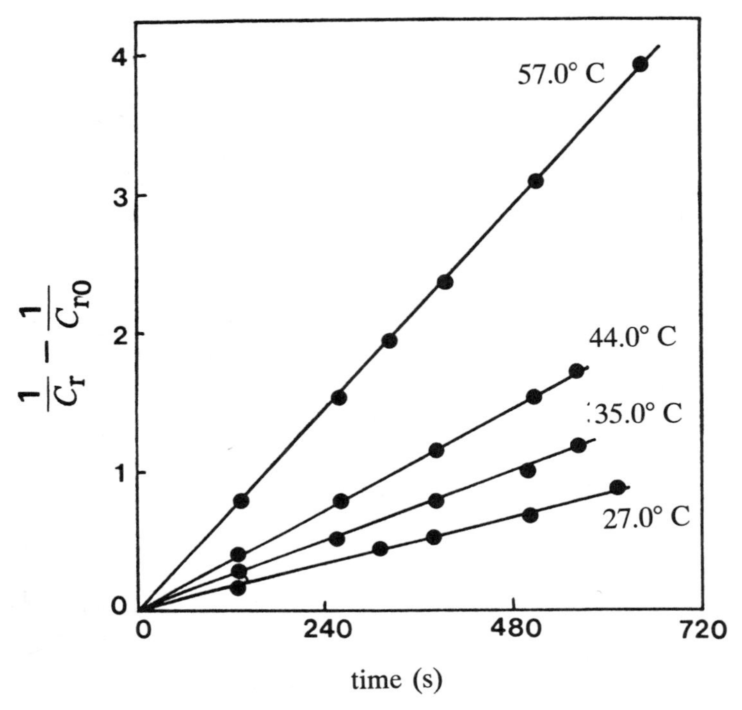

Figure 2-6. *Plot of* $(1/C_r - 1/C_{r0})$ *vs. time in the fading process in photochromism of photodimer* **2** *in benzene (Ref. 3). Concentration of* **2** *is* 2.25×10^{-5} *mol dm*$^{-3}$; C_r*: radical concentration at time* t, C_{r0}*: radical concentration at time* t = *0.*

1-2 dimer
photodimer **1**

1-4 dimer
photodimer

2-4 dimer
non-photodimer

Tanino et al. determined that the 1-2′ and 1-4′ dimers are interconvertible by UV irradiation, while the 2-4′ dimer is not affected by the light. The photodimer **1** we used for the investigation of photochromism is probably a mixture of these three dimers convertible via the triphenylimidazolyl radical formed by irradiation and/or heating. In 1966, White and Sonnenberg[9] reported that the piezo-

Table 2-4. Activation Energy in the Recombination of Triarylimidazolyl Radical in the Photochromic System in Benzene

Radical	**1R·**	**2R·**	**3R·**	**4R·**
E_a (kJ mol^{-1})	31.0	39.7	28.9	30.1

dimer is the isomer of photodimer **1** and shows thermochromism at temperatures lower than room temperatures; however, the structure of the piezo-dimer has not been determined because it is unstable and isomerizes quickly to **1** upon dissolution in benzene at room temperature.

5. Photochromic Color Change of Photodimers at Low Temperatures

Photodimers **1**, **2**, **3**, and **4** were found to show a photochromic color change not only at room temperature, but also in a wide range of temperatures, from

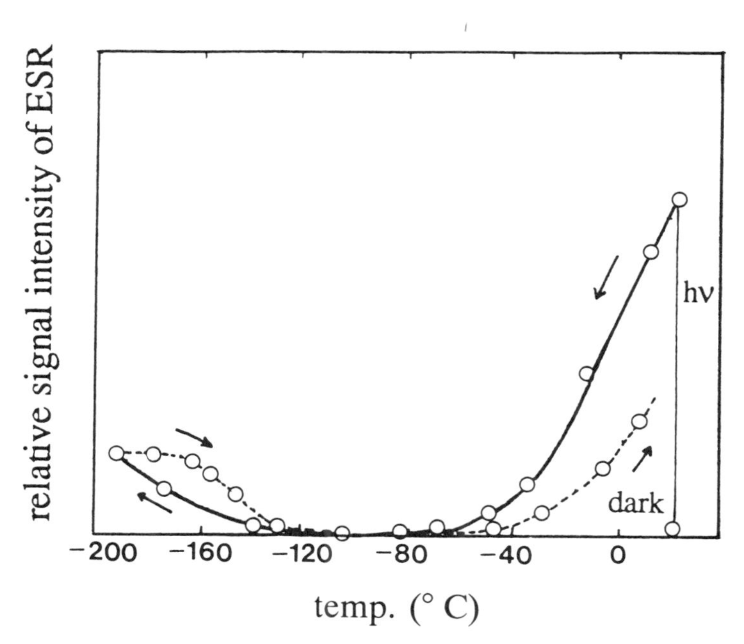

Figure 2-7. *Variation in the ESR signal intensity of photodimer* **1** *in deaerated benzene under irradiation (Ref. 11). ——: with cooling; ------: with warming. Concentration of* **1** *is* 6.8×10^{-5} *mol* dm^{-3}.

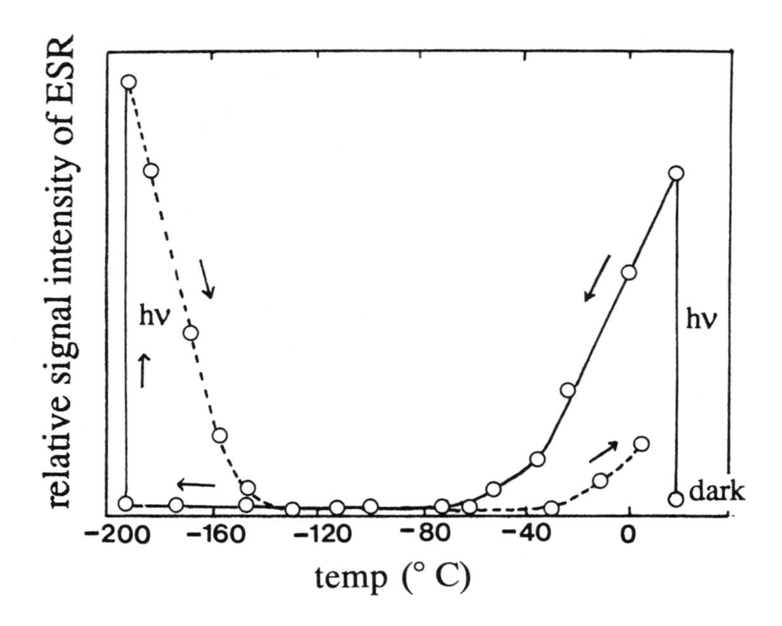

Figure 2-8. *Variation in the ESR signal intensity of photodimer **1** in deaerated benzene (Ref. 11). ——: with cooling to −196° C in the dark after irradiation at room temperature; ------: with warming to room temperature in the dark after irradiation at −196° C. Concentration of **1** is 6.8 × 10^{-5} mol dm^{-3}.*

room temperature down to −196° C. In the middle low temperature range, some variation with the type of solvent was found. We found[10] that when a purple-colored benzene solution of **1** obtained by irradiation of a pale yellow solution at room temperature was cooled in the dark, the color gradually faded and at about −20° C, the color disappeared. When the resulting colorless frozen solid was warmed by immersion in water at about 15° C in the dark, the purple color reappeared as soon as the solid began to melt. This behavior seems to be similar to that when the piezo-dimer was dissolved in benzene at room temperature: the resulting purple solution cooled immediately to below −20° C and then warmed to room temperature. On the other hand, when the purple benzene solution of **1** was cooled under irradiation to −20° C, the purple color disappeared even under irradiation. On further cooling of the resulting frozen solid to −150° C under irradiation, a bluish purple color was observed, and the color became more intense with the decrease in temperature. When the bluish purple frozen solid was warmed from −196° C to room temperature under irradiation, a color change reverse to that on cooling was observed.[11] As is shown in Fig. 2-7, the variation in the ESR signal intensity paralleled that of the color change; i.e., at −20° to −150° C, when the photochromic color disappeared, the ESR signal associated with the triphenylimidazolyl radical also disappeared. When the pale yellow benzene solution of **1** was irradiated at room temperature and then cooled in the dark to −196° C, neither a color change nor the appearance of an ESR signal was observed. Upon irradiation at −196° C, the frozen solid showed an intense bluish purple color and a strong ESR signal (Fig. 2-8). Such a temperature dependence of the photochromic behavior of **1** with a concomitant change in the ESR signal intensity was observed in other organic solvents, as well.

Variation of the absorption spectrum with the photochromic color change of **1** with a change in temperature from room temperature to −196° C was measured in EPA (ether: isopentane: ethanol = 5:5:2). An almost colorless deaerated EPA solution of **1** in a long-necked quartz cell was immersed in liquid nitrogen in a Dewar. Upon irradiation of the resulting glassy matrix from outside the Dewar, a bluish purple color appeared. Absorption spectra measured at several temperatures between −196° C and room temperature are shown in Fig. 2-9. Irradiation at −196° C gave absorption bands that were sharper and slightly red-shifted when compared to those obtained at higher temperatures. With a further increase in temperature, the intensity of the absorption bands gradually diminished. At about −130° to −60° C the bands in the visible region disappeared. Upon further warming of the solution, an absorption spectrum due to the triphenylimidazolyl radical appeared at about −50° C, and the intensity of the absorbance increased as the temperature was raised. A subtle change in color, which appeared on irradiation around room temperature and at −196° C, was also observed in other glassy-forming solvents such as 2-methyltetrahydrofuran (MTHF), 3-methylpentane, and isobutylchloride. The shift in the absorption band in the visible region was solvent dependent. For example, in MTHF, when the almost colorless glass of **1** at −196° C was exposed to UV light, the glass became purple and showed an absorption maximum for a visible band at 556 nm that corresponded to the 552-nm band at room temperature. From these observations, the temperature dependence of the photochromic behavior can be explained as follows. In solution, equilibria (a) and (b) shown in the scheme on p. 104 take place between piezo-dimer and photodimer. Equilibrium (a) lies far on the side of the radical at room temperature and shifts toward the piezo-dimer at temperatures lower than −20° C in benzene and −60° C in EPA, while the position at equilibrium (b) lies on the side of the photodimer at room temperature.

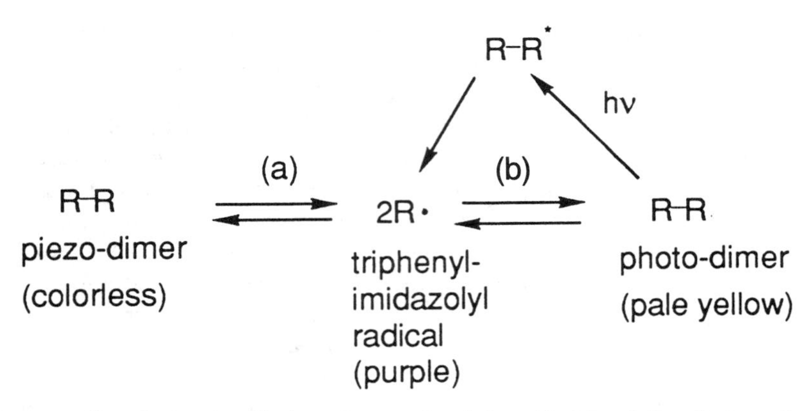

Upon cooling during irradiation, the purple triphenylimidazolyl radical produced by photodissociation of the photodimer dimerizes quickly to the piezo-dimer at intermediate low temperatures because of a low activation energy for the dimerization of the radical to the piezo-dimer. As a result, the purple color and ESR signal were not observed.

The appearance of the photochromic bluish purple color and the ESR signal at temperatures lower than ca. −150° C is due to the photodissociation of the photodimer remaining in the matrix or frozen solid. It must be difficult for the radical to surmount the activation energy barrier for dimerization to the piezo-dimer at such a low temperature. The intensity of the ESR signal observed at −196° C must depend on the concentration of the photodimer remaining in the matrix, because when the solution was cooled under irradiation (Fig. 2-7), the signal intensity was much weaker than that in the matrix that was irradiated after cooling to −196° C in the dark as shown in Fig. 2-8. The bluish purple color appearing at temperatures lower than about −150° C is attributed to a triphenylimidazolyl radical that has a more planar conformation[12] (presumed stable at −196° C) than the purple radical formed at room temperature.

Upon warming, the disappearance of the photochromic color change and ESR signal at temperatures higher than about −150° C is attributed to the dimerization of the radical to form a colorless photostable dimer: piezo-dimer, because at such temperatures the radical can surmount the barrier for dimerization to form the piezo-dimer. When the temperature was further raised to room temperature, the colorless dimer dissociated thermally to regenerate the purple triphenylimidazolyl radical in equilibrium with the photodimer.

A light yellow solid of photodimer **1** showed a weak purple color and a weak ESR signal when irradiated at room temperature, whereas upon irradiation after cooling to −196° C in the dark, the solid showed an intense bluish purple color. The color change of the solid with the change in temperature was monitored by ESR. Upon cooling of the solid of **1** in the dark after irradiation at room tem-

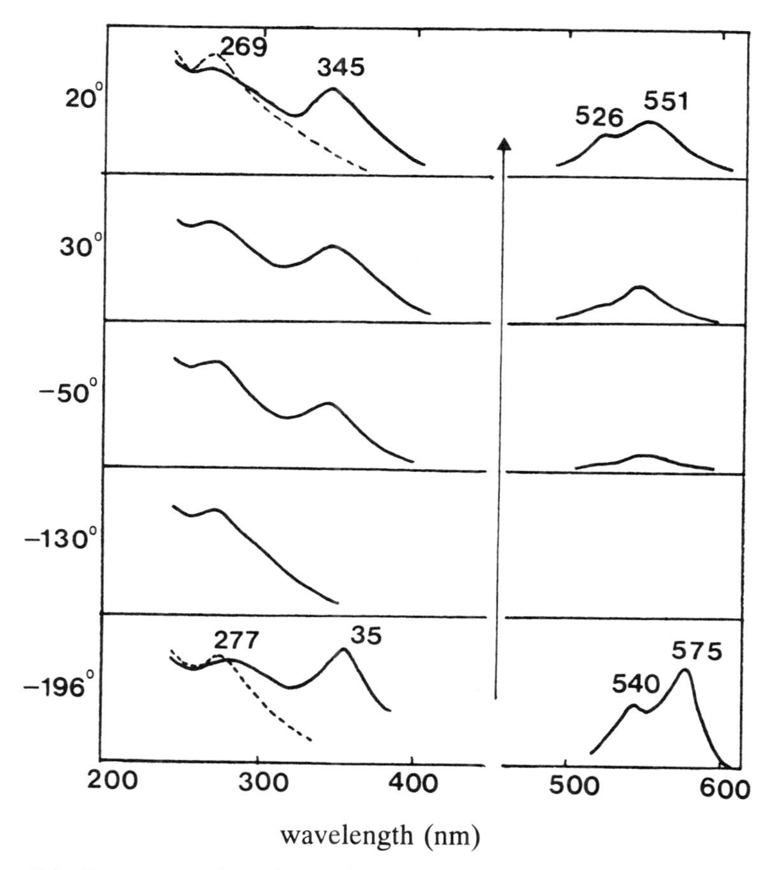

Figure 2-9. *Temperature dependence of absorption spectrum of the photocolored EPA solution of photodimer* **1** *(Ref. 11). ------: before irradiation; ——: after irradiation at −196° C. Concentrations of* **1** *are 2.54 × 10^{-5} mol dm^{-3} for λ < 400 nm and 4.48 × 10^{-4} mol dm^{-3} for λ > 400 nm.*

perature, a weak ESR signal, which appeared at room temperature, did not decrease with decrease in temperature to 196° C. Upon irradiation of the solid again at −196° C, a very strong ESR signal, corresponding to the intense bluish purple color, appeared. The signal intensity increased with irradiation time, whereas it decreased on warming in the dark as shown in Fig. 2-10. Contrary to the observations in solutions, the solid exhibited the color that did not disappear in the intermediate low temperatures, presumably because the photodimer is not transformed into the piezo-dimer due to the retarded thermal motion of

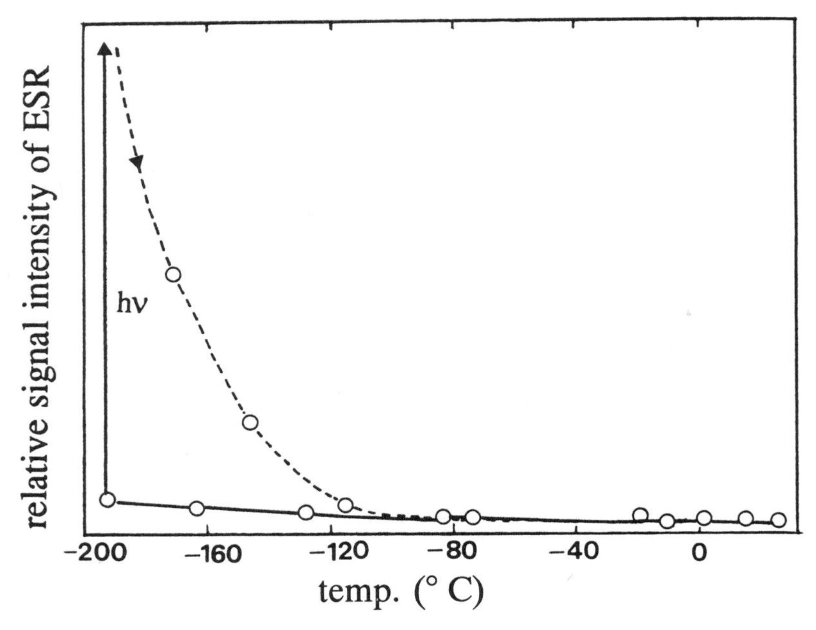

Figure 2-10. *Variation in ESR signal intensity of the solid of photodimer* **1** *with temperature (Ref. 11).* ——: *with cooling in the dark after irradiation at room temperature;* ------: *with warming in the dark after irradiation at −196° C.*

the radical. On the other hand, the large concentration of the radical corresponding to the ESR signal and deep bluish purple color, which appeared near −196° C, might be attributed to a very small rate of dimerization of the radical at such low temperatures.

References

1. B. Radziszewski, *Ber. Dtsch. Chem. Ges.*, **10**, 321 (1877).
2. T. Hayashi and K. Maeda, *Bull. Chem. Soc. Jpn.*, **33**, 566 (1960).
3. K. Maeda and T. Hayashi, *Bull. Chem. Soc. Jpn.*, **43**, 429 (1970).
4. T. Hayashi, K. Maeda, S. Shida and K. Nakada, *J. Chem. Phys.*, **22**, 1568 (1960).
5. T. Hayashi and K. Maeda, *Nippon Kagaku Zasshi*, **90**, 325 (1969).
6. T. Hayashi, K. Maeda and M. Takeuchi, *Bull. Chem. Soc. Jpn.*, **37**, 1717 (1964).
7. T. Hayashi, K. Maeda and M. Morinaga, *Bull. Chem. Soc. Jpn.*, **37**, 1563 (1964).
8. H. Tanino, T. Kondo, K. Okada and T. Goto, *Bull. Chem. Soc. Jpn.*, **45**, 1474 (1972).
9. D. M. White and J. Sonnenberg, *J. Am. Chem. Soc.*, **88**, 3825 (1966).
10. T. Hayashi and K. Maeda, *Bull. Chem. Soc. Jpn.*, **36**, 1052 (1963).

11. K. Maeda and T. Hayashi, *Bull. Chem. Soc. Jpn.*, **42**, 3509 (1969).
12. T. Shida, K. Maeda and T. Hayashi, *Bull. Chem. Soc. Jpn.*, **42**, 3044 (1969); **43**, 652 (1970).

ADVANCES IN HEXAARYLBIIMIDAZOLYL CHEMISTRY AND TECHNOLOGY (Editorial Note)

Since the first synthesis and characterization of the photochemical behavior of hexaarylbiimidazolyls (hexaphenylbiimidazoles, hexaarylbiimidazoles (HABI), etc.) by Hayashi and Maeda, hexaarylbiimidazoles continue to attract the attention of chemists. HABIs can thermally and photochemically form radicals capable of initiating acrylate polymerization and oxidation of leucodyes and other compounds. HABIs also possess very unusual photophysical properties. Hayashi and Maeda investigated HABI photochromism, electron paramagnetic spectra of triaryl on the imidazolyl radical, possibilities of HABI applications in solar energy conversion, and numerous other aspects of HABI chemistry and photochemistry.[1–7] Their discoveries stimulated extensive research by academic and industrial groups. We present below a short outline of information generated by this research. The list of references is by no means complete; the extensive review of the field lies outside the scope of this note.

Following and parallel to the work of Hayashi and Maeda described in this book, other investigations started in the middle 1960s. It was established that HABI dissociates almost exclusively from the singlet excited state, although in earlier studies the triplet state of HABI molecules was not directly observed.[8–10] It was also concluded from the investigation of HABI photochromism that the recombination of triarylimidazolyl radicals can lead to the formation of isomeric HABIs.[1–12] The imidazolyl rings in the isomers can be connected through C—N bonds or C—C bonds, and as illustrated in the previous chapter, in different positions around the imidazolyl ring.[1–12] The kinetics of photochromic and thermochromic reactions of HABI have been investigated extensively.[1–12] It was concluded that imidazolyl trimers can also form during radical recombination under certain conditions.[8–12] Extensive flash photolysis investigations of the rates of HABI photodissociation and recombination in solutions followed the first commercial applications of HABIs.[13–17] The unusual stability of triarylimidazolyl (lophine) radicals in solution was not appreciated. An investigation of the triplet state of HABI and unusual stability of the lophine radicals was conducted only recently.[22] Electron paramagnetic resonance studies at very low temperatures allowed direct observation of HABI triplet formation and confirmed the earlier conclusions[8–10] about the absence of dissociation from the triplet excited state.[22] It was demonstrated that the triplet excited state of HABI can form when the restriction of motion of triarylimidazolyls and phenyl groups in HABI exists. It was also shown that lophine recombination is sterically hindered even in solutions of low viscosity.

Industrial application of HABI-based photoinitiators led to the realization that HABI can be photosensitized with visible and infrared light.[23–25] An explanation of the photosensitization mechanism of HABI by visible light sensitizers was proposed.[26] It was alleged that it occurs through oxidation of the excited sensitizer (electron transfer). However, only recently (see Chapter 4 of this section) it was experimentally demonstrated that visible sensitization occurs through electron transfer.[27,28]

Industrial applicability of HABIs led to the interest in alternative synthetic procedures.[29] As in other radical processes, the kinetics of photodecomposition of HABI are altered when conducted in a magnetic field.[30] The magnetic field may also determine the type of HABI isomer formed during the triarylimidazolyl radical recombination.[30]

The optimization of the photopolymerization process conducted with HABI initiation resulted in the extensive work on the mechanism of HABI interactions with mercapto chain-transfer agents.[31] These investigations revealed that the electron transfer is involved in HABI reactions with mercaptans rather than through the hydrogen transfer assumed earlier.

The multiple reactive sites of biimidazolols make them a most versatile class of compounds useful in polymer chemistry and imaging applications. Some of the earlier applications of biimidazolols in photoimaging polymers were based on the ability of biimidazolol to promote photooxidation of leuco aminotriphenylmethane dyes.[16,32–37] Oxidized leuco triphenylmethane dyes are usually brightly colored and a wide range of colors could be created by substituting hydrogens in the phenyl groups. The commercially important HABI-leuco dye-based imaging systems were described in a series of Du Pont patents. Imaging material containing a substituted HABI and leuco crystal violet, giving a dark blue image upon photoexposure, was sold as Dylux ® by Du Pont for architectural design.[38,39] A thermal imaging polymeric system based on HABI and leuco dye was also developed.[40] The Du Pont patents were followed by a number of more recent patents from other chemical companies describing similar imaging materials containing HABI and hydrogen donors.[41–44] Simultaneously, with the appearance of the first Du Pont patents, it was established that substituents on the phenyl rings of HABI increase the lifetime of the triarylimidazolyl radical.[13–21] Particularly stable and active radicals could be produced by 2,2′-di(ortho-chlorophenyl)-4,4′,5,5′-tetraphenyl biimidazole (*o*-Cl-HABI).[13–21,37,38]

HABIs form long-lived radicals upon direct or sensitized photodissociation with a quantum yield ~98%. Therefore HABIs are widely used as polymerization photoinitiators. In combination with a mobile chain-transfer agent such as mercaptobenzoxazole and a suitable photosensitizer, the HABI initiation system provides efficient initiation over a wide spectral range with low oxygen sensitivity.[23–25,45–47] The HABI initiation is used in a variety of Du Pont imaging materials and holographic photopolymers.[45–47] HABI initiation is patented by

BASF, Fuji, Mitsubishi, and other major companies producing photosensitive polymeric coatings and imaging systems.[48–54]

In the majority of photocurable and photoimaging polymers the initiator consumption during imaging is less than 10%. The remaining initiator can diffuse out of the imaged polymer or form a crystalline inclusion separating from the cured polymer. Because of the large size of the substituted HABI molecules and steric restrictions, HABI retains its isotrophic distribution in the films after the imaging and does not cause undesired fogging (crystallite formation) of the polymer films.

The photophysics of hexaarylbiimidazole molecules is not fully understood, making them an intriguing object for investigations. Investigations of HABI photochemical reactions in matrix were only recently renewed.[22,55] New information appearing on HABI photochemistry will, undoubtedly, lead to new applications of HABIs in photosensitive polymers. Thus, the pioneering work of Hayashi and Maeda opened new directions in photopolymer formulations and manufacturing.

References

1. T. Hayashi and K. Maeda, *Bull. Chem. Soc. Jpn.* **33**, 566 (1960).
2. T. Hayashi and K. Maeda, *Bull. Chem. Soc. Jpn.* **35**, 2057 (1962).
3. T. Hayashi, K. Maeda, and M. Takeuchi, *Bull. Chem. Soc. Jpn.* **37**, 1717 (1964).
4. T. Hayashi and K. Maeda, *Bull. Chem. Soc. Jpn.* **37**, 1563 (1964).
5. T. Hayashi and K. Maeda, and S. Shida and K. Nakada, *J. Chem. Phys.* **32**, 1568 (1960).
6. K. Maeda and T. Hayashi, *Bull. Chem. Soc. Jpn.* **43**, 429 (1970).
7. T. Shida, K. Maeda, and T. Hayashi, *Bull. Chem. Soc. Jpn.* **43**, 652 (1970).
8. A. L. Prokhoda and V. A. Krongauz, *Khim. Vys. Energ.* **3**(6), (1969) 495, (Russ.) 449 (Engl.).
9. A. L. Prokhoda and V. A. Krongauz, *Khim. Vys. Energ.* **4**(2), (1970) 176 (Russ.) 152 (Engl.).
10. A. L. Prokhoda and V. A. Krongauz, *Khim. Vys. Energ.* **5**(3) (1971) 262 (Russ.) 234 (Engl.).
11. M. A. J. Wilks and M. R. Willis, *J. Chem. Soc.* (B), 1526 (1968).
12. M. R. Willis and M. A. J. Wilks, *Nature* **212**, 500 (1966).
13. L. A. Cescon, G. R. Coraror, R. Dessauer, E. F. Silversmith, and E. J. Urban, *J. Org. Chem.*, **36**(16), 2262 (1971).
14. L. A. Cescon, G. R. Coraror, R. Dessauer, A. S. Deutsch, H. L. Jackson, A. MacLachlan, K. Marcali, E. M. Potrafke, R. E. Read, E. F. Silversmith, and E. J. Urban, *J. Org. Chem.* **36**(16), 2267 (1971).
15. R. H. Reim, A. MacLachlan, G. R. Coraor, and E. J. Urban, *J. Org. Chem.* **36**(16), 2272 (1971).
16. A. MacLachlan and R. H. Reim, *J. Org. Chem.* **36**(16), 2275 (1971).
17. R. L. Cohen, *J. Org. Chem.* **36**(16), 2280 (1971).
18. B. S. Tanaseichuk and L. G. Resepova, *Zh. Org. Khim.* **6**, 1065 (1970).
19. B. S. Tanaseichuk, A. A. Bardina, and V. A. Maksakov, *Zh. Org. Khim.* **7**, 1508 (1971).
20. B. S. Tanaseichuk, A. Belozerov, L. G. Tikhonova, V. N. Shishkin, A. A. Bardina, and K. P. Butin, *Zh. Org. Khim.* **14**(10), 2029 (1978).

21. Yu. A. Rozin, V. F. Gryazev, V. E. Blokhin, Z. V. Pushkareva, and G. E. Martina, *Khim. Geterotsikl. Soedin.* **11**, 1536 (1974).
22. X. Z. Qin, A. D. Liu, A. D. Trifunac, and V. V. Krongauz, *J. Phys. Chem.* **95**, 5822 (1991).
23. T. E. Dueber, U. S. Patent 4,162,162 (1979).
24. T. E. Dueber, U. S. Patent 4,454,218 (1984).
25. T. E. Dueber, U. S. Patent 4,535,052 (1985).
26. R. D. Mitchell, W. J. Nebe, and W. M. Hardam, *J. Imag. Sci.* 30(5), 215 (1986).
27. A. D. Liu, A. D. Trifunac, and V. V. Krongauz, *J. Phys. Chem.* **96**, 207 (1992).
28. Y. Lin, A. D. Liu, A. D. Trifunac, and V. V. Krongauz, *Chem. Phys. Lett.* **198**(1,2), 200 (1992).
29. V. V. Nurgatin, B. M. Ginzburg, G. P. Sharnin, and V. F. Polanskii, *Khim. Geterocycl. Soed.* **8**, 1069 (1987).
30. H. Sato, K. Kasatani, and S. Murakami, *Chem. Phys. Lett.* **151**(1,2), 97 (1988).
31. D. F. Eaton, A. Gafney Horgan, and J. P. Horgan, *J. Photochem. Photobiol. A: Chem.* **58**, 373 (1991).
32. L. A. Cescon and R. Dessauer, U. S. Patent 3,585,038 (1971).
33. P. S. Strilko, U. S. Patent 3,579,342 (1971).
34. R. L. Cohen, U. S. Patent 3,563,751 (1971).
35. D. H. Fishman, U. S. Patent 3,552,973 (1971).
36. R. L. Cohen, U. S. Patent 3,554,753 (1971).
37. D. S. James and V. G. Witterholt, U. S. Patent 3,533,797 (1970).
38. R. Dessauer and C. E. Looney, in *Imaging Processes and Materials*, J. Sturge, A. Walworth, and A. Shepp, eds, Neblette's, 8th edition (Van Nostrand Reinhold, New York, (1989) pp. 263–278.
39. L. A. Cescon, Ger. Patent DE 1772534 (1972).
40. C. E. Looney, U. S. Patent 3,615,481 (1971).
41. A. Adin and S. R. Levinson, U. S. Patent 4,201,590 (1980).
42. A. Adin and S. R. Levinson, U. S. Patent 4,196,002 (1980).
43. Mitsubishi Chemical Industries Co., Ltd., Japan Kokai Tokkyo Koho JP 57/2140, (1982).
44. Oji Paper Co., Ltd. Japan Kokai Tokkyo Koho JP 60/2393 (1985).
45. W. K. Smothers, Europ. Patent 324481 A2 (1989) and ref. therein.
46. W. J. Nebe, U. S. Patent 4,009,040 (1977).
47. W. J. Chambers and D. F. Eaton, *J. Imag. Sci.* **30**(5), 230 (1986).
48. G. Bauer, G. Hoffmann, and F. Seitz, Ger. Patent DE 3,939,389 (1989).
49. H. Nagasaka, M. Masaaki, and T. Urano, Japan Kokai Tokkyo Koho JP 04,304,456 (1992).
50. Mitsubishi Kasei Corp., Japan Kokai Tokkyo Koho JP 04304456-A (1992).
51. H. Nagasaka, Europ. Patent Appl. Ep 138187 (1985).
52. Mitsubishi Chemical Industries Co., Ltd., Japan Kokai Tokkyo Koho JP 59/64840 (1982).
53. Mitsubishi Chemical Industries Co., Ltd., Japan Kokai Tokkyo Koho JP 59/56403 (1982).
54. Mitsubishi Chemical Industries Co., Ltd., Japan Kokai Tokkyo Koho JP 57/21401 (1980).
55. H. Morita and S. Minagawa, *J. Photopol. Sci.* **5**(3), 551 (1992).

3

Visible Light Photoinitiation Systems Based on Electron and Energy Transfer Processes

Tsuguo Yamaoka and Kazuhiko Naitoh

1. Introduction

Nearly 35 years ago L. M. Minsk et al. reported poly(vinyl cinnamate)[1] and E. I. du Pont de Nemours & Co. commercialized the letterpress plate made of photopolymerizing material.[2] With the current rapid development of technology, photopolymers are finding increasing applications in the fields of printing, coating, microlithography, and electronic device production.

Compared with other photosensitive materials such as silver halides and electrophotography, the sensitivity of photopolymers is not high as they require ultraviolet light with 10–100 mJ/cm^2 photon flux for the exposure.

Sensitivity enhancement of photopolymers has recently been attracting interest for application to direct laser imaging involving computer-to-plate systems and holographic recording.[3–6] Development of highly sensitive photopolymers based on photopolymerization requires photoinitiators that are sensitive to visible light in order to generate initiating free radicals with high efficiency. Some of the visible-light absorbing photoinitiators that have been reported so far are the bimolecular type, consisting of a radical generator and a sensitizing dye. Biimidazole derivatives act as visible light sensitive photoinitiators with the aid of arylketones.[7–10] Recently, the mechanism of photosensitized dissociation of hexaarylbiimidazole was reported on by A. Liu et al.[11] J. L. R. Williams et al. reported that *N*-phenylglycine, in the presence of ketocoumarin dye, shows sensitivity in the visible region up to 550 nm.[12] In addition, diphenyliodonium salts,[13–15] merocyanine dye,[16] poly(ethylene imine)-methylene blue,[17] and s-trichloromethyltriazine derivative-cyanine[18,19] have been reported to be highly sensitive initiating systems upon exposure to visible light. The efficiency of visible light photoinitiation for thioxanthones-amines[20] and thioxanthones-onium salts[21,22] has been reported. The mechanistic studies for sensitization of amino ketones by thioxanthones have been carried out as well.[23]

In printing engineering, a presensitized printing plate sensitive to 488-nm light from an argon ion laser with a sensitivity of 0.5 mJ/cm^2 has been developed and commercialized by Hoechst Co.[24]

Another category of photoinitiators consists of ionic compounds in which electron transfer occurs from counter anion to excited cation dye. D. C. Neckers et al. developed alkyl and arylborates of cyanin dyes that generate free radicals by irradiation with visible light.[25]

M. S. Kaplan proposed Eq. (1) to calculate the relation between the laser energy and the sensitivity of recording materials.[26]

$$E_g = (P \cdot T \cdot E_f)/A \tag{1}$$

where E_g is the minimum sensitivity required for the material, P is the laser energy at the material, T is the time to scan the whole area, E_f is the efficiency of the optical system, and A is the area to be exposed. Fig. 3-1 shows the relation between the laser power and sensitivity of materials to expose an area of 2700 cm^2.

2. Methods of Increasing Sensitivity

2.1. Sensitivity of Photopolymers Based on Stepwise Crosslinking

A. Reiser and E. Pihs proposed that the sensitivity of photopolymers made insoluble by stepwise crosslinking is determined by Eq. (2)[27,28] where E_g is the minimum photon energy for the gel formation(einsteins/cm^2), r is the film thickness, d is the specific gravity of the polymer, M_w is the weight average molecular weight of the polymer, ϕ is the quantum yield of crosslinking, and A is the absorption of incident light by the photosensitive group. According to Eq. (2), the limit of sensitivity of photopolymers classified by this mechanism is predicted to be 0.12 mJ/cm^2. If the crosslinking is accomplished by chain reactions such as radical polymerization, the limit of sensitivity given by Eq. (3) is 6.6×10^{-4} mJ/cm^2. Where, P_0 is the molar fraction of the crosslinking groups.

$$E_g = r \cdot d/(A \cdot M_w \cdot \phi) \tag{2}$$

$$E_g = r \cdot d/(A \cdot M_w \cdot \phi P_0) \tag{3}$$

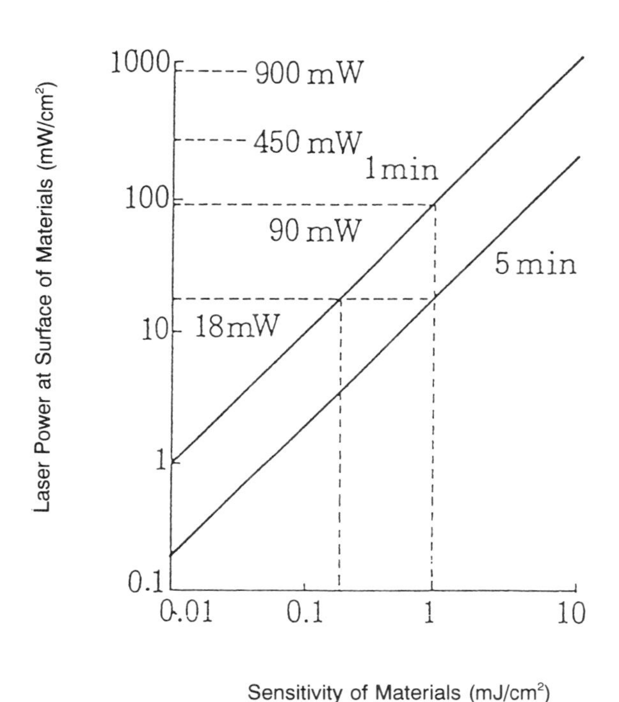

Figure 3-1. *Relation between laser power at surface of materials and sensitivity of materials required to scan-expose an area of 2700 cm² in one or five minutes.*

Several photopolymers that are sensitive to visible light have been reported, and the sensitivities of these have been expanded to visible wavelengths by using sensitizing dyes. In most cases, the sensitization is accomplished by the triplet-triplet energy transfer mechanism where the lowest triplet state energy of the dye is transferred to the triplet state of the photopolymer. Table 3-1 gives some examples of photocrosslinkable polymers and sensitizing dyes.

2.2. Sensitivity of Photopolymers Based on Photopolymerization

The mechanism of photopolymerization that is most promising for increasing sensitivity involves chain reactions of a monomeric compound. The sensitivity of the photopolymers that exhibit this mechanism may depend fundamentally on the rate of polymerization of monomeric compounds. If photopolymerization proceeds according to Eqs. (4)–(9), the rate of polymerization is given by

Table 3-1. Examples of Visible Light Sensitive Photopolymers and Sensitizers Based on Stepwise Photocrosslinking

Photopolymer	Sensitizer	Sensitivity	Ref.
$-(CH_2-CH)_n-$; $O-C(=O)-C_6H_4-N_3$	Squarylium dye: two azulenium rings (CH_3 substituents) on a squaric acid core (O, O)	10 mJ/cm^2 to 488 nm	29
	Thiopyrylium salt: OB-phenyl, two MeO-phenyl (OMe) groups, S^+, X^-	2 mJ/cm^2 to 488 nm X^-: ClO_4^-	30
$-(OC-CH=CH-C_6H_4-CH=CH-COOCH_2-C_6H_4-CH_2O)_n-$	Thiopyrylium salt: OB-phenyl, two MeO-phenyl (OMe) groups, S^+, X^-	80 mJ/cm^2 to 488 nm X^-: ClO_4^-	31
$[-C(=O)-C(CN)=CH-CH=CH-C_6H_4-CH=CH-CH=C(CN)-C(=O)-O-(CH_2CH_2O)_3-]_n$		1.5 mJ/cm^2 to 488 nm without sensitizer	32

Table 3-1. *Continued*

Photopolymer	Sensitizer	Sensitivity	Ref.
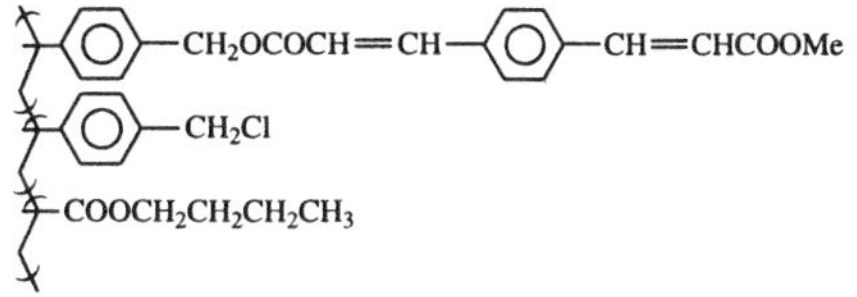 $CH_2OCOCH{=}CH$–C₆H₄–$CH{=}CHCOOMe$; CH_2Cl; $COOCH_2CH_2CH_2CH_3$	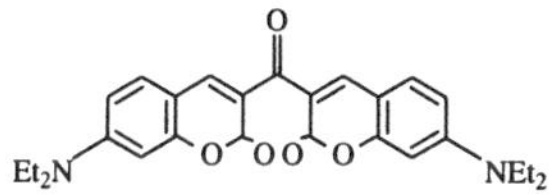 O; Et_2N; NEt_2	0.2 mJ/cm^2 to 488 nm M_w: 205,000 1 mJ/cm^2 to 488 nm M_w: 64,000	33
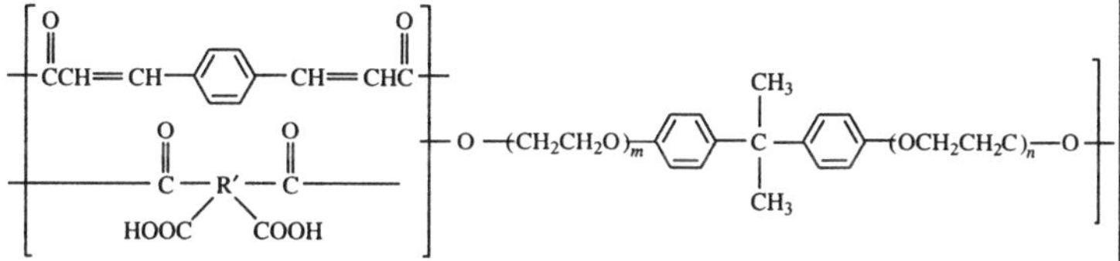 O, O; CCH=CH–C₆H₄–CH=CHC; O, O; C—R′—C; HOOC, COOH; $O\ (CH_2CH_2O)_m$; CH_3; C; CH_3; $(OCH_2CH_2C)_n$—O		Sensitive to 488 nm	34
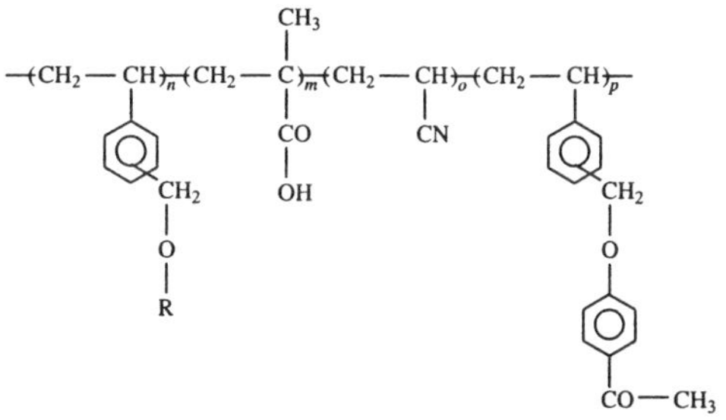 CH_3 $-(CH_2-CH)_n(CH_2-C)_m(CH_2-CH)_o(CH_2-CH)_p-$ CH_2; CO; CN; CH_2 O; OH; O R $CO-CH_3$ n = 36 mol%(R 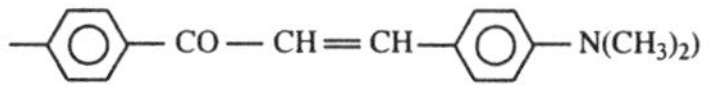 = –C₆H₄–CO—CH=CH–C₆H₄–N$(CH_3)_2$) n = 31 mol%(R 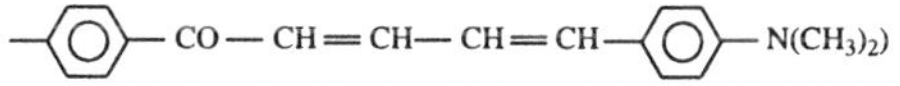 = –C₆H₄–CO—CH=CH—CH=CH–C₆H₄–N$(CH_3)_2$)		10 mJ/cm^2 to 488 nm	35

Eq. (10). When the film is very thin or very thick, Eq. (10) is approximated by Eqs. (11) or (12), respectively.

Elementary reaction			Reaction rate
In	R·	(4)	$R_i = I_{0\lambda}\phi\ [1 - \exp(-2.3_{\epsilon\lambda}\ [\mathrm{In}]L)]$
R· + nM	RMn· (Pn)·	(5)	$R_p = k_p\ [\mathrm{R}\cdot]\ [\mathrm{M}]$
Pm· + Pn·	Pm + n	(6)	$R_t = k_t'[\mathrm{P_n}\cdot]^2$
Pn· + Pn·	Pm + Pn	(7)	$R_t = k_t''\ [\mathrm{P_n}\cdot]^2$
Pn· + S	Pn + S·	(8)	$R_{tf} = [\mathrm{P_n}\cdot]\ [\mathrm{S}]$
S· + M	SM·	(9)	$R_p' = k_p'\ [\mathrm{S}\cdot]\ [\mathrm{M}]$

$$-d[\mathrm{M}]/dt = k_p/k_t^{1/2}[\mathrm{M}][10^3 I_{0\lambda}\phi f L^{-1}(1 - \exp(-2.3\epsilon_\lambda[\mathrm{In}]L))]^{1/2} \quad (10)$$

$$-d[\mathrm{M}]/dt = k_p/k_t^{1/2}[\mathrm{M}][2.3 \times 10^3 \mathrm{I}_{0\lambda}\phi f\epsilon^\lambda[\mathrm{In}]]^{1/2} \quad (11)$$

$$-d[\mathrm{M}]/dt = k_p/k_t^{1/2}[\mathrm{M}][10^3 I_{0\lambda}\phi f L^{-1}]^{1/2} \quad (12)$$

where $I_{0\lambda}$(einsteins/cm^{2}s) is the intensity of exposure light (λ in nm), [In] (in mol/L) is the concentration of the photoinitiator, ϵ_λ (in mol^{-1}L) is the molar extinction coefficient at wavelength λ (nm) of the photoinitiator, L (cm) is the film thickness, ϕ is the quantum yield of radical generation, and f is the efficiency of the initiator.

Eq. (10) shows that the rate of polymerization is governed by the concentration of the monomers, quantum yield of radical generation, reaction efficiency of radicals to monomers, and rate of propagation of monomers. The sensitivity of photopolymers can be predicted by these equations, but in practical materials, the reaction of polymerization may not proceed according to these stoichiometric reactions. For example, the rate of polymerization is proportional to the concentration monomer. Hertler et al.[36] prepared a microcrystalline matrix with *N*-vinylsuccinimide that gave a maximum concentration of monomer. This photopolymer achieved a sensitivity of 6 μJ/cm^2.

Photosensitive microgels are effective in enhancing the sensitivity. Sasa and Yamaoka and Kanda synthesized 30-nm-diameter photosensitive microgels that have acrylate monomers on the surface.[37,38] The photopolymer layer that was prepared with this microgel and a photoinitiator was made insoluble in water by exposure to visible light, and achieved a high sensitivity of 55 μJ/cm^2.

Methods that use silver halides as the photoinitiator give photopolymers with very high sensitivity. Michell et al. designed a photopolymerization system that achieves a sensitivity 100 times higher than that of conventional photopolymers.[39] Table 3-2 gives some examples of photopolymers with high sensitivity at visible wavelengths.

Table 3-2. Examples of Visible Light Sensitive Photoinitiators

Radical Generator	Sensitizer	Photopolymer	Sensitivity	Ref.
Organic peroxides 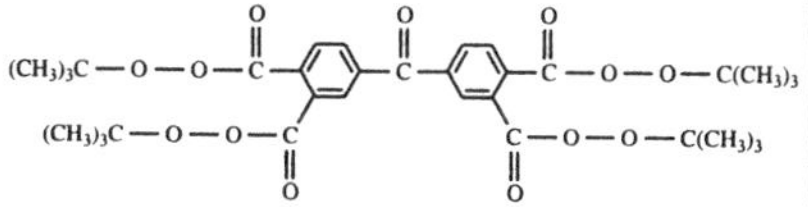	Chlorophyll	Acrylate monomer	Ar laser 514.5 nm	40
	Eosin G Ferricammonium citrate	Acrylamide/N,N′-methylene bisacrylamide	Hologram	41
	Riboflavin Methylene blue	Acrylate monomer	Visible light sensitive polymer	42
	(Thio)pyrilium salts 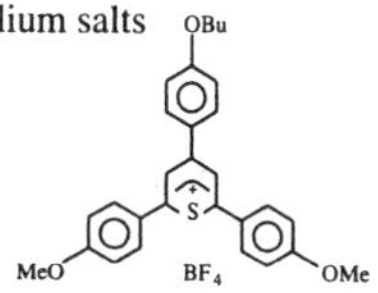	Pentaerythritol-triacrylate (PETA)/PVP	0.08 mJ/cm^2 to Ar laser 488 nm	43 44
	Merocyanine Styrylquinolone 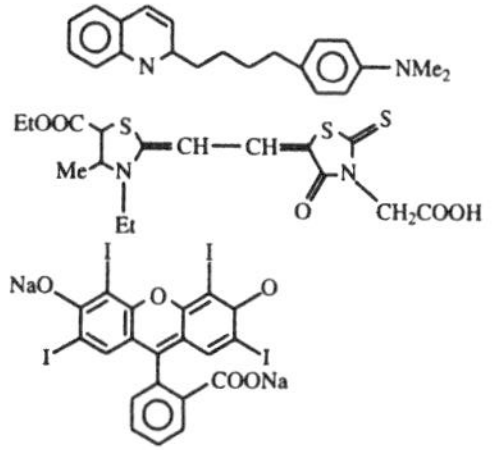	PETA/PVP	Ar laser 2.1 mJ/cm^2 to 488 nm 1 mJ/cm^2 to 514.5 nm 0.35 mJ/cm^2 to 514.5 nm	45
	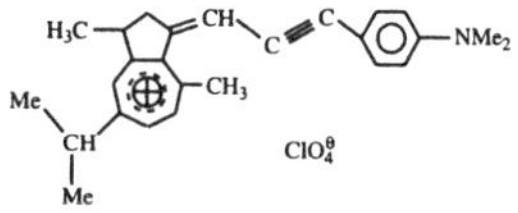	PETA/PVP	0.07 mJ/cm^2 to 488 nm	46

Table 3-2. *Continued*

Radical Generator	Sensitizer	Photopolymer	Sensitivity	Ref.
	Riboflavin tetrabutylate	PETA/P(MMA-MA-EA)	30 mJ/cm^2 to 490 nm Dry film resist	47
	ketocoumarin dye (Et$_2$N–)	PETA/PVP	Ar laser 488 nm 0.03 mJ/cm^2	48
			0.005 mJ/cm^2	49
Organic peroxides + N-phenylglycine (NPG) (Ph–NH—CH$_2$COOH)	Thioxanthene dye (OEt)	Polyfunctional acrylate monomer/Poly(MMA-MA)	0.3 mJ/cm^2 to 490 nm	48
				50
Diphenyliodonium salts (Ph–I$^+$–Ph PF$_6$$^-$)	Merocyanine dyes (Et; CH$_2$COOH; C$_7$H$_{15}$)	Urethane acrylate/PVA modified with acrolain and butylaldehyde	1.8 mJ/cm^2 to Ar laser	51
	Rhodamine dyes	PETA	A few mJ/cm^2 to Ar laser	52
	(CH$_3$)$_2$N–Ph–CH═CH—CO—CH═CH–Ph–N(CH$_3$)$_2$	Poly(methacryloyl-CMS-MMA)	0.9 mJ/cm^2 to Ar laser, 488 nm	53
	Ketocoumarin dye		10 mJ/cm^2 to Ar laser 488 nm	54
	Tetraphenylporphyline	Poly(methacryloyl-CMS-MMA)	3 mJ/cm^2 to He—Ne laser 633 nm	55
	Tetrabenzoporphyline	Poly(methacryloyl-CMS-MMA)	0.13 mJ/cm^2 (M:Cd) 0.3 mJ/cm^2 (M:Zn) to He—Ne laser	56
				57

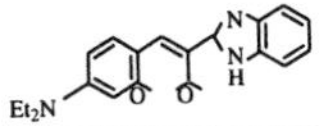

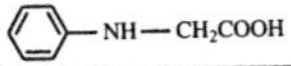

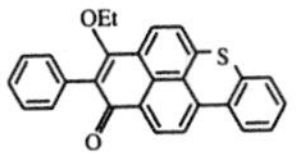

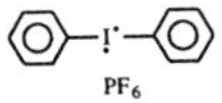

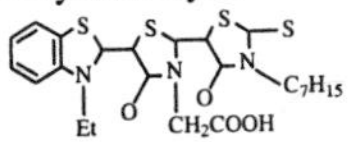

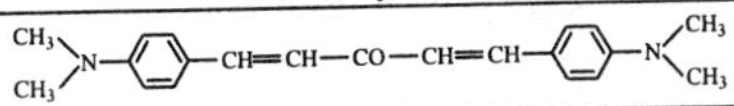

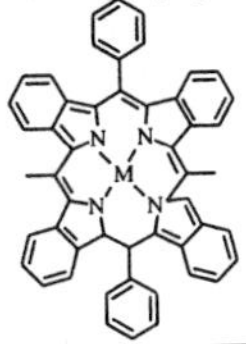

Table 3-2. *Continued*

Radical Generator	Sensitizer	Photopolymer	Sensitivity	Ref.
	Spyropyrane	Poly(methacryloyl-CMS-MMA)	Two photon system. 16.8 mJ/cm^2 to 313 nm 120 mJ/cm^2 to He—Ne laser	58
Diphenyliodonium salts + NPG	Thioxanthene dyes 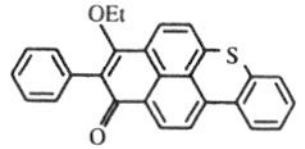	Poly(acrylate-coacrylic acid)	0.3 mJ/cm^2 to 490 nm	59
	Merocyanine dyes 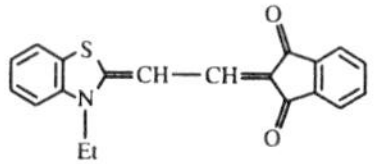	Poly(acrylate-coacrylic acid)	1.6 mJ/cm^2 to 490 nm	60
	Cyanine dyes 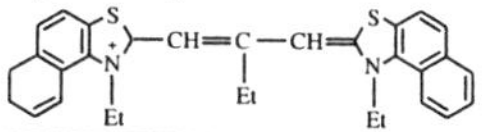	Poly(acrylate-coacrylic acid)	630 nm	61
Alkyl borate of cyanine dyes, rhodamine dyes, methylene blue dyes 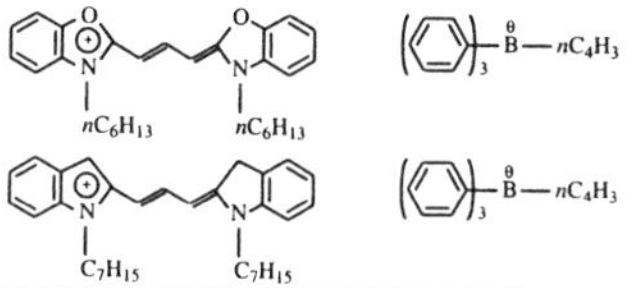		Acrylate monomer in microcapsule	Visible light	62
		Poly(allylmethacrylate)	Presensitized offset plate for visible light exposure	63
Iron arene complex 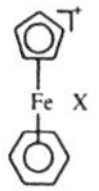 X = BF_4, PF_6, AsF_6, SbF_6		Epoxidized cresol novolak resin	10 mJ/cm^2 to 510 nm	64 65
	Ketocoumarin dyes 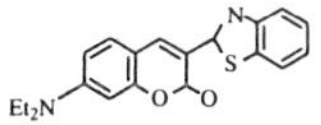	Poly(acrylate-coacrylic acid)/PETA	0.09 mJ/cm^2 to 488 nm	66

Table 3-2. *Continued*

Radical Generator	Sensitizer	Photopolymer	Sensitivity	Ref.
	Thioxanthene dyes (OEt, S, O)		0.4 mJ/cm^2 to 488 nm	66
Fluorinated titanocene (F_6, Ti, F_6)		Poly(styrene-comaleate)/poly-functional acrylate monomer	Sensitive to 560 nm 5–40 mJ/cm^2 Relief hologram	67
Bisimidazol (C=C, N—N, N=C, C=N)	Aryl-ketones (O, CH_3, CH, NEt_2)	Poly(MMA-co-MA)/poly-functional acrylate monomer	Sensitive to 600 nm	68
	Ketocoumarin dyes + thiol compound	Poly(styrene-comaleic acid ester) /polyfunctionalacrylate monomer	Sensitive to 550 nm	69 70
	(Me_2N, CN)	Poly(MMA-co-MAA)/bifunctional methacrylate monomer	Sensitive to He–Cd(440 nm) and Ar laser (488 nm)	71
NPG	Ketocoumarin dyes (O, N, O, O, NR_2)	Poly(MMA)/polyfunctional acrylate monomer	Sensitive to 560 nm	72
	Thioxanthene dyes (Cl, O, HO—C_2H_4—N, O, Cl, NEt_2)	PVP/PETA	0.2 mJ/cm^2 to 488 nm (Ar laser)	73 74

Table 3-2. *Continued*

Radical Generator	Sensitizer	Photopolymer	Sensitivity	Ref.
DITX/DMBI	Thioxanthene dyes 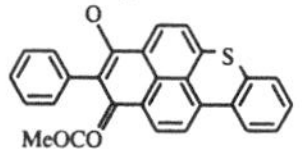	PVP/PETA	1.3 mJ/cm^2 to 488 nm (Ar laser)	74
Tris(trichloromethyl) triazine derivatives 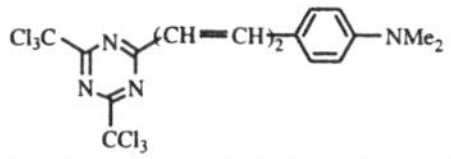		Poly(vinylbutylal)/poly-functional acrylate monomer	Sensitive to 560 nm	75
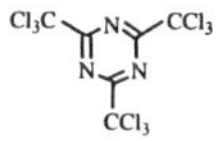	Merocyanine dye	Chlorinated polyethylene/poly-functional acrylate monomer	7 mJ/cm^2 to 488 nm	76 77
	Ketocoumarin dye		0.6 mJ/cm^2 to 488 nm	77
	Thiopyrilium salts	P(MMA-co-MA)/PETA	Sensitive to 490 nm	69 78
	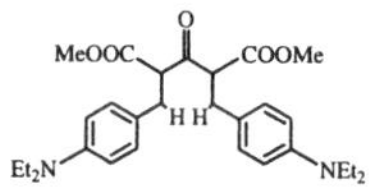			79
	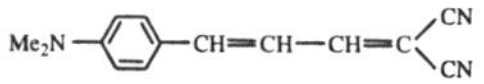			80
	Ketocoumarin dyes Thioxanthene dye 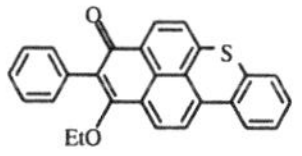	Poly(MMA-co-MAA)/PETA	0.16 mJ/cm^2 to 488 nm 0.8 mJ/cm^2 to 488 nm	81

Table 3-2. *Continued*

Radical Generator	Sensitizer	Photopolymer	Sensitivity	Ref.
Poly(ethylene imine) 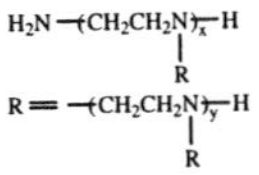 $H_2N-(CH_2CH_2N)_x-H$ (N–R) $R = -(CH_2CH_2N)_y-H$ (N–R)	Methylene blue	PVP/Li acrylate/acrylic acid/ N,N'-methylene-bisacrylamide	5 mJ/cm^2 to He—Ne laser (volume hologram)	82
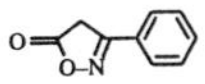	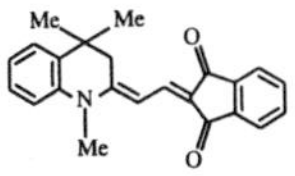	Chlorinated polyethylene/poly-functional acrylate monomer	7 mJ/cm^2 to 488 nm	77
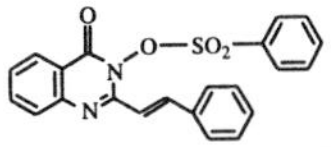	Ketocoumarin dye	Poly(methacrylate)/poly-functional acrylate monomer	Sensitive to visible light 10 m/s by 5-mW Ar laser (488 nm) focused to 25 μm	83
Aminobenzoate	Riboflavin tetrabutylate	P(MMA-co-MAA)/acrylate monomer	50 mJ/cm^2 to 457.9 nm	84

3. Photoradical Generation by Intermolecular Processes

Most bimolecular-type photoinitiators consist of a combination of a radical generator and a sensitizing dye. An electron is transferred from the excited dye to the radical generator in the ground state or from the radical generator in the ground state to the excited dye. The anion radical or the cation radical of the radical generator decomposes to generate a free radical species.

Organic peroxides alone are sensitive to light but require ultraviolet light. It is known that some electron-donating dyes sensitize the decomposition of peroxides. The combination of the ferric ion and *t*-butylhydroperoxide is used as a water processable initiator for the photopolymerization of acrylamide in gelatin. The ferric ion is photochemically reduced to a ferrous ion that reacts with hydroperoxide to generate a hydroxy radical.[85]

$$Fe^{+3} \xrightarrow{h\nu} Fe^{+2}$$

$$(CH_3)_3C\text{—}O\text{—}O\text{—}H + Fe^{+2} \rightarrow OH^- + \cdot OH$$

3.1. Electron Transfer—Peroxide/Electron Donating Dyes

Recently a tetrafunctional peroxide, 3,3′,4,4′-tetrakis(*t*-butylperoxy-carbonyl)-benzophenone (BTTB) having a relatively good stability was developed. BTTB is efficiently sensitized in visible wavelengths with electron-donating dyes to generate free radicals.[86] Bimolecular systems consisting of BTTB and dyes give highly sensitive photopolymers. Fig. 3-2 gives examples of spectral sensitivities of the photopolymers in which BTTB and the BTTB/coumarin dye bimolecular systems are used as initiators. The general composition of the photopolymers is shown in Table 3-3. When BTTB is used without sensitizing dye as the initiator, the spectral sensitivity of the photopolymer is in short wavelengths at about 300 nm. However, if the coumarin dye is used in addition to BTTB, the spectral sensitivity is shifted to visible wavelengths, ~550 nm.

The sensitivity of photopolymers to visible light also depends on the matrix polymer. The sensitivities with three kinds of matrix polymers are shown in Table 3-4. Some of these photopolymers exhibit high sensitivity of ~10μJ/cm^2 at 488 nm.

The mechanism of radical generation from the dye-sensitized BTTB system was investigated with a series of 3-substituted coumarin dyes and a thioxanthene dye, the structures of which are shown in Fig. 3-3. In Table 3-5, spectroscopic data, quantum yields, and lifetimes of these dyes, and σ_{para} values of the substituent are summarized. Among these dyes, TXD showed the longest fluorescence lifetime. For the substituted diethylaminocoumarins the fluorescence

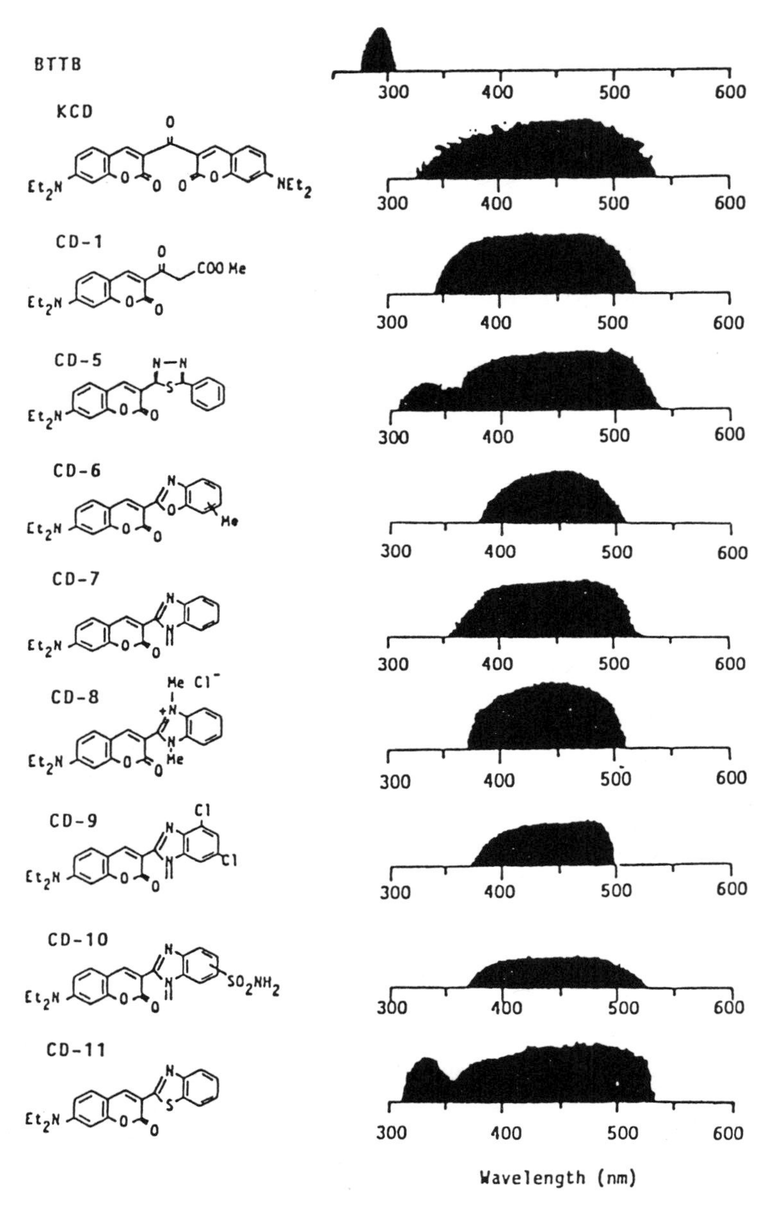

Figure 3-2. *Spectral sensitivity of the photopolymers in which BTTB or dye-sensitized BTTBs are used as photoinitiating systems.*

Table 3-3. General Composition of the Photopolymers

Matrix polymer	100 parts by weight
Polyfunctional monomer	100
BTTB	8
Sensitizing dye	8

Table 3-4. Sensitivities of the Photopolymers to 488-nm Light of an Argon Ion Laser (mJ/cm^2)

	PVP	BMCM	XL-44
KCD	0.06	0.05	0.03
CD- 1	—	—	0.9
CD- 2	18.5	—	—
CD- 3	0.8	—	—
CD- 4	0.03	0.08	0.7
CD- 5	0.06	—	—
CD- 6	0.01	0.1	0.7
CD- 7	—	—	3.6
CD- 8	—	—	3.6
CD- 9	—	—	1.2
CD-10	—	0.2	—

PVP: Poly(*N*-vinyl-2-pyrrolidone) (GAF K-90, M_w 360,000). BMCM: (0.36:0.27:0.25:0.18 copolymer of *t*-butylmethacrylate, methacrylate, cyclohexyl-methacrylate, and methacrylic acid (M_w 150,000, acid value 50). XL-44 (B.F. Goodrich, M_w 30,000, acid value 75).

lifetime increases with a decrease in the σ_{para} value of the substituent on the phenyl group, whereas the quantum yield(ϕ_f) of the fluorescence is decreased with a decrease in σ_{para}.

Fig. 3-4 shows the absorption spectrum of BTTB, which has no absorption at wavelengths longer than 410 nm. The lowest excited singlet state and the triplet state of BTTB are 72.0 kcal/mol and 61.6 kcal/mol, respectively.[87] The lowest singlet excited states of these dyes are from 53.5 (TXD) to 65.4 (PCD) kcal/mol, respectively, all of which are lower than that of BTTB. From this fact, the energy transfer mechanism from the excited dye to BTTB may be eliminated as the mechanism of sensitization. The fluorescence of the dyes is quenched by BTTB, indicating that these dyes interact with BTTB in their singlet excited states. Fig. 3-5 shows Stern-Volmer plots in which the reciprocals of the fluorescence lifetimes ($k_d^{obs} = 1/\tau_{obs}$) of the dyes are plotted as a function of the

Dye		Abbreviation
	—R	
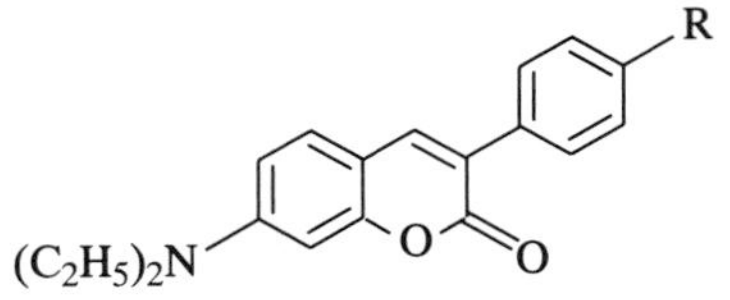 $(C_2H_5)_2N$ coumarin, 3-(4-R-phenyl)	$—NO_2$	NPCD
	—CN	CPCD
	—H	PCD
	$—OCH_3$	MOPCD
	$—NH_2$	APCD
	$—N(CH_3)_2$	DMAPCD
$(C_2H_5)_2N$ coumarin, 3-benzimidazolyl		BICD
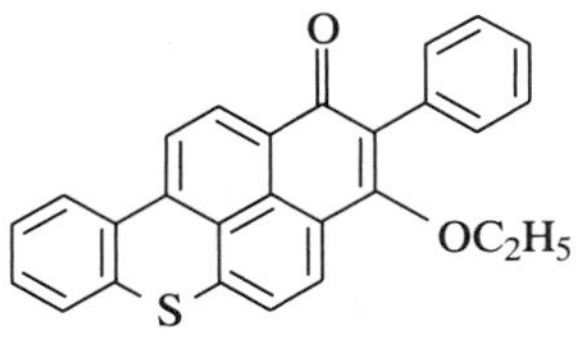 thioxanthene, OC_2H_5		TXD

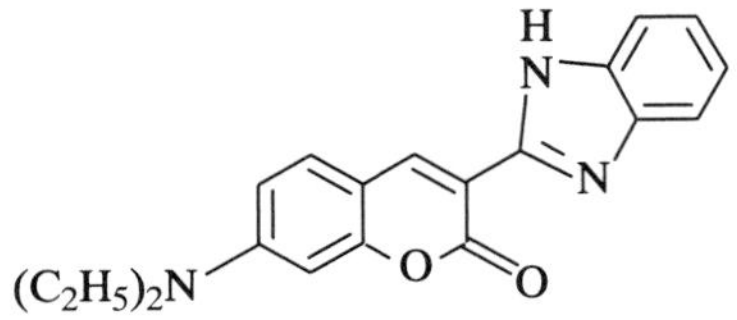

Figure 3-3. *A series of 3-substituted coumarin dyes and thioxanthene dye used for the study of the sensitization mechanism.*

Table 3-5. Spectroscopic Data, Quantum Yield (ϕ_f), Lifetime (τ) of Fluorescence and α_{para} Values for the Coumarin Derivatives and Thioxanthene Dyes

Dye	λ_{max} (nm)	ϵ 10^4 M^{-1} cm^{-1}	$^1E^*_\infty$ eV (kcal mol)	Φ_f	τ (ns)	σ_{para}
NPCD	432	3.9	2.63 (60.5)	—	—	0.78
CPCD	414	3.7	2.71 (62.5)	0.51	2.67 ± 0.01	0.66
PCD	396	3.2	2.84 (65.4)	0.71	3.31 ± 0.01	0
MOPCD	397	3.3	2.81 (64.8)	0.62	3.39 ± 0.01	−0.32
APCD	400	3.3	2.76 (63.7)	0.44	3.75 ± 0.01	−0.66
DMAPCD	405	3.2	2.68 (61.9)	0.28	4.40 ± 0.00	−0.83
BICD	434	4.4	2.64 (60.8)	0.48	2.87 ± 0.01	—
TXD	492	2.4	2.32 (53.5)	—	6.13 ± 0.01	—

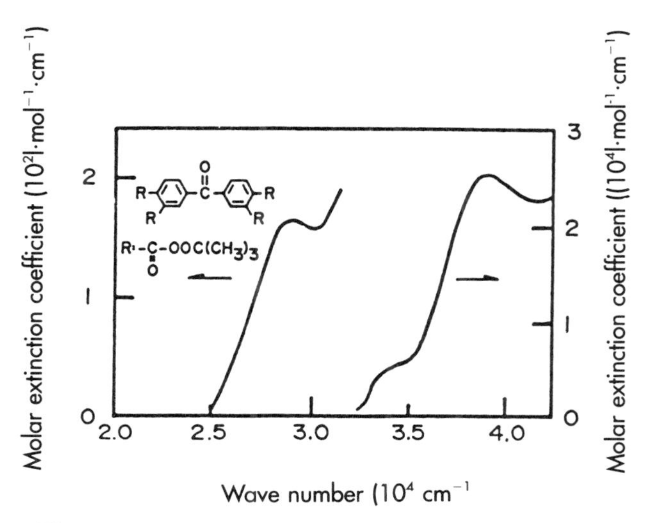

Figure 3-4. *Absorption spectrum of BTTB in acetonitrile.*

concentration of BTTB. From these plots, the quenching rate constants (k_q's) of the fluorescence caused by BTTB were determined as shown in Table 3-6. According to these k_q data, the excited singlet states of the dyes are quenched by BTTB at the diffusion-controlled rate in acetonitrile. The free enthalpy change (ΔG) accompanied by the electron transfer from the excited state of the dye to BTTB is given by the following equation:[88]

$$\Delta G = 23.06(E_{ox}^{1/2} - E_{red}^{1/2} - e_0^2/\epsilon a) - {}^1E_{00} \qquad (13)$$

Where $E_{ox}^{1/2}$ is the oxidation potential of the dye and $E_{red}^{1/2}$ is the reduction potential of BTTB. ${}^1E_{00}$ is the excitation energy (0-0 band) of the dye. $e_0^2/\epsilon a$ is the Coulombic interaction energy between the radical ions formed by electron transfer, and is generally 1–2 kcal/mol in polar solvents. This value was neglected in the present study.

BTTB shows a large reduction current at -1.0 V, suggesting that BTTB acts as an electron acceptor. Peroxides are known to be oxidative compounds and can be decomposed by nucleophilic compounds. $E_{red}^{1/2}$ of BTTB was reported to be -1.2 V for the SCE electrode. However, in the present study the cyclic voltammogram showed no oxidation current from BTTB to BTTB-, indicating that BTTB is a very unstable species that decomposes immediately after reduction. Therefore, the precise value of the reduction potential for BTTB was not determined from the cyclic voltammogram. The reduction potential of BTTB was approximated by E_{red}^{irr} as -0.85 V. If the quenching of fluorescence of the

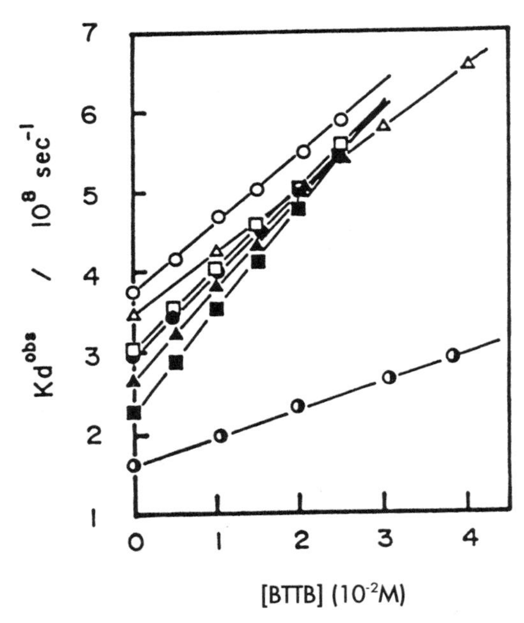

Figure 3-5. *Stern-Volmer plots for the fluorescence lifetime of coumarin dyes in acetonitrile. The reciprocal number of the lifetime* (k^{obs}) *is plotted as a function of BTTB concentration. (- ○ -) CPCD, (- △ -) BICD, (-□ -) PCD, (- ● -) MOPCD, (- ▲ -) APCD, (- ■ -) DMAPCD, (-○ -) TXD.*

Table 3-6. Rate Constants (k_q's) of Fluorescence Quenching of the Coumarin and Thioxanthene Dyes by BTTB

Dye	k_q (10^{10} M^{-1} s^{-1})
CPCD	0.854
PCD	1.03
MOPCD	1.01
APCD	1.11
DMAPCD	1.25
BICD	0.768
TXD	0.341

dye is due to electron transfer, k_q may decrease with decrease in ΔG in the electron transfer process. As $E_{red}^{1/2}$ is constant in the present case, ΔG should be proportional to $E_{ox}^{1/2} - {}^1E_{00}$.

The oxidation potentials of the dyes were successfully determined by cyclic voltammetry. The oxidation potential $E_{red}^{1/2}$ was found to be 0.98 V from the cyclic voltammogram. $E_{00}^{1/2}$ values for the dyes are summarized in Table 3-7. Figs. 3-6(a) and (b) shows plots of $E_{ox}^{1/2}$ and $E_{ox}^{1/2} - {}^1E_{00}$ as functions of σ_{para} for the substituent for a series of 3-substituted phenyl-7-diethylaminocoumarins. Both $E_{ox}^{1/2}$ and $E_{ox}^{1/2} - {}^1E_{00}$ correlate well with the σ_{para} of the substituent. $E_{ox}^{1/2}$ decreases with increase in the electron donation. In Fig. 3-7, k_q is plotted as a function of $E_{ox}^{1/2} - {}^1E_{00}$ for each dye. From these plots, it was found that k_q is closely related to the value of $E_{ox}^{1/2} - {}^1E_{00}$ for the dye. These results suggest that the quenching of the fluorescence of the dye by BTTB is due to the electron transfer from the excited singlet states of the dye to BTTB. If the value of ΔG incorporating the electron transfer process is known, k_q can be calculated. The k_q values for TXD-BTTB and BICD-BTTB were calculated as follows:[87]

$$\Delta G = \{(\Delta G/2)^2 + (\Delta G_0/2)^2 + \Delta G/2\}^{1/2} \tag{14}$$

$$k_q = 2.0 \times 10^{10}/[1 + 0.25\{\exp(\Delta G_0/RT) + \exp(\Delta G/RT)\}] \tag{15}$$

where ΔG is the activation free enthalpy and ΔG_0 is the activation free enthalpy at $\Delta G = 0$, approximated as 2.4 kcal/mol. The calculated values of k_q(cal) for TXD and BICD (Table 3-8) are 7.64×10^8 and 1.38×10^{10} $M^{-1}s^{-1}$, respectively, and are close to the experimentally determined values for k_q(obs). When the approximated value of BTTB is taken into account, the calculated values seem to be close to the experimentally determined values for k_q(obs). Consequently, the sensitization mechanism of the BTTB-dye system may be due to the following photoredox reaction.

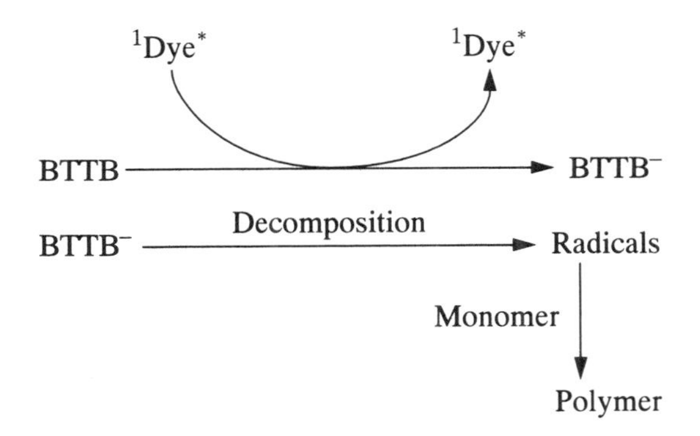

Table 3-7. Oxidation Potentials of the Coumarin and Thioxanthene Dyes

Dye	$E_{ox}^{1/2}$ (V)	$E_{ox}^{1/2} - {}^1E_{00}^*$ eV (kcal/mol)
NPCD	1.05	−1.58 (−36.4)
CPCD	1.03	−1.68 (−38.7)
PCD	0.98	−1.86 (−42.9)
MOPCD	0.92	−1.89 (−43.6)
APCD	0.68	−2.08 (−48.0)
DMAPCD	0.67	−2.02 (−46.6)
BICD	1.04	−1.60 (−36.9)
TXD	1.27	−1.06 (−24.4)

3.2. Electron and Hydrogen Transfer—NPG/TXD

Some dye-sensitized radical generators work by simultaneous transfer of electrons and hydrogen via the excited state of electron donors or acceptors.

In the combination of benzophenone and Michler's ketone, the electron transfer from the excited Michler's ketone to benzophenone and the proton transfer

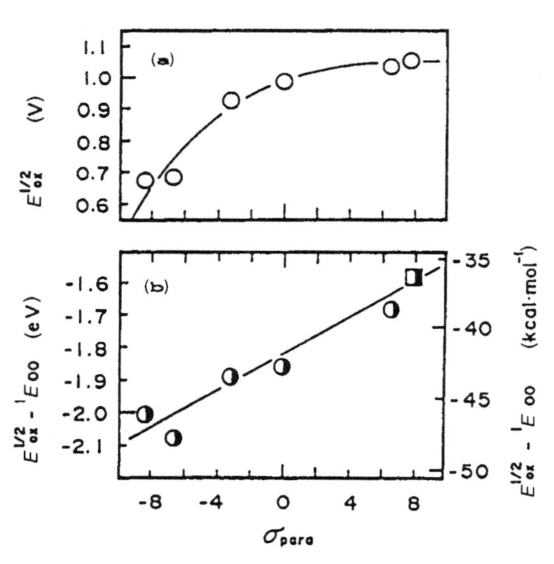

Figure 3-6. *Relationship of (a)* $E_{ox}^{1/2}$ *and (b)* $E_{ox}^{1/2} - {}^1E_{00}$ *with* σ_{para} *value of the substituent for a series of 3-substituted phenyl-7-diethylaminocoumarins.*

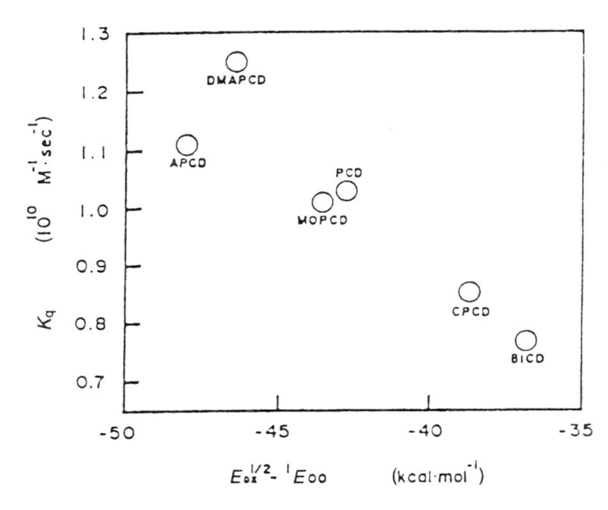

Figure 3-7. *Relationship between* K_q *and* $E_{ox}^{1/2} - {}^1E_{00}$ *for 3-substituted phenyl-7-diethylaminocoumarins.*

Table 3-8. Calculated and Experimental k_q Values for BICD and TXD

Dye	${}^1E_{00}^*$ (kcal/mol)	$1/\tau_0$ (1/s)	$k_{a(obs)}$ (1/M·s)	ΔG (kcal/mol)	ΔG^+ (kcal/mol)	$K_{q(cal)}$ (1/M·s)
TXD	53.5	1.64×10^9	3.41×10^0	0.51	2.67	7.64×10^8
BICD	60.8	3.38×10^8	7.56×10^9	−16.5	0.34	1.38×10^{10}

from Michler's ketone to benzophenone occur simultaneously via the exciplex to form ketyl radical and methyl radical. Williams et al. have reported that *N*-phenylglycine in the presence of ketocoumarin dye shows free radicals by exposure to 550-nm light.[89]

Rehm et al. studied fluorescence quenching of dyes by electron acceptors and found[88] that the flourescence of acridine is quenched by phenol with a diffusion-controlled rate, although the theoretically predicted value of ΔG accompanying the electron transfer from phenol to acridine in the lowest excited singlet state is of the order of 10^6. They proposed that the free energy change for the simultaneous occurrence of electron and proton transfer (ΔG_{HT}) should be expressed by Eq. (16).

$$\Delta G_{HT} = \Delta G + 23.06\ e^2/\epsilon a - 2.303RT(pK_{AH} - pK_{DH+}) \qquad (16)$$

It has been reported that NPG/thioxanthene dye initiator is sensitive to visible light and generates free radicals with high efficiency. This initiating system gives highly sensitive photopolymers with 400 $\mu J/cm^2$ to 488-nm argon ion laser light.[90]

$$TXD\text{-}1 \xrightarrow{h\nu} {}^1TXD\text{-}1^* \longrightarrow {}^3TXD\text{-}1^*$$

$$^3TXD\text{-}1^* + NPG \longrightarrow TXD\text{-}1^{\bar{\bullet}} + NPG^{\overset{+}{\bullet}}$$

$$C_6H_5\text{-}\dot{N}^+HCH_2C(=O)\text{-}O\text{-}H \ (NPG^{\overset{+}{\bullet}}) \xrightarrow{\text{Decarboxylation}} C_6H_5\text{-}NHCH_2^{\bullet} + CO_2 + H^+$$

$$C_6H_5\text{-}NHCH_2^{\bullet} \xrightarrow{\text{Monomer}} \text{Polymerization}$$

The sensitivities of photopolymers using NPG/thioxanthene dyes as the photoinitiator are summarized in Table 3-9.

Fig. 3-8 shows the absorption and fluorescence spectra of TXD in degassed acetonitrile. TXD has strong absorption peaks at 266 and 492 nm and an intense fluorescence peak at 555 nm.[91] Absorption of NPG is at short wavelengths near 320 nm. While TXD is very stable when exposed to visible light in acetonitrile, it shows significant change upon exposure to visible light when NPG is present in the solution. As seen in Fig. 3-9, the absorption of TXD at 492 nm was decreased by irradiation, and a new transient absorption band appeared between 380 and 450 nm. The change in the spectrum continues even after the exposure is stopped. The transient absorption with the peak at 436 nm decayed with a lifetime of 17.5 min. This photoinduced spectral change of TXD is observed with other aromatic amines having similar oxidation potential, so they are used in place of NPG. The energy diagram of the singlet and triplet states of TXD relative to the ground state is shown in Fig. 3-10. The fluorescence quantum yield and the lifetime of $^1TXD^*$ are found to be 0.36 and 6.31 ns, respectively. Pulsed laser excitation (532 nm) of an acetonitrile solution of TXD (1.6×10^{-4} M) bubbling under argon gas gave the transient absorption spectrum (Fig. 3-11) which can be attributed to T-T absorption of TXD because it has a relatively long lifetime ($\tau_T = 23.0\ \mu s$) and was efficiently quenched by oxygen. The quantum yield ϕ_{isc} of the intersystem crossing of $^1TXD^*$ has been determined to be 0.20 by a (+)-limonene photo-oxygenation method.[92] The photophysical parameters obtained from these data are tabulated in Table 3-10 and Table 3-11. The reduction potential of TXD is -0.81 V vs. SCE, and the oxidation potential of NPG is 1.01 V vs. SCE. Table 3-11 shows Hammett's values of the para-substituents of R-NPGs and their oxidation potentials in acetonitrile. A good relationship was observed between the oxidation potentials and Hammett's values

Table 3-9. Sensitivities to Ultraviolet and Visible Light of the Photopolymers in Which NPG-Xanthene Dyes and NPG-Thioxanthene Dyes Are Used as Photoinitiator. Abbreviations of the Dyes are Listed Below

		Sensitivity (mJ/cm^2)		
		Chemical lamp		Laser
	Dye Initiator	No filter	Y-43 filter	(488 nm)
XD-1	—	2.0	0.77	
	NPG	0.65	0.49	0.50
XD-2	—	150		
	NPG	17	19	
XD-3	—	6.2	2.0	8.0
	NPG	2.3	2.4	1.7
XD-4	—	8.9	6.8	
	NPG	1.2	0.44	0.37
XD-5	—	8.9	3.4	
	NPG	0.22	0.060	0.20
XD-6	—	8.9	4.8	
	NPG	0.87	0.24	0.23
XD-7	—	2.4	1.3	
	NPG	1.2	0.44	0.30
XD-8	—	49	29	
	NPG	2.4	1.3	0.40
XD-9	—	69	58	
	NPG	3.2	1.8	0.90
TXD-1	—	75	29	
	NPG	0.65	0.49	0.40
TXD-2	—	6.2	2.7	
	NPG	0.88	0.33	1.3
TXD-3	—	9.7	3.8	
	NPG	0.65	1.7	0.17

Dye no.	R_1	R_2	R_3	X	Y
XD-1	$CH_3CH(OH)CH_2$	MeO		H	O
XD-2	C_3H_7	$(Et)_2N$		H	O
XD-3	$MeO(CH_2)_3$	$(Et)_2N$		Cl	O
XD-4	$EtO(CH_2)_3$	$(Et)_2N$		Cl	O
XD-5	$HO(CH_2)_2$	$(Et)_2N$		Cl	O
XD-6	$(Et)_2N(CH_2)_3$	$(Et)_2N$		Cl	O
XD-7	$C_4H_9CH(Et)CH_2$	$(Et)_2N$		Cl	O
XD-8	$(Et)_2N(CH_2)_3$	$(Et)_2N$		Br	O
XD-9	$EtO(CH_2)_3$	$(Et)_2N$		Br	O
TXD-1*			Et		S
TXD-2*			MeCO		S
TXD-3	i-$C_3H_7O(CH_2)_3$	H		H	S

*Refers to the general formula [II]. Others refer to [I].

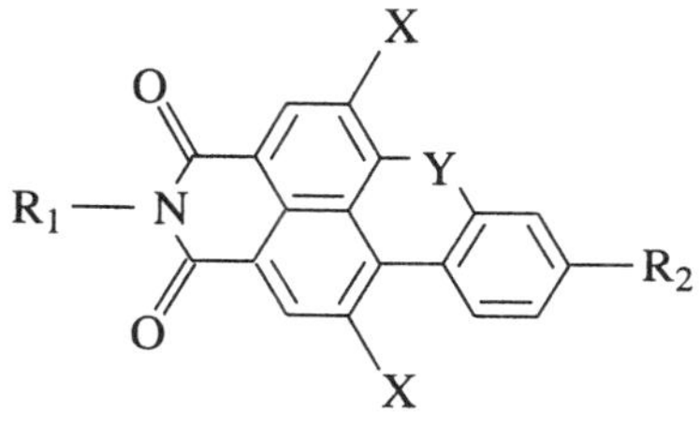

[I]

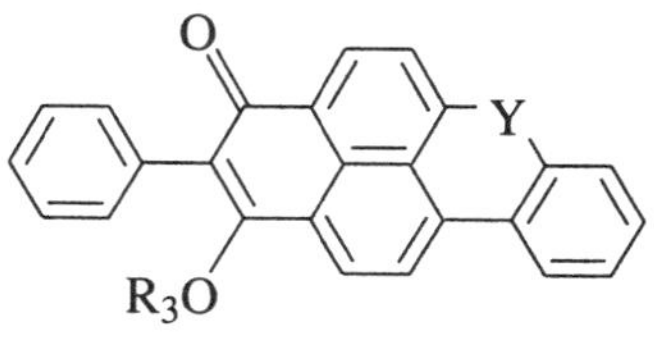

[II]

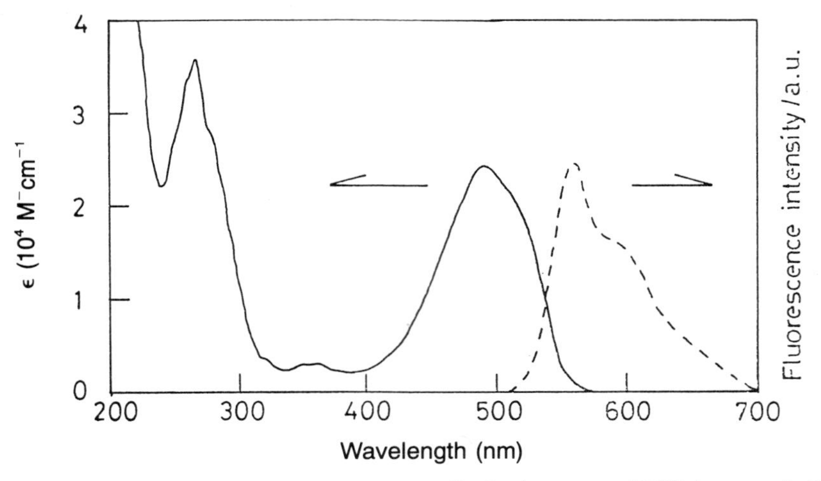

Figure 3-8. *Electronic (solid) and flourescence (broken) spectra of TXD in acetonitrile.*

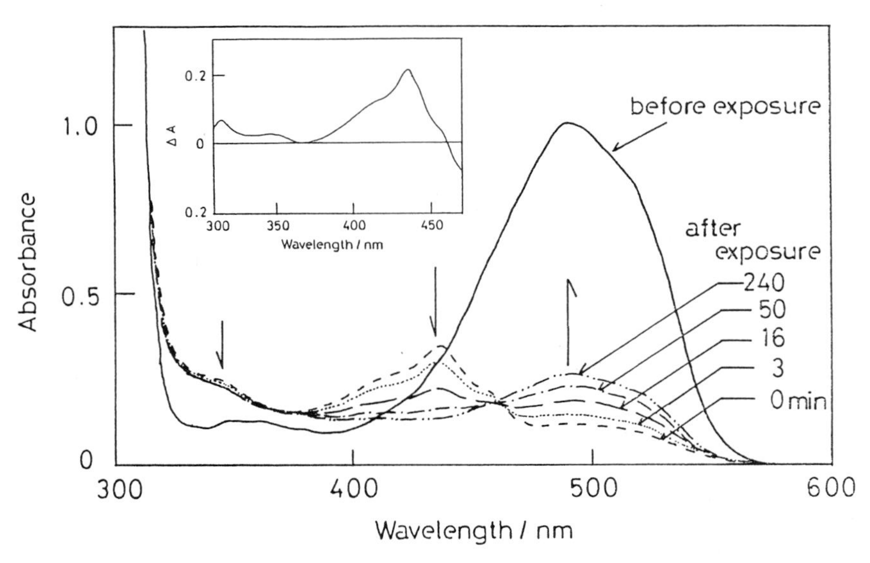

Figure 3-9. *Change of absorption spectra for TXD/NPG system in acetonitrile after irradiation with 488-nm light. [TXD] = 4.0×10^{-5} M, [NPG] = 4.2×10^{-3} M. Insert shows the difference spectrum of an acetonitrile solution of TXD/NPG before and after 240-min irradiation.*

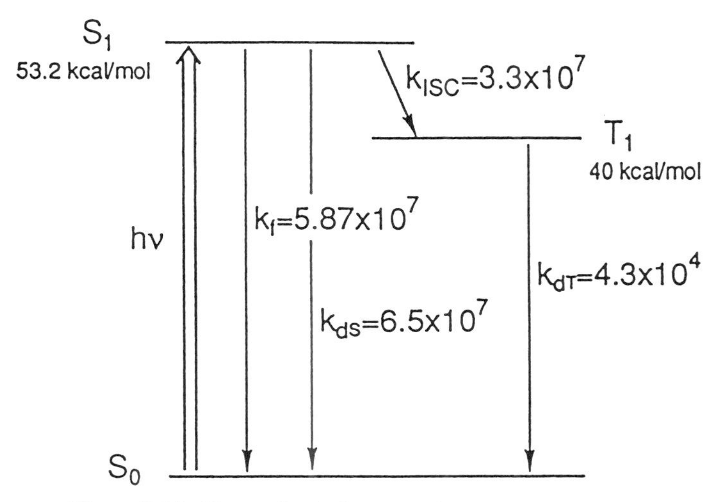

Figure 3-10. *Energy level diagram of TXD in acetonitrile.*

as seen in Fig. 3-12, suggesting that the oxidation potentials of R-NPGs can be controlled by the substituents at the para-position of the benzene ring.

The free energy changes (ΔG's) associated with electron transfer from the ground states of R-NPGs to the lowest singlet and triplet states of TXD can be calculated according to the Rehm-Weller equation shown in Eq. (17).

$$\Delta G_{S(\text{or } T)}(\text{in kcal mol}^{-1}) = 23.06[E(D/D^{+}) - E(A^{-}/A) - C] - E_{00} \quad (17)$$

The ΔG values calculated for singlet and triplet quenching of TXD by NPG are -9.4 and 3.6 kcal mol^{-1}, respectively. The calculated values are given in Table 3-11, together with the quenching rate constants. The flourescence of ^{1}TXD* is efficiently quenched by R-NPGs with a strong electron-donating substituent. Fig. 3-13 shows the decay of the fluorescence in the presence of R-NPG. The fluorescence rate constants k_q's of ^{1}TXD* by R-NPGs were determined from the following equation.

$$1/\tau' = 1/\tau + k_q[Q] \quad (18)$$

A Stern-Volmer plot of flourescence quenching of TXD by R-NPG is shown in Fig. 3-14. From this plot, k_{qS} values for several R-NPGs are summarized in Table 3-11. These values of k_{qS} indicate that electron transfer occurs from the ground state of R-NPG to the excited singlet state of TXD.

The reaction mechanism of R-NPG and TXD was investigated by laser flash photolysis. The absorption of TXD in acetonitrile is bleached by exposing to pulsed light with 5-ns half width, and simultaneously new absorptions appear

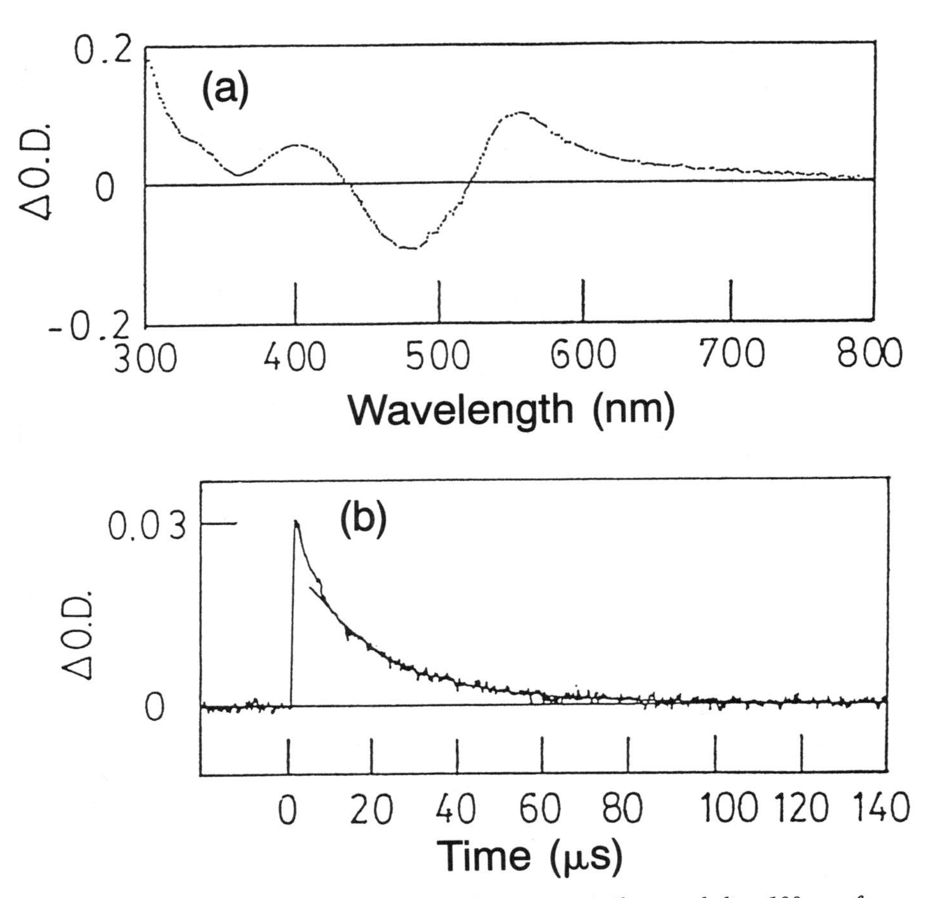

Figure 3-11. *T-T absorption spectrum of TXD in acetonitrile recorded at 100 ns after irradiation of the laser pulse. [TXD] = 1.6 × 10^{-4} M (a). Time dependence of T-T absorption decay for TXD at 300 nm in acetonitrile (b).*

at 300, 400, and 560 nm. This absorption is assigned to that of T-T absorption of TXD (Fig. 3-11). However, the irradiation of TXD by the pulsed light in the presence of NPG exhibits a different transient spectrum, as seen in Fig. 3-15(a). A typical decay trace monitored at 436 nm is shown in Fig. 3-15(b). This transient spectrum shows no appreciable decay during the laser flash experimental time scale, and the shape and the peak position are very similar to those of the transient absorption observed with exposure to continuous light.

Laser flash photolysis was investigated by Amirzadeh and Schnabel[93] (for

Table 3-10. Photophysical Parameters of TXD in Acetonitrile

λ_{max} (nm)	$\epsilon(\lambda_{max})$ (l/mol cm)	$^1E^*_{00}$	$^3E^*_{00}$ (kcal/mol)	$E^{1/2}_{RED}$ (V vs. SCE)
491.5	2.49×10^4	53.2	40	−0.81
Φ_f	Φ_{ISC}[a)]	Φ_p	τ_S (sec)	τ_T
0.36	0.2	—	6.13×10^{-9}	2.3×10^{-5}

[a)]Determined by limonene photo-oxygenation.

thioxanthene-amine photoinitiating systems), who showed that the photoinitiating capability of thioxanthene-amine systems is based on the following mechanism whereby excited triplet thioxanthone molecules are reduced by amine molecules:

$$^3\left[\rangle C{=}O \right] + AH \longrightarrow \left(^3\left[\rangle C{=}O \right]^* \cdots AH \right)$$

Exciplex

$$\longrightarrow \rangle \cdot C{=}OH + A\cdot$$

The C=O group represents the thioxanthone chromophore. These showed that

Table 3-11. Oxidation Potentials, Free Energy Changes Accompanying the Electron Transfer, and the Rate Constants for Quenching of Excited State of TXD by NPG Derivatives

R-NPG	E_{OX} (V vs. SCE)	ΔG_S (kcal/mol)	k_{qS} (M^{-1} s^{-1})	ΔG_T (kcal/mol)	k_{qT} (M^{-1} s^{-1})
CN-NPG	1.46	1.0	2.53×10^9	14.0	1.9×10^7
Ac-NPG	1.25	−3.9	3.67×10^9	9.1	2.7×10^7
Cl-NPG	1.08	−7.8	6.48×10^9	5.2	5.6×10^7
F-NPG	1.03	−9.0	6.34×10^9	4.0	1.1×10^8
NPG	1.01	−9.4	6.48×10^9	3.6	6.9×10^7
Me-NPG	0.95	−10.8	7.77×10^9	2.2	1.8×10^8
MeO-NPG	0.79	−14.5	7.83×10^9	−1.5	4.8×10^8

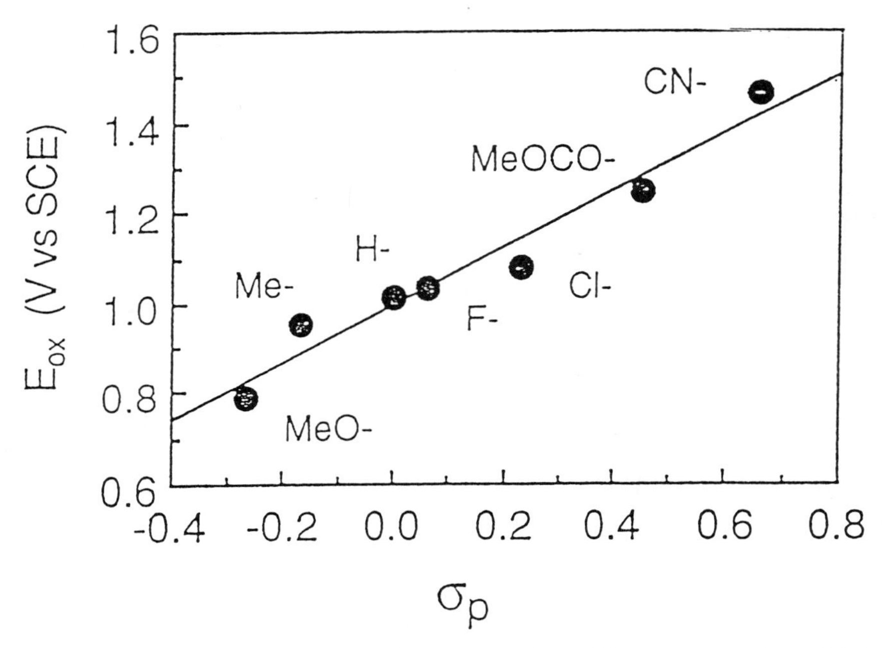

Figure 3-12. *Relationship between oxidation potentials and Hammett's σ_p values of NPG derivatives (R-NPGs).*

the radical generated on the amine molecule is mainly responsible for initiating the polymerization of the olefinic monomers. More recently, Allen et al. reported[94] that the introduction of a tertiary amine functionality into the 2-position of thioxanthene has a significant effect on the photopolymerization of acrylate monomers in comparison with a non-amine-substituted analog. They have suggested that an alkylamino radical generated through the formation of a triplet exciplex between the thioxanthone chromophore and amine moiety initiates the polymerization.

Judging from both the steady-state and transient behavior of these systems, the absorption band at 436 nm can be assigned to the TXD ketyl radical. In fact, electron transfer from amines to excited benzophenones or thioxanthines (mostly followed by proton transfer) is a well-known photochemical reaction. From these results, the generation of TXD ketyl radical may occur via electron transfer from the ground state NPG to the excited TXD followed by proton transfer from the NPG radical cation.

The decay of ^{3}TXD* (400 nm) parallels the formation of TXD ketyl radical (λ_{max} = 436 nm). Analysis of the buildup curve for the formation of TXD ketyl

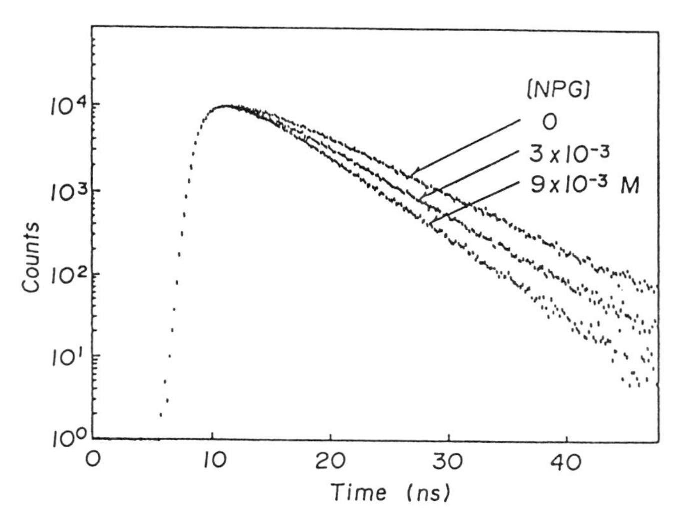

Figure 3-13. *Time dependence of fluorescence for TXD at 555 nm in the presence of several concentrations of NPG in acetonitrile.*

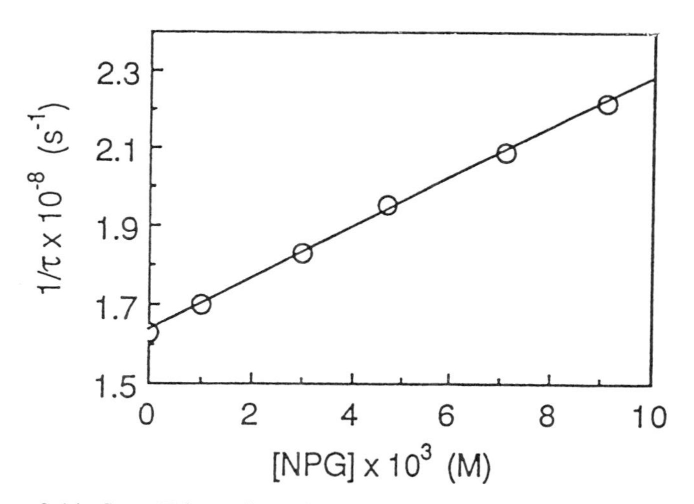

Figure 3-14. *Stern-Volmer plots of fluorescence quenching of TXD by NPG.*

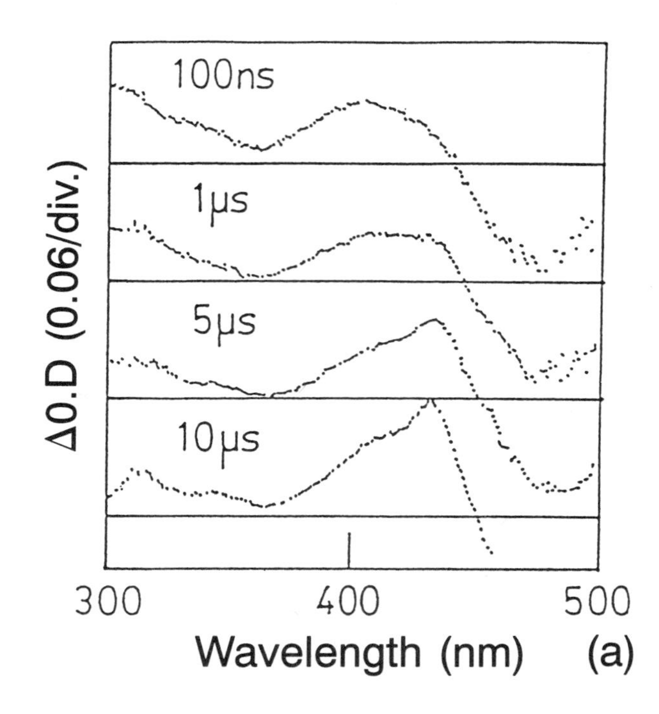

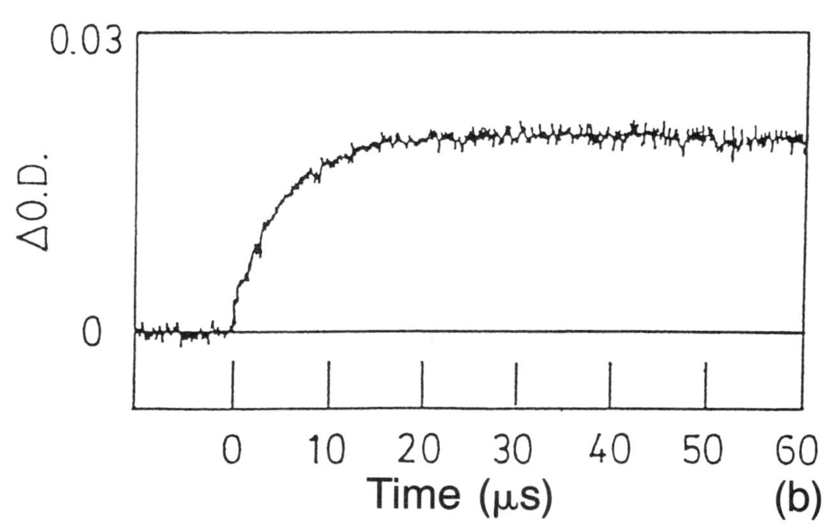

Figure 3-15. *Transient absorption spectra of TXD/NPG system in acetonitrile. [TXD] = 1.6×10^{-4} M, [NPG] = 5.0×10^{-3} M (a). Time dependence of absorption change at 436 nm for TXD/NPG system in acetonitrile (b).*

radical indicates that the apparent first-order rate constant for this process is $3.8 \times 10^4\ s^{-1}$, and is nearly the same order as that observed for the pseudo-first-order rate constant for quenching of the excited triplet state of TXD by NPG. The rate constant for the formation of ketyl radical increases with concentration of NPG. A similar time-resolved spectroscopic behavior was observed for combinations of TXD with other R-NPGs. The quenching rate constants of the excited triplet state of TXD by R-NPGs were obtained by measuring the pseudo-first-order rate constant k_d (decay of T-T absorption).

$$k_d = k_0 + k_{qT}[Q] \tag{19}$$

as a function of the R-NPG concentration, where k_0 is the decay rate constant of TXD ($= 1/\tau_T$), k_{qT} is the rate constant for quenching of the excited triplet state of TXD, and [Q] is the concentration of R-NPGs. Table 3-11 shows that these quenching rate constants appear to increase as the electron-donating ability of the substituents increases (Fig. 3-16), suggesting that the quenching mechanism of TXD by R-NPG may involve electron transfer from the ground state of R-NPG to both excited singlet and triplet TXD. When a degassed acetonitrile solution of TXD in the presence of NPG is irradiated with 488-nm light from

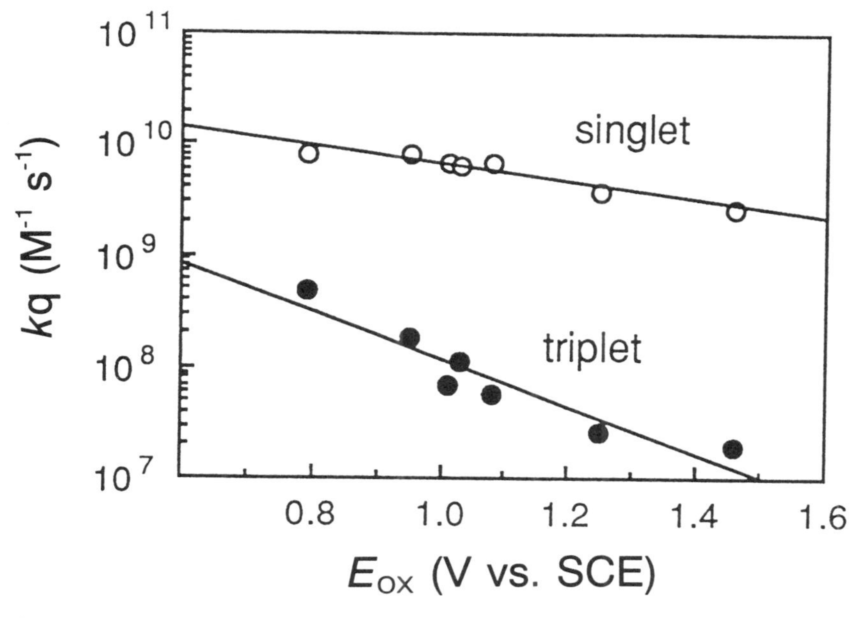

Figure 3-16. *Relationship between the quenching rate constants of the excited singlet and triplet states of TXD and the oxidation potentials of NPG derivatives.*

an argon ion laser, aniline and *N*-methylaniline are produced. Davidson et al. reported that photosensitized decarboxylation of NPG in aerated solution leads to the formation of anilinomethyl radical and successively aniline, *N*-methylaniline and formanilide.[95,96] Photochemical reaction products in the present system are similar to those of the system reported by Davidson et al.,[97,98] suggesting that TXD sensitized NPG reacts similarly to that of Davidson's system.

$$S \longrightarrow S^{*} \xrightarrow{ArNCH_2COOH} [Ar\overset{+\bullet}{N}(CH_2C(=O)OH) \cdots S^{\bar{\bullet}}]$$

S: sensitizer

$$\longrightarrow ArNCH_2^{\bullet} + CO_2 + \dot{S}H$$

The photochemical reaction of NPG in the presence of TXD is believed to proceed as the following steps. According to these reaction steps, the quantum yield ϕ_{dis} of NPG dissociation sensitized by TXD in acetonitrile is determined by Eq. (26).

$$TXD + h\nu \xrightarrow{h\nu} {}^1TXD^* \tag{20}$$

$${}^1TXD^* \xrightarrow{k_f} TXD + h\nu_f \tag{21}$$

$${}^1TXD^* \xrightarrow{k_{dS}} TXD \tag{22}$$

$${}^1TXD^* \xrightarrow{k_{isc}} {}^3TXD^* \tag{23}$$

$${}^3TXD^* + NPG \xrightarrow{k_r} \text{Photoproducts} \tag{24}$$

$${}^3TXD^* \xrightarrow{k_{dT}} TXD \tag{25}$$

$$\phi_{dis} = \left(\frac{k_{isc}}{k_f + k_{isc} + k_{dS}}\right)\left(\frac{k_r[NPG]}{k_{dT} + k_r[NPG]}\right) \tag{26}$$

$$\frac{1}{\phi_{dis}} = \frac{1}{\phi_{dis}}\left(1 + \frac{1}{k_r[NPG]}\right) \tag{27}$$

A plot of $1/\phi_{dis}$ vs. 1/[NPG] is found to be nearly linear until the concentration of NPG was 5.0×10^{-4} M (Fig. 3-17). When the concentration of NPG is increased further, the plot deviates from the linear relation, suggesting that a complicated quenching mechanism coexists. Judging from the results of laser flash photolysis mentioned above, the photoredox reaction may exist via both the excited singlet and triplet states.

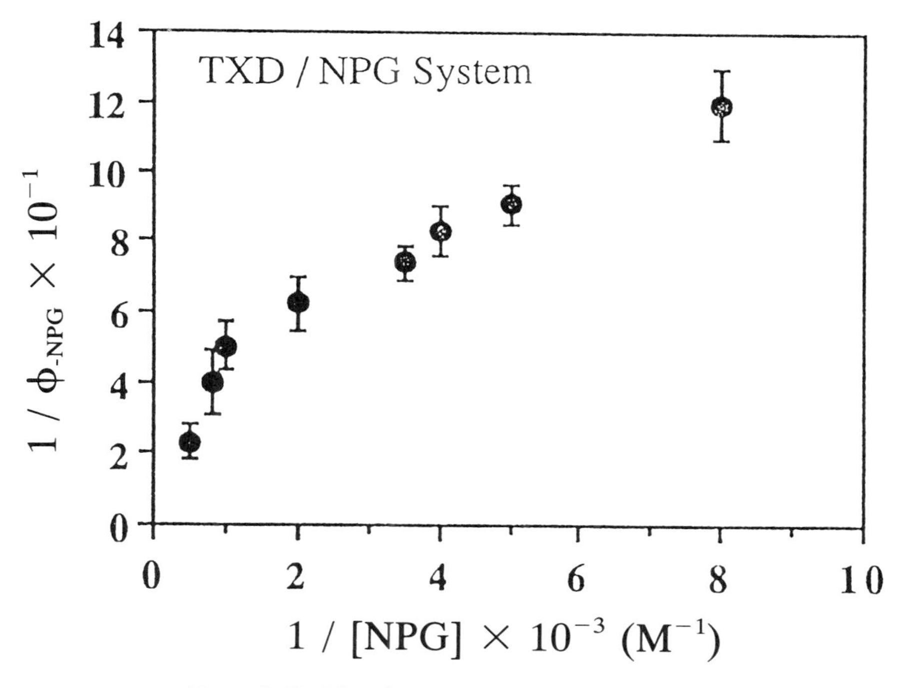

Figure 3-17. *Plot of $1/\sigma_{dis}$ as a function of $a/[NPG]$.*

3.3. Energy Transfer—Singlet Oxygen Sensitization

Ground-state oxygen molecules exist in a triplet state in which two unpaired electrons have parallel spins. Oxygen scavenges free radical species and inhibits radical polymerization. Oxygen is thus troublesome in practical use of the materials based on the radical polymerization mechanism. Singlet oxygen, however, plays an important role in obtaining highly sensitive photopolymerization systems. As the energy required to excite ground state oxygen to the first excited singlet state $^1\Delta g$ is only 22 kcal mol^{-1}, a variety of sensitizers with visible wavelength absorption bands can produce the singlet oxygen. Due to the long lifetime of the $^1\Delta g$ state, the reactions of singlet oxygen have been extensively studied, and its typical reactions have been elucidated.[99–101] Singlet oxygen undergoes an "ene-type" reaction with olefins having allylic hydrogens to form an unsaturated hydroperoxide.

$$\xrightarrow{^1O_2} \quad \text{OOH}$$

With electron-accepting compounds such as dienes or aromatic hydrocarbons, singlet oxygen behaves as a dienophile as depicted below.

$$\text{anthracene} \xrightarrow{^1O_2} \text{9,10-endoperoxide (O-O)} \qquad \text{cyclopentadiene} \xrightarrow{^1O_2} \text{endoperoxide (O-O)}$$

The third reaction, known as the singlet oxygen reaction, is 1,2-addition to olefins to form dioxetanes, which are unstable and may cleave to yield carbonyl fragments.

$$>C{=}C< \xrightarrow{^1O_2} -\underset{|}{\overset{O-}{C}}-\underset{|}{\overset{O}{C}}- \longrightarrow 2 \; >C{=}O$$

Sensitizers such as acridines, porphyrins, xanthenes (eosin, rose bengal) and thiazines (methylene blue, thionine) produce singlet oxygen with high efficiency.

The quantum yields of singlet oxygen formation by sensitizing dyes can be determined by direct and indirect methods. The former is direct analysis of singlet oxygen by luminescence measurements[102,103] or thermal lensing,[102] while the latter involves trapping of singlet oxygen by some agents such as tetramethylethylene,[104] 2,5-dimethylfuran,[105] and furfuryl alcohol.[106,107] Two novel photopolymer systems have been investigated on the basis of singlet oxygen chemistry.

As shown in the following schemes, singlet oxygen $^1O_2$ reacts with 2,5-diphenyl-3,4-benzofuran (B), which is subsequently converted to 1,2-dibenzoylbenzene (C). The carbonyl compound (C) then gives rise to free radical species (R·) under ultraviolet light irradiation. In this imaging system, the first imagewise exposure by visible light results in the formation of the latent images by (C) and the second overall exposure by ultraviolet light induces the polymerization of acrylate monomer only in those areas exposed at the first stage. The latent images thus recorded by visible light are developed by the subsequent ultraviolet light exposure. This is called a biphotonic process.[108,109]

$$\begin{cases} A \xrightarrow{h\nu} A^* \\ A^* + {}^3O_2 \longrightarrow A + {}^1O_2({}^1\Delta_g) \end{cases}$$

$$^1O_2 + B \longrightarrow [BO_2] \Longrightarrow C$$

(carbonyl compound)

$$\begin{cases} C \xrightarrow{h\nu} C^* \\ C^* \Longrightarrow R^\bullet\text{(radicals)} \\ R^\bullet + M \longrightarrow RM^\bullet \end{cases}$$

(incipient polymer)

Another system is based on the reaction of $^1O_2$ with olefins having an allylic hydrogen to form an unsaturated hydroperoxide by a so-called ene-type reaction. The photoimaging process is explained as follows. Singlet oxygen $^1O_2$ is produced by visible light with the aid of a sensitizer. The singlet oxygen reacts with olefin molecules to produce hydroperoxides that form the latent image. Subsequent heating induces a reaction between the hydroperoxides and a weak reducing agent, M^n (vanadyl naphthenate, vanadyl acetylacetonate), producing free radical species. The polymerization of acrylate monomers is initiated by these free radicals and proceeds rapidly with the aid of activation energy supplied by the heating process. After the polymerization is completed, the unreacted monomers are washed out with a solvent or an aqueous base, depending on the monomers used.[110]

Perhaps the most important requirements of this system are the discovery of efficient sensitizers for generating singlet oxygen under visible light illumination and olefin compounds for forming hydroperoxides efficiently.

Exposure (latent image formation)

$$^3O_2 \xrightarrow[\text{sensitizer}]{h\nu} {}^1O_2$$

$$\text{(olefin with allylic C–H)} + {}^1O_2 \longrightarrow \text{(ROOH)}$$

Processing (polymerization)

$$\text{ROOH} + M^n \xrightarrow{\Delta} \text{RO}^\bullet + M^{n+1} + \text{OH}^-$$

$$\text{RO}^\bullet + \text{monomer} \longrightarrow \text{X-linked polymer}$$

Development

unexposed areas removed by solvent

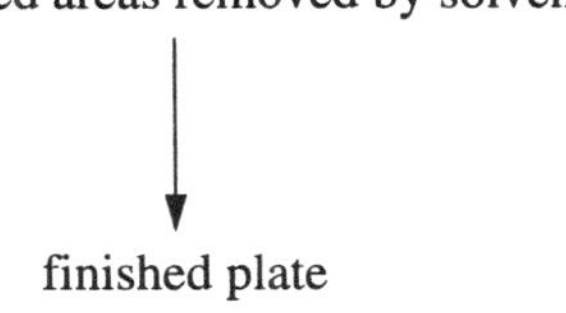

finished plate

M. Yasuike et al. found that tetrabenzoporphyrins (TBP)[111] are the efficient sensitizers for forming singlet oxygen. The combination of TBP with a dimethylfuran derivative was used. The photopolymer showed a high sensitivity to

He-Ne laser light (633 nm).

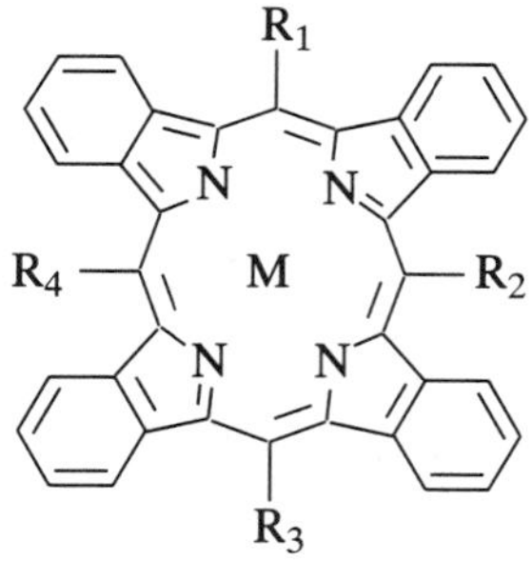

Structure of metallo-tetrabenzoporphyrins

This novel photopolymer system based on singlet oxygen chemistry is very interesting because the latent image is produced by visible light, and the ground state oxygen, which inhibits the polymerization of monomers, is used up by the sensitizer during the imagewise exposure. In addition, the thermal energy given in the heating process induces the reaction of $^1O_2$ and olefins, and the radical polymerization of acrylate monomers.

Now that C_{60} clusters[112] are available in quantities of grams,[113] several important investigations have been made possible. Electronic excitation of C_{60} in the presence of oxygen produces $^1O_2$ by energy transfer from triplet C_{60} to ground state oxygen.[114] As a result, it may be possible to use C_{60} in the present photopolymeric systems.

4. Radical Generation by Intramolecular Electron Transfer

Several dye sensitized photoinitiating systems in which photoinduced electron transfer plays an important role have been reported[115–117] and triarylalkyl borates of cationic dyes have recently been reported to be a novel category of radical generators.[118] These borates are believed to form ion pairs in nonpolar solvents, as the electron transfer between the cationic dye and the counter anion occurs efficiently despite the short lifetime of the excited singlet state of the cationic dye.[119,120] When a cyanine dye-triarylalkyl borate is exposed to visible light, the cationic cyanine dye is activated to an excited singlet state that very rapidly transfers the electron to the counter anion, producing an alkyl radical R·.

$$Cy^+Ph_3BR^- \xrightarrow{h\nu} Cy\cdot + Ph_3BR\cdot$$

$$Ph_3BR\cdot \rightarrow Ph_3B + R\cdot$$

Study of the photolysis of $Cy^+[NpCH_2B(Ph)_3]^-$ indicated that the radical formation involves the transfer of an electron from the borate anion to the cationic cyanine dye. Pulsed irradiation of a benzene solution of $Cy^{+}[NpCH_2B(Ph)_3]^{-}$ shows a transient absorption band that is bleached; the absorption bands due to the radicals of Cy· and $NpCH_2$· appear. Visible-light sensitive radical generators are studied extensively because of their possible use in photorecording materials.[121] Indolenine,[122] azulene,[123] pyrylium, and thiopyrylium[124] also act as photoradical generators by a similar mechanism. Combinations of ammonium borates with neutral dyes such as merocyanines, coumarins, or xanthene are also efficient photoinitiators.[125] When cyanine *n*-butyl-triphenylborate (TBABTB) is exposed to light (>680 nm) in the presence of tetra-*n*-butylammonium, it is converted to colorless meso-substituted leucocyanine.[126] This photoreduction reaction is dependent on the polarity of the solvent. Meso-phenyl leucocyanines are formed in polar solvents, whereas a mixture of meso-phenyl and meso-butyl leuco dyes is formed in weakly polar solvents. In nonpolar solvents, meso-butyl leucocyanine is the only product. The transfer of an electron from the borate anion to the excited state of the cyanine dye is followed by the radical cleavage of the borate anion, producing butyl or phenyl radicals. These radicals attack the cyanine dye at the meso-position to give the colorless leuco dye.

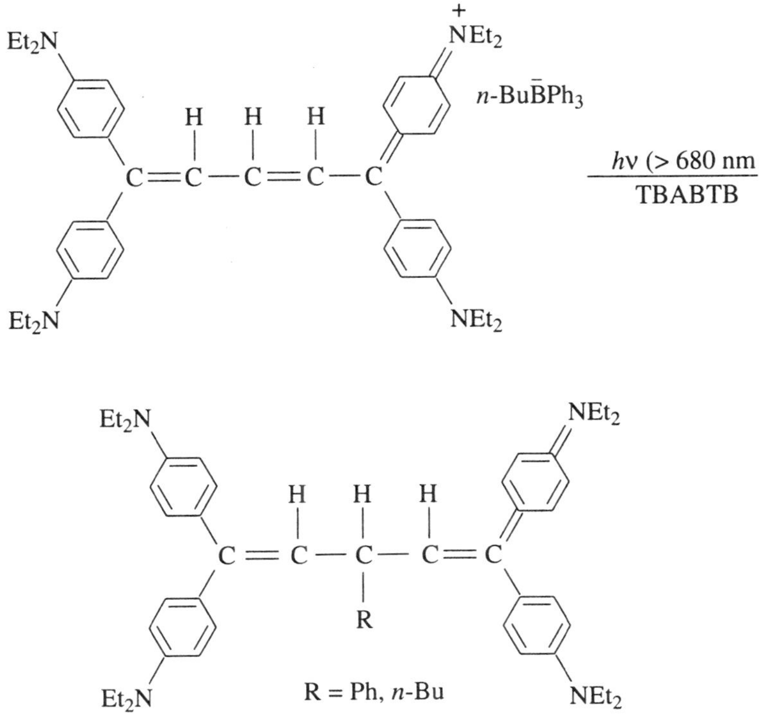

Et2N Et2N +NEt2 NEt2 H H H C=C—C=C—C n-Bu$\bar{B}$Ph$_3$

$h\nu$ (> 680 nm / TBABTB

Et2N Et2N NEt2 NEt2 H H H C=C—C—C=C R

R = Ph, *n*-Bu

A dye-aluminate complex is also reported to act as a photointiator,[127] and its mechanism seems to be similar to those of the cyanine dye borates.

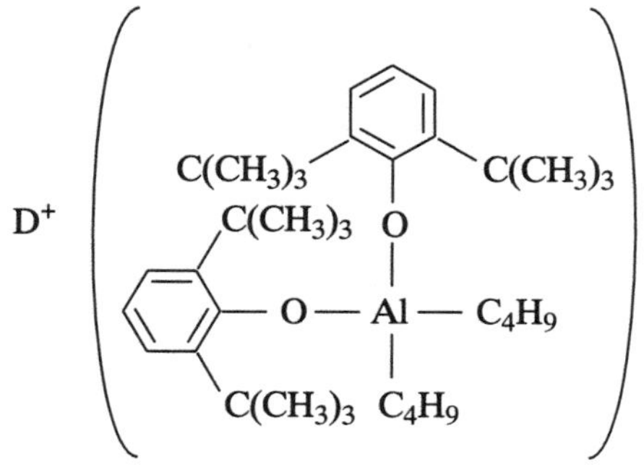

$C(CH_3)_3$ $C(CH_3)_3$

D^+ $C(CH_3)_3$ O

O — Al — C_4H_9

$C(CH_3)_3$ C_4H_9

D+ = q quaternary ammonium cation or a cationic dye

5. Acid Generation by Visible Light

Although photoinduced radical polymerization has long been studied extensively because of its industrial importance,[128–134] it is only recently that material applications of the photoinduced cationic polymerization have attracted attention. S. I. Schlesinger reported in 1970 the photoinduced cationic polymerization of epoxides with aromatic diazonium salts used as the photoinitiators. J. V. Crivello subsequently synthesized a great number of photoacid generators such as onium salts. The invention of efficient photocationic generators stimulated studies of photocationic polymerization and the acid-catalyzed modification of polymers.[135]

5.1. Ionic Molecules

The aryldiazonium salts yield Lewis acids on irradiation with visible light.[136] The spectral sensitivity of diazonium salts can be controlled by modifying the structure of aromatic moiety, as the lowest singlet excited state is mainly due to the intramolecular charge transfer from the aromatic moiety to the diazo group. The thermal instability of those salts and the generation of nitrogen gas on photodecomposition are a disadvantage.

$$Ar - \overset{+}{N_2}\, BF^-_4 \xrightarrow{h\nu} ArF + N_2 + BF_3$$

J. V. Crivello et al.[137] prepared diaryliodonium and triarylsulfonium salts that generate Brönsted acids when they are photoirradiated. The reactions of iodonium salts and sulfonium salts are described as follows.

$$Ar_2I^+X^- \xrightarrow{h\nu} [Ar_2I^+X^-]^* \rightarrow ArI\cdot^+ + Ar\cdot + X^-$$

$$ArI\cdot^+ \underset{RH}{\overset{solvent}{\rightarrow}} ArIH^+ + R\cdot \rightarrow ArI + H^+ + R\cdot$$

$$Ar_3S^+X^- + RH \xrightarrow{h\nu} Ar_2S + H^+ + R\cdot + Ar\cdot + X^-$$

The spectral sensitivities of these onium salts are at short wavelengths, around 300 nm, and the spectral sensitization based on both electron transfer or energy transfer via the triplet state has been reported.[138–141] The mechanistic studies by Hacker et al.[141] showed that the photolysis of onium salts releases acidic species in a competitive reaction of in-cage recombination and cage-escape reaction. Depending on the conditions of the photochemical reaction, triphenylsulfonium salts can undergo heterolysis, homolysis, triplet energy transfer, or electron transfer.

The aryliodonium salts can be spectrally sensitized to visible wavelengths by dyes such as acridine yellow, benzoflavin, and acridine orange.[142] Perylene efficiently sensitizes triarylsulfonium salts to 475 nm.[143]

The diphenyliodonium salts of some aromatic sulfonates have been found to react by exciting the aromatic moiety of the sulfonate. This is a novel category of spectral sensitization of iodonium cation, with the counter anion acting both as the spectral sensitizer and as the acid precursor. Diphenyliodonium 9,10-dimethoxyanthracene-2-sulfonate[144] and diphenyliodonium 8-anilinonaphthalene-1-sulfonate,[145] for example, are photochemically dissociated by visible light to give the parent sulfonic acids. This dissociation is thought to proceed by means of an intra-ion-pair electron transfer from 9,10-dimethoxyanthracene and 8-anilinonaphthalene moieties.

Saeva et al.[146] have reported that 9-anthrylmethyl-*p*-cyanobenzylsulfonium hexafluorophosphate gives a Lewis acid and other products when irradiated with visible light.

S^+ PF_6^- CH_3 CH_2 CN $\xrightarrow[CH_3CN/H_2O]{h\nu}$ CN CH_2 S CH_3 CH_3 S + CH_3–C(=O)–NH–CH_2–C$_6$H$_4$–CN + HPF_6

Another sensitizing dye for diphenyliodonium salt is 1,8-dimethoxy-9,10-bis(phenyl-ethynyl)anthracene, which absorbs the 488-nm and 514-nm light of argon ion lasers.[147] The advantage of this acid precursor is that it has no basic site like an amino group that can trap the photogenerated Lewis acid.

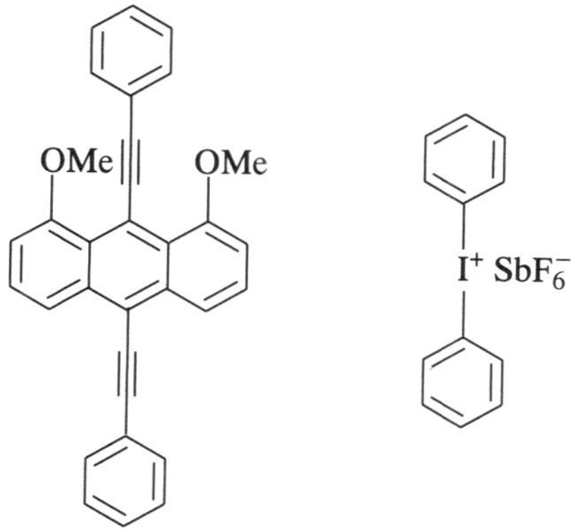

sensitizer

diphenyliodonium hexafluoroantimonate

Organometallic cationic photoinitiators like (η^6-benzene)(η^5-cyclopentadienyl)Fe(III) hexafluorophosphate are mixed-ligand arene cyclopentadienyl metal salts of complex metal halide anions.[148]

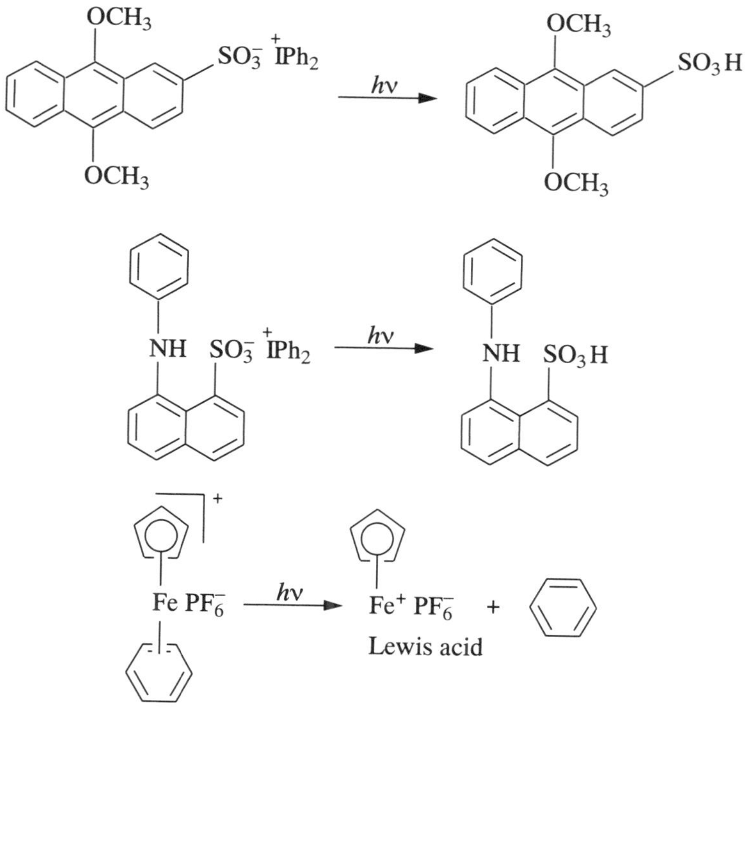

$$\text{OCH}_3,\ \text{SO}_3^- \ \overset{+}{\text{I}}\text{Ph}_2 \xrightarrow{h\nu} \text{OCH}_3,\ \text{SO}_3\text{H}$$

$$\text{NH}\ \ \text{SO}_3^- \ \overset{+}{\text{I}}\text{Ph}_2 \xrightarrow{h\nu} \text{NH}\ \ \text{SO}_3\text{H}$$

$$\text{Fe PF}_6^- \xrightarrow{h\nu} \text{Fe}^+\ \text{PF}_6^- +$$

Lewis acid

The mechanism of the photoinduced cationic polymerization of epoxides may be as follows:

$$[(\eta^5\text{-}C_5H_5)Fe(\eta^6\text{-}C_6H_6)]^+ PF_6^- \xrightarrow[\text{epoxide}]{h\nu} (\eta^5\text{-}C_5H_5)Fe^+(\text{epoxide})_3 PF_6^- + C_6H_6 \xrightarrow[\text{epoxide}]{\text{heat}} \text{polymer}$$

5.2. Nonionic Molecules

Busman et al.[149] reported that *p*-nitrobenzyl esters of alkane sulfonic acids exposed to near ultraviolet light release sulfonic acid with the aid of electron-donating sensitizers such as 9,10-dimethoxy-2-ethylanthracene. This photodissociation proceeds by means of an electron transfer from sensitizer to the sulfonate.

On the other hand, *p*-nitrobenzyl-9,10-dimethoxyanthracene-2-sulfonate is photodissociated without a sensitizer using wavelengths from the ultraviolet to 450 nm.[150] This dissociation is due to the intramolecular electron transfer from the excited 9,10-dimethoxyanthracene moiety to the *p*-nitrobenzyl moiety. Time-resolved spectroscopic study has detected the transient species, the anion radical of 9,10-dimethoxyanthracene and cation radical of the *p*-nitrobenzyl group. The intramolecular electron transfer occurs when the free energy change (ΔG) incorporating the electron transfer is small enough.

$$\text{9,10-(OCH}_3)_2\text{-anthracene-2-}SO_3CH_2C_6H_4NO_2 \xrightarrow{h\nu} \text{9,10-(OCH}_3)_2\text{-anthracene-2-}SO_3H$$

This means that various sulfonic esters other than *p*-nitrobenzyl-9,10-dimethoxyanthracene-2-sulfonate can release sulfonic acid on exposure to light if those sulfonates fulfill the above-mentioned conditions. And in fact, *p*-nitrobenzyl-5-dimethylaminonaphthalene-1-sulfonate, *p*-nitrobenzyl pyrene-1-sulfonate, and *p*-cyanobenzyl-9,10-dimethoxyanthracene-2-sulfonate also release their parent sulfonic acids when they are photostimulated.

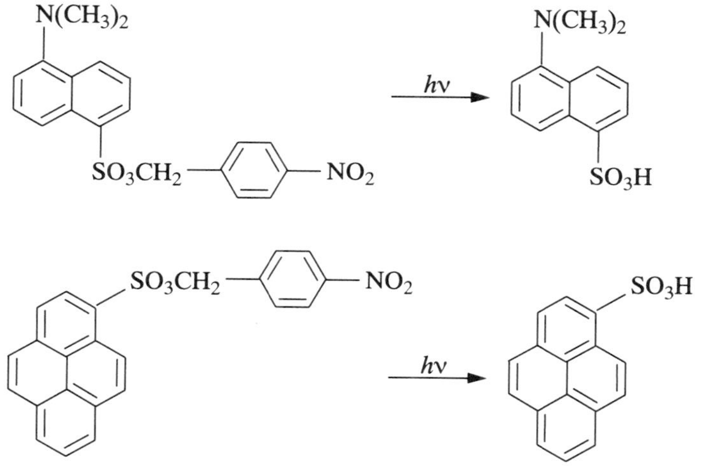

The photoinduced generation of acidic species has been studied for a long time. For example, trichloroacetophenone, tribromoethylphenyl sulfone, and desylchloride have been well studied. Bis-4,6-(trichloromethyl)-1,3,5-triazine derivatives exhibit adequate stability both in solutions and in polymer films. Bonham et al.[152] have reported that the spectral sensitivity of 4,6-bis(trichloromethyl)-1,3,5-triazine can be extended to visible wavelengths by introducing an electron-donating substituent like a dimethylamino group. From the viewpoint of photoacid generation, however, the presence of such a basic substituent in the molecule may not be a favorable because the acidic species would be neutralized. By introducing the nonbasic benzothiazoline into the 4,6-bis-(trichloromethyl)-1,3,5-triazine group, Pawlowski et al.[153] synthesized a chromophore that is sensitive to 550-nm light. This compound has good stability for storage as well as broad spectral sensitivity.

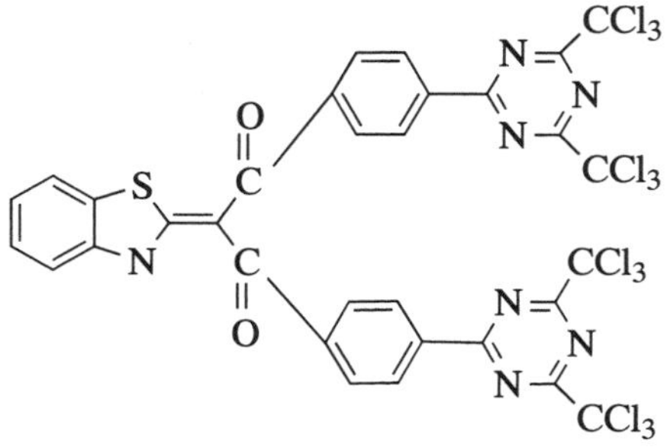

5.3. Organometallic Molecules

The photochemical reaction of titanocene complexes has also been studied, and photo-CIDNP measurement of $Ti(\eta^5\text{-}C_5H_5)_2(CH_3)_2$ shows that the cleavage of the Ti-CH_3 σ bond is the primary reaction.[154] The resulting caged radical pair rapidly decays by recombination or release of a CH_4 fragment:

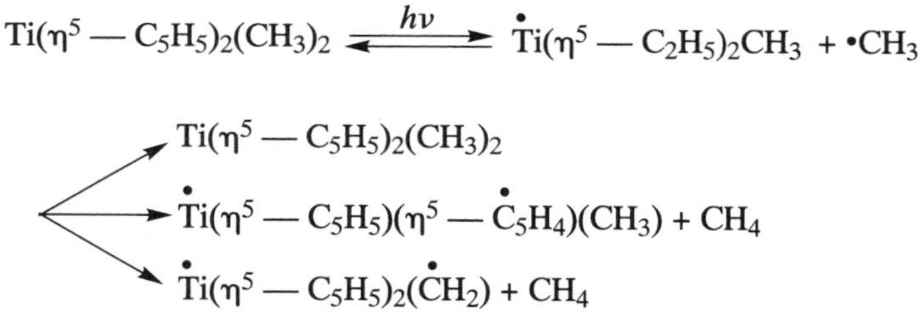

$$Ti(\eta^5 - C_5H_5)_2(CH_3)_2 \underset{}{\overset{h\nu}{\rightleftharpoons}} \dot{T}i(\eta^5 - C_2H_5)_2CH_3 + \bullet CH_3$$

$$\longrightarrow \begin{cases} Ti(\eta^5 - C_5H_5)_2(CH_3)_2 \\ \dot{T}i(\eta^5 - C_5H_5)(\eta^5 - \dot{C}_5H_4)(CH_3) + CH_4 \\ \dot{T}i(\eta^5 - C_5H_5)_2(\dot{C}H_2) + CH_4 \end{cases}$$

Despite the high photosensitivity of titanocene derivatives, these derivatives are difficult to use as photoinitiators because of their poor thermal and oxidative stabilities. It has recently been found, however, that fluorination of the aromatic rings provides thermal stability as well as resistance to oxidation by atmospheric oxygen. Bis(pentafluorophenyl)titanocene ($Ti(\eta^5\text{-}C_5H_5)_2(\eta^1\text{-}C_6F_5)_2$) is an interesting photoinitiator;[155] its photochemical reaction is quite different from what is known for the reaction of titanocene complexes.

According to a study of the fluorinated titanocene complex performed with the use of laser flash photolysis and a radical trapping agent, a highly reactive isomer is formed transiently with unit quantum efficiency. This isomer is thought to be formed by a slippage reaction of the cyclopentadienyl ring from η^5 to η^1.

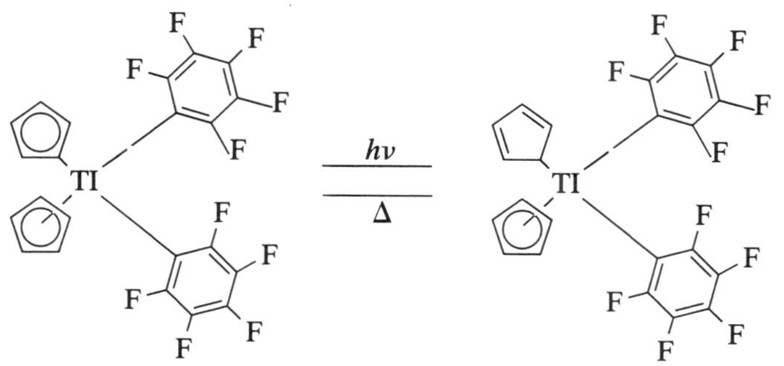

J. Finter et al. studied the initiation mechanism of the bulk polymerization of methylmethacrylate by using the fluorinated titanocene as the photoinitiator and found two transient species that are thought to be a pentafluorophenyl cyclopentenyl radical and a titanium-containing ketene acetal radical.[156]

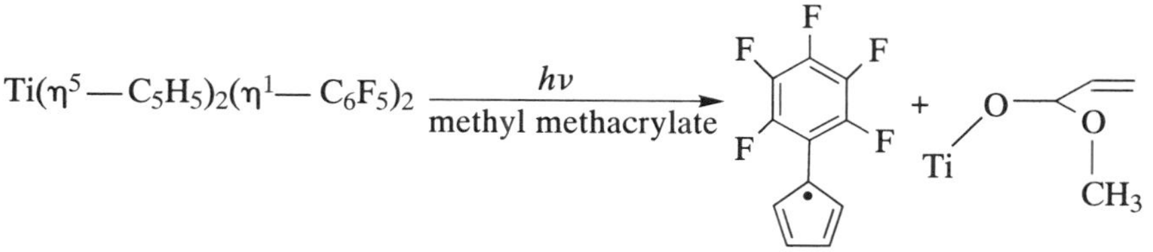

$$Ti(\eta^5-C_5H_5)_2(\eta^1-C_6F_5)_2 \xrightarrow[\text{methyl methacrylate}]{h\nu}$$

The fluorinated titanocene complex has strong absorptions in the visible region, from 400 nm to 600 nm, that are bleached by irradiating visible light (see Fig. 3-18). The bleaching property is favorable for photoinitiators because it enables the photopolymerization of thick layers.

6. Other Systems

6.1. Aluminum Complex-Silanol Systems

Some combinations of silanol and aluminum complex polymerize epoxides efficiently at room temperature.[157,158] A part of silanol reacts with aluminum complex to form compounds having Al-O-Si bonds. If the oxygen atoms in the unreacted silanols coordinate to aluminum atoms of Al-O-Si groups, the O-H groups of the silanols are strongly polarized and induce polymerization of epox-

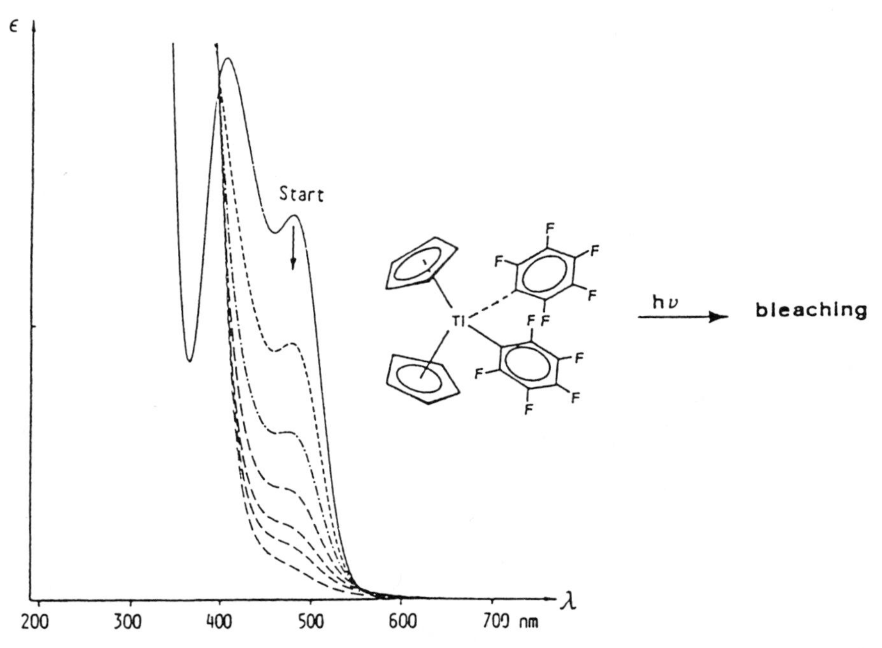

Figure 3-18. *Bleaching of absorption for fluorinated titanocene complex.*

ides in cationic polymerization.

$$\text{(Al acetoacetate chelate: } {>}Al\text{—O—C(CH}_3\text{)=CH—C(OC}_2\text{H}_5\text{)=O}\rightarrow Al) + HO-SiPh_3 \longrightarrow {>}Al-O-SiPh_3 + CH_3\overset{O}{\overset{\|}{C}}CH_2\overset{O}{\overset{\|}{C}}OC_2H_5$$

$$ {>}Al-O-SiPh_3 \xrightarrow{HOSiPh_3} Ph_3Si\overset{\delta^-}{\ddot{O}}-\overset{\delta^+}{H} \longrightarrow \text{Polymerization}$$

$$Ph_3Si\ddot{O}H \rightarrow {>}Al-O-SiPh_3$$

Because *o*-nitrobenzyl silylethers generate silanol by ultraviolet light, the mixture of *o*-nitrobenzyl silylether and aluminum complex induces the polymerization of epoxides with exposing ultraviolet light.[159,160]

$$Ph_3SiO-CH_2-C_6H_4(NO_2) \xrightarrow[(250nm \sim 400nm)]{h\nu} Ph_3SiO-CH(HO)-C_6H_4(N{=}O)$$

$$\longrightarrow Ph_3SiOH + OHC-C_6H_4(NO)$$

However, *o*-nitrobenzyl silylethers require short wavelength ultraviolet light for reaction, and the quantum yields are not high. This limitation prevents the practical use of this photocuring resin.

With photoacid generators, the sensitivity of the silylether photoinitiation system is remarkably enhanced. Some silylethers form silanols in the presence of acid. The three component system consisting of photoacid generator, silylether, and aluminum complex gives a very high speed initiator for epoxides.[161]

$$Ph_2I^+X^- \rightarrow H^+X^-$$

$$Ph_3Si—O—CH(CO_2C_2H_5)_2 + H{*}X^- \rightarrow Ph_3SiOH$$

$$\text{Epoxide} + \text{Al-complex} + Ph_3SiOH \rightarrow \text{Polymerization}$$

6.2. Photosensitive Microgels

The use of photosensitive microgels is an interesting method for photosensitivity enhancement. It has been reported that a series of crosslinked microgels (20–35-nm diameter) with quaternary ammonium ions on the surface gives highly sensitive materials. The microgels were synthesized by polymerizing chloromethylated styrene, *N,N*-dimethylvinylbenzyl ammonium chloride, and divinylbenzene in various ratios (Fig. 3-19). The microgels were further reacted with tertiary amine to quarternarize the surface. The schematic illustration of the surface modified microgels and the polymerization due to exposure to the light are shown in Fig. 3-20. The microgels are stably dispersed in water as well as in common organic solvents without surfactants and give clear film layers when coated on a substrate and dried. The photosensitive composition was prepared by adding polyfunctional monomer, photoinitiator, and photosensitizer as is shown in Table 3-12. The composition was coated on an aluminum plate to form a 2-μm-thick layer, dried, and exposed to 488-nm light of a mercury lamp or an argon ion laser. After exposure, the layer was washed with tap water to remove the unexposed areas. In Table 3-13, the compositions of microgels, the

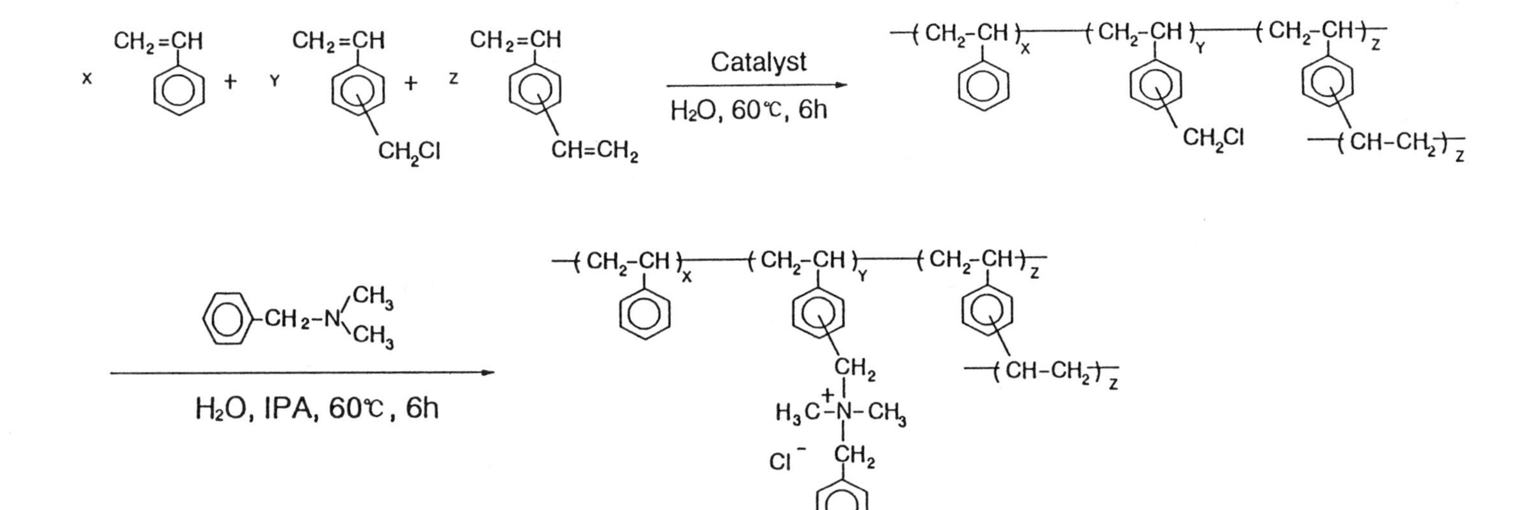

Figure 3-19. *Preparation flow-chart of microgel.*

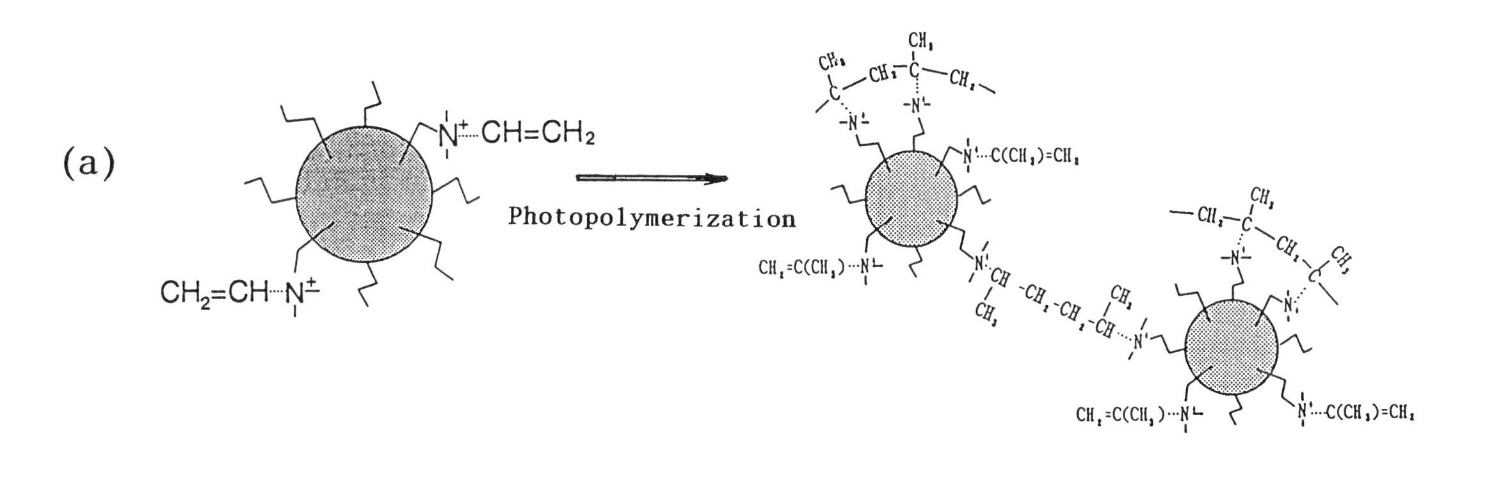

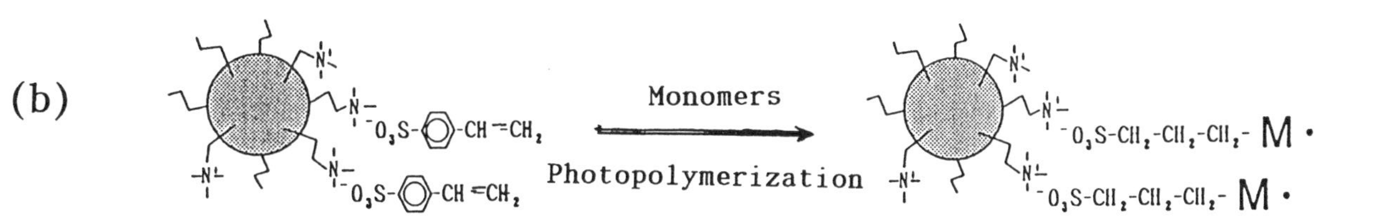

Figure 3-20. *The schematic illustration of the surface modified microgels and the polymerization due to exposure to the light. (a) Photopolymerizable monomers such as acrylates are introduced by formation of ionic bonds. (b) An example of reactive sites introduced to the microgel surfaces. During the photopolymerization of acrylate monomers, the propagating radicals are grafted on the reactive sites to crosslink the microgels.*

Table 3-12. General Composition of the Microgel-Based Photopolymers

	(Parts by weight)
Matrix polymer: Surface activiated microgel	100
Polyfunctional Monomer: Pentaerythritol triacrylate	100
Photoinitiator: BTTB	8
Photosensitizer: Ketocoumarin dye	6

Table 3-13. Composition, Average Diameter, Sensitivity, and Initial Rate of Photopolymerization for the Microgels and the Microgel-Based Photopolymers. Data for Conventional Photopolymers of Homogeneous Matrix (HSP-621 and PVP) Are Shown for Comparison

Microgel	Molar Ratio of Monomer (%)			Average Diameter (nm)	Sensitivity ($\times 10^{-3}$ mJ/cm^2) Ar Laser (488 nm)	R_p ($\times 10^{-2}$ mol/l · s)
	St.	CMS	DVB			
MG- 1	49.8	49.8	0.4	31	90	8.5
MG- 2	49.5	49.5	1.0	25	70	10.5
MG- 3	49.0	49.0	2.0	24	55	12.0
MG- 4	48.0	48.0	4.0	26	55	12.5
MG- 5	68.0	28.0	4.0	30	60	15.0
MG- 6	58.0	38.0	4.0	25	85	11.5
MG- 7	38.0	58.0	4.0	21	90	9.5
MG- 8	28.0	68.0	4.0	32	85	9.0
MG- 9	46.0	46.0	8.0	62	80	9.0
MG-10	40.0	52.0	8.0	65	75	16.5
MG-11	44.0	44.0	12.0	65	70	13.0
HSP-621					250	2.5
PVP					140	5.5

R_p: Initial rate of monomer polymerization in the matrices.

photosensitivity, and initial photopolymerizing rates are summarized. As can be seen from the table, these microgel photopolymeric layers exhibit high sensitivity from 55 to 90 μJ/cm^2. These sensitivities are about ten times higher than those of conventional homogeneous photosensitive layers. Besides the sensitivity, the microgel sensitive layers give water developability, high resolution, and a high oleophilic property.

Another series of the microgels is made from photopolymerizable acryloyl groups introduced into the microgel surface. The introduction of acryloyl groups gives higher light sensitivity with the use of photoinitiators.[162,163] A pattern formed by an argon ion laser scanning and the SEM picture of 25 μm line and space pattern are shown in Fig. 3-21.

7. Summary

With the growing application of photopolymers to various fields, demands on their performance are becoming more complex and stringent. For example, LSI and other electronic devices require quarter-micron patterns and high thermal stability, as well as low dielectric constants. Printing engineering, where the pre-press process is now computerized and interest in computer-to-plate systems is increasing, needs highly sensitive photopolymers that can be exposed to visible laser light. Holographic recording is another field that requires photopolymers that are highly sensitive to visible laser light. The following are several approaches for improving the sensitivity of photopolymers.

1. *Radical photopolymerization.* Among the primary reactions used in photopolymers, photopolymerization is one of the most attractive candidates because of its potential for high sensitivity. Areas to investigate include developing photoinitiators that generate free radicals with a high quantum efficiency, monomers that have a high catalytic chain length, and matrix polymers that accelerate the rate of monomer polymerization. Another, more sophisticated, mechanism for increasing sensitivity is the application of silver halide reactions to the photoinitiating system. Similarly, the application of the silver halide photography development mechanism to the crosslinking reaction of photopolymers has been reported to be successful for the sensitivity enhancement of the photopolymer.
2. *Cationic photopolymerization.* This polymerization mechanism is attractive because it can eliminate the unfavorable effects of atmospheric oxygen encountered in radical polymerization. Accumulation of basic data will be helpful for finding mechanisms to improve the sensitivity of photopolymers.
3. *Chemical amplification.* Chemical amplification is based on the mechanism where a catalyst is generated in the polymer on exposure to light, and this catalyst accelerates the polymer reactions assisted by thermal energy released after light exposure. Reactions such as crosslinking, photodepolymerization, and photodeblocking that use photogenerated acid as the catalyst are being studied extensively.
4. *Photoreactive microgels.* The surface of microgels are modified by photosensitive substituents to make them sensitive to light. For example, microgels with acryloyl groups on their surfaces were prepared. The acryloyl groups on the surfaces of microgels polymerize on exposure to light and result in

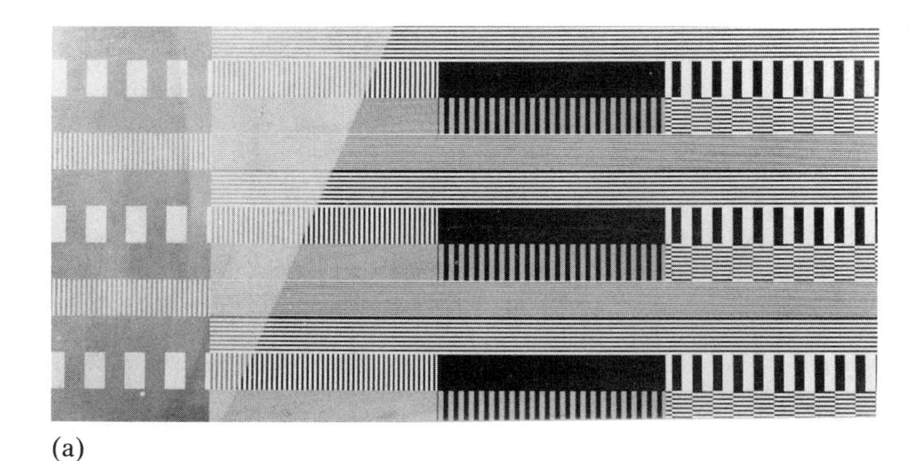

(a)

25 μm

(b)

Figure 3-21. *A pattern formed by an argon ion laser scanning and the SEM picture of 25 μm line and space pattern. (a) The microgel photosensitive layer coated on a grained aluminum plate was scan-exposed by 488 nm light of an argon ion laser, washed with tap water and dried. (b) A part of the pattern was observed with a SEM. A 25 μm line and space pattern on the grained aluminum surface is clearly resolved.*

Table 3-14. Examples of the photopolymers which are highly sensitive to 488 nm light of an argon ion laser.

Sensitivity/mJ cm^{-2}

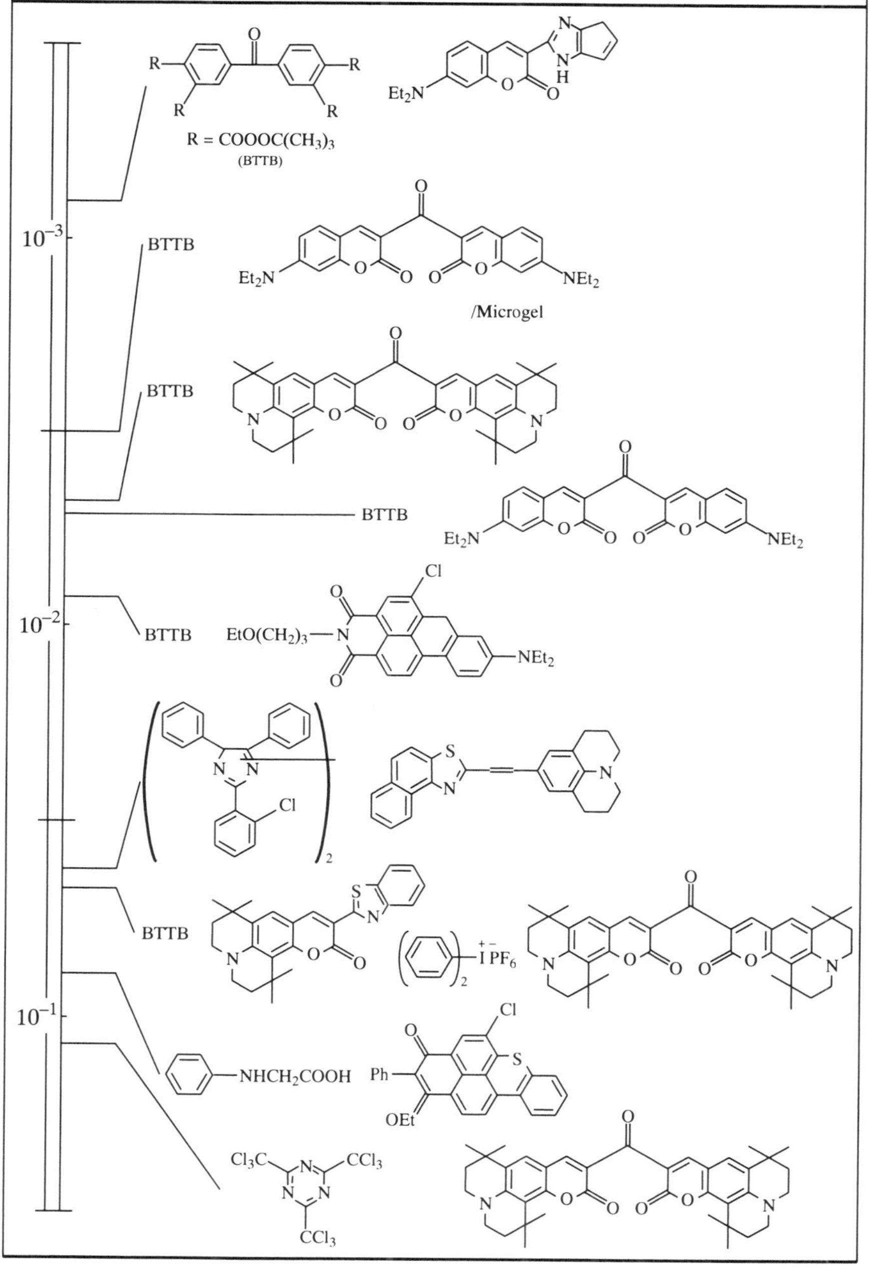

10^{-3}

10^{-2}

10^{-1}

R = $COOOC(CH_3)_3$
(BTTB)

Et_2N

BTTB

Et_2N NEt_2

/Microgel

BTTB

BTTB

Et_2N NEt_2

BTTB $EtO(CH_2)_3$—N Cl NEt_2

Cl

BTTB IPF_6

Ph OEt Cl S

—$NHCH_2COOH$

Cl_3C CCl_3 CCl_3

crosslinking of microgels. This is an interesting way to achieve sensitivity enhancement because a large mass of polymer molecules is crosslinked instead of the individual polymer molecules. It was reported that the sensitivity of a microgel photopolymer was increased by up to 1000 times that of conventional photopolymers with a similar composition.

Several examples of highly sensitive photopolymers are shown in Table 3-14. These photopolymers are not necessarily suitable for practical use because some of them do not satisfy the characteristics other than high sensitivity that are required for practical materials. More basic studies must be undertaken to develop photopolymers that have the properties and high sensitivity necessary for practical materials.

References Section 1

1. (a) L. M. Minsk, J. G. Smith, W. P. Van Deusen, and J. F. Wricht, *J. Applied Polym. Sci.* **2**, 302 (1959) (b) E. M. Robertson, W. P. Van Deusen, and L. M. Minsk, *J. Applied Polym. Sci.* **2**, 308 (1959). USP 2,610,120.
2. Brit. Patent 1,128,850 (1968). USP 3,469,982 (1969).
3. H. W. Vollmann, *Angew. Chem., Int. Edn.* **19**, 99 (1980).
4. S. M. Shahbazian, *J. Photogr. Sci.* **32**, 1111 (1984).
5. M. C. J. Twaalfhorven, *Electronic Packaging and Production* (1985), p. 64.
6. Jap. Kokai Tokkyo Koho, 54-155, 292 to E. I. Du Pont.
7. Jap. Kokai Tokkyo Koho, 61-123, 603 to E. I. Du Pont.
8. Jap. Kokai Tokkyo Koho, 61-170, 747 to E. I. Du Pont.
9. R. D. Michell, W. J. Nebe, and W. M. Hardam, *J. Imag. Sci.* **30**, 215 (1986).
10. U. S. Patent 3652275 (E. I. Du Pont).
11. A. Liu, A. D. Trifunac, and V. V. Krongauz, *J. Phys. Chem.* **96**, 207 (1992).
12. J. L. R. Williams, D. P. Specht, and S. Farid, *Polym. Eng. Sci.* **23**, 1022 (1983).
13. J. V. Crivello and J. H. Lam, *J. Polym. Sci., Polym. Symp.* No. 56, 383 (1976).
14. J. V. Crivello and J. H. Lam, *J. Polym. Sci., Polym. Chem. Ed.* **16**, 2441 (1978).
15. J. V. Crivello and J. H. Lam, *J. Polym. Sci., Polym. Chem. Ed.* **17**, 1059 (1979).
16. U. S. Patent 4,304,923 to 3M.
17. PCT WO85/01127. Jap. Kokai Tokkyo Koho 60-502125 to Polaroid.
18. Jap. Kokai Tokkyo Koho 48-36281.
19. Jap. Kokai Tokkyo Koho 54-151024.
20. J. P. Fouassier and S. K. Wu, *J. Appl. Polym. Sci.* **44**, 1770 (1992).
21. U. S. Patent 4,868,092 (Nippon Paint).
22. G. Manivannan, J. P. Fouassier, and J. V. Crivello, *J. Polym. Sci., Polym. Chem. Ed.* **30**, 1999 (1992).
23. G. Rist, A. Borer, K. Dietliker, V. Desorby, J. P. Fouassier, and D. Ruhlmann, *Macromolecules* **25**, 4182 (1992).

24. Jap. Kokai Tokkyo Koho 4-219756 to Hoechst.
25. Jap. Kokai Tokkyo Koho 62-143044, 62-150242 to The Mead.
26. M. S. Kaplan, *TAGA Proc.* **90** (1977).

Section 2

27. A. Reiser and E. Pitts, *Photogr. Sci. Eng.* **20**, 225 (1976).
28. A. Reiser and E. Pitts, *J. Photogr. Sci.* **29**, 187 (1981).
29. Jap. Kokai Tokkyo Koho 61-262737 to Canon.
30. K. Koseki, H. Echigo, T. Yamaoka, and T. Tsunoda, *J. Soc. Photogr. Japan* **46**, 99 (1983).
31. K. Koseki, T. Yamaoka, T. Tsunoda, S. Shimizu, and N. Takahashi, *J. Chem. Soc. Japan*, 1983, 798.
32. (a) K. Iwata, T. Hagiwara, and H. Matsuzawa, *J. Polym. Sci., Chem. Ed.* **23**, 2361 (1985). (b) *J. Polym. Sci., Part A, Polym. Chem.* **24**, 1043 (1986).
33. K. Ichimura and Y. Nishio, *J. Polym. Sci., Part A, Polym. Chem.* **25**, 1579 (1987).
34. (a) T. Nakamura and H. Shoue, The 78th Spring Conf., Preprints p. 94, Tech. Assoc. Graph. Arts, Jpn. (1987). (b) N. Sakaguchi and H. Syoue, The 78th Spring Conf., Preprints p. 98, Tech. Assoc. Graph. Arts, Jpn. (1987).
35. Y. Namariyama, H. Washio, A. Kinoshita, K. Nakamura, T. Asano, and G. Nagamatsu, *J. Chem. Soc. Japan*, 1983, 798.
36. W. H. Hertler, P. J. McCartin, J. R. Merrill, G. R. Nacci, and W. J. Nebe, *Photogr. Sci. Eng.* **23**, 297 (Sep./Oct. 1979).
37. N. Sasa and T. Yamaoka, *Polym. Adv. Technol.* (in press).
38. K. Kanda, Radcure '86, Tech. Paper FC86-820, p. 5–31, Sep. 9–11, 1986.
39. R. D. Michell, W. J. Nebe, and W. M. Hardam, *J. Imag. Sci.* **30**, 215 (1986).
40. G. Oster and E. H. Immergut, *J. Am. Chem. Soc.* **76**, 1393 (1954).
41. Jap. Kokai Tokkyo Koho 51-103424 to NTT.
42. Jap. Kokai Tokkyo Koho 49-121885 to ICI Amer.
43. K. Koseki, S. Miyaguchi, T. Yamaoka, E. Yamada, and Y. Goto, *J. Chem. Soc. Japan*, 1985, 119.
44. Jap. Kokai Tokkyo Koho 60-76503 to Nippon Oil & Fats.
45. K. Koseki, S. Miyaguchi, and T. Yamaoka, *J. Chem. Soc. Japan*, 1986, 1234.
46. Jap. Kokai Tokkyo Koho 61-233736 to Canon.
47. Jap. Kokai Tokkyo Koho 62-156103 to Yunichika.
48. T. Yamaoka, H. Maede, and Y. Zhang, *Polym. Preprints Jpn.* **35**, 477 (1986).
49. K. Koseki, S. Muraoka, T. Shirosaki, and T. Yamaoka, *Circuit Technol.* **2**, 150 (1987).
50. M. Kawabata and Y. Takimoto, 76th Spring Conf., Preprints p. 159, Tech. Assoc. Graph. Arts, Jpn. (1986).
51. U. S. Patent 4,304,923 to 3M.
52. Jap. Kokai Tokkyo Koho 60-76740.
53. Jap. Kokai Tokkyo Koho 60-78443.

54. Jap. Kokai Tokkyo Koho 60-88005.

55. O. Ohno and K. Ichimura, *Polym. Preprints Jpn.* **43**, 467 (1985).

56. K. Ichimura, M. Sakuragi, O. Ohno, and M. Yasuike, *Polym. Preprints Jpn.* **36**, 529 (1987).

57. K. Ichimura, M. Sakuragi, H. Morii, M. Yasuike, H. Tanaka, and O. Ohno, *J. Photopolym. Sci. Technol* **1**, 204 (1988).

58. K. Ichimura and M. Sakuragi, *J. Polym. Sci., Polym. Lett.* **26**, 18 (1988).

59. M. Kawabata and Y. Takimoto, *Chem. Express* **1**, 619 (1986).

60. M. Harada, M. Kawabata, and Y. Takimoto, The 78th Spring Conf., Preprints p. 63, Tech. Assoc. Graph. Arts, Jpn. (1987).

61. M. Kawabata, M. Harada, and Y. Takimoto, The 79th Fall Conf., Preprints p. 81, Tech. Assoc. Graph. Arts, Jpn. (1987).

62. Jap. Kokai Tokkyo Koho 62-143044 and 62-150242 to The Mead.

63. Jap. Kokai Tokkyo Koho 64-13139 to Fuji Photogr. Film.

64. K. Meier and H. Zwefel, *J. Rad. Curing* **13**, 26 (1986).

65. K. Meier and H. Zwefel, *J. Imag. Sci.* **30**, 174 (1986).

66. K. Koseki, K. Kaku, Y. Nakamura, and T. Yamaoka, The 81st Fall Conf., Preprints p. 12, Tech. Assoc. Graph. Arts, Jpn. (1988).

67. H. Angerer, V. Desobry, M. Riediker, H. Spahni, and M. Rembold, Proc. Conf. Rad. Curing Asia (CRCA '88) p. 461 (1988 Tokyo).

68. Jap. Kokai Tokkyo Koho 54-155292 to E. I. Du Pont.

69. Jap. Kokai Tokkyo Koho 61-123603 to Mitubishi Chemicals.

70. S. Shimizu, *TAGA Proc.* 232 (1986).

71. Jap. Kokai Tokkyo Koho 61-270747 to Toyo Bouseki.

72. J. L. R. Williams, D. P. Specht, and S. Farid, *Polym. Eng. Sci.* **23**, 1022 (1983).

73. K. Koseki, S. Miyaguchi, T. Yamaoka, and T. Shirosaki, 1984 Annual Meeting, Preprints, p. 115, Soc. Photogr. Sci., Jpn.

74. Jap. Kokai Tokkyo Koho 60-221403 to Nippon Chemicals.

75. Jap. Kokai Tokkyo Koho 48-36281 to 3M.

76. Jap. Kokai Tokkyo Koho 54-151024 to Fuji Photogr. Film.

77. A. Umehara, S. Kondo, K. Tamoto, and A. Matsufuji, *J. Chem. Soc. Japan*, 1984, 192.

78. Jap. Kokai Tokkyo Koho 58-40302 to Mitsubushi Chemicals.

79. Jap. Kokai Tokkyo Koho 62-161802 to Dainippon Printing Ink and Chemicals.

80. Jap. Kokai Tokkyo Koho 61-180359 to Toyobo.

81. K. Kaku, K. Koseki, and T. Yamaoka, The 80th Spring Conf., Preprints, p. 59, Tech. Assoc. Graph. Arts, Jpn. (1988).

82. PCT WO85/01127, Jap. Kokai Tokkyo Koho 60-502125 to Polaroid.

83. Jap. Kokai Tokkyo Koho 61-97650 to Fuji Photogr. Film.

84. Jap. Kokai Tokkyo Koho 60-57832 to Hitachi Chemicals.

Sections 3–6

85. F. W. H. Mueller, H Evans, and E. Cerwonka, *Photogr. Sci. Eng.* **6**, 227 (1962).
86. Y. Goto and E. Yamada, *J. Chem. Soc., Japan*, 1987, 1027.
87. T. Yamaoka, Y. Nakamura, K. Koseki, and T. Shirosaki, *Polym. Adv. Technol.* **1**, 287 (1990).
88. S. Shima, K. Naitoh, and T. Yamaoka, *Chem. Mater.* (in press).
89. Y. Zhang, K. Koseki, and T. Yamaoka, *J. Applied Polym. Sci.* **38**, 1271 (1989).
90. J. L. R. Williams, D. P. Specht, and S. Farid, *Polym. Eng. Sci.* **23**, 1022 (1983).
91. D. Rehm and A. Weller, *Israel J. Chem.* **8**, 259 (1970).
92. Y. Toba, M. Yasuike, M. Shirosaki, K. Koseki, and T. Yamaoka, The 60th Annual Meeting of the Chem. Soc. Jpn. 1990 Abstr. 3F-331, Hiroshima.
93. G. Amirzadeh and W. Schnabel, *Makromol. Chem.* **182**, 2821 (1981).
94. N. S. Allen, D. Mallon, and I. Sideridou, A. Green, A. Timms, and F. Catalina, *Eur. Poly. J.* **28**, 647 (1992).
95. A. A. Lamola and H. D. Roth, *J. Amer. Chem. Soc.* **96**, 6270 (1974).
96. R. S. Davidson and P. R. Steiner, *J. Chem. Soc. Perkin*, 1682 (1971).
97. D. R. G. Brimage and R. S. Davidson, *J. Chem. Soc. Perkin I*, 496 (1973).
98. R. S. Davidson, K. Harrison, and P. R. Steiner, *J. Chem. Soc.* (C), 3480 (1971).
99. C. S. Foote, *Chem. Rev.* **1**, 104 (1968).
100. D. R. Kearns, *Chem. Rev.* **71**, 395 (1971).
101. H. H. Wasserman and R. W. Murray, eds., *Singlet Oxygen* (Academic Press, New York, 1979).
102. K. Gollnick, *Adv. Photochem.* **6**, 1 (1968).
103. See for example, R. G. Zepp, N. L. Wolfe, G. L. Baughman, and R. C. Hollis, *Nature* **264**, 421 (1977).
104. P. Murasecco-Suardi, E. Gassmann, A. M. Braun, and E. Oliveros, *Helv. Chim. Acta* **70**, 1760 (1987).
105. A. M. Braun, F. H. Frimmel, and J. Hoigne, *Intl. J. Environ. Anal. Chem.* **27**, 137 (1986).
106. R. W. Redmond and S. E. Braslawsky, *Chem. Phys. Lett.* **148**, 523 (1988).
107. E. Oliveros, P. Suadri-Murasecco, A. M. Braun, T. Amimian-Saghafi, and H. J. Hansen, *Helv. Chim. Acta* **74**, 79 (1991).
108. C. Carre, D. Ritzenthaler, D. J. Lougnot, and J. P. Fouassier, *Optics Lett.* **12**, 646 (1987).
109. D. J. Lougnot, D. Ritzenthaler, C. Carre, and J. P. Fouassier, *J. Appl. Phys.* **63**, 4841 (1988).
110. D. S. Breslow, D. A. Simpson, B. D. Kramer, R. J. Schwarz, and N. R. Newburg, *Ind. Eng. Chem. Res.* **26**, 2144 (1987).
111. M. Yasuike, T. Yamaoka, O. Ohno, M. Sakuragi, and K. Ichimura, *Inorg. Chim. Acta* **184**, 191 (1991).
112. W. Kratschmer, L. D. Lamb, K. Fostiropoulos, and D. R. Huffman, *Nature* **347**, 354 (1990).
113. H. W. Kroto, J. R. Heath, S. C. O'Brien, R. F. Curl, and R. E. Smalley, *Nature* **318**, 162 (1990).
114. J. W. Arbogast, A. P. Darmayan, C. S. Foote, Y. Rubin, F. N. Diederich, M. M. Alvarez, S. J. Anz, and R. L. Whetten, *J. Phys. Chem.* **95**, 11 (1991).

115. D. F. Eaton, in *Advances in Photochemistry*, D. H. Volman, G. S. Hammond, and K. Gollnick, eds., vol 13 (Wiley-Interscience, New York, 1986), pp. 427–487.

116. D. F. Eaton, in *Topics in Current Chemistry*, J. Mattay, ed., vol. 156 (Springer-Verlag, Berlin, 1990), pp. 199–225.

117. D. F. Eaton, *Pure Applied. Chem.* **56**, 1191 (1984).

118. P. Gottschalk, D. C. Neckers, and G. B. Schuster, U. S. Pat. 4772530 (1988); *Chem. Abstr.* **107**, 187434n (1987).

119. S. Chatterjee, P. Gottschalk, P. D. Davis, and G. B. Schuster, *J. Am. Chem. Soc.* **110**, 2326 (1988).

120. S. Chatterjee, P. D. Davis, P. Gottschalk, M. E. Kurz, B. Sauerwein, X. Yang, and G. B. Schuster, *J. Am. Chem. Soc.* **112**, 6329 (1990).

121. J. S. Arney and J. A. Dowler, *J. Imaging Sci.* **32**, 125 (1988).

122. J. Yamaguchi, F. Shinozaki, M. Okazaki, and K. Adachi, U. S. Pat. 4952480 (1990); *Chem. Abstr.* **111**, 222119v (1989).

123. J. Yamaguchi, M. Okazaki, and T. Hoiki, U. S. Pat. 4902604 (1990); *Chem. Abstr.* **111**, 244332m (1989).

124. K. Kawamura and Y. Okamoto, U. S. Pat. 4971891 (1990); *Chem. Abstr.* **111**, 222173h (1989).

125. N. Kita and M. Koike, U. S. Patent 4,937,161 (1990); *Chem. Abstr.* **110**, 183029y (1989).

126. M. Matsuoka, T. Hikida, K. Murobushi, and Y. Hosoda, *J. Chem. Soc., Chem. Commun.* 299 (1993).

127. H. Matsumoto, J. Yamaguchi, N. Yanagihara, and H. Yamamoto, E. P. Appl. 90119767 (1990).

128. C. G. Roffey, *Photopolymerization of Coatings*, C. G. Roffey, ed. (John Wiley & Sons, New York, 1982), p. 1.

129. S. P. Pappas, *UV Curing—Science and Technology*, S. P. Pappas, ed. (Technology Marketing Corp., Norwalk, 1985), p. 1.

130. H. J. Hageman, *Photopolymerization and Photoimaging Science and Technology*, N. S. Allen, ed. (Elsevier Applied Science, London, 1989), p. 1.

131. E. Reichmanis and L. F. Thompson, *Polymer in Microlithography—Materials and Process*, E. Reichmanis, S. A. MacDonald, and T. Iwayanagi, eds. (ACS Symposium Series 412, American Chemical Society, Washington, D.C., 1989) p. 1.

132. A. Reiser, *Photoreactive Polymers, The Science and Technology of Resists*, A. Reiser, ed. (John Wiley & Sons, New York, 1989), p. 1.

133. Y. Shirota, *Photoinduced Electron Transfer, Part D—Applications*, M. A. Fox and M. Chanon, eds. (Elsevier, Amsterdam, 1988), p. 441.

134. H. J. Timpe, *Topics in Current Chemistry 156, Photoinduced Electron Transfer I* (Springer-Verlag, Berlin, 1990), p. 167.

135. H. Itoh and C. G. Wilson, *Polym. Eng. Sci.* **23**, 1012 (1983).

136. S. I. Schlesinger, *Photogr. Sci. Eng.* **18**, 387 (1974).

137. J. V. Crivello and J. H. Lam, *J. Polym. Sci., Polym. Symp.* No. 56, 383 (1976).

138. R. J. Devoe, M. R. V. Sahyun, and E. Schmidt, *Can. J. Chem.* **66**, 319 (1988).

139. R. J. Devoe, M. R. V. Sahyun, and E. Schmidt, *J. Imaging Sci.* **33**, 39 (1989).

140. J. P. Fouassier, D. Burr, and J. V. Crivello, *J. Photochem. Photobio.* **A49**, 317 (1989).

141. K. M. Welsh, J. L. Dektar, M. A. Garcia-Garibaya, N. P. Hacker, and N. J. Turro, *J. Org. Chem.* **57**, 4179 (1992).

142. J. V. Crivello and J. H. Lam, *J. Polym. Sci., Polym. Chem. Ed.* **16**, 2441 (1978).

143. J. V. Crivello and J. H. Lam, *J. Polym. Sci., Polym. Chem. Ed.* **17**, 1059 (1979).

144. K. Naitoh, T. Yamaoka, and A. Umehara, *Chem. Lett.* 1869 (1991).

145. K. Naitoh, K. Ishii, T. Yamaoka, and T. Omote, *J. Photopolym. Sci. Technol.* **5**, 339 (1992).

146. F. D. Saeva and D. T. Breslin, *J. Org. Chem.* **54**, 712 (1989).

147. G. W. Wallraff, R. D. Allen, W. D. Hinsberg, and C. G. Willson, *J. Imaging Sci. Technol.* **36**, 468 (1992).

148. K. Meler and H. Zweifel, *J. Imaging Sci.* **30**, 174 (1986).

149. S. Busman and J. E. Trend, *J. Imag. Technol.* **11**, 191 (1985).

150. K. Naitoh, K. Yoneyama, and T. Yamaoka, *J. Phys. Chem.* **96**, 238 (1992).

151. K. Naitoh and T. Yamaoka, *J. Chem. Soc. Perkin Trans.* **2**, 663 (1992).

152. J. A. Bonham and P. C. Petrellis, U. S. Patent 3,987,037 (1976).

153. G. Pawlowski, R. Dammel, K. J. Przybilla, and W. Spiess, *J. Photopolym. Sci. Technol.* **4**, 389 (1991).

154. P. W. N. van Leeuwen, H. van der Heijden, C. F. Roobeek, and J. H. G. Frijns, *J. Organomet. Chem.* **209**, 169 (1981).

155. M. Riedieker, M. Roth, N. Buehler, and J. Berger, U. S. Patent 4,910,121 (1990).

156. J. Finter, M. Riedieker, O. Rohde, and B. Rotzinger, *Makromol. Chem., Macromol. Symp.* **24**, 117 (1989).

157. S. Hayase, T. Ito, S. Suzuki, and M. Wada, *J. Polym. Sci., Polym. Chem. Ed.* **19**, 2185 (1981).

158. S. Hayase, T. Ito, S. Suzuki, and M. Wada, *J. Polym. Sci., Polym. Chem. Ed.* **20**, 3155 (1981).

159. S. Hayase, Y. Onishi, S. Suzuki, and M. Wada, *Polym. Preprints, Japan* **33**, 190 (1984).

160. S. Hayase, Y. Onishi, S. Suzuki, and M. Wada, *Macromol.* **18**, 1799 (1985).

161. Jap. Kokai Tokkyo Koho 61-291621.

162. N. Sasa and T. Yamaoka, *Polym. Adv. Technol.* (in press).

163. N. Sasa and T. Yamaoka, *Chem. Materials* **5**, 1434 (1993).

4

Investigation of Electron Transfer Between Hexaarylbiimidazole and a Visible Sensitizer

Yi Lin, Andong Liu, Alexander D. Trifunac, and Vadim V. Krongauz

1. Introduction

Hexaarylbiimidazole (HABI) is a widely used photopolymer photoinitiator (Structure I). Triarylimidazolyl (lophyl ≡ L) radicals recombine in the dark and form HABI reversibly.[1,2] The L radical generated in the photodecomposition can interact with chain-transfer agents either by direct hydrogen abstraction or by electron transfer to produce an initiating species capable of being added to the monomer molecule.[2,3] Because of the development of inexpensive visible and IR light sources, it is advantageous to develop materials sensitive in these spectral regions. Therefore, there has been considerable interest in producing L radicals with visible light excitation. It was observed that the polymerization processes may be initiated by a visible light excitation ($\lambda > 480$ nm) if certain visible absorption dyes such as JAW (Structure II) are added to the solution.[4–7]

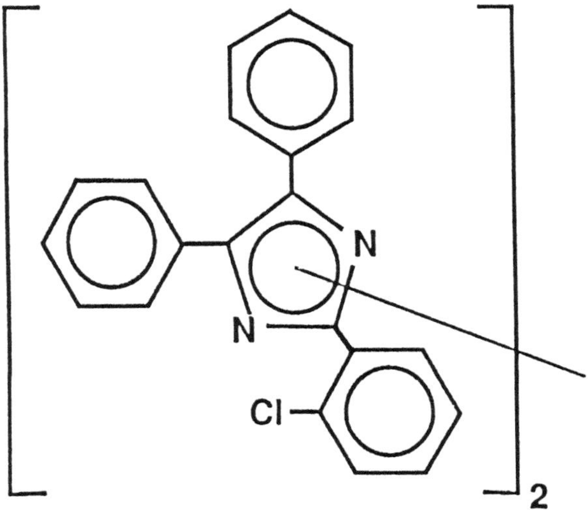

N

N

Cl

2

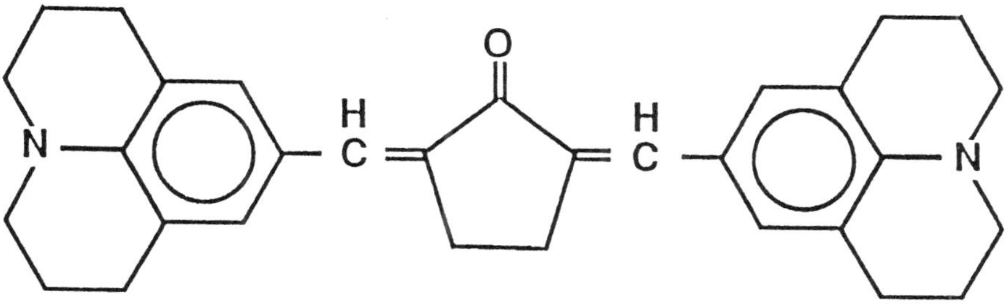

The purpose of this study was to determine the mechanism of the visible photoinitiation and unfold the dynamics of the interaction between the sensitizer and HABI molecules. The photochemical and photophysical properties of the sensitizer molecule (JAW) were studied extensively in our laboratory.[8,9] The lowest electronic excited state of the JAW molecule is characterized by the intramolecular charge-transfer (ICT) interaction between the aniline and carbonyl moieties. In this work, we studied the dependence of the relative fluorescence yield of the sensitizer on the HABI concentration in solution. The fluorescence yield was found to decrease exponentially as a function of HABI concentration. The time evolution of the fluorescence decay was also measured. When the solution contains the sensitizer molecule alone, the fluorescence decays as a single exponential function with a lifetime of 1.5 ns. The addition of the HABI molecules to the solution causes the fluorescence to decay faster. As the concentration of HABI molecules increases, the fluorescence decays faster and becomes extremely nonexponential. We discuss several dynamic models in light of our experimental observations. Among them are: short-range electron exchange model,[10] diffusion-controlled electron transfer,[11] and electron exchange in a diffusive media.[12] Only the model of electron transfer in diffusive media was consistent with all our observations. The transient species was also probed by the laser flash photolysis experiment. We observed the transient absorption of the L radicals in the mixed solution of the HABI and sensitizer system, following the optical excitation of the sensitizer molecules (JAW).

Based on the above experimental observations, the reaction mechanism shown in Eq. (1) is proposed: The L radicals are formed by dissociation of the HABI radical anion that results from the electron transfer interaction between HABI and the sensitizer.

$$\mathrm{Dye}^* + \mathrm{L}_2 \rightarrow \mathrm{Dye}^+ + \mathrm{L}_2^-$$

$$\mathrm{L}_2^- \rightarrow \mathrm{L}\cdot + \mathrm{L}^- \tag{1}$$

2. Experiments

o-Cl-HABI and the dye sensitizer JAW were synthesized at Du Pont according to the known procedures.[13] The sensitizer concentrations used in the experiments were around 10^{-4} to 10^{-5} M, and the concentrations of the *o*-Cl-HABI were varied from 0 to 0.15 M. All samples were degassed by bubbling with argon for at least 30 minutes, after which the cuvette was capped and sealed with Parafilm. Absorption spectra were taken before and after the laser photolysis to ensure that no significant photochemical degradation of the sample had occurred.

Steady-state fluorescence spectra were acquired with a Spex Fluorolog, and absorption spectra were recorded on a Perkin-Elmer Lambda 4B spectrophotometer.

In the time-dependent fluorescence experiments, a CW:Nd^{+3} YAG laser (Coherent model 76s) was actively mode locked at 76 MHz. The output of the IR pulse train (20 W) was frequently doubled to produce 1.5-W, 532-nm green light by a KTP crystal. This green light was used to synchronously pump a rhodamine-6G/DODCI cavity-dump dye laser (Coherent model 702-1). At 570 nm the output was 30 nJ/pulse with an autocorrelation trace of 1 ps. This output was used to excite the sample at a right angle. The repetition rate of the system was typically 1 MHz. The kinetic decay was obtained by a time-correlated single-photon counting apparatus. The detector was a microchannel plate photomultiplier tube (Hamamatsu model R1564) with an instrument response function of 70 ps. Fluorescence wavelength selection was achieved with an ISA HR-320 monochromator with a spectral resolution of 0.5 Å. The fluorescence data were collected by a multichannel analyzer and stored on a Macintosh IIx computer.

In the time-resolved fluorescence experiment, the fluorescence decay shape was determined to be independent of the laser intensity and the laser repetition rate. The excitation light intensity was low enough to ensure that two-photon processes did not occur, and only samples with low optical densities (~0.1) were used to ensure that reabsorption of the fluorescence had little effect on the observations. The overall time response of the system (70 ps) was determined by placing a scatterer solution in the sample holder and by measuring the pulse shape of the scattered light. This impulse response was recorded and used for convolution with the theoretical calculation to permit an accurate comparison with the data.

In the transient absorption experiment, the excitation pulse was provided by a Lambda Physik dye laser pumped by 308-nm light from a Questek excimer laser. The output of the dye laser was about 480 nm with a pulse width of 10 ns and a pulse energy of 30 mJ. The optical probe was a pulsed xenon lamp. A Tektronix (DSA 601) digitizer was used to record the transient absorption signal with a sampling rate of 1 GHz. The time resolution of the digitizer was approximately 100 ps. The data was collected and stored on a Macintosh IIx computer through a GPIB interface.

3. Results

We first examined the optical properties of the sensitizer molecules. The steady-state absorption and emission spectra of the JAW molecule in methylene chloride are shown in Fig. 4-1. The major absorption peak appears between 400 and 600 nm, which was compatible with our laser output wavelength. The fluorescence emission of the dye molecules excited at 600 nm is plotted in Fig. 4-1(b). The decay as shown in Fig. 4-4 (with Ca = 0.0) fits well with an exponential function with a time constant of 1.53 ns.

The optical absorption and emission of the *o*-Cl-HABI occurs further in the blue compared to the sensitizer molecule as shown in Fig. 4-2. The center of the *o*-Cl-HABI spectrum is approximately 270 nm, with an extinction coefficient of $3 \times 10^4\ M^{-1}\ cm^{-1}$. The absorption curve extends to 450 nm. At an excitation wavelength longer than 550 nm *o*-Cl-HABI molecules are not excited.

The interaction between the sensitizer and *o*-Cl-HABI is best monitored through the time-dependent fluorescence quenching and relative fluorescence yield measurement. In the experiment, we selectively excited the sensitizer molecules at 560 nm, as *o*-Cl-HABI molecules do not absorb at this wavelength. We first measured the sensitizer fluorescence yield at concentrations of *o*-Cl-HABI ranging from 0.0–0.166 M (Fig. 4-3). As the concentration of the *o*-Cl-HABI increases, the logarithm of the relative fluorescence yield decreases linearly with a slope equal to $0.23\ M^{-1}$.

Time-resolved fluorescence quenching was monitored at 600 nm for HABI concentrations ranging from 0.0–0.166 M (Fig. 4-4). With increasing concentrations of the *o*-Cl-HABI molecules, the sensitizer fluorescence decays faster, and the fluorescence becomes extremely nonexponential. This strongly suggests that interaction between HABI and the sensitizer occurs in the singlet state of the molecules and over a very short distance range.

To further investigate the kinetics of the transient species, a pump probe experiment was conducted in a nanosecond laser flash photolysis system. The sensitizer molecules were excited at 480 nm, and the transient absorption signal was probed by a pulsed xenon lamp from 300 nm to 700 nm. In the *o*-Cl-HABI and JAW mixture, the L radical was the only species absorbing at around 550 nm. Fig. 4-5 shows a typical absorption spectrum obtained 5 μs after the laser flash. One can clearly see an absorption peak around 550 nm, which was taken as the evidence for the presence of L radicals.

The extinction coefficient of the L radical is known to be $\epsilon_{550} = 2800\ M^{-1}$

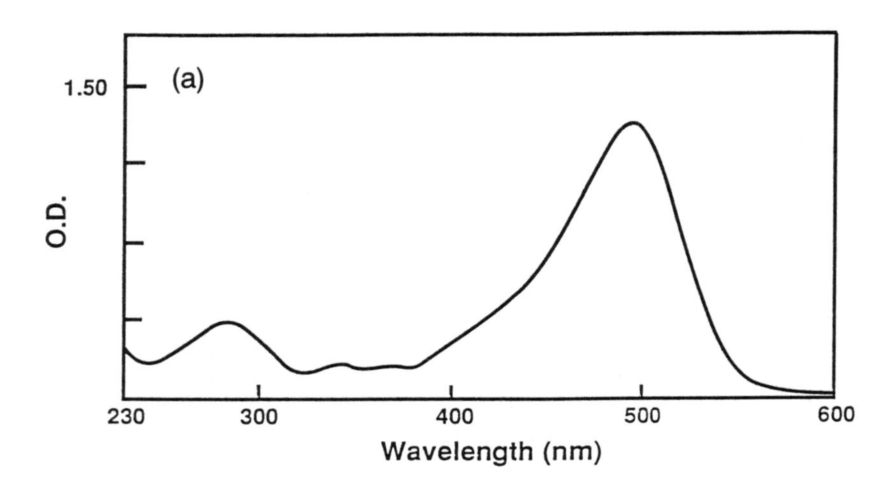

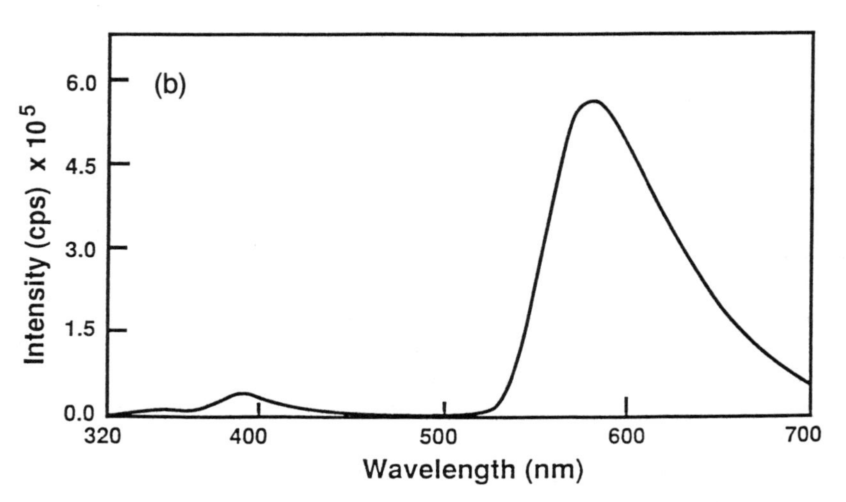

Figure 4-1. *(a) Optical absorption spectrum of JAW in dichloromethane. (b) Steady-state fluorescence spectrum of JAW in dichloromethane. The sample was excited at 600 nm.*

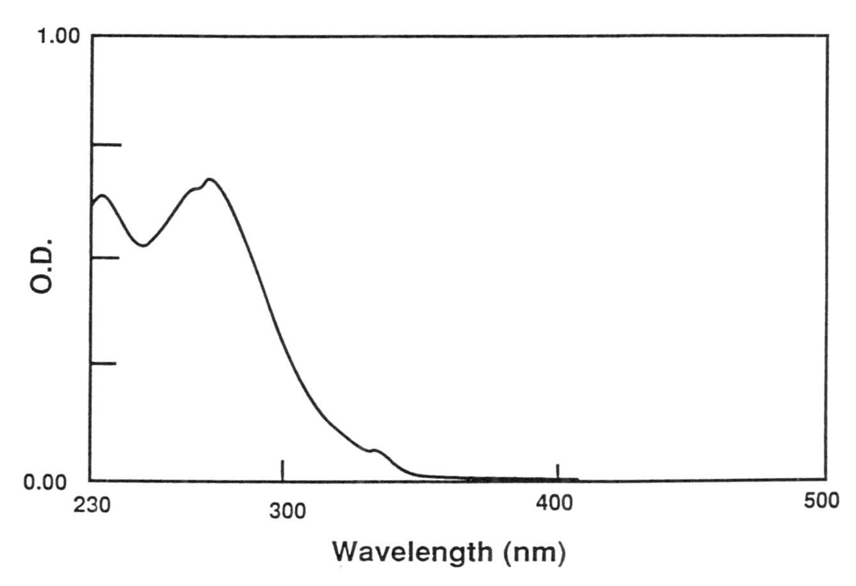

Figure 4-2. *Optical absorption spectrum of o-Cl-HABI in dichloromethane.*

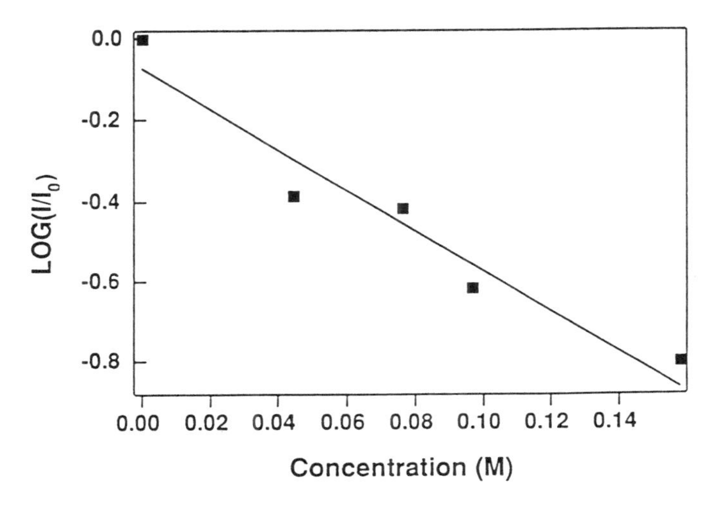

Figure 4-3. *The natural log of the relative flourescence yield of the sensitizer is plotted as a function of the* o-*Cl-HABI concentration. The points are the experimental data and the line is the linear fit. The slope is related to* R_0 *by* $R_0 = 7.346(\text{-slope})^{1/3}$. *In this case* $R_0 = 12.0$ Å.

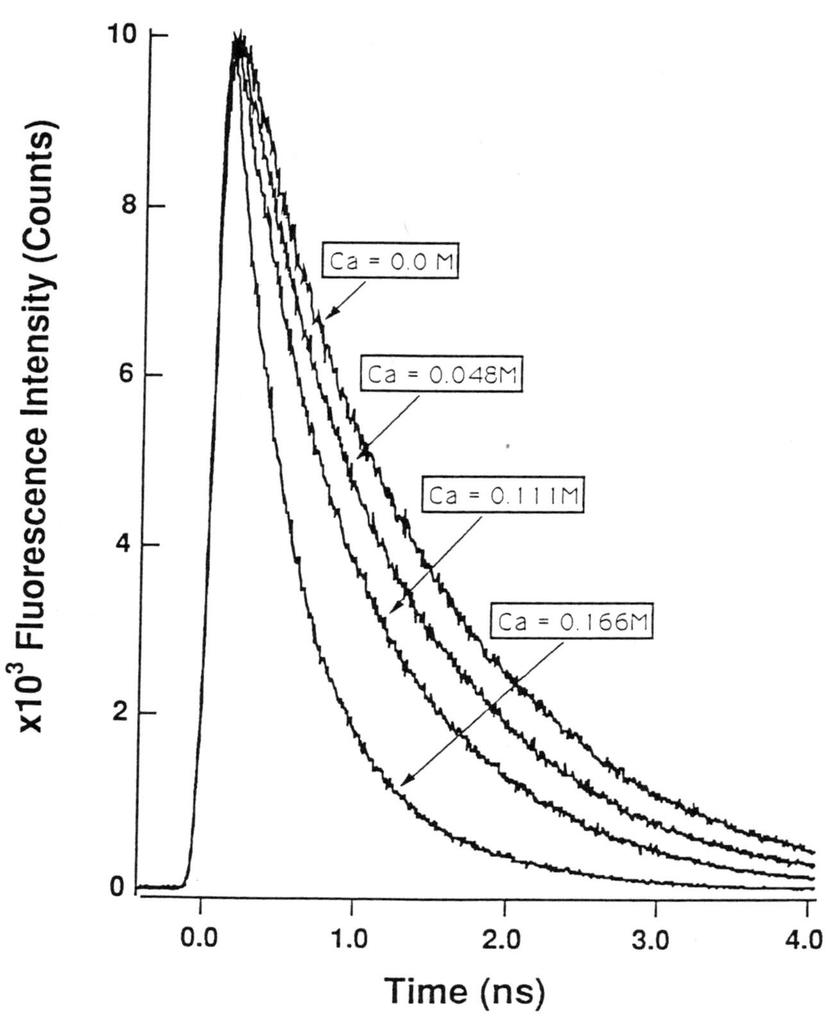

Figure 4-4. *Sensitizer time-resolved fluorescence decay at various* o-*Cl-HABI concentrations. The lifetime is obtained from* Ca = *0* M *samples and* τ *is 1.5 ns.*

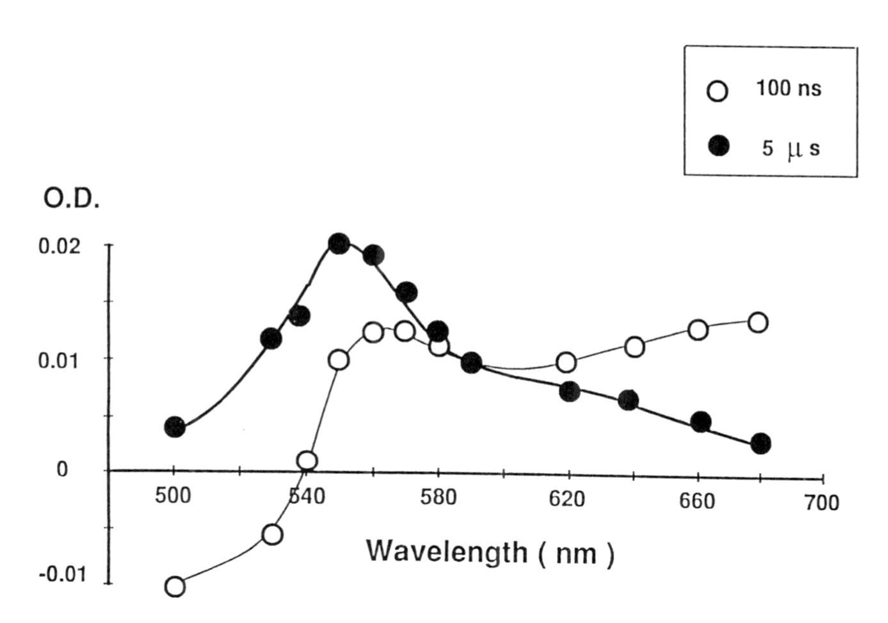

Figure 4-5. *Transient absorption spectra of the* o-*Cl-HABI and JAW mixed system observed at 100 ns and 5 μs after 480-nm laser excitation. The concentrations of* o-*Cl-HABI and JAW were 0.1 M and* 10^{-5} *M, respectively.*

cm^{-1} from previous work.[13] With this knowledge, the measured optical density (*OD*) can be readily converted to the absolute L radical concentration:

$$[\mathrm{L}] = \frac{OD(\mathrm{HABI} + \mathrm{dye}) - OD(\mathrm{HABI}) - OD(\mathrm{dye})}{\epsilon l} \tag{2}$$

where l is the optical cell size and *OD* is measured through the flash photolysis experiment.

Fig. 4-6 shows the production of the L radical vs. *o*-Cl-HABI concentration in the sample, along with the result of a set of reference experiments in which HABI solution was used without a sensitizer (JAW). In a wide range of concentrations of HABI, the presence of a small amount of sensitizer (JAW, 10^{-5} M) enhances the production of L radicals two- to threefold.

4. Discussion

The observed change of the fluorescence (Figs. 4-3 and 4-4) indicates that HABI interacts strongly with the sensitizer molecules in its singlet state. Furthermore,

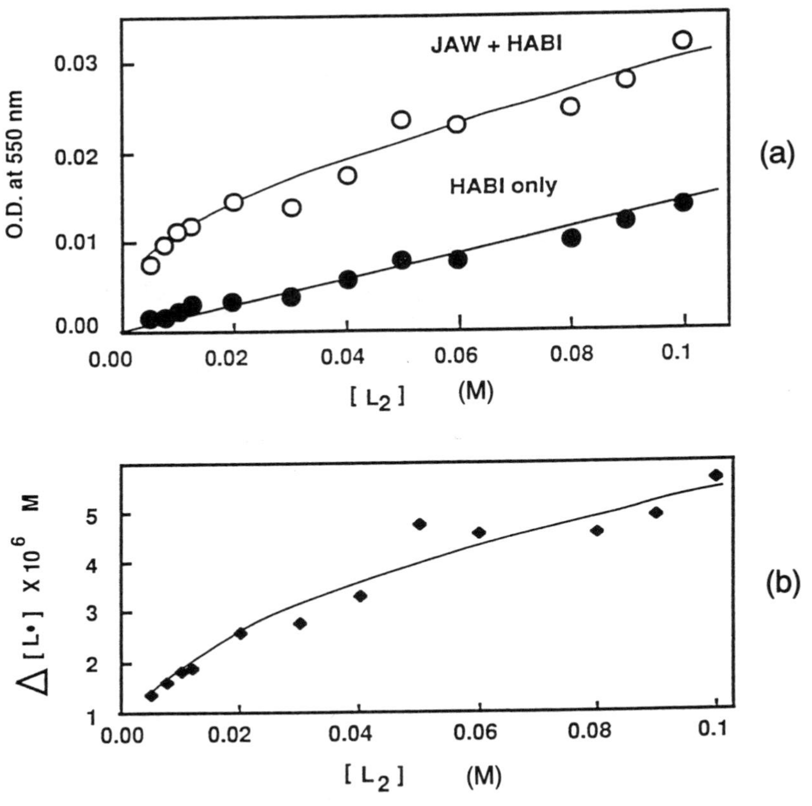

Figure 4-6. *The production of the L radical vs. the* o-*Cl-HABI concentration. Excitation was at 480 nm and probe was 550 nm. (a) shows the enhancement of the L radical yield and (b) shows the net L radical concentrations which was derived from equation (2).*

the concentration of the sensitizer (JAW) used was much lower than that of HABI so that the interaction between sensitizer and HABI molecules should dominate, and the sensitizer-sensitizer multistep energy transport or electron transfer was not likely to occur. However, the solvent molecules could participate in the interaction between HABI and the sensitizer. They could act as bridges in the electron transfer process, and their effect was included in our transfer parameters, which were obtained by comparing the theory with experimental data.

The possible interactions between the HABI and sensitizer molecules that could occur in the singlet states are singlet-singlet electronic energy transfer or electron transfer. The energetics illustrated by the UV-VIS absorption spectra

(Figs. 4-1 and 4-2) indicate that the energy transfer is not an energetically feasible process, because the HABI molecule does not absorb at the wavelength longer than the excitation wavelength (560 nm). Therefore, the observed quenching of the sensitizer fluorescence must be due to the electron transfer process.

The electron transfer rate is usually an exponential function of the donor-acceptor separation distance. It can be written as[14–16]

$$K^{T} = \frac{1}{\tau}\exp\left(\frac{R_0 - R}{a_0}\right) \tag{3}$$

where R is the donor-acceptor (center-to-center) separation, R_0 is used to parameterize the distance scale of the transfer, a_0 characterizes the fall off (or decrease) of the electronic wave function between the neutral donor and acceptor levels, and τ is the fluorescence lifetime of the donor. Conventionally, the value of R_0 was obtained by fitting the theory with the steady-state relative fluorescence yield data, whereas the other electron transfer parameter a_0 was determined through the time-dependent donor fluorescence quenching.

We first analyzed the steady-state relative fluorescence yield data according to the method of Inokuti and Hirayama (IH).[10] The data and fitting are plotted in Fig. 4-3. The natural log of the data plotted vs. HABI concentrations yields a straight line as predicted by the theory. The slope, which is related to R_0, gives $R_0 = 12.0$ Å. Because of the short-range nature of the exchange interaction, a_0 was not sensitive in determing the steady-state fluorescence yield.

The time-dependent fluorescence quenching was modeled with IH theory. The calculated curve was then convolved with the instrument response function and compared to the experimental data. Only one parameter, a_0, was adjusted to obtain the best fit. The measured fluorescence lifetime of 1.5 ns of JAW was used in the calculation. As can be seen in Fig. 4-7, even the best fit substantially deviated from the experimental data. The experimental data always decayed faster than the calculated curve, and this behavior was observed in other concentrations as well.

The observed failure of the IH theory lies in the basic assumption of the model. The IH theory assumes that donor and acceptor molecules are distributed in a fixed configuration. However, in a system such as methylene chloride, molecules are diffusively moving in the solution. The diffusion will increase the chance for the donor-acceptor contact. As a consequence, diffusion increases the electron transfer probability.

Inclusion of the donor-acceptor relative diffusive motion into the electron transfer theory has been previously carried out by many authors.[12,17,18] The solution to this problem is complicated and involves a partial differential equation for which only a numerical solution is possible. For simplicity, we used the approximate solution derived by Allinger and Blumen.[12] The excited state pop-

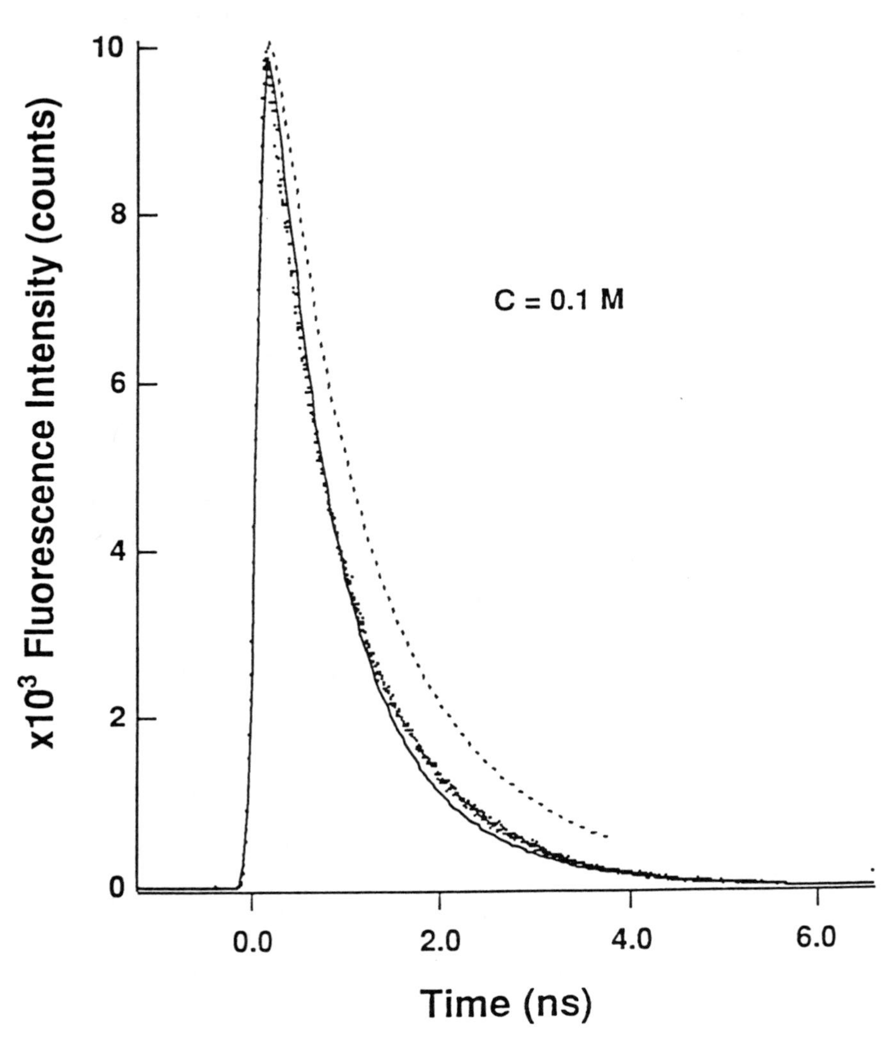

Figure 4-7. *Time-resolved fluorescence data and theoretical fits are shown for the 0.1 M HABI concentration. The dots are the experimental data and the dashed line shows the fit to IH theory with* a_0 = *1.9 Å. The solid line shows the fit to the theory incorporating diffusion with* a_0 = *0.8 Å and D* = *8.5* $\times$ *10^{-6} cm^2 s^{-1}.*

ulation decayed as

$$\Phi(C, t) = \exp(-t/\tau)\exp\left[-\frac{C}{C_0}\gamma^{-3}g_3\left(e^r\frac{t}{\tau}\right) + 3P(a_0^2 - DT)g_2\left(e^r\frac{t}{\tau}\right)\right] \quad (4)$$

where $g_m(u)$ is the generalized Inokuti-Hirayama function and is defined as[19]

$$g_m(u) = m\int_0^\infty [1 - \exp(-ue^{-y})]y^{m-1}d_y \quad (5)$$

C_0 is the critical transfer concentration ($C_0 = 1/(4/3\pi R^3{}_0)$), γ is given by R_0/a_0, and D is the relative diffusion coefficient between the donor and acceptors. $P(x)$ is a diffusion related function and is given by

$$P(x) = \sqrt{x}e^{\pi^2}/x \ \mathrm{erfc}\left(\frac{\pi}{\sqrt{x}}\right) \quad (6)$$

where erfc(z) is the complementary error function [Eq. (7.12) of Ref. 20].

The solution of Eq. (4) was evaluated numerically, and convolved with the instrument response function before comparison with the experimental data. The wavefunction overlap parameter a_0 and diffusion coefficient D were adjusted to obtain a good git. The best fit was obtained with $a_0 = 0.8$ Å and $D = 8.5 \times 10^{-6}$ cm^2 s^{-1}, and the result is shown in Figs. 4-7 and 4-8. It should be emphasized that the two parameters in the theory were obtained by fitting the data for one concentration ($C = 0.1$ M). With these parameters determined, the theory was able to fit correctly the decay curves for samples with other acceptor concentrations.

The diffusion coefficient is similar to that expected for large molecules in solvents of low viscosity. For comparison purposes, the diffusion coefficient in water is 10^{-5} cm^2 s^{-1}. The value of parameter a_0 is within the range obtained for other systems.[17-20] Therefore the assumption of HABI sensitization by JAW through electron transfer seems to fit our experimental data well. Because the energetics of the systems exclude all other paths, we can conclude that the visible light sensitization of *o*-Cl-HABI by JAW occurs by the electron transfer mechanism.

5. Summary

We have studied the electronic interaction between *o*-Cl-HABI and JAW molecules in liquid solution under visible light excitation. A time-correlated single photon counting technique was used to study the fluorescence decay of the JAW molecules. The quenching of the JAW fluorescence was attributed to electron

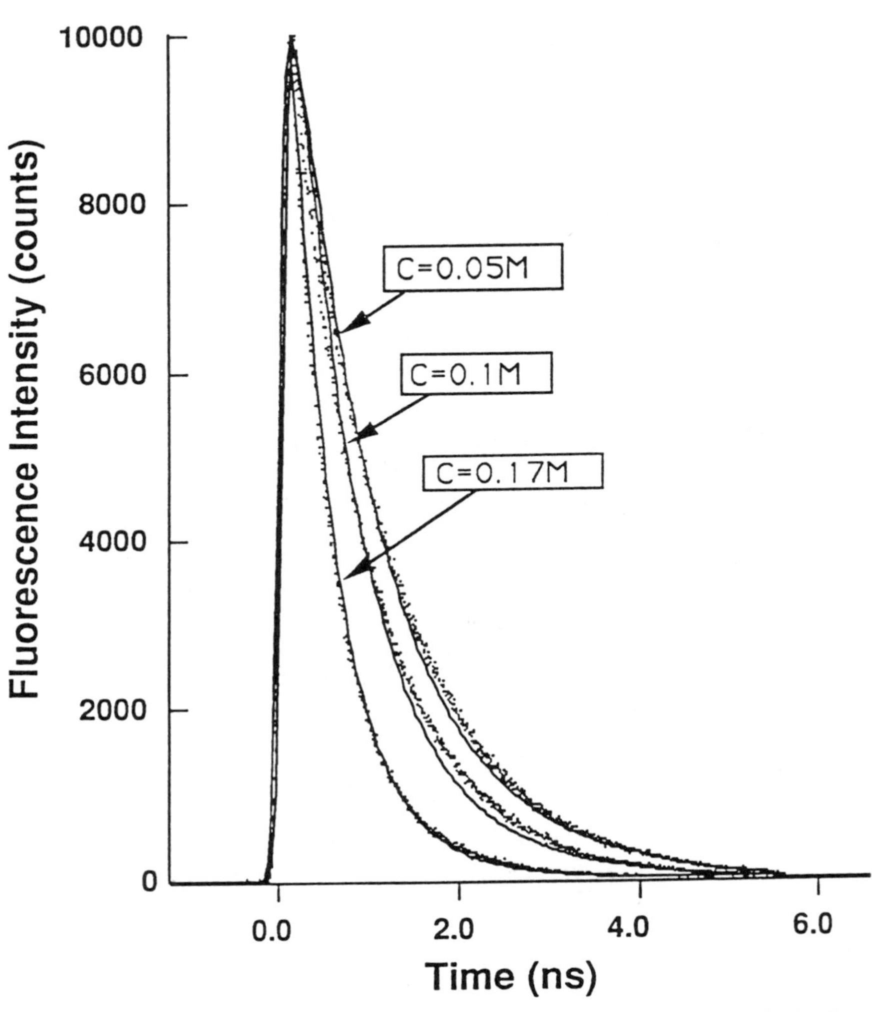

Figure 4-8. *The time-resolved fluorescence data and the theoretical curves obtained using the diffusion model are shown for three concentrations. The dots are the experimental data, and the lines are the theoretical simulation curves with* $R_0 = 12.0$ Å, $a_0 = 0.8$ Å, *and* $D = 8.5 \times 10^{-6}$ cm^2 s^{-1}.

transfer from JAW to *o*-Cl-HABI. A laser flash photolysis technique was also applied to monitor the reaction intermediates, L radicals. The growth of the yield confirmed the electron transfer interaction between HABI and the sensitizer.

Exponential HABI concentration dependent behavior was observed for the steady-state sensitizer relative fluorescence yield. From this dependence we determined the critical interaction distance for the electron transfer process to be 12.0 Å.

In determining the dynamics of the electron transfer, two models were calculated and compared with the experimental data. The Inokuti-Hirayama theory cannot be reconciled with the experimental data because it lacks the proper description of the diffusive motion of the solute molecules. Inclusion of the diffusion motion into the electron transfer model gives a good theoretical simulation of the data when electron transfer parameters have the value of R_0 = 12 Å, a_0 = 0.8 Å, and relative diffusion coefficient $D = 8.5 \times 10^{-6}$ cm^2 s^{-1}. It will be interesting to extend this study to the mechanism of similar reactions in a polymer matrix, where HABI is commercially used as the initiator for photopolymerization.

References

1. A. B. Cohen and P. Walker, *Imaging Processes and Materials*, Neblette's 8th edition, J. M. Sturge, V. Walworth, and A. Shepp, eds. (Van Nostrand Reinhold, New York, 1989), pp. 226–262.
2. G. R. Coraor, A. Maclachlan, R. H. Riem, and E. J. Urban, *J. Org. Chem.* **36**, 2272 (1971).
3. D. F. Eaton in *Topics in Current Chemistry*, Vol. 156 (Springer-Verlag, Berlin, 1990).
4. G. A. Delzenne, *Advances in Photochemistry*, **11**, 1 (1979).
5. T. E. Dueber, 1979 U.S. Patent 4,162,162.
6. T. E. Dueber, 1979 U.S. Patent 4,454,218.
7. T. E. Dueber, 1979 U.S. Patent 4,535,052.
8. A. Liu, A. D. Trifunac, and V. V. Krongauz, *J. Phys. Chem.* **96**, 207 (1992).
9. M. Barnabas, A. Liu, A. D. Trinfunac, V. V. Krongauz, and C. T. Chang, *J. Phys. Chem.* **96**, 212 (1992).
10. M. Inokuti and F. Hirayama, *J. Chem. Phys.* **43**, 1978 (1965).
11. D. F. Calef and J. M. Deutch, *Ann. Rev. Phys. Chem.* **34**, 493 (1983).
12. K. Allinger and A. Blumen, *J. Chem. Phys.* **75**, 2762 (1981).
13. L. A. Cescon, G. R. Coraor, R. Dessaur, E. F. Silversmith, and E. J. Urban, *J. Org. Chem.* **36**, 2262 (1971).
14. D. L. Dexter, *J. Chem. Phys.* **21**, 836 (1953).
15. N. R. Kestner, J. Logan, and J. Jortner, *J. Phys. Chem.* **78**, 2148 (1974).
16. R. K. Huddleston and J. R. Miller, *J. Phys. Chem.* **86**, 200 (1983).
17. R. C. Dorfman, Y. Lin, and M. D. Fayer, *J. Phys. Chem.* **94**, 8007 (1990).

18. B. Sipp and R. Voltz, *J. Chem. Phys.* **79**, 434 (1983).
19. A. Blumen, *J. Chem. Phys.* **72**, 2632 (1980).
20. M. Abramowitz and J. A. Stegun, eds., *Handbook of Mathematical Functions*, (Dover, New York, 1968).

5

Diffusion in Polymer Matrix and Anisotropic Photopolymerization

Vadim V. Krongauz

1. Introduction

Photopolymers are used in printing and electronic industries, and their use is expanding to encompass holography, data storage and processing, optical waveguides, and compact disks. A variety of new compositions is being sought to cater to new applications and novel methods of photoexposure. The photopolymers we will be discussing here consist of a plasticized polymer matrix and low molecular weight reagents dissolved in the plasticizer. In most applications (excluding three-dimensional imaging) this reactive mixture is coated from a solution onto an inert support such as a glass plate or a polymer sheet. After solvent evaporation the photopolymer film can be covered with an inert transparent cover for mechanical integrity. The manufacturing process usually does not exclude air, and oxygen dissolved in a plasticizer prior to or during the coating operation plays an active role in the imaging photopolymerization. In typical applications the illumination applied from one side, perpendicular to the surface, initiates a chemical process that records the incident light pattern as a variation of polymerized and unpolymerized regions (Fig. 5-1). We will examine some peculiarities of polymerization in the photopolymer that affect the image quality and resolution. We will also consider how the photopolymers should be formulated and exposed in order to achieve optimal photospeed and image quality.

It is common knowledge that the rate of chemical reactions in viscous media is influenced by the rates of reagent diffusion toward each other. Often radical polymerization reaction kinetics are diffusion controlled, even in nonviscous solutions.[1,2] Growing polymer chains acquire substantial size and, in a viscous

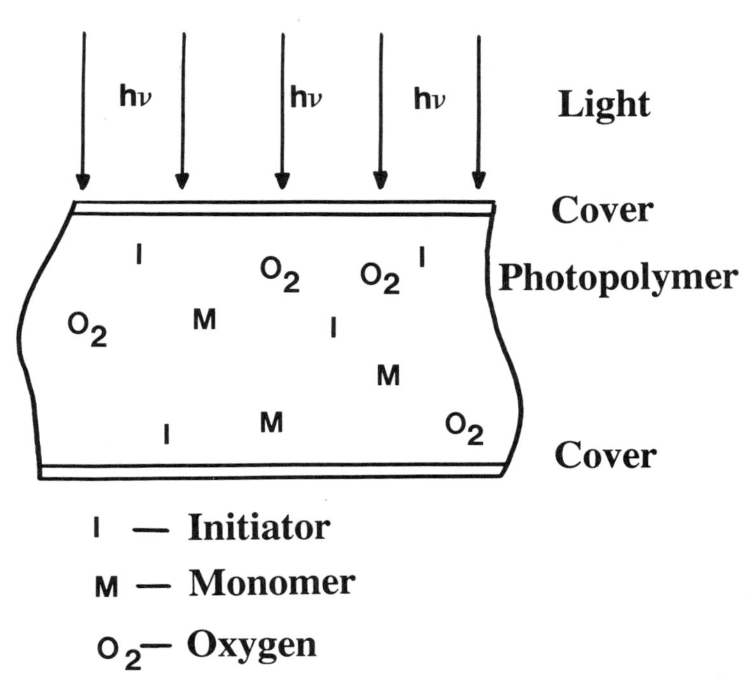

Figure 5-1. *Diagram of the unidirectionally illuminated photopolymer film containing reactive photopolymerizable formulation within the polymer binder.*

fluid or a polymer matrix, lose their mobility in the course of the polymerization reaction. However, monomeric molecules continue to diffuse more or less freely. Thus, in the course of the reaction, the ratio of chain growth and termination rates changes. A number of unique diffusion-stipulated effects can occur in a photopolymer film that would not be observed during photopolymerization in a low viscosity solution, where the polymer matrix restricting molecular mobility is absent. These effects are enhanced when a difference of several orders of magnitude develops between the diffusivities of monomeric and polymeric moieties in a plasticized matrix. For example, diffusivities of ''small'' molecules (benzene, carbazole, water, etc.) range from 10^{-6} to 10^{-10} cm^2/s, whereas values ranging from 10^{-11} to 10^{-13} cm^2/s are usually reported for polymeric species.[3–10] In addition to variations in the reagent diffusion rates, the mechanism of photopolymerization in a photopolymer film is complicated by nonuniform initiation rates resulting from the illumination from one side and the light attenuation by the photopolymer film. This spatially uneven initiation in conjunction with substantial differences between the mobility of reactants and products leads to an anisotropy in the distribution of polymer molecules forming within the photopolymer film during exposure to light. The nonuniformity in the product distribution also leads to spatial variation in properties of the imaged regions relative to the unexposed regions of the photopolymer. These variations can be successfully used for image recording. Thus, photopolymer films can be used in holographic recording, because variations of the density resulting from the diffusion of the monomer during photopolymerization from the regions of lower light intensity to the regions of higher light intensity lead to corresponding variations in the refractive index.[11–18] In proofing and printing applications of photopolymers requiring post exposure wash-off of portions of the film for image development, the inhomogeneity in polymer formation may produce undesirable effects such as the "undercut."[19] These effects of nonuniform photopolymerization become more pronounced in thicker photopolymer films such as those used in flexographic printing. In multilayered photopolymer films,[20,21] where the peel-apart method of the exposed and unexposed portions of the film is used for the final image formation, an uneven distribution of products can lead to difficulties with layer separation. It is also quite common to observe a deviation of the image on the photopolymer from the size of the initial image projected on the photopolymer film. This difference is partially stipulated by the unequal mobility of reagents in the photopolymer. In addition to nonuniformity in the distribution of products and resulting defects in image formation by wash-off or peel-apart methods, preferential diffusion of the low molecular weight reagents to the regions in which low mobility polymeric products are formed results in morphological changes in the photopolymer films. The lack of product outflow from the reaction regions during localized light exposure results in swelling (and corresponding shrinking), formation of mechanical stresses, and wrinkling of the exposed photopolymer films.[22] These structural changes in exposed films should be taken into account in optical applications such as waveguides and optical digital recording.[23,24] Until the recent interest in image recording in photopolymers using lasers, photopolymerization in photopolymer films was treated by analogy to that of photocurable coatings. However, the effects caused by uneven illumination or higher drying rates at the surface of the coating are well described and investigated, while the mechanistic investigations of photopolymer films almost completely neglected this aspects of coating behavior. The eddy currents formed in the drying solvent based coating are, to some extent, analogous to the effects caused by migration of the monomer in the photopolymer towards the exciting light, though the uniformity of product distribution is not as crucial in the coatings.[25] The migration of the photopolymer components during photoimaging resulting in anisotropy in polymer formation and other diffusion-related phenomena will be discussed in this chapter.

An experimental investigation of the inhomogeneity in the formation of transients and in the product distribution during polymerization in a matrix is a difficult task. Even modeling of the photoinduced polymerization in a matrix is a complicated process, as was demonstrated by the modeling of the local fluctuations in the density of the forming polymer by Kloosterboer and co-workers.[26,27] Their comprehensive model provided the explanation for the formation of inhomogeneities in polymer distribution; however, it did not take into account the mobility of the reactants and light intensity changes during the polymerization process. Some of the experimental data of Kloosterboer and Litjen,[28] on the other hand, indicated the importance of monomer diffusion in determining polymerization kinetics in a matrix. They reported that the rate of photopolymerization exhibited a sudden decrease when the free monomer, able to migrate in the polymer matrix, was exhausted. The shorter range (relative to molecular diffusion) of electron or hydrogen transfer radical generation could account for the decrease in the crosslinking rate reported by Kloosterboer. Monte Carlo analysis of photopolymerization directly pertaining to the processes in photopolymer films was presented recently by Gupta.[29] In this work the inhomogeneous gelation and the dependence of the size of forming polymer molecules on distance from the light source were computed. Although details of the mechanism of photopolymerization initiation and chain propagation were omitted, these qualitative results provide a convincing argument for the importance of concentration gradients resulting from activating light attenuation and monomer diffusion. In this chapter we will concentrate on the classic kinetic modeling of the photopolymerization mechanism.

The classic kinetic analysis of photopolymerization within a thin liquid layer, with consideration of light attenuation by the photopolymerizable mixture, was recently presented.[30] The model of photopolymerization in thin layers of liquid can describe the three-dimensional or "solid imaging," where the part is "grown" layer by layer from a liquid photopolymerizable solution.[31] The initial stage of photopolymerization of liquid photoresists applied to the mask as a thin layer of liquid[32] can also be described using kinetic schemes derived for polymerization in liquid layers. However, it is not applicable to the description of the processes in photopolymer films. The effect of unidirectional illumination on photoresist imaging was considered in detail by Dill and co-workers.[33] The kinetics of photopolymerization induced by a strongly absorbed light were considered. The consumption of an inhibitor as a function of the distance from the illuminated surface of the resist and the exposure energy was evaluated. These results can be successfully applied to the immobile initiator consumption in photopolymer films as well. The author also considered oxygen effects on resist imaging and gradients in oxygen consumption due to an exponential dependence of the initiating light intensity on the distance from the illuminated surface. These studies emphasize the particular aspects of photopolymerization not observed in nonviscous solutions. However, to describe the photopolymerization kinetics and the mechanism in photopolymer films containing mobile reagents, both the initiating light attenuation by the film and differences in the mobilities of reagents must be considered. A comprehensive analysis of such a system was absent prior to the work described below.

In numerous studies of photopolymerization mechanisms, the photopolymerization is conducted in a dilute solution where the polymeric molecules have a mobility similar to that of small molecular species.[2,34,35] In a solvent, reagents and proucts mix at a substantial rate and, therefore, an exponential drop in the light intensity with distance from the illuminated wall of the reaction vessel does not have a significant effect on the mechanism of the process. The diffusional effects in solution polymerization were considered mostly in regard to the chain termination processes.[2] A lack of appreciation of the difference between photoinduced polymerization in solutions and solid systems was, often prevalent in research on photopolymers.[36] Even when investigators explicitly based their assignment of products on diffusion-controlled photopolymerization in a polymer matrix (holographic photopolymers, flexographic printing, etc.), they failed to address the consequences of a nonuniform illumination and the difference in mobilities of reactive species and products.[37] Despite the existence of experimental data pointing to the photopolymer performance dependence on factors influencing diffusion, such as the free volume in the matrix[38] and the viscosity of the plasticizer,[39] diffusion was not considered in many recent publications describing photopolymerization mechanism in a polymer matrix.[40] The neglect of this subject was not universal, and the swelling stipulated by the unidirectional monomer diffusion and absence of polymer outflow has recently been modeled and applied in the production of optical elements.[23,41]

Time-modulated illumination was proposed for the recording of holographic images in photopolymers[42] in order to fully utilize the effect of monomer diffusion towards the illuminated regions and, thus, maximize the polymer yield and create an optimum gradient in the refractive index. The diffusion of monomer in photopolymer systems was not directly observed or proven. In the investigations acknowledging diffusion control of polymerization kinetics and unidirectional diffusion of the monomer towards the activating light,[43] the proof that diffusion actually occurs was absent. During the hologram recording in a photopolymer the refractive index change was detected, and the diffusion mechanism was postulated.[36,43] The increase of the refractive index change when recording a hologram at higher temperatures was used as a proof of diffusion, in the absence of the kinetic analysis of the data.[36] However, the rates of reactions are known[1] to increase with the temperature, and the temperature dependence of the polymerization rate and the polymer yield is not proof of the monomer diffusion taking place, but rather an indication of the possibility of such an event.

After surveying the literature, we concluded that in spite of more than a decade of claims regarding the diffusional nature of the holographic recording, no serious attempt has been made to investigate the monomer diffusion control of photopolymerization in a matrix, or to measure the rate-controlling diffusivities of the reacting species, or to estimate the direction and the magnitude of the mass transport during the photoimaging.

Several years ago we initiated the study of photoimaging mechanism and kinetics, to examine experimentally diffusion-controlled photopolymerization, and devised experimental and numerical models showing the practical consequences of molecular migration in photopolymers during imaging. We have also examined why the diffusion restrictions of photodecomposition make one particular initiator, hexaarylbiimidazole, an initiator of choice in photopolymer systems. This concerted effort to analyze the diffusional effects in photopolymer films is reviewed in the following sections.

2. Fluorescent Monomer in High Optical Density Photopolymer Film

2.1. Photopolymer

The most direct way to investigate diffusion-controlled photopolymerization in the photopolymer film is to analyze the material that was designed to maximize the differences between the illuminated and the unexposed regions produced by the alleged monomer diffusion towards light. Formulations similar to those used for holographic image recording and optical waveguide production and optimized for the highest change of refractive index upon exposure to light[25,36,37,44–47] were ideal materials to study diffusion-controlled photopolymerization.

These holographic photopolymers, designed at E. I. Du Pont de Nemours & Co.[11,36,37,44–47] contained the fluorescent monomer, *N*-vinylcarbazole (NVC), the initiator, *o*-chlorohexaaryl-biimidazole (HABI), the chain transfer agent, 2-mercaptobenzoxazole, the visible light sensitizers, the photoreactive plasticizer-diluent, phenoximethylacrylate, and, in some instances, a less reactive glycol plasticizer. Poly(vinyl acetate) (PVA), cellulose acetate butyrate (CAB), and poly(vinyl butyrate) (PVB) were typically selected for a matrix material (binder) compatible with the photopolymerizable mixture. The formulation was made in a methylene dichloride solution and further coated on a polymer or glass substrate. Knife and spin coatings of methylene dichloride solution of photopolymer formulation were used to form 0.5- to 50-μm thick films on an optical quartz slide. After deposition of the coating formulation, the coated substrate was dried (in experiments described below under ambient conditions) to ensure the absence of the methylene dichloride solvent and the formation of a consistent and reproducible image after exposure to light. Similar photopolymerizable compositions were patented by several companies including Polaroid, BASF, Japan Synthetic Rubber Company, Fuji, and others.

Industrial research and development practices and traditions demand that an investigation of material performance be conducted using only commercial formulations. Accordingly, the techniques discussed below were developed to have low sensitivity to impurities and to allow photopolymerization kinetics monitoring in real time using easily accessible instrumentation. These experimental and computational methods are applicable to numerous photopolymer systems, and the obtained results and conclusions are rather general.

2.2. Patterned Illumination

Mobility in thin films could be monitored by the fluorescence photobleaching recovery kinetics[48–52] and the induced transient (holographic) grating relaxation[53–59] methods. These methods monitor the destruction of the tracer in the illuminated regions and the diffusion-controlled redistribution of the remaining tracer throughout the polymer film. These techniques were successfully used in nonreactive systems. However, the application of these techniques to monitor the diffusion-controlled kinetics in reactive photopolymer films has never been attempted.

We needed nondestructive monitoring methods, which do not alter the structure and the behavior of systems under investigation and which permit real-time observation of the process.

The DuPont holographic photopolymer that was investigated contained a fluorescent monomer, *N*-vinyl carbazole (NVC).[11,36,37,44–47] The corresponding polymer, poly(*N*-vinyl carbazole) (PVCA), has a substantial yield of fluorescence (Fig. 5-2). We used the modified patterned photobleaching technique based on monitoring fluorescence emission. Several holographic studies[11,36,37,43] claimed that local change of the refractive index during the imaging is caused by the migration of the monomer from the shadow to the light followed by conversion to the immobile polymer during the patterned imaging. We considered that if both polymer and monomer are fluorescent, migration from the shadow to the light region should lead to an increase in flourescence intensity during patterned photopolymerization induced by the illumination through the grating. The fluorescence must be monitored at the isoemissive point, and the excitation must be applied at the isobestic point of monomer, NVC, and polymer, PVCA. We hoped to use the pattern excitation method to monitor the diffusion of the monomer. The pattern photoexcitation method worked and we found, as described below, a simplification of it, as well.

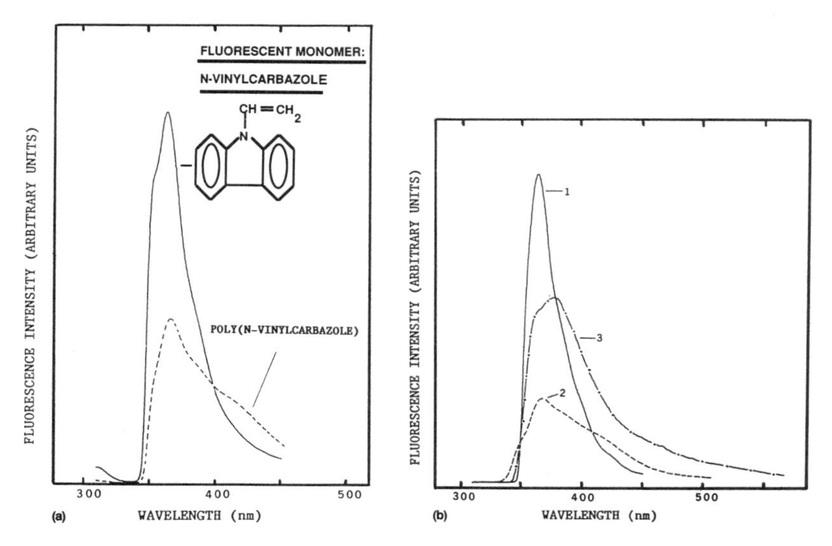

Figure 5-2. *(a) Emission spectra of the equimolar solutions of NVC and PVCA in methylene dichloride; (b) emission spectra of the poly(vinylacetate) films containing equimolar amounts of NVC (1) and PVCA (2) and the spectrum of the exposed photopolymer film (3).*

2.3. Apparatus and Experimental Procedure

Time dependence of emission intensity in the photopolymer film containing flourescent photopolymerizable monomer *N*-vinyl carbazole was monitored using a conventional experimental set-up (Fig. 5-3).[7] A 150-W xenon arc lamp (Oriel Co.) was used for the photopolymerization initiation and the excitation of fluorescence emission. The majority of photoinitiators can also initiate polymerization upon heating. Thermal initiation can be substantially reduced by placing a water filter in the path of the excitation light. After passing through the water filter, interference filter, monochromator, focusing quartz lenses, and 0.1-ms mechanical shutter (Vincent Assoc.), the excitation light is focused by quartz lenses onto a plate coated with the photopolymer film.

In some of the experiments the photopolymer was sandwiched between two quartz plates to ensure that both photopolymer film surfaces were optically equivalent. In the pattern illumination experiments, the photopolymer was coated directly on a photomask with periods 0.5, 5, and 50 μm (electron beam etched grating in chromium coating). In the initial experiments using the grating and in the experiments imitating the conditions of the experiments using a grating mask, the coating was located on the side of the slide directed away from the

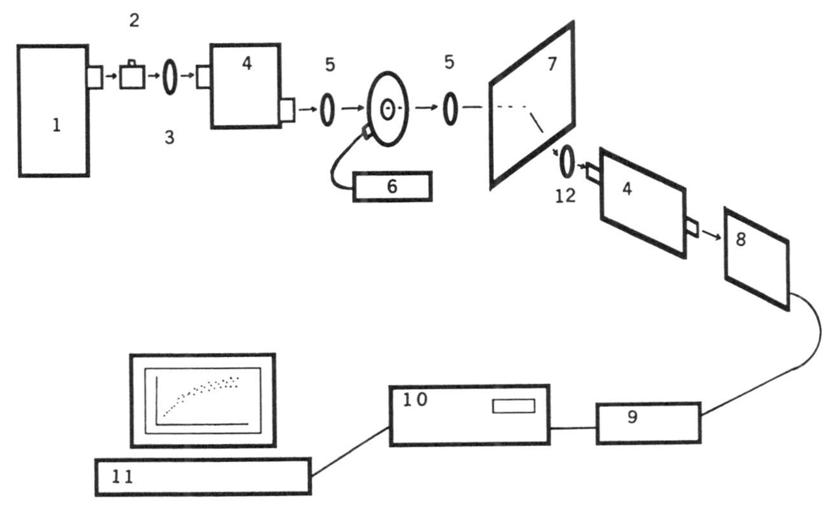

Figure 5-3. *A diagram of the experimental set-up: 1-xenon arc light source; 2-water filter; 3-bandpass interference filter; 4-monochromator; 5-quartz lens; 6-electromechanical shutter; 7-quartz slide coated with the photopolymer film; 8-photomultiplier tube, housing, and shutter; 9-preamplifier; 10-photon counter 11-computer; 12-high pass optical filter.*

xenon arc lamp. However, in all of the experiments in the controlled atmosphere, the photopolymer-coated side of the quartz plate was illuminated. To control the atmosphere, the quartz plate coated with the photopolymer film was placed in a stainless steel vacuum chamber made from a six-way pipe connector (Varian Associates). It was equipped with inlet valves and three sapphire windows (Varian) (two windows along the path of illuminating light, and one at the right angle to it) to illuminate the sample and to monitor the emission and the absorption of light by the polymer film. A vacuum system with a mercury diffusion pump was used to evacuate the chamber. To ensure that the changes in emission intensity do not arise from the conversion of the monomer into the polymer, the excitation has to be conducted at the isobestic point of fluorescent monomer and the resulting polymer and the detection must be carried out at their isoemissive point. For *N*-vinyl carbazole and poly(*N*)-vinyl carbazole), the isobestic point lies at 295 nm; fluorescence was monitored near the isoemissive point of NVC and PVCA, at 400 nm (Fig. 5-2). To reduce the scattering light interference the detection of fluorescence was conducted in a transmission mode at a 90° angle to the incident light and the coated plate positioned at 45° relative to the incident light direction. The emitted light was filtered and monochromated. A head-on

2-inch photomultiplier tube was used without a cryogenic jacket system. A Stanford Research Systems preamplifier and a photon counter interfaced with the personal computer provided the signal recording and processing. The data was further transferred for processing to a mainframe computer system.

2.4. Fluorescence Detected Photopolymerization Kinetics

The migration of the fluorescent monomer into the illuminated regions was monitored. Illumination of the film through the photomask with the spacing from 50 to 0.5 μm initiated photopolymerization with the consequent conversion of the mobile fluorescent monomer in the illuminated regions into the immobile fluorescent PVCA. Because the monomer was consumed in the illuminated regions of the film, the monomer concentration gradient developed between the shadowed and the illuminated regions. We observed an increase in fluorescence emitted by the photopolymer film illuminated through the mask. This increase in emission intensity presented direct proof of the monomer migration toward the illuminated regions of the film. This was the first recorded direct observation of material movement toward the light-irradiated regions in photopolymers.

The patterned illumination method yields itself to numerical modeling. When patterns with different spacings are used, the modeling can be simplified. The results of our model computations differed substantially from the experimental data. The experimentally measured rise of the fluorescence was faster and higher than could be expected when only lateral diffusion from the shadow to the light is considered. When the photopolymer film was illuminated by UV light without the photomask, the emission increase was still observed, although it was not as substantial as the relative increase observed in the presence of the mask. The observation of this increase in the fluorescence yield led to the direct photopolymerization kinetics monitoring, described below. However, because of the interest in imaging where photopolymerization is taking place under the patterned illumination of the film, we will return to the effects associated with pattern illumination later in this chapter.

2.5. Direct Monitoring of Diffusion-Controlled Photopolymerization in Films

Let us consider what can cause the deviation of the experimentally observed kinetics of fluorescence emission increase during the illumination through the grating compared to the computed effect. In most applications the photopolymer films are exposed to UV light (below 350 nm). We used 295-nm excitation. The DuPont holographic photopolymer that we investigated has been designed to be activated by an argon ion laser at 488 nm. Holography requires visible light for image recording, as the images must be visible to the human eye. Long wavelength initiation is more uniform because of the low optical density of the films in this spectral region. The efficiency of activation at this wavelength is low because the HABI initiator does not absorb significantly in this region and sensitization by electron transfer from the visible sensitizer dyes is also low.[60,61] To improve the initiation rate the HABI initiator is added in high quantities, up to 4%.[11,44,45] Consequently, optical density of the photopolymer at 295 nm is high; for 25-μm film, for example, it exceeds 14 at 295 nm.[7] The 295-nm excitation light is completely absorbed within a fraction of a micron from the illuminated surface of the photopolymer film. As a result, the monomer is consumed more quickly near the exposed surface of the film. This creates a monomer concentration gradient between the surface and the bulk of the film. The monomer can diffuse within the plasticized matrix; it moves toward the illuminated surface. The produced polymer, PVCA, has several orders of magnitude lower diffusivity in the plasticized matrix, resulting in immobilization and accumulation of fluorescent carbazyl groups near the illuminated surface. Accumulation of carbazyl groups near the illuminated surface increases the fraction of light absorbed by the fluorescent carbazyl groups, and thereby the intensity of emitted fluorescence increases as well. The photopolymer usually contains some fluorescent quenchers dispersed throughout the film. There is no mechanism for the quencher migration during polymerization. As the carbazyl groups in the NVC monomer migrate during photopolymerization the luminophore and quencher separation will contribute to the increase in the fluorescence intensity. The extent of the luminophore-quencher separation effect is also proportional to the concentration change of the illuminated carbazyl. Both factors, the stronger light absorption at the surface and the decrease in the relative concentration of the quencher per carbazyl group, can be treated as a total effect indicating the extent of the diffusion from the bulk of the film. When the photopolymer is illuminated with a strongly absorbed light through the grating, the lateral diffusion parallel to the film surface contributes to the monomer migration. However, the major contribution comes from the diffusion from the depth of the film towards the illuminated surface. That is why the pattern illumination produced a higher increase in fluorescence intensity than was expected, and the kinetics of this increase deviated from those one would expect in the case of purely lateral diffusion.

This conclusion was verified by experimentation with the photopolymer film illuminated without a mask, where the possibility of lateral diffusion from the regions of the film that were not illuminated was excluded. To ensure complete film illumination, a small square (5 × 5 mm) was outlined on the film surface, and all of the film outside the square boundaries was removed. Uniform 295-nm light illumination of the photopolymer film containing fluorescent monomer (*N*-vinyl carbazole) led, as was expected, to an increase in the intensity of film fluorescence emission (Fig. 5-4). This was a direct demonstration of the monomer migration from the depth of the high optical density film to its illuminated surface during photoinitiated polymerization.

When the 40% solution of the photopolymer formulation in methylene chloride was squeezed between two quartz slides separated by a 25-μm Mylar gasket, photopolymerization occurred without any fluorescence intensity increase. The fluorescence intensity remained constant, because the forming polymer and the reactants were mixed throughout the solution volume, and the concentration of the illuminated carbazyl groups remained constant during the exposure. Several other phenomena that prove the monomer migration mechanism of the fluorescence increase have been enumerated.[7] However, the most direct demonstration of the extent of the monomer depletion was obtained when the photopolymer film was flipped over after the initial exposure. A small rectangle of the film was squeezed between two quartz slides and illuminated until the emission intensity reached the limiting value. Then, it was flipped and exposed from the opposite side of the film. It was observed that the exposure of the opposite (flip) side of the same area of the film to the excitation light yields

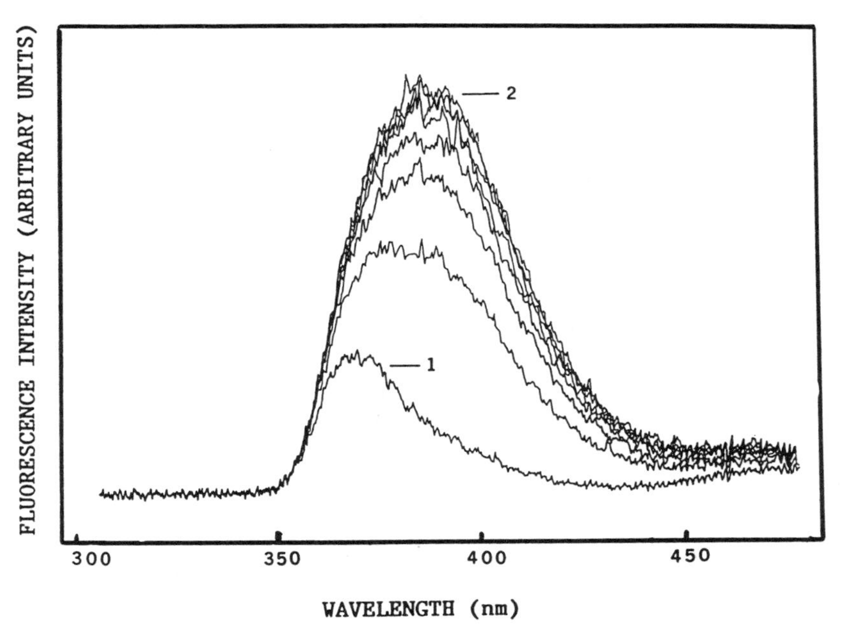

Figure 5-4. *Dependence of the photopolymer emission spectrum on the time of exposure: 1-at the beginning of the exposure to 295 nm UV light; 2-close to the end of the photopolymerization reaction (600 seconds later).*

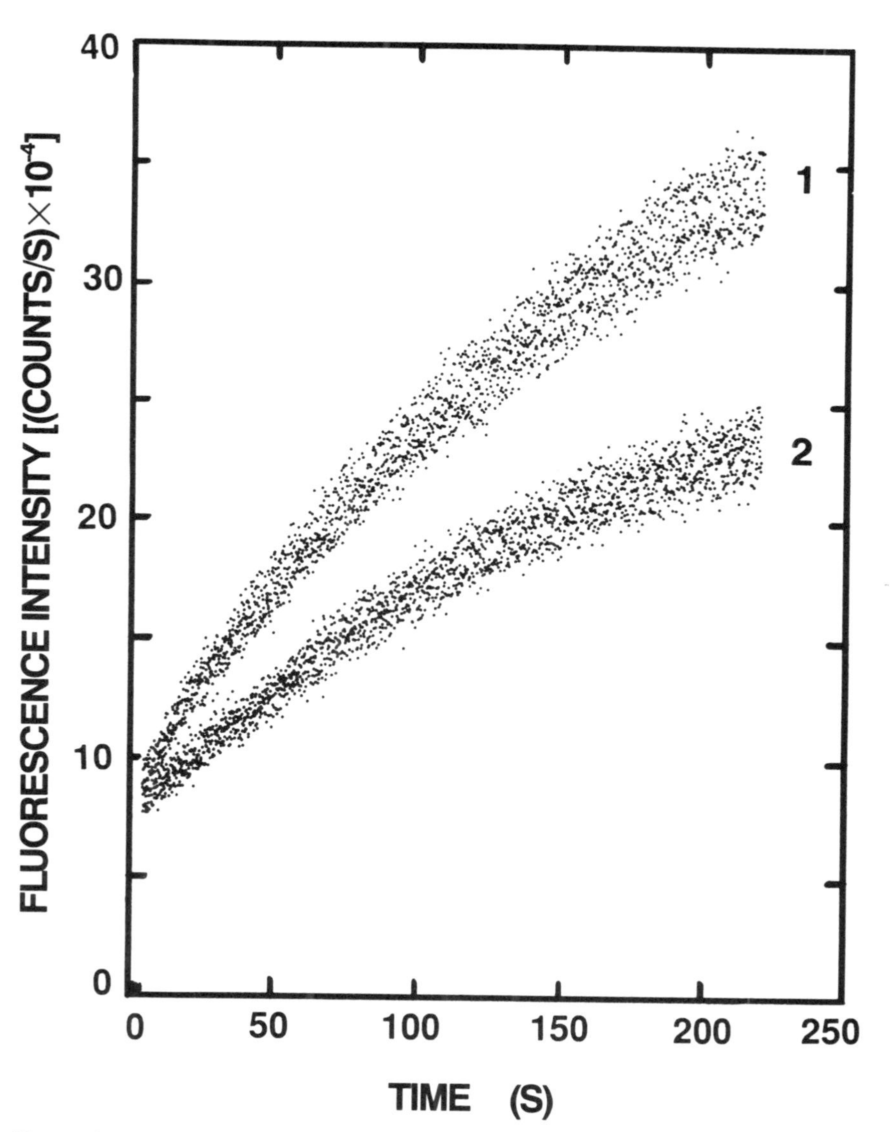

Figure 5-5. *Time dependence of the fluorescence intensity during the photopolymerization in 24.5-μm film. 1-front surface exposure, 2-flip surface exposure.*

lower initial fluorescence intensity, a slower increase in emission, and a lower ratio of final to initial fluorescence (Fig. 5-5). These results provided direct proof that the diffusion of the monomer to the side of the photopolymer film exposed to activating light had occurred, thereby depleting the overall amount of the free unreacted monomer in the film. The yield of the polymer, as detected by the fluorescence intensity increase, was higher for the thicker films having larger amounts of monomer free to migrate. When the film was thin so that its optical density allowed more uniform illumination, all the monomer polymerized after the first exposure. The flip side of the film yielded a fluorescence intensity independent of time (Fig. 5-6). Some of the monomer, even in thin film, had migrated to the surface illuminated first; and quencher-fluorophore ratio also increased, so that the final fluorescence of the initially illuminated film was higher than that of the flip side of the thin film (Fig. 5-6).

The data presented above have demonstrated that the monomer migrates toward the illuminated regions of the film during photopolymerization and that photopolymerization is diffusion controlled.[7] This detection of the monomer migration towards the illuminated surface of the photopolymer film by monitoring the rise in film fluorescence intensity, in conjunction with the simple model used to deduce the diffusion coefficients of a monomer in this photopolymer film, can be used to optimize the illumination regime for hologram recording.[7,42] We also observed an improvement of polymer yield (as detected by fluorescence intensity increase) when the illumination was conducted in a stepwise fashion with the interruption periods allowing more time for the monomer to migrate towards the light.

2.6. Monomer Diffusivity

The monomer diffusion coefficients must be measured in order to understand and quantify the diffusion-controlled kinetics of photopolymerization in the photopolymer film. As with any kinetic parameter, the diffusion coefficient depends on the model of the process selected for its computation.[1–3,62] Selection of the model depends on the computational ability of the researchers and on the underlying chemistry of the process. The process of the monomer diffusion towards the illuminated surface can be described by the analogy with the diffusing substance evaporating from the surface of the film, because the monomer conversion during the photopolymerization reaction occurs in a very thin layer of the film. This is especially true for the early stages of photopolymerization. For the fluorescent monomer converting into an immobile fluorescent polymer, the fluorescence intensity increase is proportional to the amount of monomer removed from the bulk of the film by photopolymerization. The mathematical expression describing the loss of the diffusing substance by surface evaporation (in the

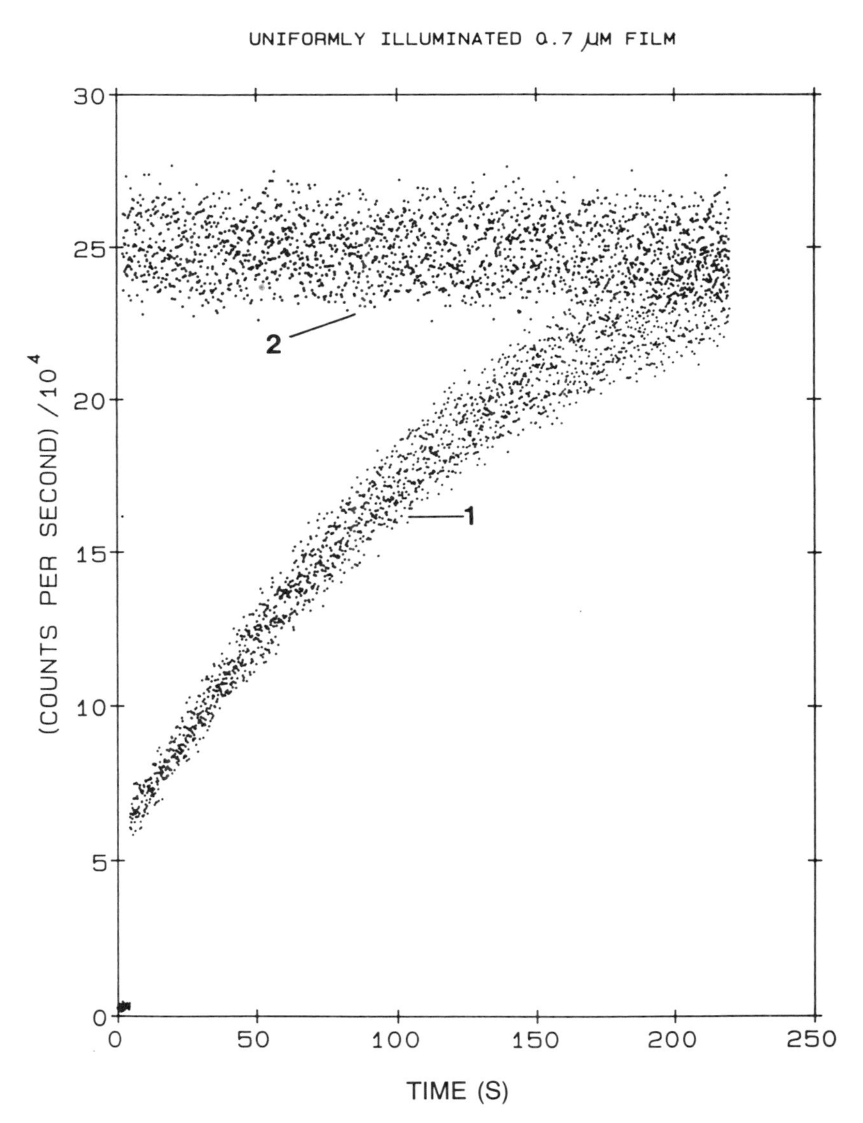

Figure 5-6. *Time dependence of the fluorescence intensity during the photopolymerization in 0.7-μm film. 1-front surface exposure, 2-flip surface exposure.*

present case surface polymerization) was presented by Crank[3,62] (Eq. (1)).

$$[M(t) - M(0)]/[M(\infty) - M(0)] = 1 - \sum_{n=1}^{\infty} \left\{ \left[2L^2 \exp\left(\frac{-\beta_n^2 D t}{x^2}\right) \right] / [\beta_n^2 (\beta_n^2 + L^2 + L)] \right\} \quad (1)$$

Here β_n are the positive roots of $\beta \tan\beta = L$; $L = x\,\alpha/D$; x is the film thickness in cm; D is the diffusion coefficient (cm^2/s) used as a parameter in calculations, and α is a fraction of the molecules that can evaporate. It was known from the current fluorescent measurements (comparison of the front and "flip" side emission)[7] and from the chromatographic data, that no more than 10% of the monomer is converted to the polymer by the initial exposure of the 24.5-μm-thick films. Thus, $\alpha = 0.1$. The diffusion coefficient for the plasticized film measured for other molecules lies in the range 10^{-7} to $10^{-10} cm^2/s$.[3,4,8,9] Therefore, $L > 100$ was expected, and β_n were chosen correspondingly using the tabulated data[62] (Fig. 5-7). The curves computed using the monomer diffusion coefficient as a parameter in Eq. (1) were compared with experimental data on kinetics of the fluorescence intensity increase. The diffusion coefficient giving the best overall fit was selected. Thus, the *N*-vinyl carbazole diffusion coefficient in plasticized cellulose acetate butyrate (CAB) matrix was determined to be: $D_{CAB} = 6 \times 10^{-9}$ cm^2/s. This represents a real-time, real-composition, no-tracer determination of the rate of the monomer diffusion during photopolymerization. The diffusion coefficient is within the range of diffusion coefficients determined for other aromatic molecules in a plasticized polymer matrix.[3,4,8,9]

3. Kinetic Models of Anisotropic Photopolymerization in Films

3.1. Why Compute?

Steady-state approximation for the radical concentrations[1-3] does not necessarily apply in matrix reactions. Because of the low mobility of growing chains, radical recombination is also not as significant a factor in chain termination as in solution. The analytic solution of the kinetic equations becomes difficult in the absence of a uniform reaction. To describe the photopolymerization kinetics in a polymer matrix under a unidirectional illumination, the equations of mass balance, rate of polymerization, etc., must be solved for each location within the photopolymer.

Light attenuation is an important limiting factor in the formation of photocured adhesives, paints, and coatings.[33,34,63-65] Even in a well-stirred tank polymerization, the attenuation of light leads to complications in the photopolymerization chemistry.[66,67] A combined effect on the photopolymerization of the

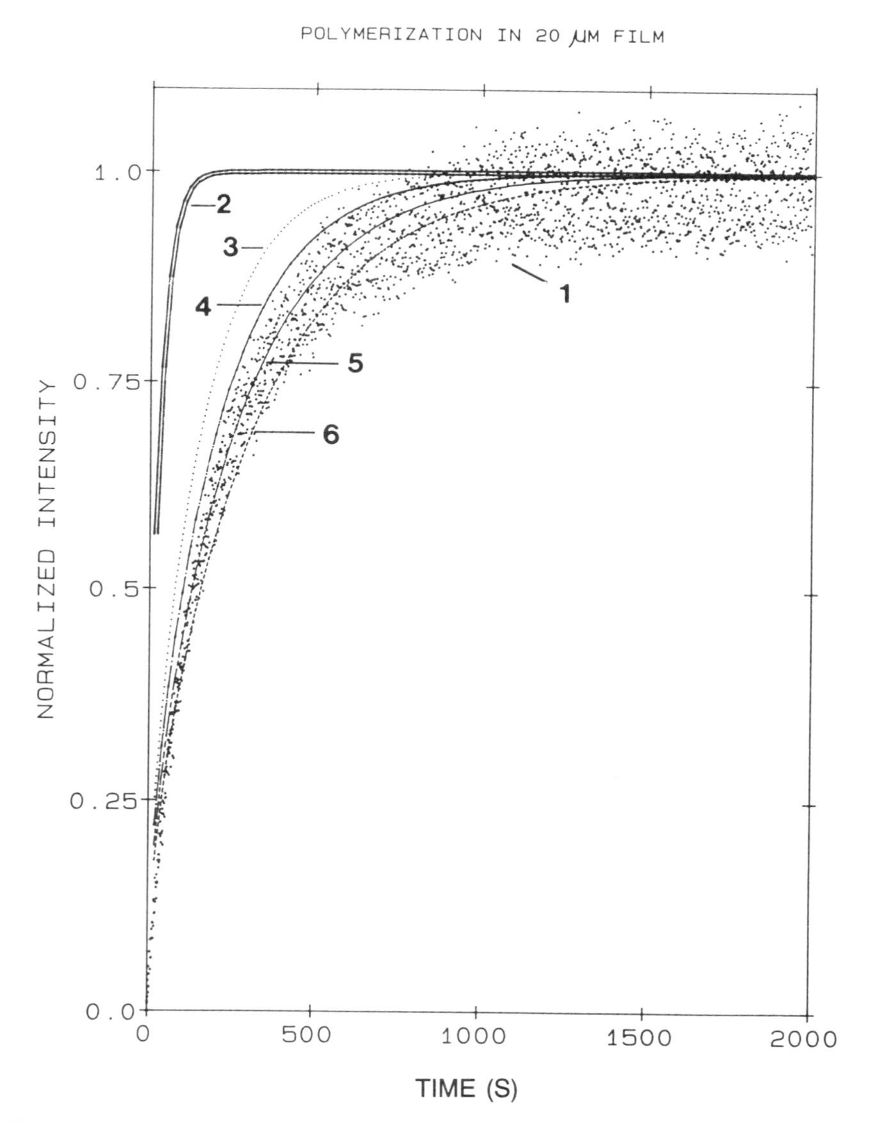

Figure 5-7. *Deduction of the monomer diffusivity from the fluorescence detected photopolymerization kinetics: 1-experimental data for 20-μm film normalized to 1; 2-computations using diffusion coefficient* 5×10^{-8} cm^2/s; *3-*1×10^{-8} cm^2/s; *4-*7×10^{-9} cm^2/s; *5-*6×10^{-9} cm^2/s; *6-*5×10^{-9} cm^2/s.

unidirectional illumination, the limited and varying mobilities of species in the polymer matrix, the effects caused by oxygen, the mobile and immobile initiators, and the migration of the reactive species towards the illuminated regions of the film was considered in our modeling.[68–72] It is prohibitively difficult to evaluate the influence of the mentioned effects on the kinetics of photopolymerization, using an analytic solution of kinetic equations. Below we present the numerical description of the photopolymerization in a unidirectionally illuminated polymer matrix with the consideration of limited mobilities of reactants, transients, and products. Only a few of the assumptions made here are stipulated by the particular composition of the DuPont holographic photopolymers used in the experiments, and the results of the modeling are rather general and apply to many photopolymer systems. The modeling revealed and explained several interesting aspects of the reactions in a matrix not observed in the solution photochemistry.

3.2. Kinetic Model of Photopolymerization in Matrix

The photopolymerization reaction proceeds through several well-established steps. It starts with the light absorption resulting in excitation of the photoinitiator. It then proceeds through initiation of polymerization by charge, energy transfer, or radical reaction; propagation; and termination of the chain.[1,2]

Reaction	Step
$S + h \rightarrow S^* \rightarrow S\cdot$	Initiator photoactivation, radical formation
$S\cdot + M \rightarrow M\cdot + S$	Chain initiation
$S\cdot + O_2 \rightarrow SOO\cdot$	Initiator radical deactivation by oxygen
$S\cdot + C \rightarrow C\cdot + S$	Initiator radical termination by chain-transfer
$M\cdot + M \rightarrow M_2\cdot$	Chain propagation
$M_2\cdot + M \rightarrow M_3\cdot$	
....	
$M_{k-1}\cdot + M \rightarrow M_k\cdot$	

$M\cdot + O_2 \rightarrow MOO\cdot$ Radical scavenging by oxygen

....

$M_k\cdot + O_2 \rightarrow M_kOO\cdot$

$M\cdot + C \rightarrow M + C\cdot$ Chain termination

....

$M_k\cdot + C \rightarrow C\cdot + M_k$

Here S is a photoinitiator, M is a monomer, $M_k\cdot$ is a radical containing k monomeric units, C is a chain transfer agent acting in the present scheme as a chain-terminating agent, and O_2 is oxygen. The mobility of growing polymer radicals in a matrix is usually two orders of magnitude lower than that of monomeric molecules. Therefore, the radical-radical chain termination is absent in the kinetic scheme. The data described below demonstrated that photopolymerization is significantly slower in the presence of oxygen. Thus, oxygen is considered as a prime chain-terminating agent. The possibility for coinitiation by peroxy radicals as seen in other systems is reportedly substantially slower than the primary process and is not considered in this kinetic model.[74,75]

3.3. Photoinitiation

The polymerization process begins with the absorption of light by the photoinitiator and the formation of radicals. The rate of this process at a particular point within the photopolymer film depends on the extinction coefficient ϵ_s, the local concentration of the photoinitiator [S], and the local light intensity I:

$$\left\{\frac{\partial[S]}{\partial t}\right\} = -\epsilon_s[S]I \tag{2}$$

The diffusivity of an o-chloro hexaarylbiimidazole (HABI) molecule used as an initiator is considered negligible relative to that of the monomer, the oxygen, and the chain transfer agent. Indeed, the size of the HABI molecule approaches that of the oligomer M_6. Two radicals are generated by each dissociated molecule of the HABI photoinitiator. The nature of triarylimidazolyl (lophyl) radicals is discussed below, and their generation is described in detail in other chapters of this book. It is sufficient to note here that these radicals are stable in solution and even more so in a polymer matrix. Because of the steric restrictions discussed below,[73] the recombination of the initiator radicals is less likely than their reaction with the monomer, the oxygen, and the chain-transfer agent. The mobility of triarylimidazolyl radicals is also relatively low. Modeling the photopolymerization kinetics with an inclusion of reasonable values for lophyl radicals and HABI diffusivities demonstrated that neither the kinetics nor the computed reactant and product distributions change perceptibly relative to the values computed for fully immobile HABI. The behavior of the photopolymer containing a highly mobile and reactive photoinitiator instead of HABI is discussed below.[68]

In the DuPont photopolymer, the HABI initiation reaction is assisted by "chain-transfer" agents, such as mercaptobenzoxazole, although direct initiation by HABI also occurs.[76] The mercapto chain-transfer agents are activated by electron transfer.[76] They react further with a monomer producing radical species by electron or hydrogen transfer or by attachment to the double bond of the monomer. The overall effect is an addition to the polymer chain with the transfer of the radical to another growing chain (here the mercapto additive acts as another monomer moiety), or chain termination.[32] The parallel reaction of the coinitiation by the mercapto additive is incorporated into the initiation kinetics by ascribing it to the higher reactivity of the HABI initiator. Any consideration of the monomer-like capability of the mercapto additive is unnecessary because of its low contribution (one or two terminal groups). However, the chain-terminating function of this mercapto-additive is significant and was handled as described below.

The introduction of a mobile coinitiator unable to absorb light by itself into the kinetic scheme does not alter the qualitative results of the computations. The slow degradation of the peroxide radical with the delayed initiation of the polymerization[74,75] could also be neglected. The rate of initiation is given by an expression:

$$\text{Initiation Rate} = -2\left\{\frac{\partial[\mathrm{S}]}{\partial t}\right\} \tag{3}$$

3.4. Propagation

The monomer addition to the radicals of the "live" and growing polymer chains can be represented by a conventional expression for the rate of the second-order reaction:

$$\text{Propagation Rate} = K_{\mathrm{p}}[\mathrm{M}]\left\{\sum_{j=1}^{\infty} r_j^*\right\} \tag{4}$$

In this equation K_p is the second-order rate constant, and [M] r_j^*; are the local concentrations of the monomer and the radicals of the chain length j, respectively. The local concentration of the monomer is reduced by the propagation. However, monomer diffusion from regions rich in the monomer substantially affects the monomer distribution in the course of photopolymerization. Using monomer diffusivity, which is one of the most important features of our model, the rate of the monomer concentration change is expressed as:

$$\frac{\partial[\mathrm{M}]}{\partial t} = D_m\left\{\frac{\partial^2[\mathrm{M}]}{\partial z^2}\right\} - K_p[\mathrm{M}]\left\{\sum_{j=1}^{\infty} r_j^*\right\} \tag{5}$$

The distance z from the illuminated surface of the film is used as a spatial variable. D_m is the monomer diffusivity in the plasticized polymer matrix. In the photopolymer films there is no mass transport by convection or turbulence as occurs in stirred tank systems.[66,67] The only source of polymer molecules is chain propagation, and neither polymer molecules nor polymer radicals migrate perceptibly over the course of the reaction, so we can write for polymer concentration [P]:

$$\frac{\partial[\mathrm{P}]}{\partial t} = K_p[\mathrm{M}]\left\{\sum_{j=1}^{\infty} r_j^*\right\} \tag{6}$$

There should be substantial variations in the molecular weights of polymer molecules forming in the regions with different light intensity.[29] However, the molecular weights and their distribution are not the subject of the current investigation; consequently, chain transfer and coinitiation are omitted in the model.

3.5. Termination

It was previously established experimentally that in the absence of scavenging species, radicals can remain in the polymer matrix almost indefinitely.[27,77] The data indicate the low rate of bimolecular recombination of radicals in the polymer matrix. Our preliminary computations showed that the concentration of radicals is low even near the illuminated surface. Further modeling confirmed a negligible rate for bimolecular recombination. Therefore, the bimolecular recombination can be excluded from the computational model of polymerization in the matrix. Oxygen can form a low reactivity peroxy-radical upon reaction with the live chain, thus terminating the chain growth. Although the HABI initiator system is reported to have low sensitivity to the presence of oxygen,[78,79] oxygen does react with lophyl radicals (included in the total radical count) as was shown by Hayashi and Maeda.[80,81] The reason for low oxygen effects on the HABI system is not related to the radical reactivity and is discussed below.[73] Taking into account Hayashi and Maeda data on the lophyl radical reactions with oxygen,[80,81] lophyl radicals are considered as vulnerable to oxygen attack as other radicals in the system. The chain-transfer agent is considered as a chain terminator analogous to oxygen. Both oxygen and the chain-transfer agent are free to diffuse within the matrix. The chain-termination rate is then expressed as:

$$\text{Termination Rate} = \{K_T[O_2] + K_{CT}[C]\}\left\{\sum_{j=1}^{\infty} r_j^*\right\} \tag{7}$$

where K_T and K_{CT} are the second-order rate constants of the chain termination by oxygen, O_2, and a chain-transfer (chain-terminating) agent, respectively. Oxygen is always present in the photopolymer film. The diffusivity of oxygen D_{O_2}, is higher than that of the monomer; thus the polymerization would not occur in the unprotected photopolymer film, if the oxygen solubility in the film were high. Fortunately, usual photopolymer oxygen solubility is <90 ppm (3×10^{-6} mol/cm^3). The rate of concentration change of oxygen and the chain-terminating agent are expressed by similar equations, although the initial and the boundary conditions are, of course, different:

$$\frac{\partial[O_2]}{\partial t} = D_{O_2}\frac{\partial^2[O_2]}{\partial z^2} - K_T[O_2]\left\{\sum_{j=1}^{\infty} r_j^*\right\} \tag{8}$$

$$\frac{\partial[C]}{\partial t} = D_C\frac{\partial^2[C]}{\partial z^2} - K_{CT}[C]\left\{\sum_{j=1}^{\infty} r_j^*\right\} \tag{9}$$

3.6. Radicals

The radical concentration evolution during photopolymerization is traditionally handled using the stationary-state hypothesis; that is, assuming that the rates of radical generation and disappearance are equal. However, in a spatially nonuniform, low mobility system the validity of such an approach is doubtful. Considering the Kloosterboer results[77] and our data on low diffusivity of the monomer and oxygen in the photopolymer,[7,70–72] the stationary-state approximation is not valid for the photopolymer system description. Therefore, the local radical concentration time dependence was computed explicitly by the equation:

$$\frac{\partial R^*}{\partial t} = 2\left\{\frac{\partial[S]}{\partial t}\right\} - \{K_T[O_2] + K_{CT}[C]\}\left\{\sum_{j=1}^{\infty} r_j^*\right\} \tag{10}$$

where $R^* = \left\{\sum_{j=1}^{\infty} r_j^*\right\}$

It has been reported that the HABI initiated system has a low sensitivity to oxygen.[78,79] Hayashi and Maeda[80,81] reported that lophyl radicals do react with oxygen. To the best of our knowledge, the experimental data on the lifetime of triarylimidazolyl radicals in the presence of oxygen and in the inert environment are not available. Because the lophyl radicals formed by the initiator dissociation react with oxygen as do all other radicals present, they were included in the radical count and are treated here as any other radical within the system. Consequently, the integration of Eq. (10) from the initial condition of the absence of radicals automatically includes any delay in chain propagation caused by oxygen scavenging during the initiator radical oxygen reaction. Actually, at high light intensities the polymer matrix degrades, generating methyl and other radicals.[73] This generation of the reactive species must be taken into account in the design of the imaging systems used with high light intensities (for example, Du Pont proofing systems such as Cromalin, Cromacheck, etc.).[37,43,82] In order to bypass the matrix photodegradation problem in our experiment, light intensity was low enough to reduce the photodegradation to levels below detection limit.[7,73] The tests were done by monitoring photobleaching of the dyes.

3.7. Local light intensity

Light is absorbed in the photopolymer film by the initiator, the monomer, and the polymer. As a result, because reactions and diffusion change the distribution of the absorbing species, the local light intensity is changed in the course of imaging. Because the overall optical density is very high, the changes may not be substantial. Nevertheless, we consider the local light intensity within the photopolymer film as a variable changing in time and with the distance from the illuminated surface, as described by the equation:

$$\frac{\partial I}{\partial t} = - I \left\{ \epsilon_s[S] + \epsilon_m ([M] + [P]) + \sum_k (\epsilon_k[Q_k]) \right\} \tag{11}$$

where $[Q_k]$ is the concentration of the kth component of the reactive mixture, I is the local time-dependent light intensity, and ϵ_k is the extinction coefficient of the kth component. The modeling of the inhomogeneous photopolymerization in a high optical density system was conducted previously for well-stirred and unstirred reactors[66,67] in UV-curable adhesive polymers[64] and photoresist systems.[33,63] Our modeling extended the range of the considered light attenuation effects. The results of our computations using the kinetic scheme presented above indicated that relations between the yield of the polymer, the distance from the illuminated surface of the film, and the concentration of reactive species are more complicated than one would expect from the exponential Bouguer-Lambert-Beers law of light attenuation.[68–72]

3.8. Fluorescence intensity

The only experimentally observed kinetic variable was fluorescence emitted by the monomer, NVC, and the resulting polymer, PVCA (Figs. 5-4 through

5-6).[68–72] The intensity of fluorescence is proportional to the amount of light absorbed by the fluorescent groups. Because of intramolecular quenching, the monomer radical fluorescence yield is significantly lower than that of nonradical species.[83] In model computations only dimeric and higher molecular weight radicals are treated as fluorescent species.[70–72] The photopolymer film fluorescence excitation and detection were conducted in isobestic and isoemissive points, respectively.[68–71] Therefore, the same extinction coefficient and fluorescence yield were used in computations for NVC and PVCA. The spatial derivative of the fluorescence intensity is expressed as:

$$\frac{\partial(I)_{\text{fluorescence}}}{\partial z} = Iqf\epsilon_m\{[M] + [P]\} \tag{12}$$

where q is the quantum yield of fluorescence in the isoemissive point of NVC and PVCA at 400 nm, and ϵ_m is an extinction coefficient of carbazyl groups in the monomer and the polymer in an isobestic point at 295 nm.[84] The factor f is a spatially dependent adjustment parameter designed to account for the presence of fluorescence quenchers in the photopolymer mixture. The fluorescence yield is inversely proportional to the quencher concentration,[83,85] and directly proportional to the number of illuminated carbazyl groups. The parameter f is equal to the local fraction of fluorophore in a fluorophore-quencher mixture and depends on time and the distance from the illuminated surface. The quencher is considered to be immobile. Introduction of the f factor into the model makes it more flexible, allowing consideration of the variation of the total quantum yield of fluorescence in the course of the reaction. The omission of this factor, of course, does not substantially change in the qualitative behavior of the system, only slightly reducing the computed rise in the fluorescence intensity.

The expression for the fluorescence intensity evolution provides a basis for comparison of the experimental data with the model computations. If the kinetic model is reasonable, the results of computations should qualitatively reproduce the experimental data. The experimental data can, in turn, be explained if the observed and computed dependence of fluorescence intensity on the time of film illumination change similarly when reaction conditions are modified.

3.9. Computational Parameters

Some values for computational parameters, such as diffusion coefficients, were experimentally determined. Rate constants used in the computations were found in the literature. Modeling of photopolymerization kinetics was done for the photopolymer system used in the experiments; that is, for the 24.5-μm thick plasticized polymer film initially containing 8 weight % of the monomer, 2.3% of the chain-terminating agent, and 2% of the initiator. An equilibrium oxygen level of 90 ppm (3×10^{-6} mol/cm^3) was used as the oxygen concentration in a glassy polymer matrix or a viscous organic liquid at room temperature.[86,87] The rate constants of 1.2×10^{10} cm^3/mol s for chain propagation, 1.5×10^{11} cm^3/mol s for the chain-terminating oxygen reaction with radicals, and 4×10^8 cm^3/mol s for the mercaptobenzoxazole chain-terminating agent reaction with radicals were based on average values found in the literature for a wide variety of radical reactions in the gas phase. The extinction coefficient for NVC and PVCA was 3×10^7 cm^3 (mol cm)$^{-1}$ at 295 nm.[84] The extinction coefficient of HABI was 10^6 cm^3/mol cm. The total optical density of invariant components such as the binder and sensitizing dyes (used for long wavelength imaging, and inactive here) was measured to be 2.5×10^4 for 1 cm. The intensity of the excitation light corresponding to the experiments was 2.5×10^{-8} einsteins. The diffusion coefficient for the monomer was approximated in our experiments to be 6×10^{-9} cm^2/sec for CAB matrix.[7] Because the size of the chain-transfer agent molecule, 2-mercaptobenzoxazole, was similar to that of the monomer, the diffusion coefficient of 10^{-8} cm^2/sec was used in computations for the chain-transfer agent diffusion. This value was close to those reported for other molecules with analogous cross-sections.[88] The diffusivity of oxygen in the plasticized polymer matrix used in the modeling was computed as an average of those in the nonplasticized polymer matrices[86–89] and in organic liquids.[90] Although the computational parameters were selected for the particular DuPont holographic photopolymer, the values listed above are well established and can be used with a minor modification in modeling of the kinetics in other photopolymer systems.

3.10. Initial and boundary conditions

Initially, prior to illumination by the activating light, all the ingredients of the photopolymer formulation and the oxygen dissolved in it are distributed uniformly within the film. The phase separation reported for polymerizing photopolymers[39,91,92] containing more than 40% of the binder does not occur in the present case, as was determined upon investigation of polymerization. The degree of conversion of the investigated monomer dissolved in the plasticizer did not exceed 10%. Holographic photopolymer films are formulated with the intent to avoid phase separation after polymerization of the reactive plasticizer.

For most of the photopolymeric applications it can be assumed that all of the ingredients, with the exception of oxygen (when the film surface is not protected), are confined to the film. Therefore, the concentration gradients vanish at the surfaces. The oxygen gradient vanishes at the surfaces in the presence of an impermeable barrier layer.

The incident light is assumed to be of constant intensity. Because experimentally the sample can be illuminated from both sides just by flipping the film, a provision was made in the computations to step-turn the incident beam by 180° at the prescribed time for modeling the consequent imaging on the "flip-side."

3.11. Calculations

The numerical solution of the system of differential equations describing a diffusion-limited photopolymerization reaction in space is found by a method of points using a finite difference approach.[62] The volume of film in such a technique is divided into a set of nonuniform finite elements, with the smallest elements near the illuminated surface of the film and, in the case of patterned illumination, near the boundary of the illuminated regions, where abrupt and rapid changes in reagent concentrations are anticipated. When the photopolymer film surface is illuminated uniformly, diffusion occurs in one direction towards the light, and only one geometric variable z (distance from the illuminated surface) is needed. Partial derivatives for the z variable are approximated by three-point formulas in the computations presented below. For uniform illumination the finite element grid is symmetric relative to the midplane of the film for ease of processing of flip-side exposure. For patterned illumination we used a two-dimensional diffusion model, in which diffusion towards the illuminated region occurs not only from the depth of the film toward the illuminated surface, but also from the shadowed regions in the direction parallel to the film surface. A more detailed description of this two-dimensional diffusion is presented below. Time derivatives of all dependent variables are integrated by a stiff differential equation solver (Gear algorithm).[93,94] Results are recorded at selected fractional times of the overall exposure period. Numerical modeling of such magnitude requires substantial computational power. Here it was conducted using a Cray computer. Data was further processed on a mainframe system and plotted with the help of a personal computer.[68–72]

4. Modeling, Experiments, and Photopolymerization Anisotropy

4.1. Confirmation of Experimental Data

The kinetic scheme of polymerization and the corresponding mathematical model of photopolymerization in a plasticized polymer matrix were constructed using almost no traditional simplifications such as steady-state radical concentration or narrow distribution of the transients within the system. Even mobilities of the polymeric species and the HABI photoinitiator were removed from the computations only after it was established that these low values do not affect the results of the modeling. The main feature of our model is an explicit acknowledgment of the difference between mobilities of the reagents and their products (often diffusivity is hidden within the selected rate constants). It is clear that the computed change in fluorescence intensity during photopolymerization of the hypothetical fluorescent monomer with properties resembling those of NVC does not depend on the experimental data obtained with the DuPont holographic photopolymer.

The model computations qualitatively reproduced an experimentally observed increase in fluorescence intensity upon exposure to light of the high optical density photopolymer film containing a fluorescent mobile monomer, polymerizing into an immobile fluorescent polymer (Fig. 5-8). As in the experimental observations, the increase of fluorescence intensity is higher upon first exposure of the film. After photopolymerization initiated by the front surface exposure ceased, the exposure of the flip surface yielded a lower initial emission and a lower overall increase in fluorescence intensity over the time of the exposure than that of the front surface. As was explained above, this occurs because some monomer is consumed during the front surface exposure and fewer fluorescent monomer molecules contribute to the fluorescence and polymer formation during the subsequent exposure of the flip side of the film.

The model computations have also reproduced several other experimentally observed effects, resulting from unidirectional exposure of photopolymer films. One of these effects is the dependence of the fluorescence-detected photopolymerization kinetics on the film thickness. It is well known that the kinetics of polymer film photocuring depend on film thickness and on the nature of the light-absorbing components of the formulation. A substantial amount of experimental work has been carried out on the effects of light absorption by the photoinitiator.[29,64,95–98] However, the reported experimental data are contradictory. In some reports the initiator concentration giving optimal photopolymerization yield should be reduced when the film thickness is increased;[95–97] in others the opposite is recommended.[98] However, this change occurs not only because of the variation in the initiator concentration and consumption, but also as a result of monomer mobility. Indeed, as the monomer is consumed near the illuminated surface, the monomer concentration within the photopolymer film decreases. Because the depth of light penetration is the same in both thick and thin films, the monomer consumption in the thick film can continue longer and the yield of polymer is higher than in the thin film. Moreover, the amount of free monomer still remaining in the film after the front surface polymerization is also higher (Figs. 5-5, 5-6, 5-9, 5-10). The "thickness of the film" must be understood as being relative to the light absorbency. Indeed, low optical density material will behave as a "thin" film with respect to the amount of free monomer remaining in the film after the front surface exposure is completed. In a "very thin film" in which complete polymerization is possible (in our experiments

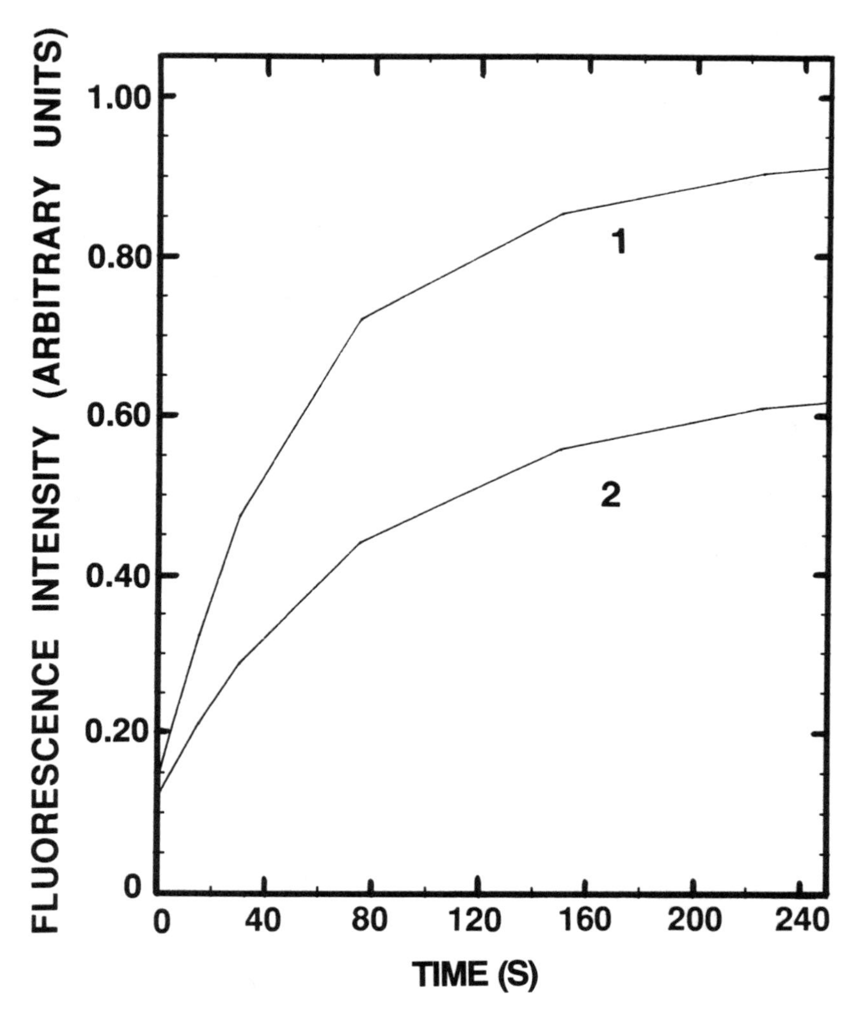

Figure 5-8 *Model calculations of the fluorescence detected kinetics of photopolymerization taking place in consecutive UV exposure of the two sides of 24.5-μm photopolymer film: 1-exposure of the front, oxygen impermeable, surface; 2-exposure of the unprotected, flip side of the film.*

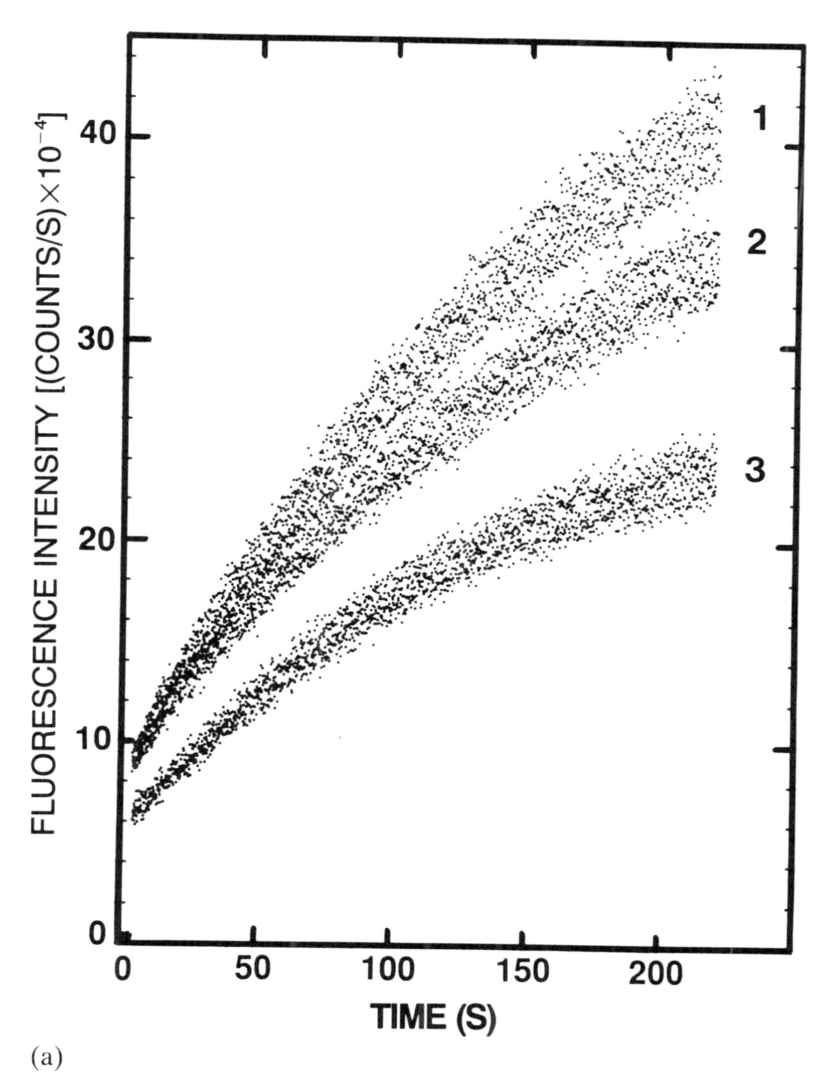

(a)

Figure 5-9. *(a) Experimentally observed film thickness dependence of photopolymerization kinetics: 1-39.2-μm film; 2-24.5-μm film, and 3-0.7-μm film. The photopolymer was exposed from the air impermeable side. (b) Photopolymerization kinetics in the 24.5-μm film observed at different intensities of activating light: 1: 100%; 2, 70% of initial intensity; and 3, 20% of initial intensity.*

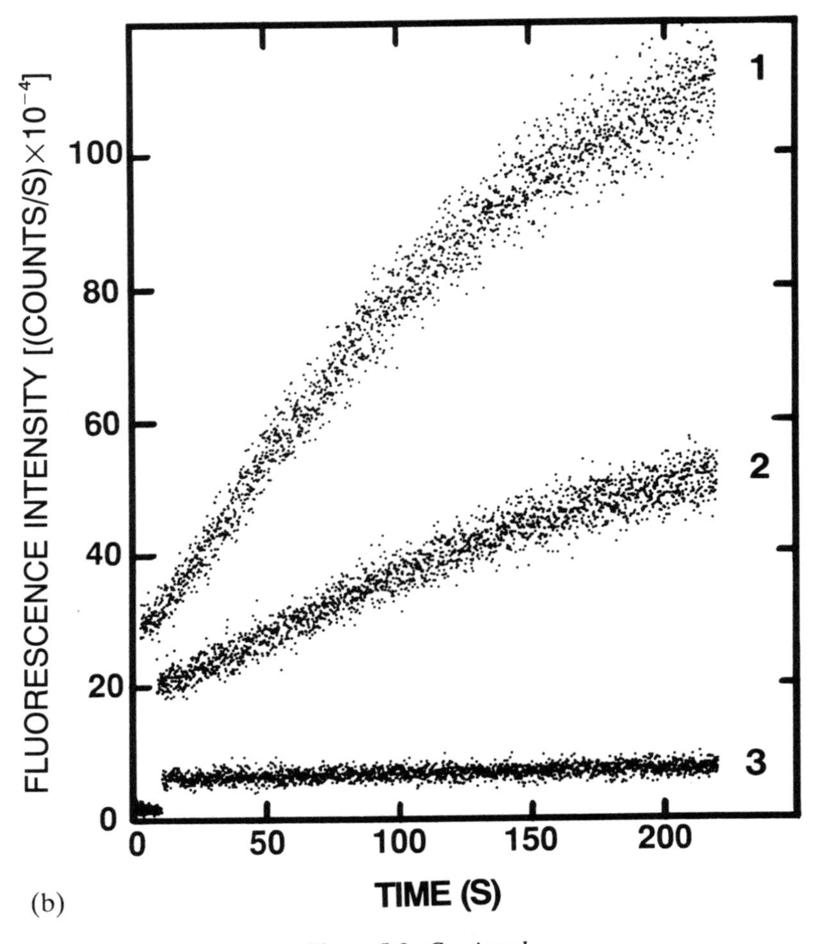

Figure 5-9. *Continued*

>0.7 μm) after the front surface exposure, the exposure of the flip side of the film does not result in any fluorescence increase (Fig. 5-6). Variation of the activating light intensity affects the photopolymerization kinetics in film similarly to the change of the film thickness, because the depth of light permeation into the film increases with the intensity increase [Fig. 5.9(b)]. The change in light intensity also affects the number of radicals generated at the illuminated surface, therefore the result of increasing light intensity on photopolymerization kinetics is clearly not identical to that of reducing the film thickness [Fig. 5.9(a,b)]. The model computations qualitatively reproduce the observed thick-

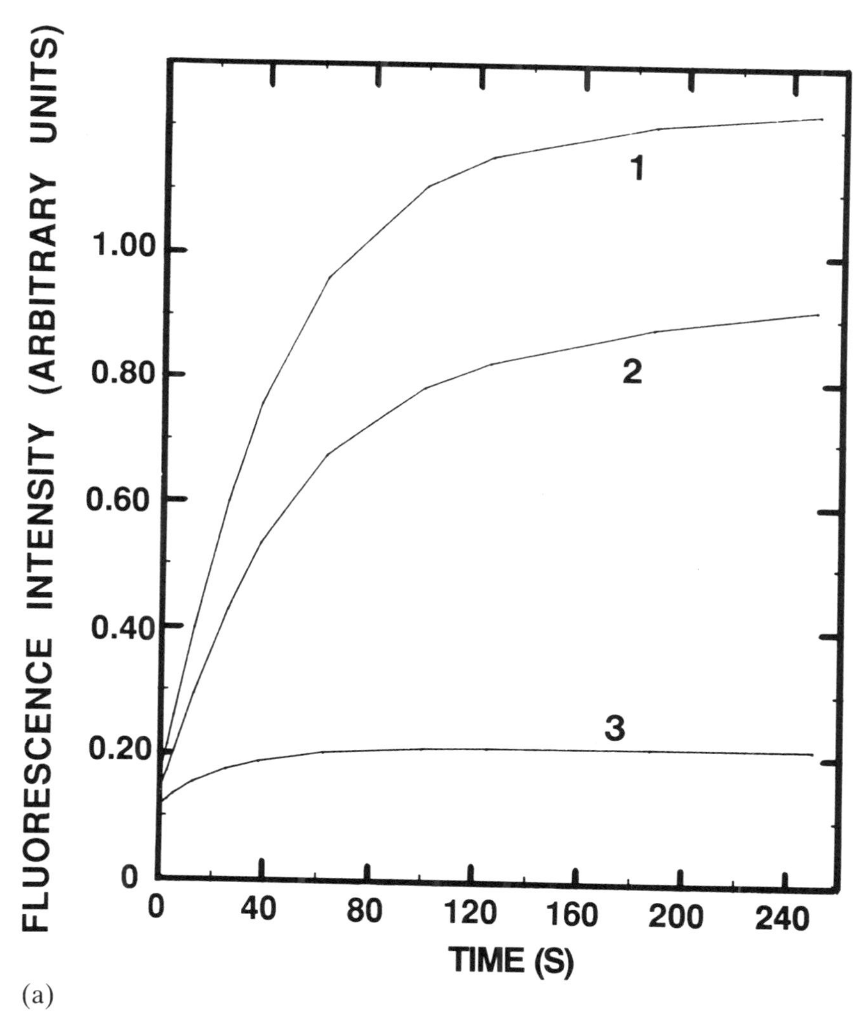

(a)

Figure 5-10. *(a) Computed thickness dependence of the photopolymerization kinetics: 1-39.2-μm film; 2-24.5-μm film, and 3-0.7-μm film. The surface of the photopolymer protected from the contact with oxygen was exposed to light. (b) Computer simulation of the photopolymerization kinetics in 24.5-μm film at different light intensities: 1, 100%; 2, 70% of initial intensity; and 3, 10% of the initial intensity.*

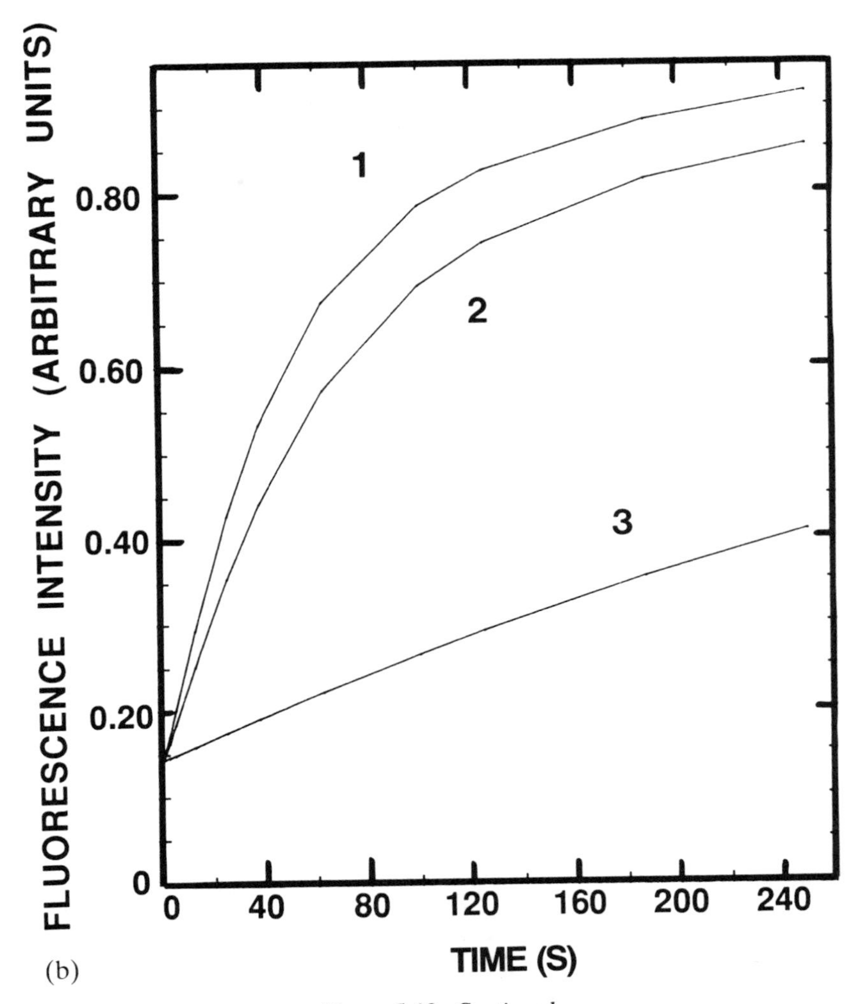

Figure 5-10. *Continued*

ness and light intensity dependence of the photopolymerization kinetics (Fig. 5-10 (a,b)).

4.2. Evolution of Spatial Distribution of Reagents in Photopolymer Films

Attempts to measure reactant and product distribution in the photopolymer systems were tried, unsuccessfully, in the past using a spin-echo technique.[36,43] More recently NMR imaging was employed for determination of the polymer distribution after photocuring of an acrylate-based photopolymer composition.[99] NMR mapping seems to be a promising approach, and substantial progress has been achieved in the mapping of the polymer during different stages of photopolymer exposure, although the spatial resolution of the method is relatively low. Constraints in the sample preparation for NMR analysis precluded exact reproduction of commonly used conditions of the photopolymer imaging. Progress in the determination of the spatial distribution of light-absorbing species in polymer films was made recently using photopyroelectric spectroscopy.[100] The demand for depth uniformity of imaged photoresists, proofing materials, and digital recordings led to repeated attempts to visualize spatial behavior of the photopolymer images by numerical modeling.[19,29,33,34] Attention was also given to modeling of the deterioration of image uniformity caused by optical effects observed when the submicron image resolution is sought.[101] Correlation between the image resolution and quality and diffusion of photopolymer components clearly exists. Recently the resolution of photopolymer imaging was studied as a function of the rate of photoinhibitor diffusion and the intensity of the exposing light.[101,102] In the absence of direct monitoring methods, kinetic modeling is the most useful tool for understanding the evolution of spatial distribution of reagents and products during photoimaging.

We computed the change in species distribution as a function of exposure time. Some modeling results were intuitively obvious. For instance, because the excitation light falls from one direction, the product formation and reagent consumption are not isotropic. Thus, in the high optical density photopolymer film considered here, most of the radical species are produced and concentrated within a fraction of a micron from the illuminated surface. In the surface layers of the film, the immobile photoinitiator distribution changes faster with time than in the deep layers of the film. As the photopolymerization progresses, the initiator is consumed farther away from the illuminated surface (Fig. 5-11). In a film fully isolated by a substrate and a cover sheet from further oxygen uptake, oxygen will behave as other reagents confined in the photopolymer film and not replenished after consumption. Oxygen and the monomer molecules are also consumed more rapidly near the illuminated surface; however, oxygen and the monomer are free to diffuse. Even though the reaction does not occur deeply below the illuminated surface, oxygen and the monomer are depleted by diffusion into a photoactive region of the film (Figs. 5-12, 5-13). In the beginning of photopolymerization the change in the monomer concentration occurs mostly near the surface, and only later the monomer is depleted by diffusion from the depth of the film towards the light. The ''information'' about the loss of the monomer is propagated through the film at the rate of the monomer diffusion. Because oxygen diffusivity is higher than that of the monomer, depth equilibration and the consumption of oxygen in the photo-

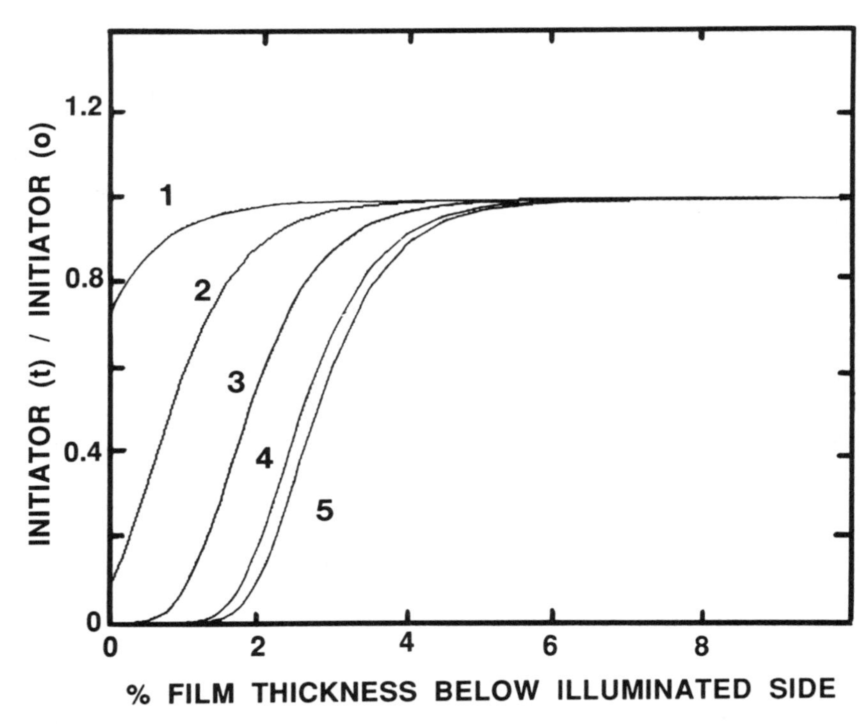

Figure 5-11. *The immobile initiator consumption computed as a function of distance from the illuminated surface of the film: 1-13 s, 2-100 s, 3-500 s, 4-1500 s, and 5-2000 s from the beginning of the exposure. Both surfaces of the film were protected from contact with air.*

polymer film isolated from air occur faster than for the monomer (Figs. 5-12, 5-13). The kinetics of a chain-terminating agent consumption closely resembles that of diffusion-controlled consumption of the monomer and will not be discussed in detail here.

4.3. Oxygen Effects on Photopolymerization Kinetics in Films

The absence of a pronounced induction period in the fluorescence detected accumulation of polymers is one of the peculiarities of the matrix photopolymer-

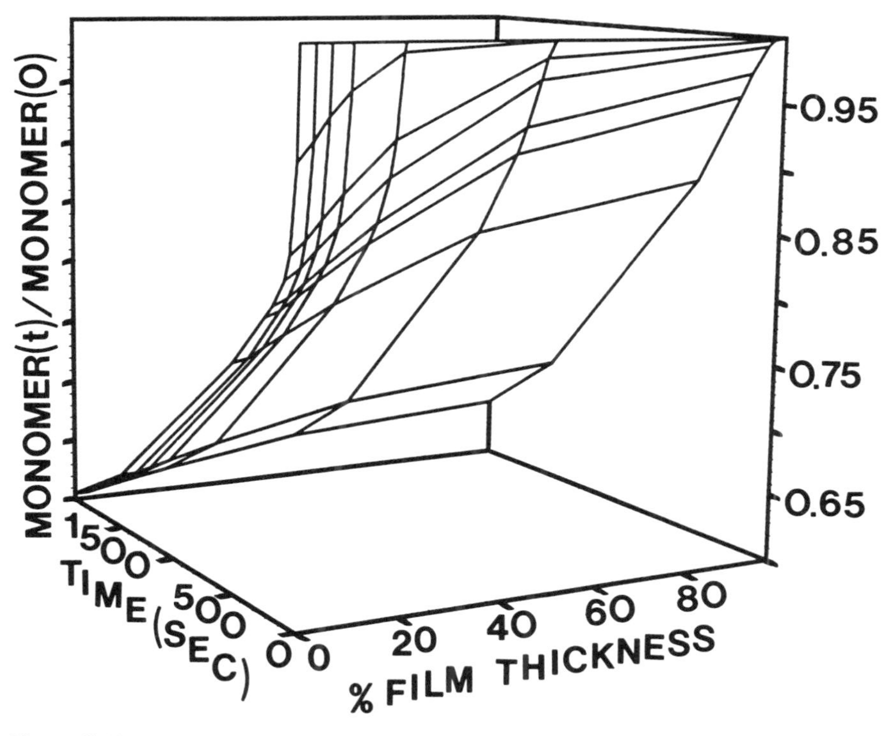

Figure 5-12. *Monomer concentration computed as a function of distance from the illuminated side of the film and the time of exposure to light. Both sides of the film were assumed to be protected from contact with air.*

ization. The induction period is observed in the solution polymerization because the rate of the radical reactions with oxygen is substantially higher than the rate of the chain propagation.[2] Solubility of oxygen in the polymer matrix or in plasticizers is 3×10^{-6} mol/cm^3, which is significantly lower than that of the monomer. The oxygen diffusivity in the matrix is only one or two orders of magnitude higher than that of the monomer, and the rates of the radical reactions here are controlled by diffusion of the reagents. Therefore polymer chain growth in the matrix reaction can successfully compete with chain termination by oxygen. Thus, the induction period is not observed, even though the presence of oxygen slows the polymerization rate and reduces the yield of polymer.

Oxygen does not quench *N*-vinyl carbazole or poly(*N*-vinyl carbazole) fluorescence in the polymer matrix. Consequently, the changes in fluorescence-detected

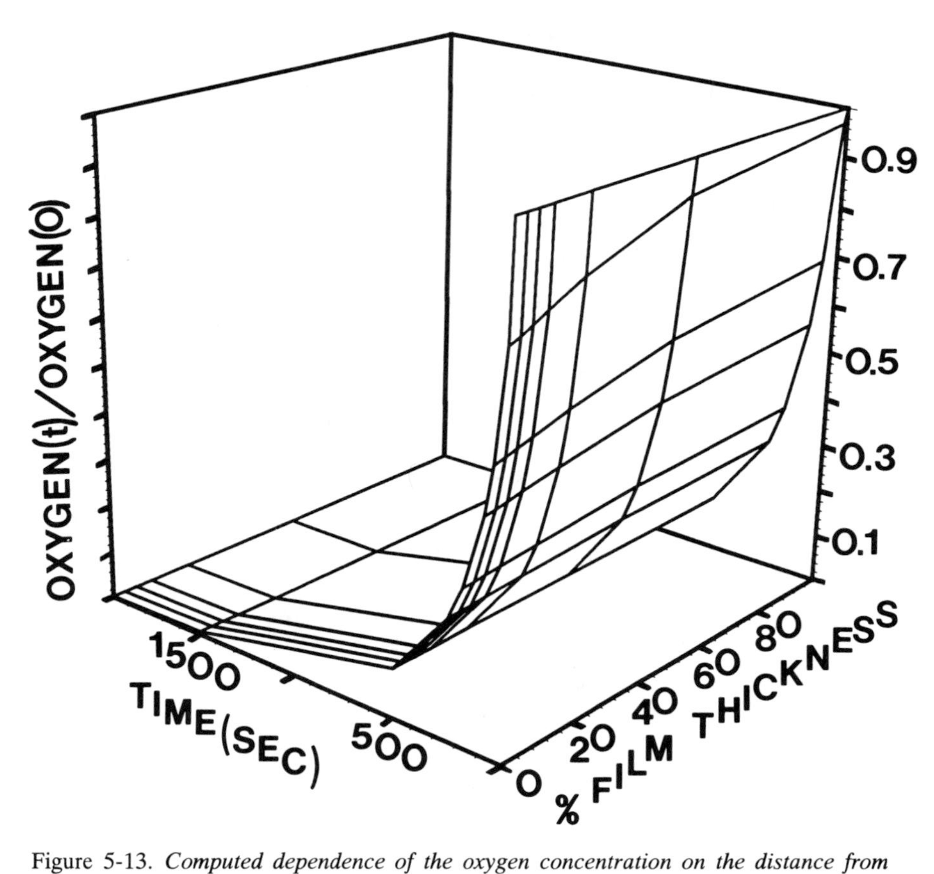

Figure 5-13. *Computed dependence of the oxygen concentration on the distance from the illuminated surface and on the exposure time. Both sides of the film were assumed to be covered by oxygen impermeable film.*

kinetics in the presence and absence of oxygen are not caused by oxygen-induced quenching.[71,72] Because oxygen effects were in question, a vacuum chamber in which the photopolymerization could be studied in different atmospheric environments was used.[71,72] The exposure to the chamber atmosphere was always from the illuminated side of the film. The samples were degassed for 20 minutes at 10^{-5} mm of Hg to reduce the amount of oxygen dissolved in the photopolymer film prior to introduction of the oxygen-nitrogen mixture to the chamber.

When photopolymerization was conducted in the oxygen-nitrogen mixture

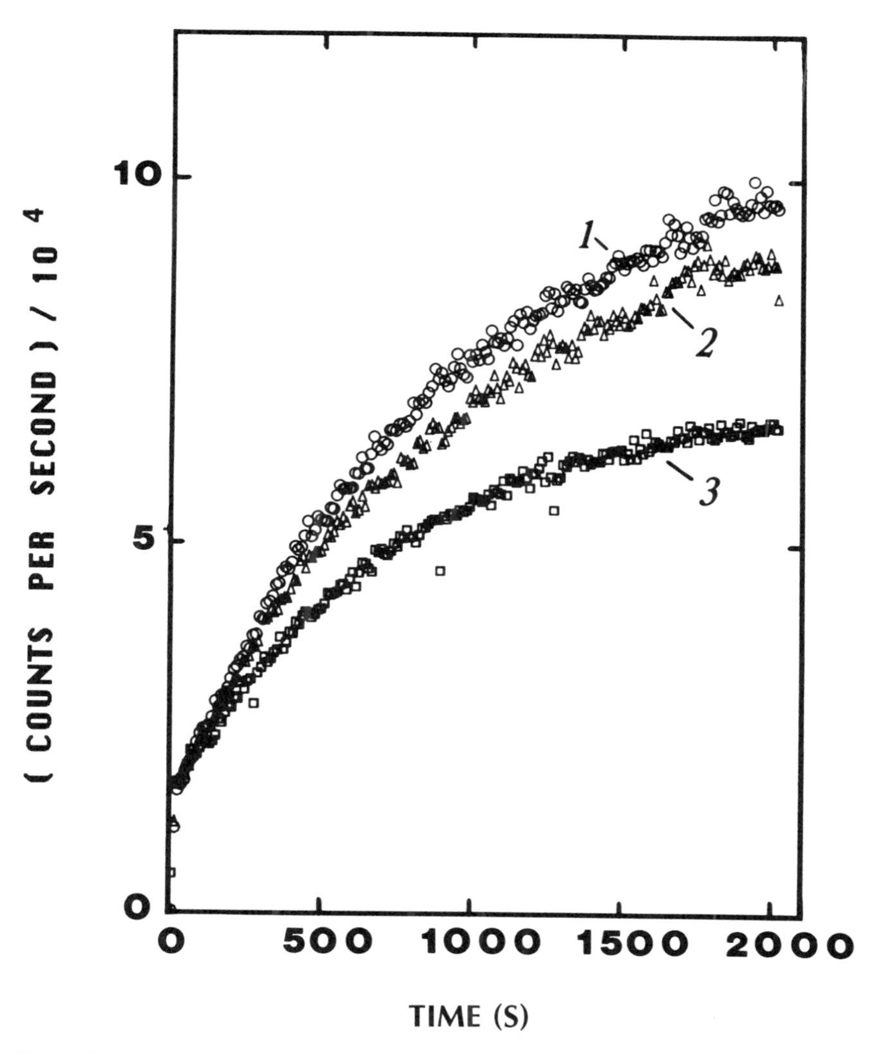

Figure 5-14. *Kinetics of photopolymerization in photopolymer film monitored by the change in the film fluorescence intensity at various oxygen concentrations in the atmosphere surrounding the film: 1-neat nitrogen, 2-0.05% oxygen in nitrogen, and 3-0.5% and higher concentrations of oxygen in nitrogen. The illuminated side of the film was in contact with the surrounding atmosphere.*

containing various concentrations of oxygen, the yield of polymer increased when oxygen contents in the nitrogen-oxygen mixture was lower (Fig. 5-14). Kinetic modeling confirmed the experimental data, indicating that when the diffusivity of oxygen was comparable to that of the monomer, the induction period in photopolymerization was not observed (Fig. 5-15).

4.4. Matrix and Oxygen Effects on Photopolymerization Kinetics

Model computations predict that kinetics of photopolymerization in a film exposed to air will change as a result of the change in the monomer diffusion rate (Fig. 5-16). The most direct way to alter the diffusivities without altering other reagents in the photopolymer is to change the polymer matrix. The change of the polymer matrix is often used in developing holographic and proofing applications of photopolymers.[36,37,43,103] Most formulations are intentionally made in such a way that the matrix is saturated with a liquid monomer, a plasticizer, etc. (usually >50% of formulation). This is done to avoid the phase separation observed in monomer-poor systems.[43,91,92] In monomer rich-systems, the polymer matrix reduces reagent mobilities and provides reaction sites. The polymer matrix influence on radical chemistry and molecular mobility are still not well investigated areas.

Progress in the selection of photopolymer components based on their diffusion coefficients in the polymer matrix was made recently and resulted in improved resolution under patterned exposure.[101,102] In order to vary molecular mobility we used three inert polymeric matrices: cellulose acetate butyrate (CAB), poly(vinyl butyrate) (PVB) and poly(vinyl acetate) (PVA). The coatings were studied in a dry nitrogen environment after degassing in vacuum. There was some alteration of formulation relative to that described in the section on monomer diffusivity. In the absence of oxygen the photopolymerization kinetics are different in different matrices (Fig. 5-17). It was possible to deduce the monomer diffusivities comparing these data with the results of the computations using Eq. (1).[68,71,72] The monomer diffusion coefficient in CAB was $D_{CAB} = 1.5 \times 10^{-9}$ cm^2/sec, and in PVA and PVB, $D_{PVA,PVB} = 8.5 \times 10^{-10}$ cm^2/sec. The values of the monomer diffusivities are within the range of those measured by other techniques.

In the presence of oxygen the rate of monomer migration does not change. However, the polymer yield and the total amount of the monomer migrating to the illuminated surface decline as a consequence of radical scavenging by oxygen (Fig. 5-18), as expected (Figs. 5-14, 5-15). We encountered a remarkable oxygen effect on the photopolymer film performance. In a matrix in which diffusivity of the monomer and other small molecules is the highest, the polymer yield in the presence of oxygen is the lowest. Several patents and papers claim-

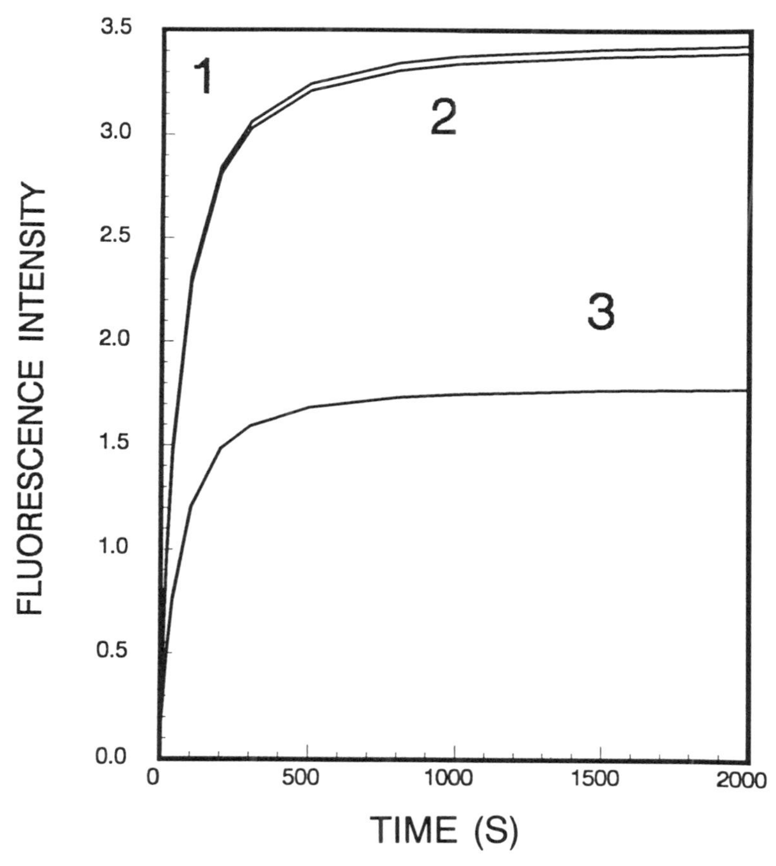

Figure 5-15. *Computed kinetics of photopolymerization in photopolymer film monitored by the change in the film fluorescence intensity at various oxygen concentrations in the atmosphere surrounding the film: 1-neat nitrogen, 2-0.05% oxygen in nitrogen, and 3-0.5% and higher concentrations of oxygen in nitrogen. The illuminated side of the 24.5-μm film was in contact with the surrounding atmosphere.*

ing unexpected properties in the holographic photopolymers were written after the authors substituted one matrix for another.[103,104] To explain these "unusual" effects it was even suggested that a crosslinked polymer matrix can migrate out of the illuminated region of the film.[104]

The reason for the decline of the polymerization yield is, most likely, the scavenging of the radicals by oxygen. In a matrix with a high diffusion rate the oxygen diffusion into the film from the atmosphere is faster. Replenishing of

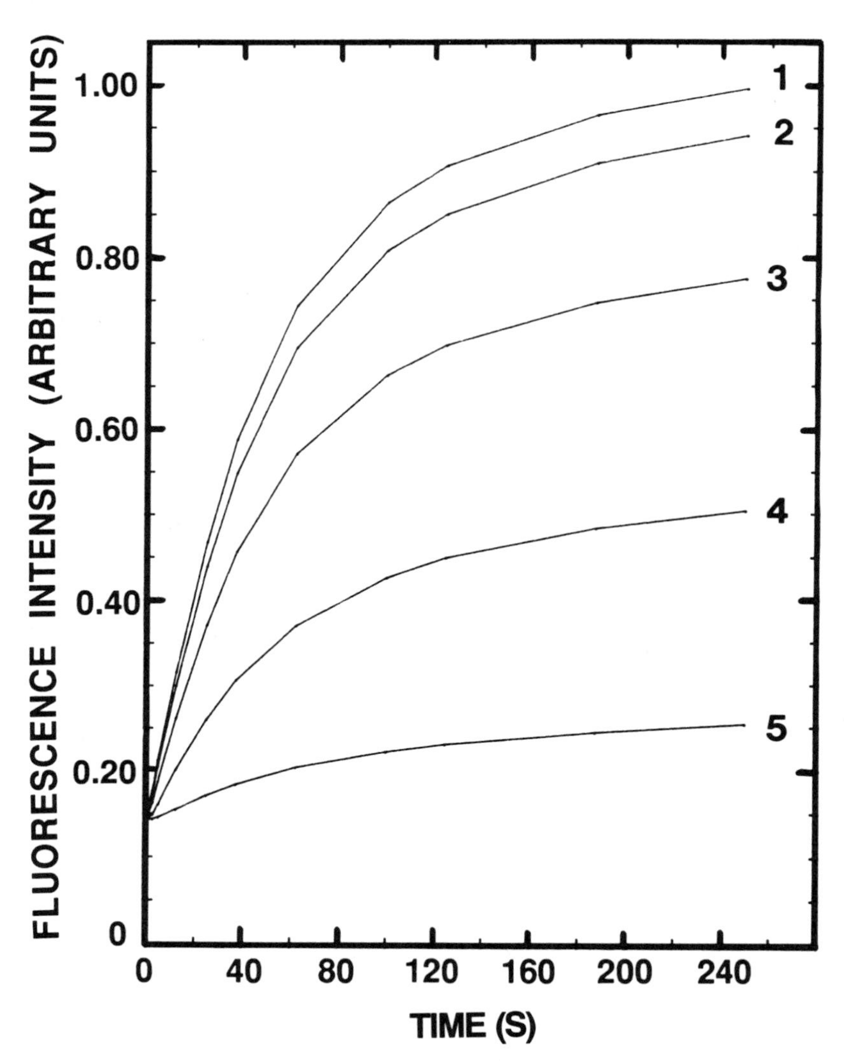

Figure 5-16. *Modeling results illustrating the effect of the change in the monomer diffusion coefficient on photopolymerization kinetics detected by the fluorescence intensity change. The 24-5-μm film isolated from the air contact on the illuminated side was considered. Diffusion coefficients of the monomer* (cm^2/s) *used in computations were: 1-*10^{-7}*; 2-*10^{-8}*; 3-*10^{-9}*; 4-*10^{-10}*, and 5-*10^{-11}.

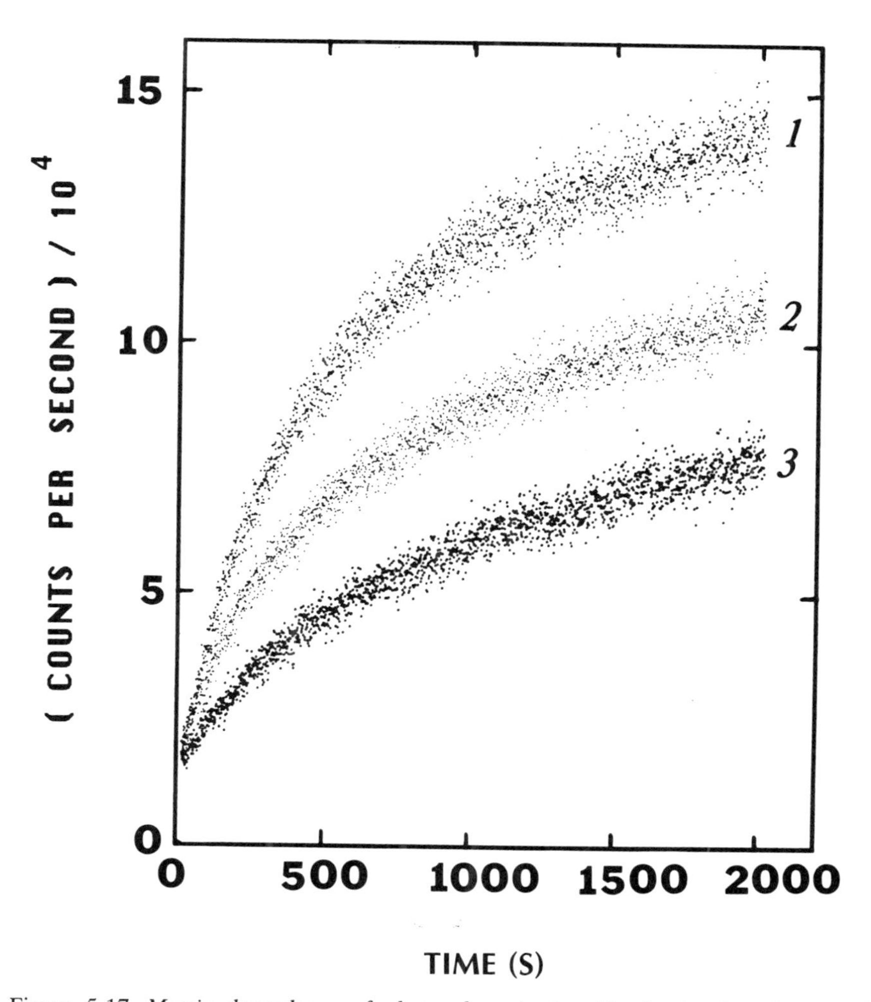

Figure 5-17. *Matrix dependence of photopolymerization kinetics in the absence of oxygen. the samples were degassed prior to the illumination in nitrogen atmosphere. The kinetics were measured in the following matrices: 1-cellulose acetate butyrate; 2-poly(vinyl butyrate); 3-poly(vinyl acetate).*

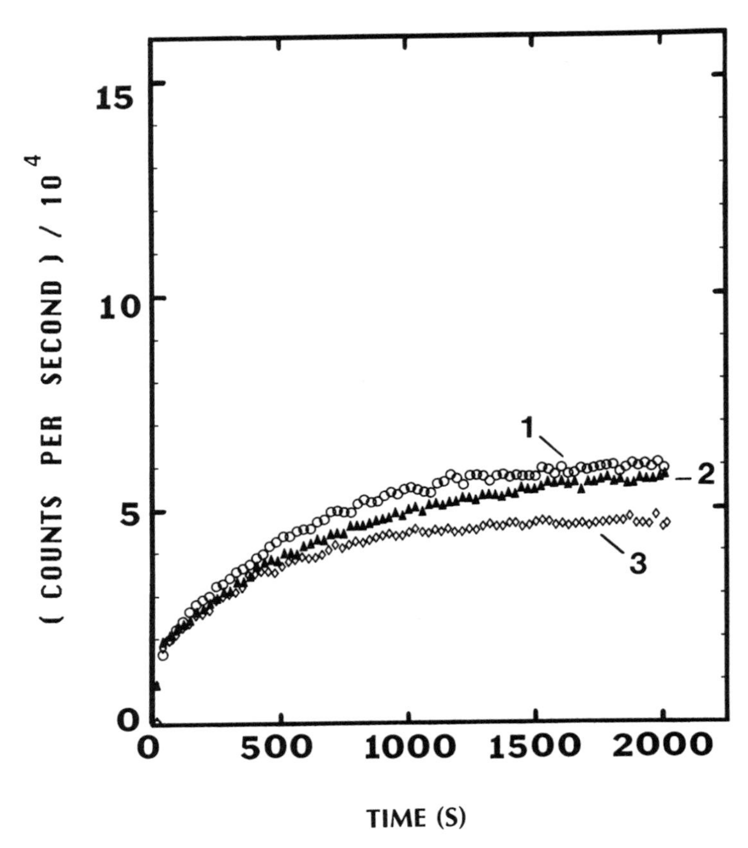

Figure 5-18. *Kinetics of photopolymerization in various matrices in the presence of oxygen. The illuminated side of the film was in contact with the oxygen atmosphere. The following matrices were investigated: 1-poly(vinyl butyrate); 2-poly(vinyl acetate); 3-cellulose acetate butyrate.*

reacted oxygen and the reduction of the radical concentration are more efficient. The radical concentration, Eq. (6), becomes substantially lower and cannot be compensated by the increase in monomer diffusion rate. Thus, the matrix with the increased rate of molecular diffusion is not necessarily the best matrix for imaging in the presence of oxygen.

Let us now consider in more quantitative terms the behavior of the system exposed to oxygen. The relative change of the oxygen diffusion coefficient with the change of the polymer matrix could be expected to be similar to the relative change in the monomer diffusion coefficient.[105,106] The monomer diffusion coefficient in CAB was determined above to be equal (within the formalism of the surface "evaporation" model) to 1.5×10^{-9} cm^2/sec and in PVA to be 0.9×10^{-9} cm^2/sec. Therefore, the selected oxygen diffusion coefficient would change from 10^{-7} cm^2/sec to 6×10^{-8} cm^2/sec with the change from CAB to PVA matrix. The diffusivity of the chain-terminating agent would be similarly affected. Using these diffusion coefficients in the kinetic modeling, we see (Fig. 5-19) that, indeed, the inversion of the polymer yield ratios occurs when the polymerization in two matrices proceeds in the presence and absence of oxygen. The matrix in which diffusion of the monomer is higher yields less polymer upon irradiation in the presence of oxygen (Fig. 5-19). For comparison with the experimental data, the computed fluorescence intensity dependence on the exposure time is presented (Fig. 5-19).

4.5. Oxygen Effects on the Polymer Spatial Distribution

The reduction of spatial inhomogeneities in polymer formation during anisotropic photopolymerization is of particular interest. Despite the absence of experimental data on the film photopolymerization, we learned from simple experiments on the acrylate photopolymerization in a beaker that the homogeneity of the polymerization is destroyed by the presence of oxygen. The visible gelation due to polymerization occurred in the depth of the reaction vessel and near the walls, where radical scavenging by oxygen is minimal. The open surface of the polymerizing mixture remained liquid long after the rest of the mixture had solidified. Similar behavior was detected for thinner polymerizing layers by NMR imaging.[99] The "undercut" observed in the developed photoresists and in the flexographic printing plates was also attributed to the presence of oxygen in the system.[19] In the absence of experimental data, the analysis of various theories of oxygen effects on the spatial distribution of molecules in the photopolymer film must be carried out by kinetic modeling. We expanded earlier efforts[19] to construct a comprehensive model of photopolymerization in the film[69–72] and used it to deduce the spatial distribution of the reactants and products in the photopolymer.

It was proposed some time ago that by covering the surface of the photopolymer with a transparent oxygen impermeable cover one may be able to reduce oxygen effects and increase the polymer yield during photopolymerization in films.[107,108] The cover would prevent oxygen diffusion into the photopolymer during imaging. Let us consider what effect such an oxygen protection would have on the image uniformity. When the photopolymer film in the model is exposed to oxygen only during manufacturing, it contains 90 ppm of dissolved oxygen and is isolated from the contact with air by the cover sheet and the substrate, the computed distribution of the polymer formation under unidirec-

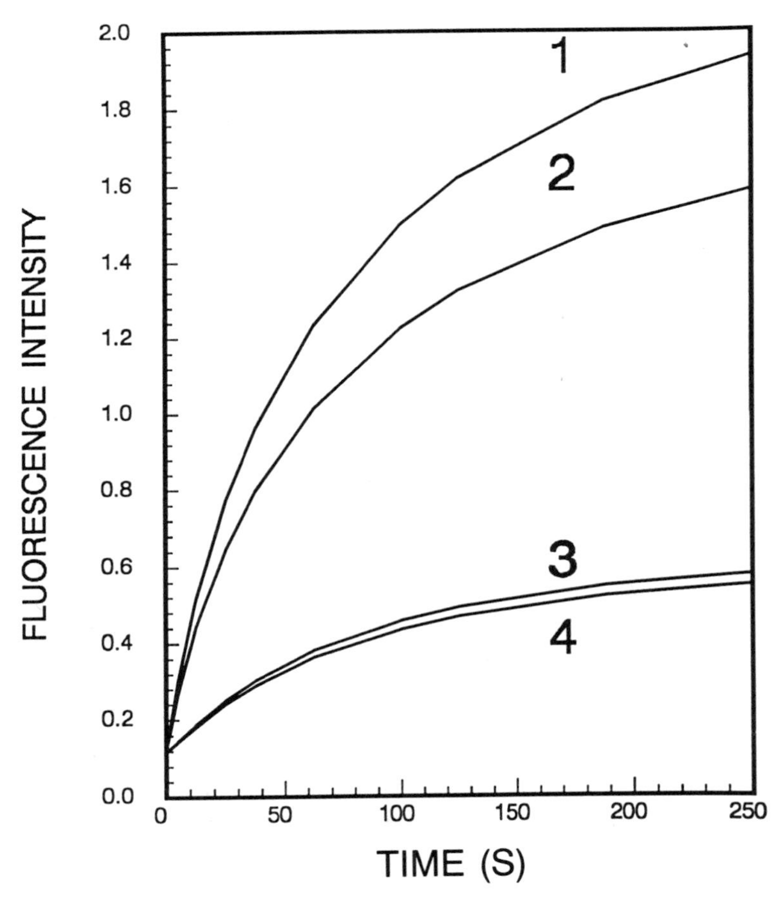

Figure 5-19. *Computed kinetics of photopolymerization in the different matrices: 1-high diffusivity matrix,* $D_{MONOMER} = 1.5 \times 10^{-9}$ *cm²/s, no oxygen; 2-low diffusivity matrix,* $D_{MONOMER} = 0.9 \times 10^{-9}$ *cm²/s, no oxygen; 3-high diffusivity matrix opened to oxygen,* $D_{MONOMER} = 1.5 \times 10^{-9}$ *cm²/s,* $D_{OXYGEN} = 1 \times 10^{-7}$ *cm²/s; 4-low diffusivity matrix opened to oxygen,* $D_{MONOMER} = 0.9 \times 10^{-9}$ *cm²/s,* $D_{OXYGEN} = 0.6 \times 10^{-7}$ *cm²/s.*

tional exposure to light is inhomogeneous (Fig. 5-20). Inhomogeneity is expected and predicted even by simple modeling.[19] However, the maximum in the polymer yield that may occur according to our computations[68–72] at longer exposure times some distance away from the illuminated surface (Fig. 5-20) requires some discussion.

The concentration of radicals is highest near the illuminated surface of the film. Thus, with short exposure (at the beginning of the reaction), the polymer yield is highest near the illuminated surface and follows an exponential decrease of the excitation light intensity within the film. The consumption of oxygen and the monomer is also initially higher near the illuminated surface (Figs. 5-12, 5-13). As oxygen is more mobile than the monomer, the larger fraction of radicals initially reacts with oxygen. This leads to faster consumption and depletion of oxygen in the photopolymer (Fig. 5-13), while there is still a large fraction of monomer left deeper within the film (Fig. 5-12). The initiator and the growing chain radicals have low mobility, and their consumption cannot be compensated by their diffusion (Fig. 5-11). Formation of radicals farther away from the pho-

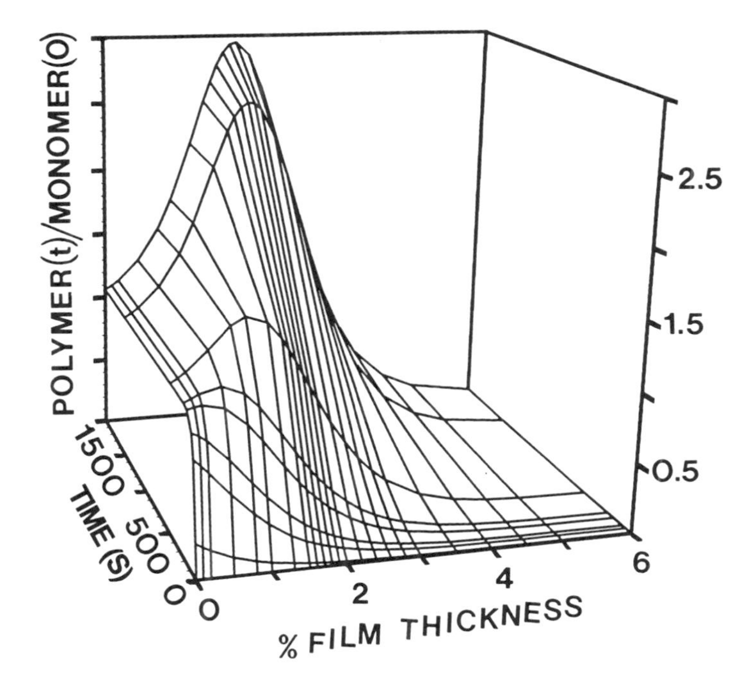

Figure 5-20. *Dependence of the concentration of the forming polymer on the distance from the illuminated surface and on the time from the beginning of the photoexposure. Both surfaces of the film were isolated from contact with the atmosphere.*

topolymer film surface, where the excitation light intensity is low, occurs more slowly than near the surface. As a result, when most of the oxygen is consumed, radicals forming deeper within the film survive and have a higher probability of reacting with the monomer diffusing from the depth of the film where the light intensity and, hence, radical concentration are low. Therefore, according to our modeling of a photopolymer film protected from the uptake of oxygen by a cover sheet, one may observe a maximum in polymer yield some distance away from the illuminated surface (Fig. 5-20). The distance of the maximum yield from the illuminated surface would be determined by the light intensity, the optical density of the film, and the relative mobility and concentration of oxygen and monomer.

Oxygen induced inhomogeneity of polymerization in photopolymer films is rather troublesome in the imaging and the electronic applications of photopolymers. It particularly affects the quality of the systems where the imaged and the not-imaged areas are differentiated by the wash-off or the peel-apart methods.[37] It is impractical to conduct a large-scale coating operation in the absence of oxygen in an inert environment. As was demonstrated by the model computations (Fig. 5-20), the use of an oxygen impermeable cover is not sufficient to prevent the lack of uniformity in the polymer yield profile. One of the means to reduce the oxygen content immediately prior to the imaging exposure is based on the utilization of higher oxygen diffusivity and reactivity relative to the monomer. Thus, the oxygen concentration in high optical density films can be lowered by pre-exposing one side of the film to light and, after that, using the flip side of the film for image recording.[19,109] A similar approach to the reduction of oxygen content in a photopolymerizing reaction mixture by several laser pulses was used in the quantitative analysis of the oxygen diffusivities and the oxygen inhibition of photopolymerization.[110,111]

Our model computations revealed a considerable decrease of oxygen content after one-sided exposure of the isolated photopolymer film (Fig. 5-13). In a relatively short time, even at the low light intensity used in modeling and in experiments, the oxygen content becomes uniformly low throughout the film. The computations confirmed that the consequent exposure of the reverse side of the isolated film would result in a significantly higher polymer yield and an improved image uniformity (Fig. 5-21(a)). The yield of the polymer near the illuminated surface of the isolated film is more than two times higher after the preexposure of the flip side than it is after single exposure. The polymer yield after long exposure becomes almost independent of the distance from the illuminated surface (Fig. 5-21(a)). These results confirm the validity and illustrate the benefits of oxygen removal by reverse side pre-exposure of the photopolymer film protected on both sides from the oxygen uptake.

The image uniformity through the depth of the film can sometimes be achieved by exposing the films to light while keeping the oxygen concentration

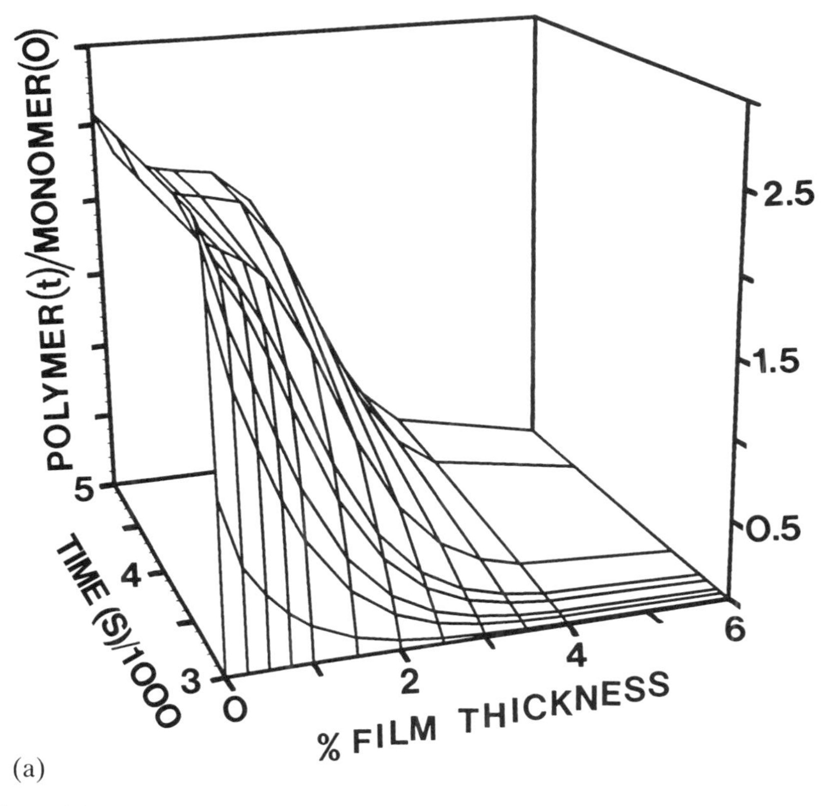

Figure 5-21. *Computed dependence of the spatial distribution of the forming polymer on time of the illumination. (a) Distribution of polymer after the second exposure, which "started" immediately after the reverse side of the film was pre-exposed to light for 3000 s. Both surfaces of the film were protected from contact with oxygen. (b) Computed results of the first exposure of the film with the illuminated surface opened to contact with oxygen.*

constant. The loss of oxygen within the film due to the photochemical reaction can be replenished by keeping the illuminated surface of the film open to air. Considering the high oxygen concentration gradient between the film and atmosphere and the relatively high oxygen diffusion coefficient, the rate of diffusion of oxygen into and within the film is high relative to the monomer diffusion-controlled chain growth. In the course of the reaction, when the illuminated surface of the photopolymer is open to air, the radicals formed near the illuminated surface and deeper inside are exposed to a roughly constant oxygen

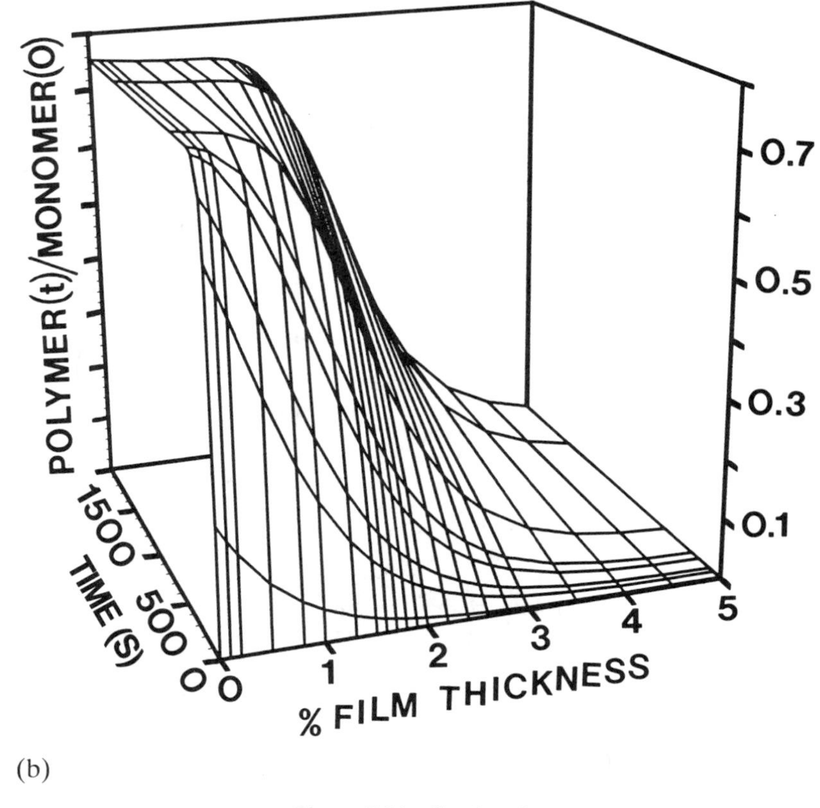

(b)

Figure 5-21. *Continued*

concentration. Computations demonstrate that the concentration of the forming polymer will be uniform at longer exposure times throughout the photopolymer film, when the illuminated photopolymer film surface is open to air (Fig. 5-21(b)). Of course, the maximum yield of the polymer will be lower than in the protected film, and the forming polymer chains will be shorter than in the protected film (see also Figs. 5-14, 5-15, 5-17–5-19). The computed nonuniformity of the polymer formation throughout the film contradicts the expectation that, in the presence of oxygen, the imaging is always nonuniform.[19] The modeling results described here indicate that, in those applications in which a high-depth uniformity of the image is required (for example, in the flexographic plate preparation),[112, 113] the uniform image may be obtained by opening the illuminated surface to air rather than by protecting it from contact with air by a cover film. Perhaps the use of sensitizing dyes forming active peroxides with oxygen[108, p. 122 and reference therein] would improve the polymer yield while maintaining a uniform image profile. Oxygen effects on photopolymerization are considered in numerous publications;[19,68–72, 110–115] however, direct experimental observations of the oxygen effects on spatial distribution of products in bulk photopolymerization are not available.

4.6. Patterned Exposure of Photopolymer Films

Photopolymers are designed for the reproduction of patterns in electronic and imaging applications. The pattern features recorded in photopolymers are usually comparable in their dimensions to the thickness of the photopolymer film. The thickness of the lines projected for imaging on the photopolymer surface can vary from 0.2 μm in the holographic and compact disc applications to 50 μm and more in waveguides and flexography. It was mentioned above that under patterned illumination of the film, the monomer, oxygen, and other reagents will diffuse into the illuminated area not only from the depth of the film perpendicular to the illuminated surface, but also from the shadowed regions of the projected image, moving parallel to the film surface. The relative contribution of these two processes is determined by the relationship between the area and the perimeter of the illuminated region.

Numerical analysis of the kinetics of localized polymer imaging with the consideration of two-dimensional diffusion was conducted by the finite element techniques using a Cray computer. The diffusion of every component was assumed to be nonzero in the kinetic scheme presented above.[68–72] Two types of illumination patterns are prevalent in the photopolymer applications: stripes with the length substantially exceeding the width, and circular dots. Expressed in cylindrical coordinates, the problem of diffusion towards the circular illuminated dot is equivalent to the problem of diffusion toward an infinitely long (negligible end effects) illuminated stripe. We modeled the photopolymerization kinetics in the high optical density photopolymer film exposed to a series of alternating dark and illuminated stripes. Exposures with the sinusoidal variation of light intensity with distance were also analyzed. Diffusion in cases of randomly exposed regions can be visualized based on the results of the regular pattern exposure.

Let us consider a 25-μm-thick photopolymer film protected from contact with air by a cover sheet with a composition similar to that described above, which is used in holography, proofing, etc. Let us assume that it is exposed to light in a pattern of dark and light stripes with an incomplete Gaussian distribution of light intensities across the illuminated region (Fig. 5-22). The results of the model computations for the short exposure duration (relative to the total reaction time) indicate that the polymer yield distribution follows the light intensity pattern within the film; that is, the yield is higher at the center of the stripe and

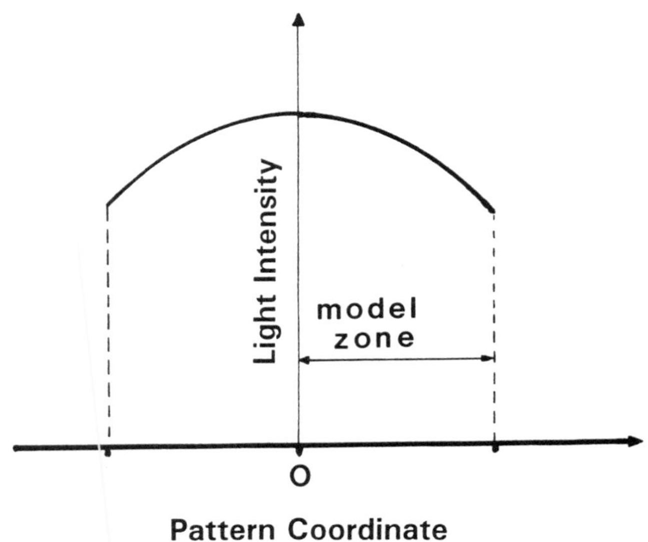

Figure 5-22. *The distribution of intensity of the exposing light (used in model computations) as a function of the distance from the center of the stripe in the light pattern.*

close to the illuminated surface (Fig. 5-23). At longer exposures, when the oxygen concentration is reduced by reaction with radicals, the monomer consumption within the illuminated stripe is compensated by the diffusion from the depth of the film and from the shadowed regions of the pattern. As a result of monomer diffusion and the fast, non-compensated depletion of oxygen, the yield of polymer starts increasing at the edges of the illuminated stripe and some distance away from the illuminated surface (Fig. 5-24). At advanced stages of photopolymerization, the yield of the polymer at the edges of the illuminated stripe and away from the illuminated surface exceeds that in the center of the illuminated region and at the surface of the film (Fig. 5-25). These computational results can be verified experimentally. Indeed, although it is difficult to detect the difference in polymer yield within the depth of the film, the changes taking place on the film surface can be observed.

4.7. Exposure Induced Swelling of Photopolymer Films

The influx of the monomer and the accumulation of the forming polymer in the illuminated regions should lead to stretching of the matrix and swelling of

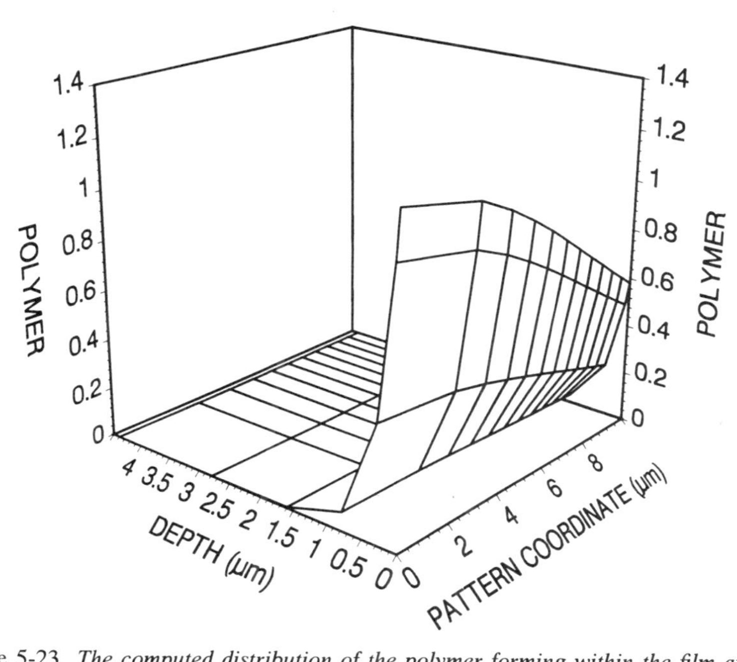

Figure 5-23. *The computed distribution of the polymer forming within the film after 10 seconds of the patternwise exposure to light. Both film surfaces were assumed to be isolated from the contact with oxygen.*

the illuminated regions of the photopolymer film. Such a swelling was observed in the holographic photopolymers.[23,72] When the exposure is short, the swelling shape follows the incident light intensity distribution (Fig. 5-26). At longer exposure times, the edge of the pattern swells more strongly [72] (Fig. 5-27), as was predicted by the computations (Fig. 5-25). The detection of the expected swelling of the exposed regions of the photopolymer film and the similarity of observed and predicted shapes of the swelled regions provided confirmation of the diffusion control of the photopolymerization reactions.

The correlation between the exposure induced monomer diffusion and the resulting photopolymer swelling was noticed only recently.[22,41,72] Utilization of the photo-induced swelling of the photopolymer for information storage and compact disc recording was reported,[41] as well as the detailed analysis of the swelling mechanism.[22,72] The shrinkage observed in the oligomer crosslinking based photocurable systems is well understood.[116] Results presented here demonstrate that swelling of the illuminated regions can be expected in the monomer diffusion based photoactive polymers.

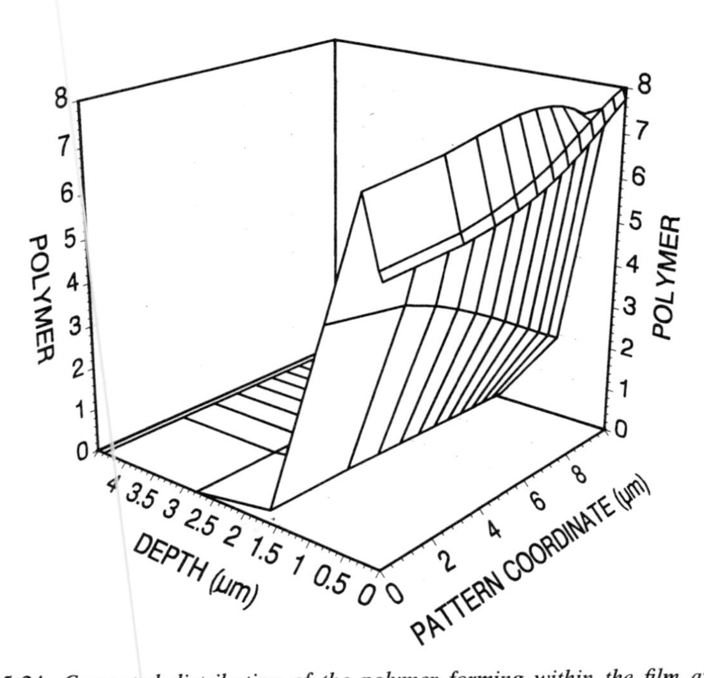

Figure 5-24. *Computed distribution of the polymer forming within the film after 350 seconds of patternwise exposure to light. Both film surfaces were assumed to be protected from contact with oxygen.*

4.8. Image Resolution and Swelling

One of the most important issues in photopolymer applications is image resolution. In optics and electronics the term ''resolution'' is defined quite clearly and quantitatively. However, in photoresist and photopolymer applications, as well as in printing and proofing, the definition is lost in practical traditions. Thus, the resolution is most frequently quantified by looking at the smallest line of a standard pattern visible after projection through the pattern mask on the photopolymer or the photoresist.[108] The typical ''Military Standard''pattern consists of parallel lines of various sizes.

The swelling of photopolymers is related to the issue of a resolution achievable in photopolymer imaging, because a change in the polymer surface morphology would change the results of the postprocessing such as toning and metallizing. A uniform grating of the alternating transparent and metallized regions etched by an electron beam in the chromium layer of the metallized quartz plate was used as a photomask for investigating resolution limits of a

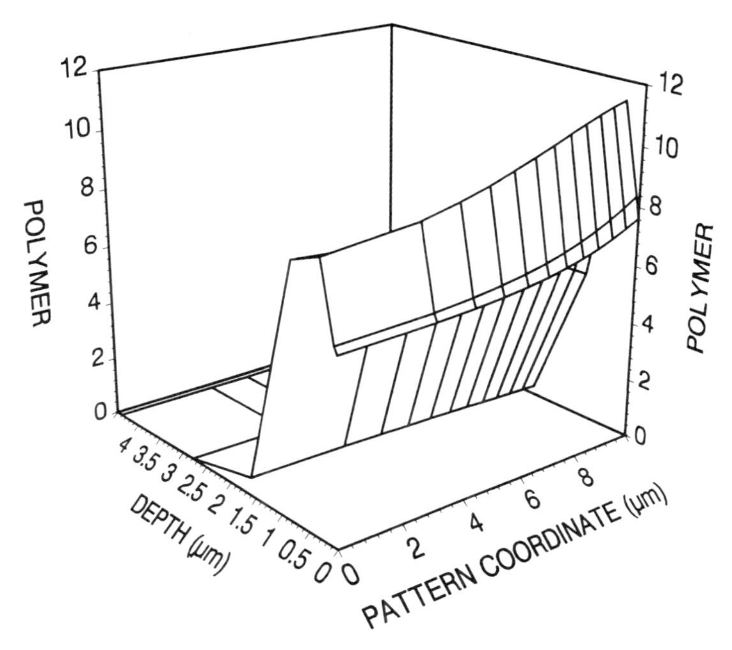

Figure 5-25. *The computed distribution of the polymer forming within the film after 500 seconds of the patternwise exposure to light. Both film surfaces were assumed to be isolated from oxygen contact.*

holographic photopolymer. An image of the photomask placed on the photopolymer was investigated. To avoid embossing, the photomask was separated from the photopolymer film surface by a thin gasket.

The image of a mask with a 50-µm line size was viewed between crossed polarizers in reflected light using a differential interference contrast technique.[117] The swelling of the illuminated portions of the film was clearly visible (Figs. 5-28, 5-29). Apparently, the alignment of the polymer molecules perpendicular to the film surface and the boundary of the illuminated regions resulted in the dichroism around these photopolymer regions. Because monomer diffusion towards the illuminated regions also occurs in the same directions, it can be concluded that the observed dichroism resulting from anisotropic polymer growth can be used as another direct experimental confirmation of the monomer diffusion towards the illuminated regions during photoexposure.

To reconfirm the diffusional origin of the image formation in photopolymers and to simplify the imaged area's visualization, the liquid crystalline mixture was added to the formulation,[22] as liquid crystalline phases and dichroic dyes

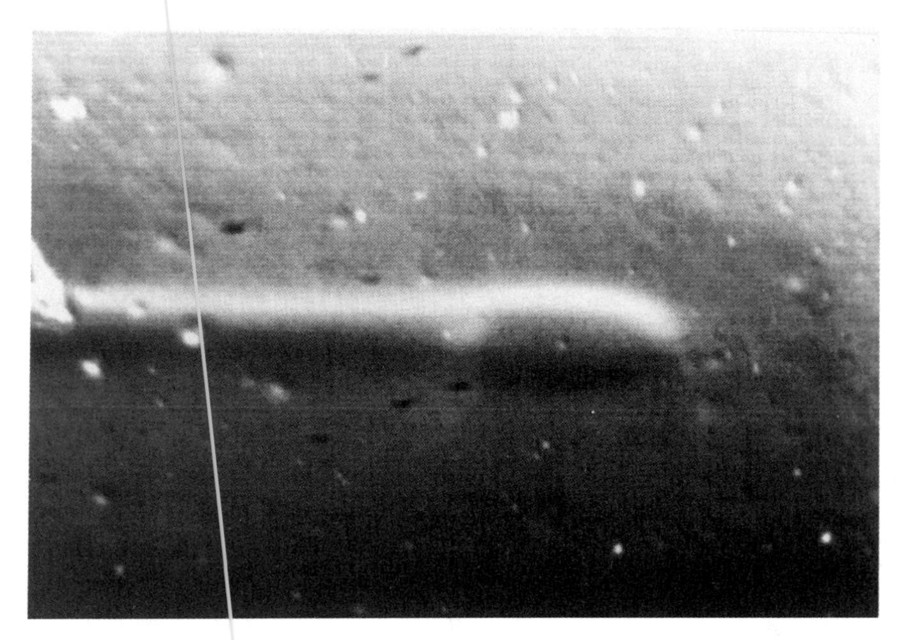

Figure 5-26. *Observed swelling of the illuminated region after a short exposure of the photopolymer to light.*

have the ability to align along the flow and polymer stretch direction.[118] The formation of the parallel orientation of the polymer molecules around the swollen region observed in polarized light becomes substantially more visible in the presence of orientable phases. (Fig. 5-30). During photopolymerization, liquid crystals and dichroic dyes align perpendicular to the film surface and boundary of the illuminated region, as can be seen using differential interference contrast microscopy[117] (Fig. 5-30). This provides unequivocal experimental confirmation that monomer diffuses towards light.

The larger the shadowed regions separating the illuminated regions, the larger is the pool of the monomer available for polymerization. The mechanical performance of the materials is also dependent on the size of the area subjected to stress. When high-resolution imaging of photopolymer film is attempted, these factors must be considered. When the holographic photopolymer was exposed through the photomask with a 5-μm period grating pattern, the swelling of the individual illuminated lines was lower than in the case of 50-μm grating (Fig. 5-31). When the distance between the illuminated regions was further reduced (0.5-μm grating), no swelling of individual lines was observed, although the entire area of the exposed pattern was raised over the surrounding unexposed film surface (Fig. 5-32 (a)). The image

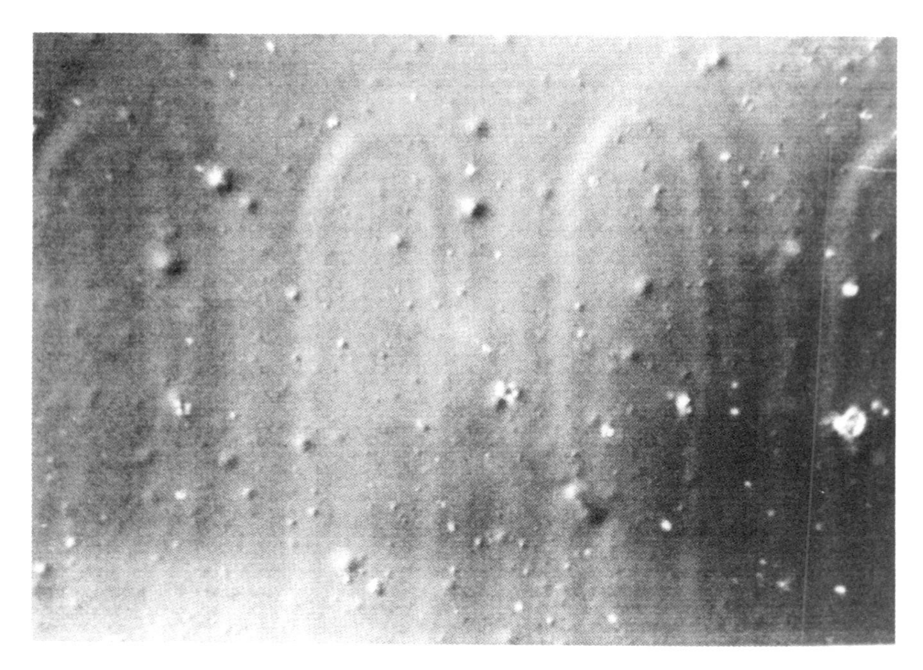

Figure 5-27. *Observed swelling of the illuminated regions of the photopolymer after prolonged exposure to light.*

of the 0.5-μm lines in the pattern can be observed only in transmitted light, confirming that image recording by monomer diffusion has indeed occurred without individual swelling of the illuminated regions (Fig. 5-32(b,c)). Apparently the elasticity of the polymer matrix (PVA, CAB, or PVA) is not sufficient to allow individual stretching on a submicron scale. The pattern size dependence of swelling explains why, in hologram recording where the pattern size is small (0.2–0.5-μm patterns), the swelling and mechanical image distortion are not noticeable.

The swelling and the polymer alignment in the imaged areas can be successfully used in the production of optical waveguides. However, when the photopolymer is used as a master (for example, in electrostatic printing), the difference in the extent of swelling of various size halftone dots or lines can cause distortions of the print and variations in the dot-gain. Variations in swelling can also affect its application in digital recording.[41]

4.9. Self-Focusing and Monomer Diffusion

Photobleachable, highly light-absorptive dyes are added to photopolymer[119] and photoresist[120] formulations to reduce light scattering and diffraction effects.[101]

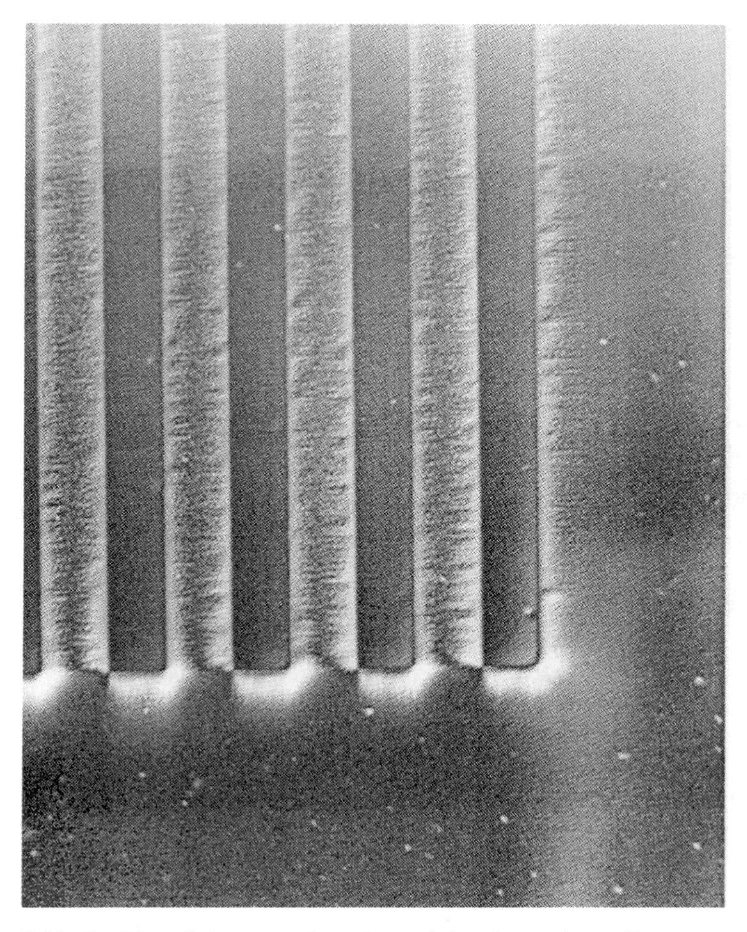

Figure 5-28. *Swelling of the exposed portions of the photopolymer film as seen under the reflected light and external side illumination. The sample was placed between two crossed polarizers. The objective was equipped with a Nomarski prism. The size of the exposed line was 50 μm.*

When the photopolymer containing such a dye is imaged through a submicron photomask, light is initially absorbed in a very thin film layer, preventing depth scattering and diffraction. As the exposed portion becomes transparent, light is diffracted most strongly in the direction normal to the film surface, creating a self-focusing effect.[119,120] The resulting image has side walls normal to the illuminated photoresist surface. These effects are rather interesting, considering

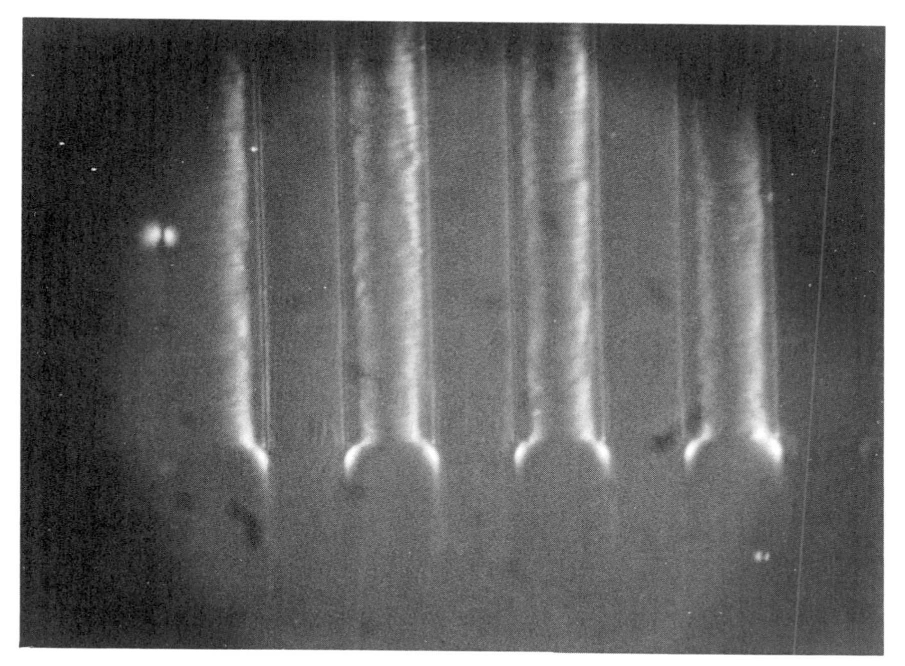

Figure 5-29. *The same pattern as in Fig. 5-28 seen in polarized reflected light through the Nomarski prism-equipped objective: (1) The lines in the pattern were perpendicular to the polarization plane of the first polarizer; (b) Lines in the pattern were at a 45° angle to the planes of polarization; (c) Lines were turned 90° relative to those of (a).*

that one can obtain a pattern image with dimensions below 0.5 μm with a very high depth uniformity. A substantial amount of work was devoted to the development of contrast enhancing coatings[121,122] and the theoretical analysis of self-focusing in the pattern reproduction in photoresists.[120,123] It is difficult to establish the exact composition of industrial materials, therefore it is not clear whether there was any possibility of molecular migration during the time of the exposure.

Self-focusing was also observed in the refractive index-changing materials such as photopolymers used in holography and proofing (Fig. 5-32).[124] These materials have a high optical density at the wavelength used for exposure; no drop of optical density during exposure was detected.[124] We observed formation of a uniform image of 0.5-μm bars of the mask through the entire thickness of the 25-μm-thick photopolymer film.[124] The intensity of light within the investigated photopolymer film is very low in the absence of photobleaching. There-

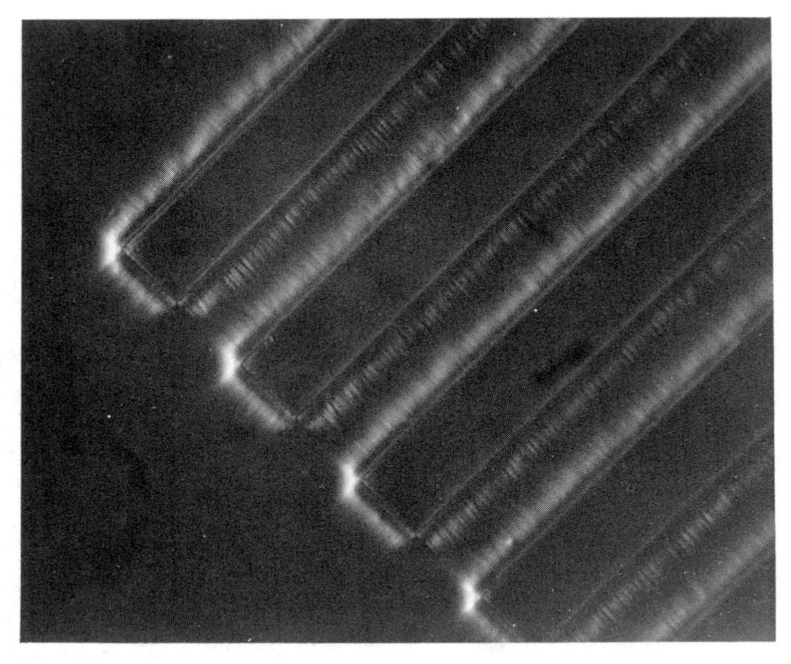

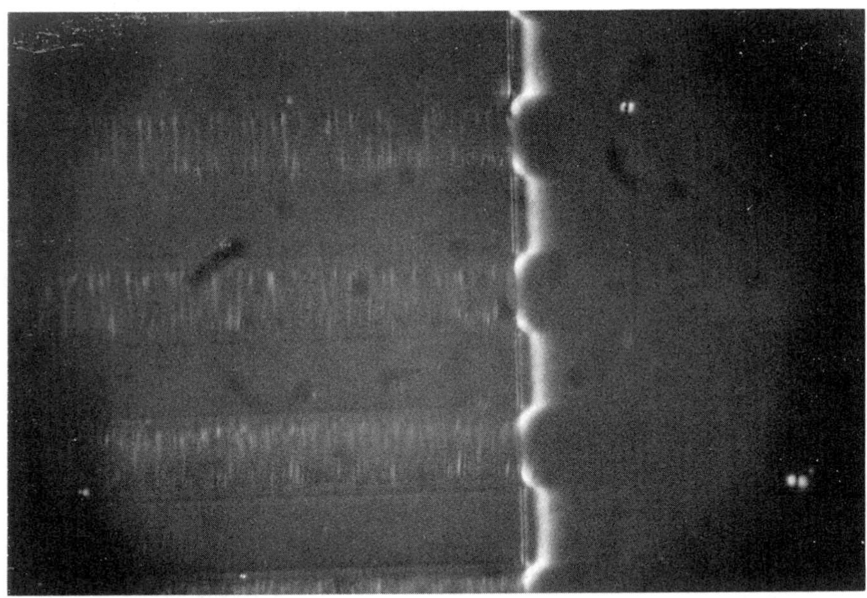

Figure 5-29. *Continued*

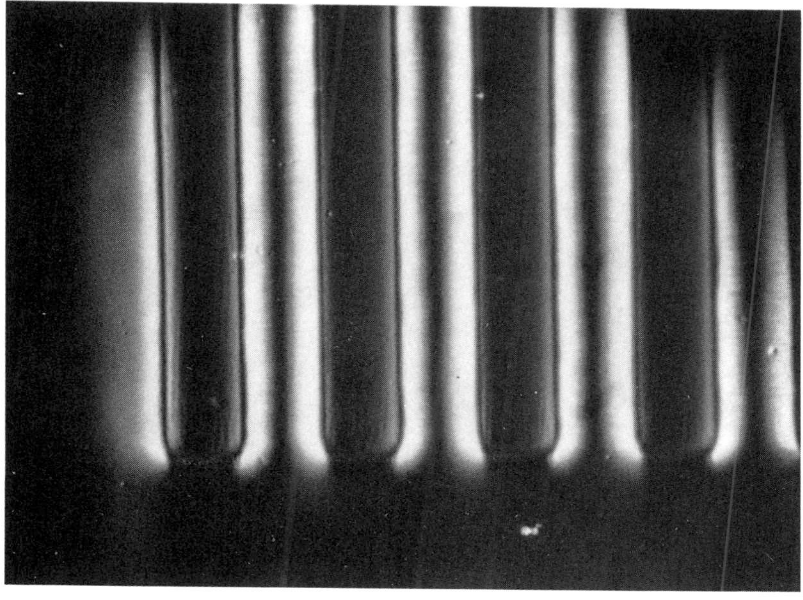

Figure 5-30. *Swelling of the exposed regions of the photopolymer containing 0.1% of the inert liquid crystal phase in formulation. Pattern period was 50 μm. The sample was placed between two crossed polarizers and was viewed in reflected light through the objective equipped with a Nomarski prism: (a) The lines in the pattern were perpendicular to the polarization plane of the first polarizer; (b) Lines in the pattern were at a 45° angle to the planes of polarization; (c) Lines were turned 90° relative to those of (a).*

fore, formation of the uniformly polymerized narrow regions has to be caused by effects other than the optical self-focusing. Our computations indicated that an almost uniform polymer yield can be achieved throughout the film thickness as long as the diffusivity of the initiator is substantially lower than that of the monomer.[124] We conclude the diffusion of the monomer, oxygen, and other reagents toward the light can explain the self-focusing effects observed in the photopolymer films containing an immobile photoinitiator.

4.10. Mobile Initiator

In a system with an immobile initiator, even weak illumination in the depth of the photopolymer film eventually leads to polymerization throughout the depth of the film. In the case of a mobile initiator, the initiation rate in the darker regions of the film is not sufficient to compete with the rate of the initiator

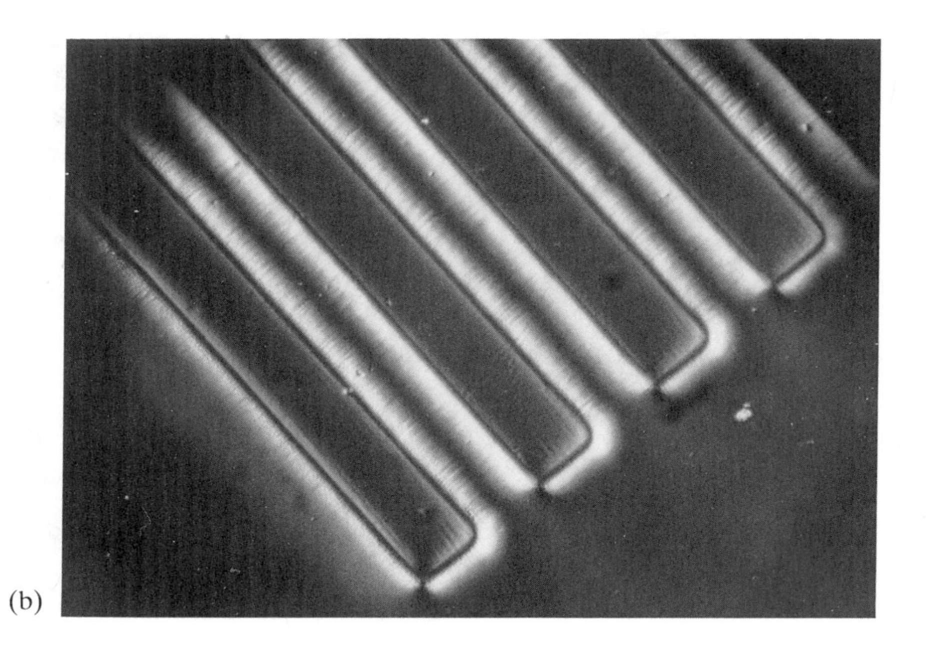
(b)

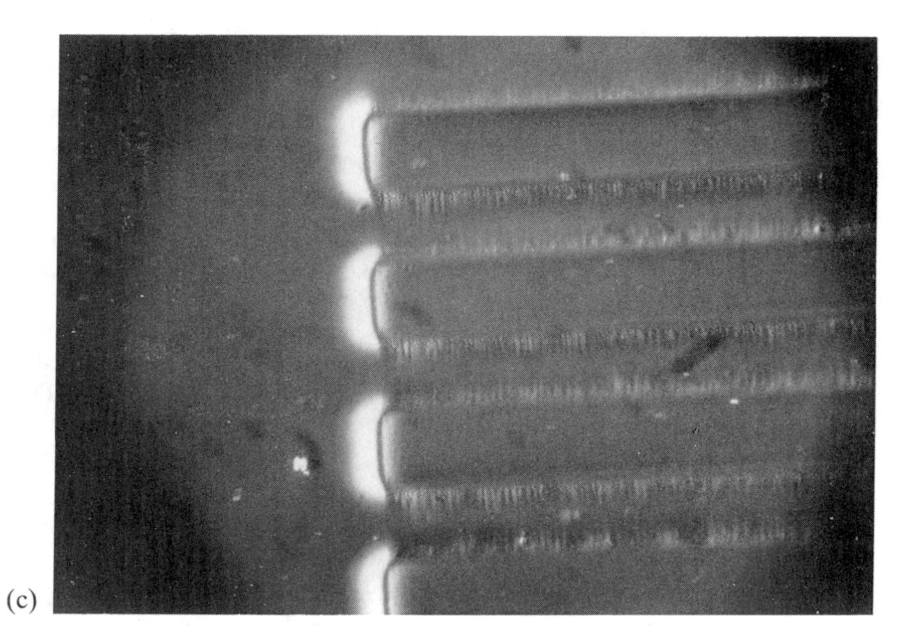
(c)

Figure 5-30. *Continued*

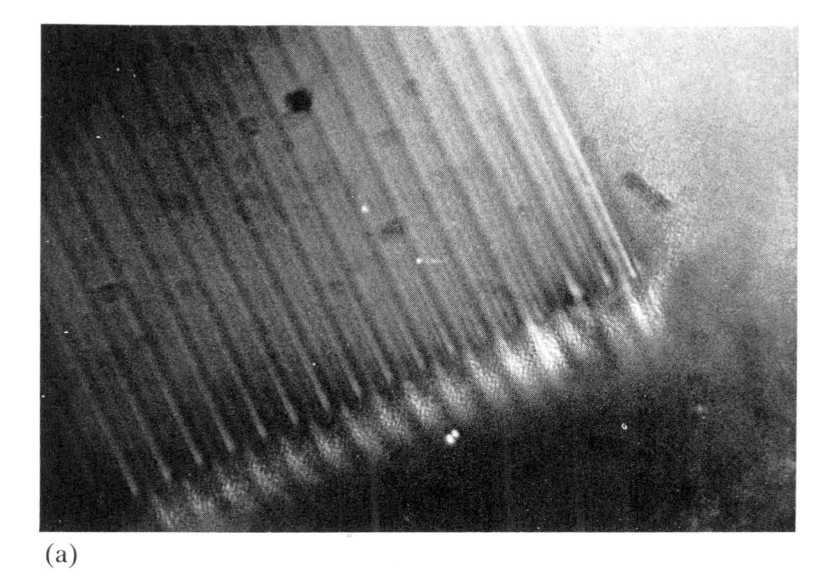

(a)

(b)

Figure 5-31. *Photopolymer exposed through 5-μm mask. The sample was viewed as in Figs. 5-28 and 5-29. Reduction in swelling of the individual lines was observed. Swelling of the entire pattern was visible: (a) Photopolymer film contained no liquid crystals; (b) Photopolymer formulation contained 0.1% of the liquid crystal phase.*

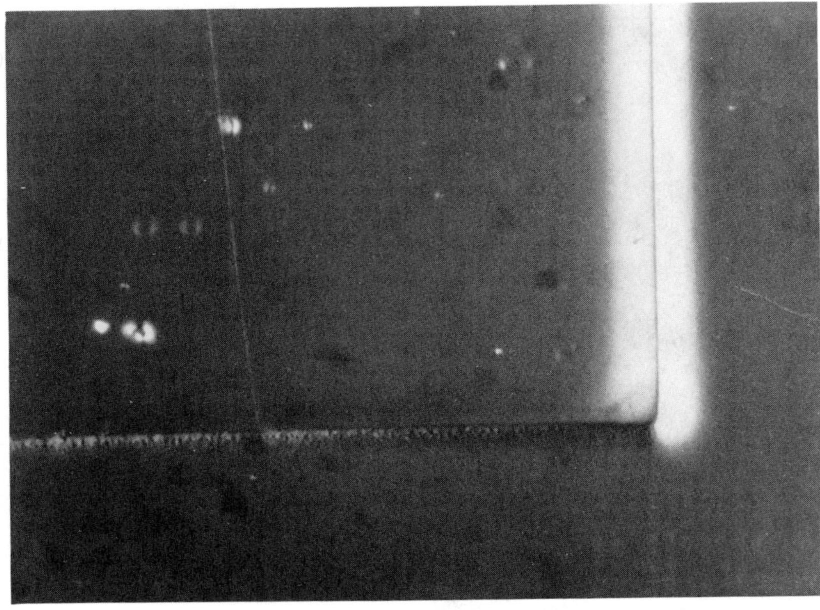

(a)

Figure 5-32. *Photopolymer containing 0.1% of liquid crystal mixture exposed through the 0.5-μm mask: (a) The exposed film was viewed in the reflected light as in Figs. 5-29 through 5-31. No swelling of the individual lines in the pattern was detected. Overall swelling of the exposed region was visible. (b) (c). The same exposed film viewed between two crossed polarizers in the transmitted light using Nomarski prism-equipped objective. Recorded 0.5-μm pattern was clearly visible.*

consumption at the surface and consequent initiator migration out of the depth. The initiator is removed from the depth of the film and photolyzed before the monomer can diffuse towards light. Numerical modeling confirms the intuitive analysis of the system with the mobile photoinitiator. When the diffusion coefficient of the initiator is close to that of the monomer (in our system around 10^{-9} cm^2/s), the polymer yield is low and closely follows the distribution of the light intensity within the film. When the rate of the initiator migration exceeds that of the monomer ($D_{\text{initiator}} = 10^{-8}$ or 10^{-7} cm^2/s), photopolymerization does not proceed in the depth of the film to any extent. The initiator decomposition occurs faster than the migration of most of the monomer.

To support model computations, photopolymer mixtures formulated with three different photoinitiators, *O*-chlorohexaarylbiimidazole, 1-hydroxycyclohexyl phenyl ketone, and 2,2′-azobis(2-methylpropionitrile), were tested. The

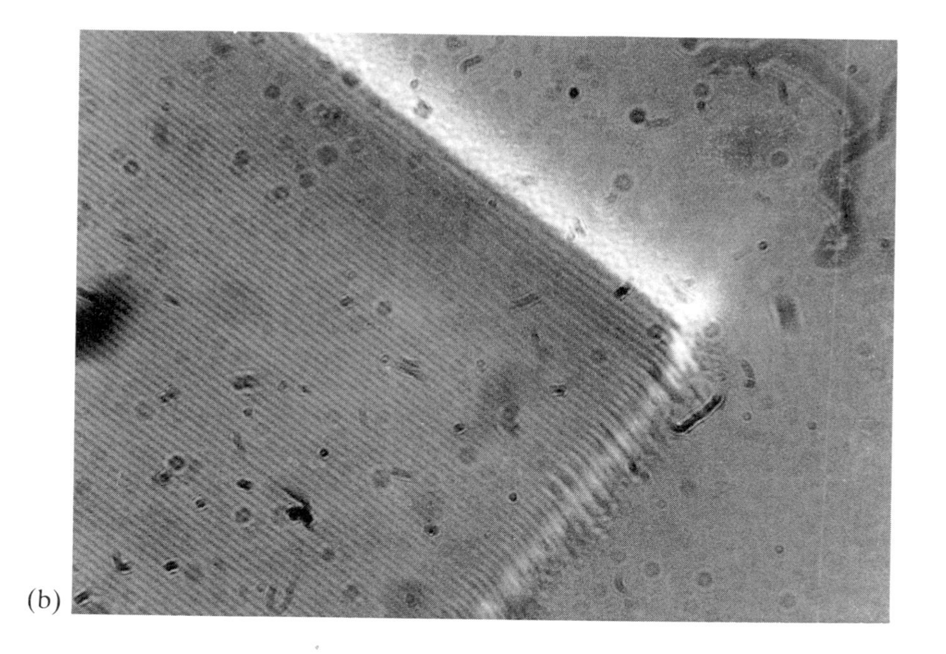
(b)

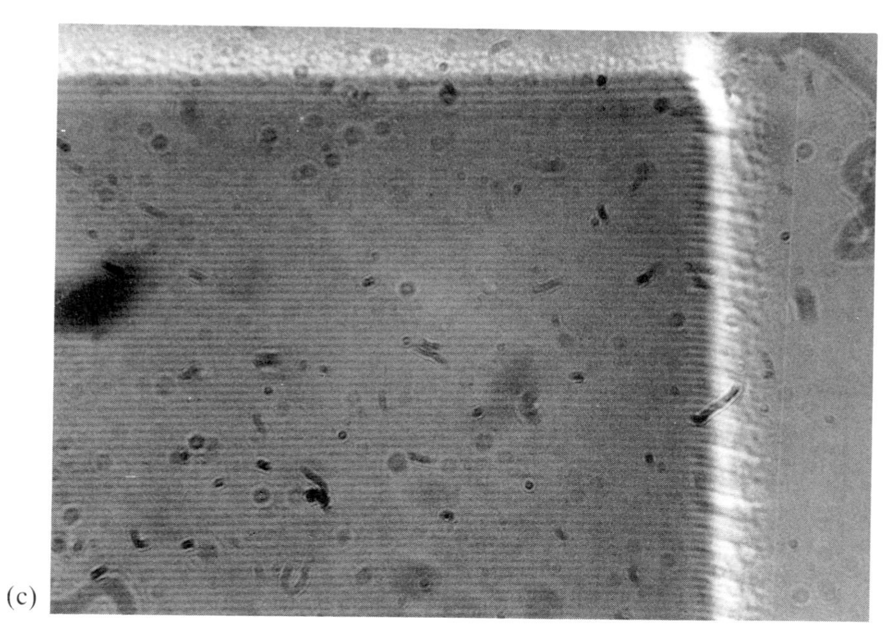
(c)

Figure 5-32. *Continued*

experiments were conducted in a nitrogen atmosphere after careful degassing of the photopolymer samples. The photopolymerization kinetics detected by the monomer fluorescence differed drastically for these three formulations (Fig. 5-33).[68] The photoinitiators studied quench *N*-vinyl carbazole fluorescence with different efficiencies.[125,126] Although the initial fluorescence of these formulations may differ, the relative change of the fluorescence intensity during photopolymerization provides information on the photopolymerization mechanism. When the initiator has mobility similar to that of a fluorescent monomer, it migrates towards activating light and is consumed at a rate close to that of the monomer. Thus, the extent of monomer migration towards the surface is lower than when the monomer is immobile. This is reflected in a lower rise in the intensity of fluorescence during photopolymerization (Fig. 5-33). Migration of the initiator towards light decreases the light intensity within the film and increases the ratio of the concentration of the quencher (initiator) to that of the fluorophore (monomer, polymer). This depresses the rise of fluorescence intensity during the time of the exposure even further. When an initiator is more mobile than a monomer, such as 2,2′-azobis(2-methylpropionitrile), migration of the photoinitiator towards activating light occurs prior to migration of the monomer. This is reflected in the kinetics of the film fluorescence change. When the initiator is highly mobile, the fraction of the light absorbed by the initiator and quencher to fluorophore concentration ratio increases near the illuminated surface because monomer migration to light is slower than that of the initiator. As a result, the fluorescence of this photopolymer film declines with the exposure (Fig. 5-33).[68] These results support computations showing that photopolymerization with a highly mobile initiator does not give a high yield of polymer. Further, more detailed investigations of initiator mobility effects are required.

4.11. Stability of the Initiator Radicals, Rotational Diffusion in Matrix

After the discovery and first characterization of hexaarylbiimidazole[80,81] the mechanism of the photodissociation and recombination of HABI was not adequately known.[80,81,127–129] Several isomers with C—N vs. C—C bonds between the imidazolyl moieties and the formation of imidazolyl trimers at certain temperatures were reported.[81,127–129] The unique stability of the lophyl radicals was attributed to a high radical stabilization by resonance.[130,131] Resonance was also claimed (without experimental proof) as a reason for the unusual insensitivity of the HABI photochemistry to the presence of oxygen. It was never explained why and how this resonance stabilization would retard the triarylimidazolyl (lophyl) radical recombination within the polymer cage.

Recent laser flash-photolysis experiments confirmed that the lifetime of triarylimidazolyl (lophyl) radicals in solution is disproportionately high (Fig.

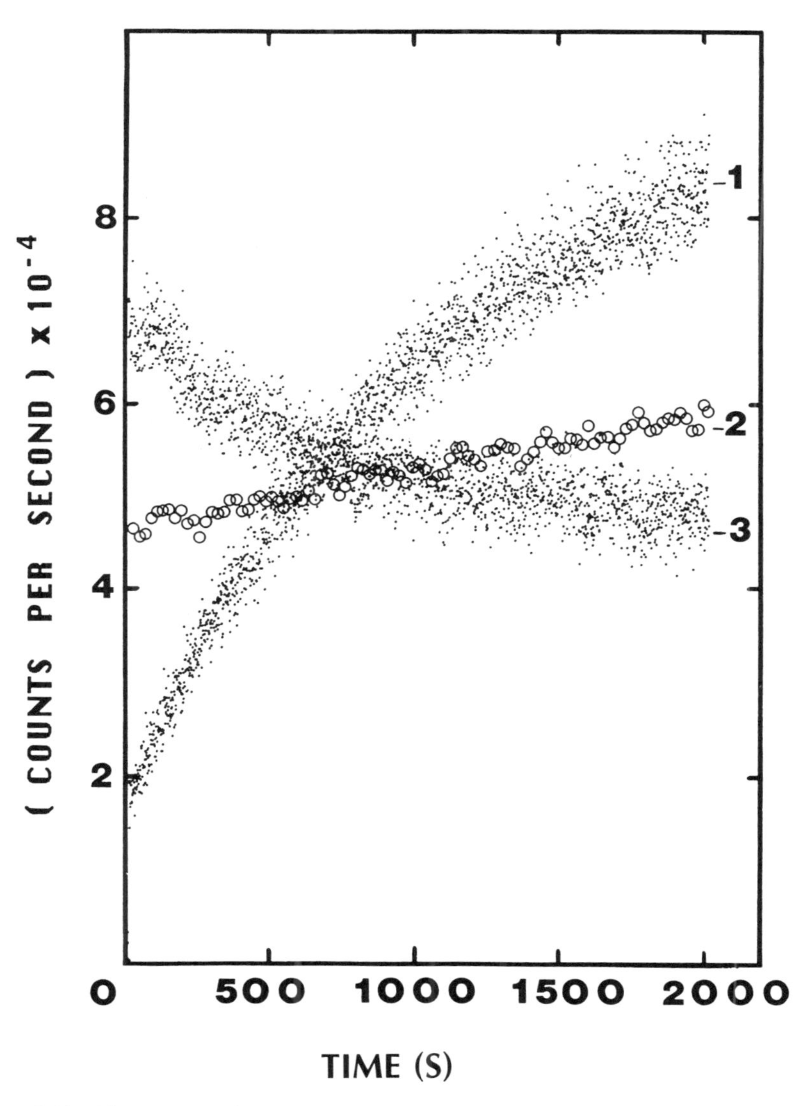

Figure 5-33. *Fluorescence detected kinetics of photopolymerization of the photopolymer formulations containing the different mobility initiators: 1-o-chloro hexaarylbiimidazole; 2-1-hydroxycyclohexyl phenyl ketone; 3-2,2′-azobis(2-methylpropionitrile).*

5-34).[73] Straightforward computations reveal that in the lophyl radical the phenyl and imidazolyl rings lie in a plane, whereas in the HABI molecule the phenyl and imidazolyl groups are not planar (Fig. 5-35).[73] Photodissociation of the "porcupine-like" HABI molecule leads to formation of a planar system of lophyl radicals, and recombination of lophyl radicals to form the HABI molecule can occur only through the torsion or rotation of the phenyl groups. The recombination of lophyl radicals also requires a shift of two imidazolyl rings relative to each other. Such a rearrangement occurs with the rate of the torsional and wagging vibrations and the "rotational diffusion" of these groups. Hindrance of the torsional motions and the low rate of rotational diffusion in a plasticized polymer matrix is responsible for the high stability and the low cage recombination probability of lophyl radicals. The HABI photodecomposition required for the polymerization initiation would also depend on the rigidity of the matrix.

To understand the mechanism of photoinitiation by HABI and to analyze its media rigidity dependence, HABI photodecomposition was conducted in methylene dichloride and toluene matrices at 5 to 8 K and in a poly(vinyl acetate) matrix at temperatures ranging from 8 K to room temperature using a 308-nm pulsed output of an excimer laser or continuous irradiation by a xenon arc lamp.[73] The concentration of HABI (3% by weight) was similar to one used in the photopolymer composition. The formation of lophyl radicals was monitored by the electron paramagnetic resonance (EPR), while the HABI-containing sam-

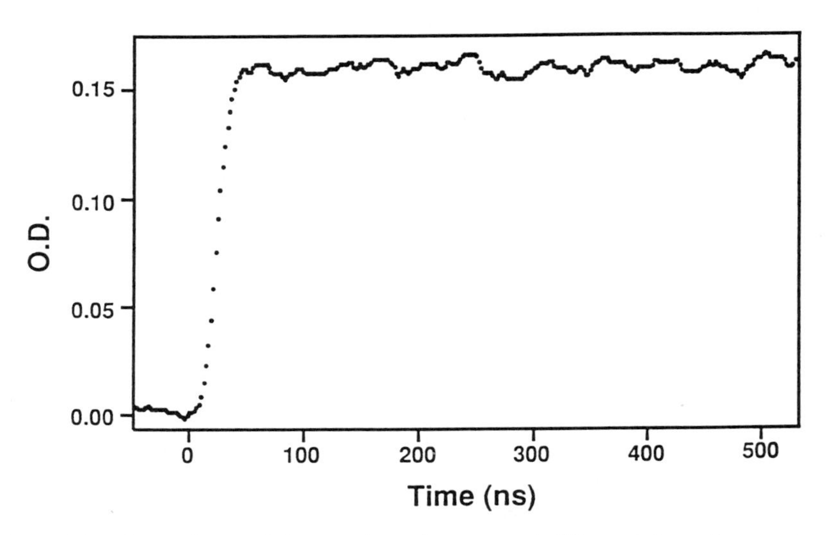

Figure 5-34. *Dependence of the transient absorption at 500 nm (absorption maximum of lophyl radicals) on time after the irradiation of 10^{-5} M solution of* o-*Cl-HABI in CH_2Cl_2. The solution was irradiated by 308-nm light at room temperature.*

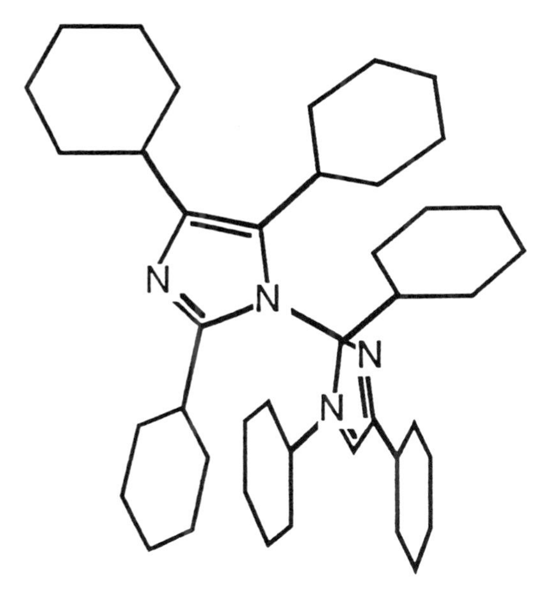

Figure 5-35. *Diagram of the computed nonplanar HABI structure.*

ples were irradiated directly within the EPR cavity.[73] The observed decomposition of HABI was definitely dependent on the temperature and consequently, on the matrix rigidity.

At very low temperatures, such as 5 K, no free lophyl radicals were formed. Excitation of the HABI molecule at low temperatures led to formation of the triplet state instead of decomposition (Figs. 5-36, 5-37). The triplet signal was shown to form from the excited hexaarylbiimidazole molecule rather than from two lophyl radicals trapped in a case after photodissociation.[73] The increase of temperature led to a change in the HABI photochemistry. At higher temperatures the matrix becomes softer, the torsional and rotational motion less hindered, and some decomposition of HABI into lophyl radicals can be observed (Fig. 5-38). Simultaneously, the concentration of the HABI triplets decreases when the temperature is increased (Fig. 5-38).

After the HABI triplet was observed and identified it was demonstrated that the lophyl radicals could be formed only from the singlet excited state of hexaarylbiimidazole.[73] These findings were in agreement with the results of earlier investigations demonstrating that hexaarylbiimidazole decomposition cannot be sensitized using a triplet sensitizer.[128] The absence of the lophyl radical formation from the HABI triplet excited state makes HABI substantially different from other photoinitiators for which such a path for radical formation is com-

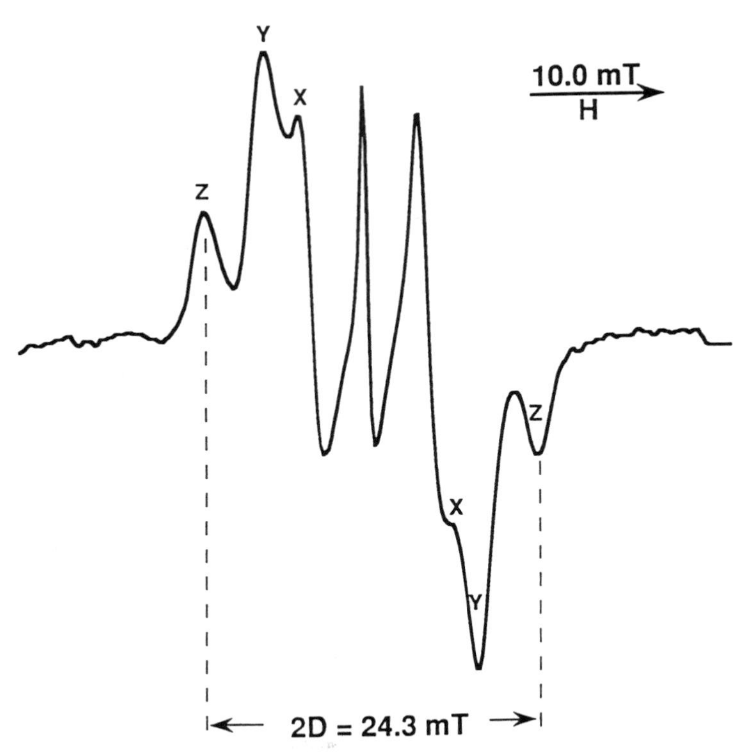

Figure 5-36. *Electron paramagnetic resonance spectrum recorded after 308-nm irradiation of the HABI solution in toluene at 6 K. The features attributed to the HABI triplet state were clearly distinguishable.*

mon.[130, p. 106 and references therein] The absence of the triplet photosensitization of the lophyl radical formation naturally precludes use of the triplet sensitizing dyes in the HABI-based systems. The mechanism of the dye photosensitization of biimidazole is discussed in detail in Section II, Chapter 4 of this monograph.[132] As the HABI triplet is not involved in the lophyl radical formation, the oxygen induced deactivation of the HABI triplet excited state cannot possibly reduce the yield of the lophyl radicals. This explains, to some extent, the low sensitivity of the HABI initiation to oxygen present in the photopolymer coatings.

In a photopolymer film the restrictions for the HABI photodissociation and the lophyl radical recombination are less severe than in rigid glass. However, the recombination is still sterically hindered and must proceed through the rotational and torsional rearrangement. Substitution of aromatic rings in the ortho-position increases radical lifetimes,[129] apparently because of further hindrance of

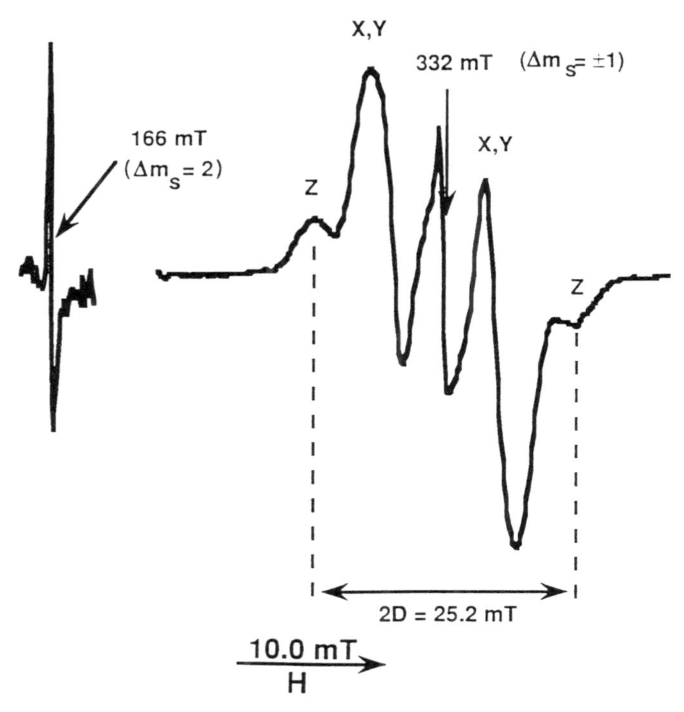

Figure 5-37. *EPR spectrum recorded after the 308-nm irradiation of 3% solution of* o-*Cl-HABI in poly(vinyl acetate) at 8 K. The HABI triplet state features were detected.*

the rotational and torsional reorientation. The increased radical lifetime is the main reason for the use of *o*-chloro-hexaarylbiimidazole in a number of photopolymer-coating formulations. The relatively long lophyl radical lifetime increases initiation efficiency and allows the chain-transfer agent or a monomer to diffuse to the distance required for activation.

5. Conclusion

In this chapter we have discussed molecular mobility effects on the photopolymer film performance. The emphasis was placed on directly monitored photopolymerization kinetics. The extensive computer modeling of the photopolymerization kinetics[70–72] described in this review was designed to illustrate the combined effects of the diffusion-dependent photopolymerization kinetics and

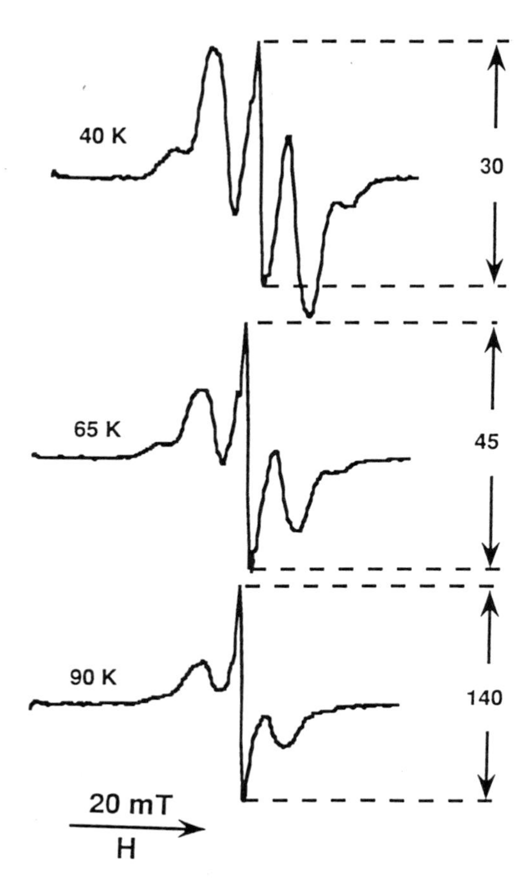

Figure 5-38. *The EPR spectra recorded after 308-nm irradiation of 3% solution of* o-*Cl-HABI in poly(vinyl acetate) at 40 K, 65 K, and 90 K. The numbers on the right indicate the relative intensities of the lophyl radical signal.*

the effects caused by anisotropic illumination[133] of the photopolymers that led to the unidirectional reactant migration during photopolymer imaging.[68–72]

It is often assumed that photopolymers and other photocurable materials consist of a number of immobile reactive centers that interact only by charge or energy transfer. In this review we demonstrated that this is not so. The processes occurring in the photopolymer films result from the combination of unidirectional illumination and diffusivity of the reactants, making photopolymer films a unique and complex system. The results reviewed above indicate the opportunities for creation of the new photopolymers and investigation of one- and two-dimensional phenomena with many versatile practical applications.

Acknowledgment

I would like to thank Professor Valeri A. Krongauz of The Weitzmann Institute of Science for his helpful comments and encouragement of this work.

References

1. S. W. Benson *The Foundations of Chemical Kinetics* (McGraw-Hill Book Co., New York and London, 1960).
2. Kh. S. Bagdasarian, *Theory of Free Radical Polymerization*, Israel Program for Scientific Translations, Jerusalem, IPST Cat. No. 2197, 1968.
3. J. Crank and G. S. Park, eds., *Diffusion in Polymers* (Academic Press, London and New York, 1968).
4. N. Lakshminarayanaiah, *Transport Phenomena in Membranes* (Academic Press, New York and London, 1969).
5. L. H. Sperling, *Introduction to Physical Polymer Science* (John Wiley and Sons, New York, Chichester, Brisbane, Toronto, Singapore, 1986).
6. V. V. Krongauz and D. Reddy, *Polymer Com.* **32**(1), 7 (1986).
7. V. V. Krongauz and R. M. Yohannan, *Polymer* **31**(6), 1130 (1990).
8. J. Comin, ed., *Polymer Permeability* (Elsevier, London, 1988).
9. J. M. Vergnaud, *Liquid Transport Process in Polymeric Materials, Modeling and Industrial Applications* (Prentice-Hall, Englewood Cliffs, NJ, (1981).
10. D. J. Meier, *Molecular Basis of Transitions and Relaxations* (Gordon and Breach Research Science Publishers, London, New York, Paris, 1978).
11. E. F. Haugh, U. S. Patent 3,658,526 (1972).
12. D. G. Howe, H. T. Thomas, and J. J. Wrobel. *J. Photogr. Science. Eng.* **23**, 97 (1979).
13. T. Ishitsuka and A. Yamagishita, Japanese Patent 60/227280 A2[85/227280] (1985).
14. T. Kurokawa and N. Takato, Japanese Patent 60/46690 B4[85/46690] (1985).
15. Fujitsu Ltd., Japanese Patent 57/31550 A2[82/34550] (1982).
16. (a) M. Yu. Bazhenov, Yu. M. Bardash, A. A. Kostyuk, N. G. Kuvshinskii, S. I. Kudinova, N. G. Nahodkin, V. A. Pavlov, N. I. Sokolov, and E. E. Sirotkina, USSR Patent 840786 (1981): (b) Otkrytiya. Izobret. Prom. Obraztsy, Tovarnye Znaki, **23**, (1981), 198.
17. K. Morimoto, A. Minobe, and M. Kuroda, Japanese Patent 49/106585 [74/106585] (1974).
18. T. Yamaoka and S. Namai, Nissan Motor Co., Japanese Patent 104183 [WO-9220016-A1] (1992).
19. D. K. Smith, *Photogr. Sci., Engin.* **12**(5), 263–266 (1968).
20. A. B. Cohen and R. N. Fan, U.S. Patent 4,174,216 (1976).
21. R. B. Held, U. S. Patent 3,854,950 (1974).
22. V. V. Krongauz and C. C. Legere-Krongauz, *Polymer* **34**(17), 3614-3619 (1993).

23. D. G. Howe, H. T. Thomas, and J. J. Wrobel, *Photogr. Sci., Engin.* **23**(6), 370–374 (1979).

24. B. L. Booth, in *Polymers for Electronic and Photonic Applications*, C. P. Wong, ed. (Academic Press, Inc., Harcourt Brace Jovanovich, Publ., Boston, San Diego, New York, London, Tokyo, Toronto, 1993), pp. 549–599.

25. E. W. Orr, in *Handbook of Coatings Additives*, L. J. Calbo ed., (Marcel Dekker, Inc., New York, Basel, Hong Kong, 1987), p. 51.

26. J. G. Kloosterboer, G. M. M. van de Hei, and H. M. J. Boots, *Polym. Comm.* **25**, 354 (1984).

27. J. G. Kloosterboer and G. J. M. Lippits, *J. Imag. Sci.* **30**, 177 (1986).

28. J. G. Kloosterboer and G. F. C. M. Lijten, in *Cross-Linked Polymers, Chemistry, Properties, and Applications,* ACS Symposium Series 367, R. A. Dickie, S. S. Labana, and R. S. Bauer, eds. (ACS, Washington, D. C. 1988), pp. 409–426.

29. A. M. Gupta, *J. Phys. II France* **3**, 407–409 (1993).

30. (a) Yu. G. Medvedevskikh and V. V. Sirnonenko, *Zh. Fiz. Khim.* **66**(5), 1432–1435 (1992). (b) *Ibid.*, **66**(6), 652-8 (1992).

31. D. C. Neckers, *Polym. Eng., Sci.* **32**(20), 1481 (1992).

32. C. G. Roffey, *Photopolymerization of Surface Coatings* (John Wiley & Sons, Chichester, New York, Brisbane, Toronto, Singapore, 1982).

33. (a) F. H. Dill, *IEEE Trans. Electron. Dev.* ED-22, 440 (1975). (b) F. H. Dill, W. P. Hornberger, P. S. Hauge, and J. M. Shaw, *IEEE Trans. Electron. Dev.* ED-22, 445 (1975).

34. S. V. Babu and Srinivasan, *Proc. SPIE* **539**, 36 (1985).

35. J. F. Rabek, *Mechanism of Photophysical Processes and Photochemical Reactions in Polymers. Theory and Applications* (John Wiley & Sons, New York, 1987).

36. B. L. Booth, *Appl. Optics*, **26**(6), 593–601 (1975).

37. A. B. Cohen and P. Walker, in *Imaging Processes and Materials*, J. M. Strurge, ed.) (Van Nostrand Reinhold, New York, 1989), pp. 226–278 and references therein.

38. M. F. Molaire, *J. Pol. Sci., Pol. Chem.* Ed. **20**, 847–861 (1982).

39. C. T. Chang, L. Galloway, and M. Grossa, Proc. of SPSE's 41st Annual Conference, May 22–26, (1989), 85–87.

40. R. T. Ingwall and M. Troll, *Opt. Eng.* **28**(6), 586–591 (1989).

41. T. Suzuki, Y. Todokoro, and K. Komenou, U.S. Patent 4,877,717 (1989).

42. D. J. Lougnot and C. Turck, *Pure Appl. Opt.* **1**, 269–279 (1992).

43. (a) B. M. Monroe, SPSE Proc. Photochem. Imag. Sys. Symp., A. Herbert, ed., pp. 89–100, Springfield, VA (1988). (b) B. M. Monroe in *Radiation Curing: Science and Technology*, S. P. Papas ed. (Plenum Press, New York, 1992).

44. W. K. Smothers, U.S. Patent 88-144281L, (1988); EP 89-100496 (1989).

45. W. K. Smothers, T. J. Trout, A. M. Weber, and D. J. Mickish, IEE Conf. Publ., 311 (Int. Conf. Hologr. Syst.. Compon. Appl., 2nd) (1989) 184–189.

46. T. Yamaoka and K. Koseki, JP 02216180 A2 (1990).

47. N. Ikeda, Y. Yamagishi, T. Ishizuka, and M. Tani, JP 01300287 A2 (1989).

48. D. Axelrod, D. E. Koppel, J. Schlessinger, E. Elson, and W. W. Webb, *Biophys. J.* **16**, 1055–1069 (1976).

49. B. A. Smith and H. M. McConnell, *Proc. Nat. Acad. Sci. USA* **75**(6), 2759–2763 (1978).

50. D. E. Koppel and M. P. Sheetz, *Biophys. J.* **43**, 175–181 (1983).

51. D. E. Koppel in *Fast Methods in Physical Biochemistry and Cell Biology*, R. I. Shaafi and S. M. Fernandes, eds. (Elsevier Science Publ., Amsterdam, 1983) pp. 339–367.

52. B. R. Ware, *Am. Lab.* April (1984), 16–28.

53. J. G. Kirkwood and J. Riseman, *J. Chem. Phys.* **16**, 565 (1948).

54. J. G. Kirkwood, *J. Polym. Sci* **12**, 1 (1954).

55. D. G. Miles, Jr., P. D. Lamb, K. W. Rhee, and C. S. Johnson, Jr., *J. Phys. Chem* **87**, 4815–4822 (1983).

56. M. Antoniety, J. Coutandin, R. Gruttar, and H. Sillescu, *Macromolecules* **17**, 798–802 (1984).

57. J. A. Wesson, I. Noh, T. Kitano, and H. Yu, *Macromolecules* **17**, 782–792 (1984).

58. D. Lougnot, C. Carre, and J. P. Fouassier, *Macromol. Chem., Macromol. Symp.* 24 (Eur. Symp. Polym. Mater., Pt. 3) 209–216 (1987).

59. C. Carre, D. J. Lougnot, and J. P. Fouassier, *Macromolecules* **22**(2), 791–799 (1989).

60. A. Liu, A. D. Trifunac, and V. V. Krongauz, *J. Phys. Chem.* **96**, 207 (1992).

61. Y. Lin, A. Liu, A. D. Trifunac, and V. V. Krongauz, *Chem. Phys. Lett.* **198**(1,2), 200–206 (1992).

62. J. Crank, *The Mathematics of Diffusion*, 2nd edition, (Clarendon Press, Oxford 1975).

63. C. Decker, in *Radiation Curing Science and Technology*, (S. P. Pappas, ed. (Plenum Press, New York and London, 1992), pp. 135–179.

64. J. Guthrie, M. B. Jeganathan, M. S. Otterburn, and J. Woods, *Polym. Bul.* **15**, 51–58 (1986).

65. X. Zhang, I. N. Kochetov, J. Paczkowski, and D. C. Neckers, *J. Imag. Sci. Technol.* **36**(4), (1992), 322–327.

66. E. A. Lissi and A. Zanocco. *J. Polym. Sci., Polym. Chem. Edn.* **21**, 2197 (1983).

67. A. R. Shultz and M. G. Joshi, *J. Polym. Sci., Polym. Phys. Edn.* **22**, 1753 (1984).

68. V. V. Krongauz and R. M. Yohannan, SPIE OE/Lase Conference Proceedings, Photopolymer Device Physics, Chemistry and Applications, Los Angeles, U.S.A., 17–19 January, Vol. 1213, (1990), 174–183.

69. V. V. Krongauz and R. M. Yohannan, *Mol. Cryst. Liq. Cryst.* **183**, 495–503 (1990).

70. V. V. Krongauz, E. R. Schmelzer, and R. M. Yohannan, *Polymer* **32**(9), 1654–1662 (1991).

71. V. V. Krongauz and E. R. Schmelzer, *Polymer* **33**(9), 1893–1901 (1992).

72. V. V. Krongauz and E. R. Schmelzer, SPIE Conference Proceedings. Photopolymer Device Physics, Chemistry and Applications, San Diego, U.S.A., 24–26 July, Vol. 1559, (1991), 354–376.

73. X.- Z. Qin, A. Liu, A. D. Trifunac, and V. V. Krongauz, *J. Phys. Chem.* **95**(15), 5822–5826 (1991).

74. G. Oster, U.S. Patent 2,850,445 (1958).

75. J. D. Margerum, L. J. Miller, and J. B. Rust, *Photogr. Sci. Eng.* **12**, 177 (1968).

76. D. F. Eaton, A. G. Horgan, and J. P. Horgan, *J. Photochem. Photobiol. A: Chem.* **58**, 373 (1991).

77. J. G. Kloosterboer, G. F. C. M. Lijten, and F. J. A. M. Greidanus, *Polym. Commun.* **27**, 268 (1986).

78. (a) L. A. Cescon, G. R. Coraor, R. Dessauler, E. F. Silversmith, and E. J. Urban, *J. Org. Chem.* **36**(16), 2262 (1971). (b) *Ibid.* 2267 (1971).

79. (a) R. H. Reim, A. MacLachlan, G. R. Cori, and E. J. Urban, *J. Org. Chem.* **36**(16), 2272 (1971). (b) *Ibid.* 2275 (1971).

80. T. Hayashi and K. Maeda, *Bul. Chem. Soc. Jpn* **33**, 565 (1960).

81. T. Hayashi and K. Maeda, *J. Chem. Phys.* **32**, 1568 (1960).

82. D. F. Eaton, *Top. Cur. Chem.* **156**, 199 (1990).

83. C. A. Parker, *Photoluminescence of Solutions* (Elsevier Publishing Co., Amsterdam, London, New York, 1968).

84. L. P. Elinger, *Polymer* **5**(1), 559 (1964).

85. O. Stern and M. Volmer, *Phys. Z.* **20**, 183 (1919).

86. J. Y. Moisan, in *Polymer Permeability*, J. Comin, ed. (Elsevier, London, 1988), p. 127.

87. D. Y. Chu, J. K. Thomas, and J. Kuczynski, *Macromolecules* **21**, 2094 (1988).

88. N. C. Billingham, P. D. Calvert, and A. Uzuner, *Polymer* 31, 258 (1990).

89. H. J. Timpe, B. Basse, F. W. Muller, and C. Muller, *Europ. Polym. J.* **23**(12), 967–971 (1987).

90. I. M. Krieger, G. W. Mulholland, and C. S. Dickey, *J. Phys. Chem.* **71**(4), 1123 (1967).

91. S. B. Maerov, *J. Imag. Sci.* **30**, 235 (1986).

92. S. B. Maerov, *J. Appl. Polym. Sci* **30**, 1499 (1985).

93. C. W. Gear, in *Information Processing* 68, A. J. H. Morrell, ed. (North Holland, Amsterdam, 1969), pp. 187–193.

94. C. W. Gear, *Comm. ACM* **14**, 176 (1971).

95. J. Hutchison and A. Ledwith, *Polymer* **14**, 405 (1973).

96. J. Woods, *Radcure Europe, 85, Conference Proceedings, FC85-* 414 (May 6–8, 1985).

97. R. J. Holman and H. Rubin, *J. Oil Col. Chem. Assoc.* **61**, 189 (1978).

98. S. Clarke and R. A. Shanks, *Polym. Photochem.* **1**, 103 (1981).

99. K. Albert, U. Gunther, M. Ilg, E. Bayer, and M. Grossa, *Magnetic. Res. Microscopy* (1992), 277.

100. M. C. Prystay and J. F. Power, *Polym. Eng. and Science* **33**(1), 43 (1993).

101. V. M. Treshnikov, S. A. Esin, N. A. Kuritsyna, B. P. Kalashnikov, L. L. Pomerantseva, and A. V. Oleinik, *Zh. Nauchn. Prikl. Fotogr. Kinematogr.* **32**(5), 340 (1987).

102. (a) C. A. Mack, *Optical Eng.* **27**(12), 1093 (1988). (b) M. J. Bowden, *J. Electr. Soc.* **128**(5), 195C (1981).

103. W. K. Smothers, B. M. Monroe, A. M. Weber, and D. E. Keys, *Proc. SPIE-Int. Soc. Opt. Eng., Practical Hologr. IV*, **1212**, (1990), 20.

104. B. M. Monroe, *J. Imag. Sci.* **35**, 25 (1991).

105. J. Crank and G. S. Park, *Diffusion in Polymers* (Academic Press, London and New York, 1968).

106. J. O. Herschelder, C. F. Curtis, and R. B. Bird, *Molecular Theory of Gases and Liquids* (John Wiley & Sons, New York, Chichester, Brisbane, Toronto, 1954).

107. G. Oster, *Nature* **173**, 300 (1954).

108. W. S. DeForest, *Photoresist: Materials and Processes* (McGraw-Hill Book Co., Inc., New York, Sidney, St. Louis, San Francisco, 1975).

109. E. Leberzammer and R. P. Held, E. I. Du Pont Imaging Research and Development, private communications.

110. C. E. Hoyle, R. D. Hensel, and M. B. Grubb, Proc. 8th Int. Conf. Radiat. Curing, Soc. Manuf. Eng., Dearborn, Mich. (1984), p. 13.

111. C. E. Hoyle, R. D. Hensel, and M. B. Grubb, *Polym. Photochem.* **4**, 68 (1984).

112. A. D. Kuchta, *Electron. Manufact.* **34**, 8 (1988).

113. T. Omote, T. Yamaoka, and K. Koseki, *J. Appl. Polym. Sci.* **38**, 389 (1989).

114. (a) F. R. Wight, *J. Polym. Sci., Polym. Lett., Ed.* **16**, 121 (1978). (b) F. R. Wight and J. A. Ors, in *Polymers for Electronic and Photonic Applications*, C. P. Wong, ed. (Academic Press, Inc., Harcourt Brace Jovanovich, Publ., Boston, San Diego, New York, London, Tokyo, Toronto, 1993), pp. 387–434.

115. J. G. Kloosterboer and G. F. C. M. Lijten, *Polym. Commun.* **28**, 2 (1987).

116. P. Karrer, S. Corbel, J. C. Andre, and D. J. Lougnot, *J. Polym. Sci. A: Pol. Chem.* **30**, 2715 (1992).

117. L. C. Sawyer and D. T. Grubb, *Polymer Microscopy* (Chapman and Hall, University Press, Cambridge, 1987).

118. V. N. Tsvetkov, E. I. Rjumtsev, and I. N. Shtennikova, in *Liquid Crystalline Order in Polymers*, A. Blumstein, ed. (Academic Press, New York, 1978), pp. 50–52.

119. D. G. Howe, H. T. Thomas, and J. J. Wrobel, *Photogr. Sci. Eng.* **23**(6), 370 (1979).

120. S. V. Babu and E. Barouch, *J. Imag. Science* **33**(6), 193 (1989).

121. B. F. Groffing and P. R. West, *Solid State Tech.* **28**, 152 (1985).

122. D. C. Hofer, C. G. Wilson, A. R. Neureuther, and M. Makey, *Proc. SPIE* **334**, 196 (1982).

123. S. V. Babu, E. Barouch, and B. Bradie, *J. Vac. Sci. Technol.* **B6**, 564 (1988).

124. V. V. Krongauz, unpublished results.

125. R. G. Jones and R. Karimian, *Polymer* **21**(7), 832 (1986).

126. V. D. McGinniss, *J. Rad. Curing*, January (1975), 3.

127. (a) M. A. J. Wilks and M. R. Willis, *Nature* **212**, 500 (1966); (b) Ibid. *J. Chem. Soc.*, (B) 1526, (1968).

128. (a) A. L. Prokhoda and V. A. Krongauz, *Khim. Vys. Energ.* **3**(6), 495 (1969); (b) Ibid. **4**(2), 174 (1970). (c) Ibid. 176 (1970). (d) **5**(3), 262 (1970).

129. (a) L.A. Cescon, G. R. Coraror, R. Dessauer, E. F. Silversmith, and E. J. Urban, *J. Org. Chem. 36*(16), 2262 (1971), (b) Ibid. 2267 (1971). (c) Ibid. **36**(16), 2272 (1972).

130, A Reiser, *Photoreactive Polymers: The Science and Technology of Resists* (John Wiley & Sons, New York, Chichester, Brisbane, Toronto, Singapore, 1989.

131. R. Dessauer and C. Looney, *Photogr. Sci. Eng.* **13**, 287 (1979).

132. Y. Lin, A. Liu, A. D. Trifunac, and V. V. Krongauz, *Chem. Phys. Let.* **198**(1,2), 200 (1992).

133. T. Omote, T. Yamaoka, and K. Koseki, *J. Appl. Polym. Sci.* **38**, 389 (1989).

6

Benzoin Ether Photoinitiators Bound to Acrylated Prepolymers

Kwang-Duk Ahn

1. Characteristics of Macromolecular Photoinitiators

Oligomers that contain photoinitiating groups are of considerable interest for practical applications in UV-curable formulations. These photoinitiators (PIs) are called macromolecular PIs, oligomer(ic) PIs, polymeric PIs, or polymer-bound PIs depending on the chemical structure and molecular weight. The main advantages expected from the high molecular weight photoinitiators are:

- Good compatibility with ingredients in the formulation
- Low migration tendency both in the uncured formulation and in the UV-cured product
- Low volatility, hence reduced odor problems
- Reduced yellowing effect due to the low migration of the residual unreacted PI and photoreacted PI fragments to the film surface.

Although not all of these expectations are fulfilled by the oligomeric PIs known today, these potential advantages have motivated growing research efforts involving the macromolecular PIs.[1]

One of the significant problems with photoinitiators is their incompatibility with a viscous formulation that contains ingredients such as multifunctional monomers, a reactive diluent, reactive oligomers, and a photosensitive prepolymer. In addition, being low molecular weight compounds, photoinitiators are apt to bleed out of the system in which they are incorporated both before and after curing. The incompatibility of photoinitiators in the UV-curing formulation leads to insufficient curing, leaving unreacted ingredients and results in adverse effects on the physical properties of the cured system. The PIs remain as undesirable small molecular inclusions in the finished products and as photoreacted PI fragments. The use of macromolecular photoinitiators is expected to eliminate the problems encountered with conventional PIs such as volatility, odor, and mixing difficulty.

Good compatibility and high photoreactivity have always been major requirements for a practical photoinitiator. Another important requirement for applications such as optical fiber and lens coatings, furniture and houseware coatings, and food packaging, is a low migration tendency. This minimizes potential hazards, odor, and deterioration caused by the various low molecular weight components contained in the cured product.

Macromolecular photoinitiators can be divided into two categories: the oligomeric PI compounds containing a small number of repeat units and photoinitiating groups per molecule, and the polymeric PI molecules, which possess many repeat units and photoinitiating groups with high average molecular weights. Oligomeric PIs differ from monomeric compounds containing the same photoreactive groups in physical properties such as solubility and diffusion rates; they do not usually exhibit a pronounced difference in their photocuring ability. Their application characteristics can be extrapolated from that of low molecular weight PI model compounds.

Polymeric PIs often have photochemical properties that differ significantly from those of their corresponding low molecular weight PIs. This has been called a ''polymer effect,'' because it occurs due to features of the polymeric structure. The high rigidity of the polymer chain is known to affect various photochemical parameters, and it is postulated that it is responsible for some specific polymer effects. Energy migration along the polymer chain, through interactions between excited and ground state photoreactive moieties fixed in a favorable geometry, may produce another polymer effect.

2. Polymer-Bound Benzoin Ether Photoinitiators

Benzoin ethers are an important class of commercial PIs utilized in UV-curing formulations. Benzoin alkyl ethers (BAE) photochemically undergo α-cleavage to produce benzoyl (B) and benzyl ether (E) radicals. The role of these two radicals B and E in photopolymerization is a source of controversy, but both radicals are known to be effective in photocuring compositions containing high concentrations of acrylates and methacrylates.[2]

The PIs of benzoin ethers exhibit poor storage stability (shelf life) in the presence of reactive monomers, resulting in premature polymerization even in dark storage. This instability has been attributed to the benzylic hydrogen of BAE molecules, which is readily abstracted by adventitious radicals such as peroxy radicals. Various radical photoinitiators that do not possess a benzylic hydrogen have been developed and commercialized. The substitution of the benzylic hydrogen atom in BAE photoinitiators brings about an improved storage stability.[3]

Some of the polymer-bound α-hydroxy acetophenones are shown in Fig. 6-1. An oligomeric PI of 2-hydroxy-2-methylpropiophenone (Fig. 6-1(a)) was reported to be very efficient in acrylate-containing coatings in which it displayed high photoreactivity and generated nonyellowing and low odor photolysis prod-

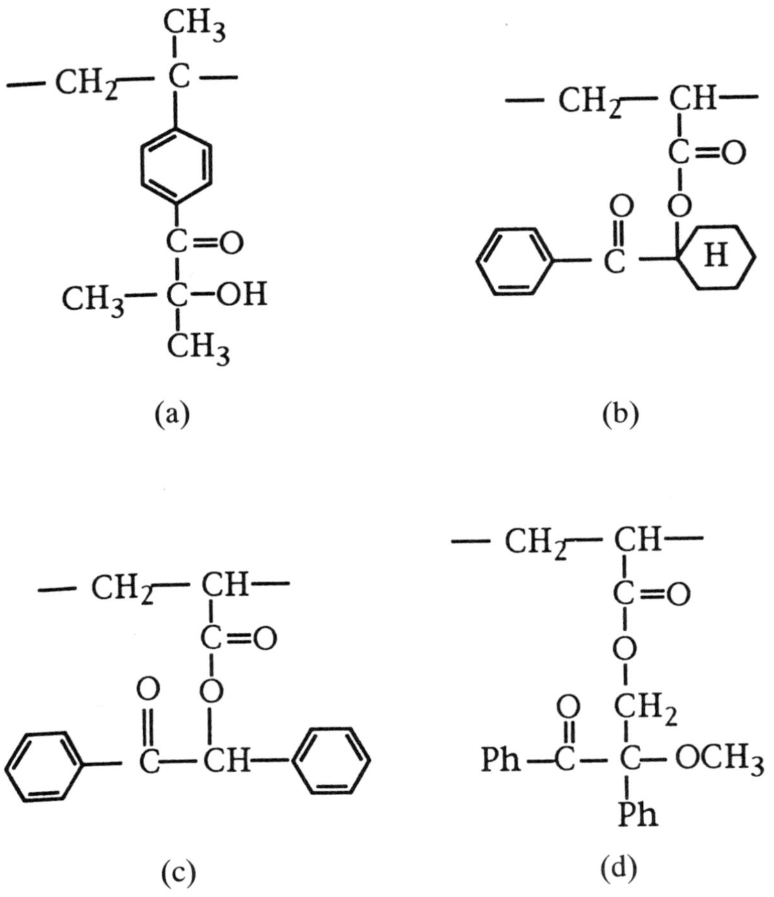

(a)

(b)

(c)

(d)

Figure 6-1. *Polymer-bound α-hydroxyacetophenone and α-methylolbenzoin methyl ether photoinitiators.*

ucts.[4,5] Both 1-hydroxycyclohexyl phenyl ketone (Fig. 6-1(b)) and benzoin (Fig. 6-1(c)) were incorporated via ester bonds with polyacrylates.[6] The photosensitivity of these polymeric PIs was remarkably reduced in comparison with that of the corresponding low molecular weight PIs. This reduction in the reactivity of these polymer-bound PIs was explained by the presence of electron-withdrawing substituents (ester bonds) instead of α-OH groups.

The photosensitivity of benzoin ether PIs is expected to be enhanced by modifying the α-position benzylic hydrogen. Ahn et al.[7–9] designed the polymer-bound BAE PIs with improved storage stability, compatibility, and high photoreactivity. Utilizing the reactive benzylic hydrogen of BAE, several useful functional groups such as carboxylic acid, methylol, or isocyanate were introduced, as shown in Fig. 6-2.

α-(2-Carboxyethyl)benzoin methyl ether (BAE-CO_2H in Fig. 6-2) was prepared, and the carboxylic acid was then linked to the epoxy groups of a copolymer of glycidyl methacrylate and methyl methacrylate to make a polymer-bound BAE.[7] A photocrosslinkable polymer containing both BAE and acryloyl groups in one polymer molecule was synthesized by reacting the BAE-CO_2H and acrylic acid (AA) with the copolymer of glycidyl methacrylate and methyl methacrylate, as shown in Fig. 6-3. The polymeric BAE photoinitiator was less reactive than the low molecular weight BAE-CO_2H or other BAE PIs in photopolymerization of styrene in solution, but the photocrosslinkable polymer exhibited a substantially enhanced crosslinking ability compared to a photocurable system consisting of low molecular weight BAE-CO_2H.[7] The stronger tendency to photocrosslinking was ascribed to higher homogeneity and compatibility of both units, viz., the photoinitiator BAE and the crosslinkable acryloyl groups in one polymer molecule.

Benzoin ethers were transformed into α-methylolbenzoin alkyl ether (BAE-OH in Fig. 6-2) and reacted with acryloyl chloride to make α-methylolbenzoin alkyl ether acrylates (BAE-AA in Fig. 6-2) which is a polymerizable photoinitiator monomer. Homopolymerization and copolymerization of BAE-AA yielded polymer-bound BAE as shown in Fig. 6-1(d).[8,10] Similarly, α-vinyloxymethyl-

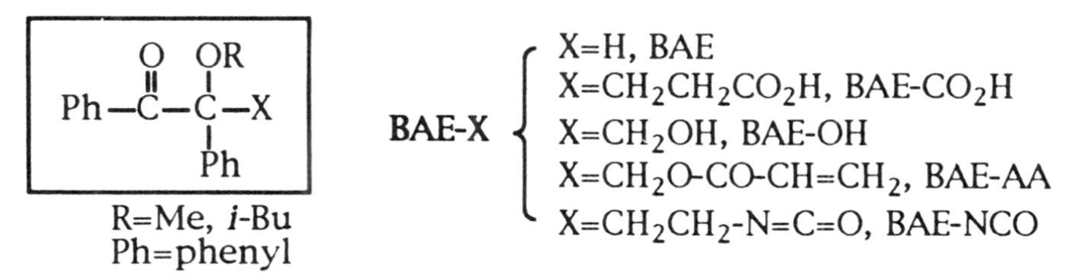

Figure 6-2. *Functional benzoin alkyl ether photoinitiators BAE-X with α-substituents.*

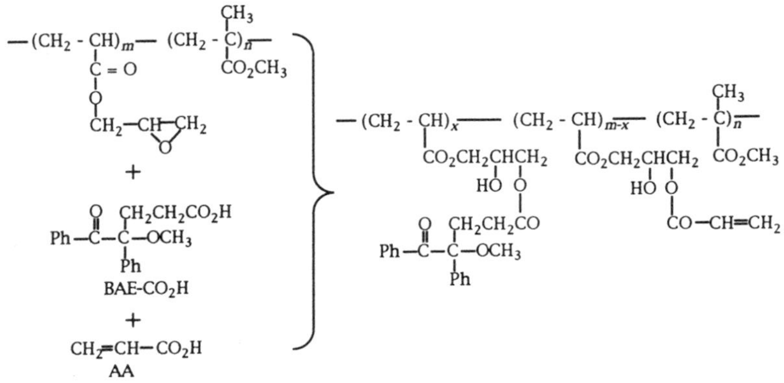

Figure 6-3. *Synthesis of a photocrosslinkable polymer containing benzoin methyl ether (BAE) and acryloyl groups.*

benzoin methyl ether was prepared to make polymeric PI by cationic polymerization by Angiolini et al.[10,11] The polymeric acrylate BAEs were more reactive toward photopolymerization than the corresponding low molecular weight BAE photoinitiators.[8] This polymer effect was attributed to the reduced mobility of the polymer-bound radicals, preventing radical coupling in the polymer chains without appreciably affecting their capability to initiate polymerization.

3. Oligomer Photoinitiators Bound to Prepolymers

Reactive oligomers or prepolymers are the most important components in UV-curable formulations because they usually bestow all the required physical properties such as hardness, flexibility, durability, strength, adhesion, thermal stability, and weatherability as well as chemical resistance and appearance. Therefore, good compatibility of the employed PIs with the rather viscous prepolymers is of the utmost importance for photocuring. Ahn et al.[9] designed a variety of oligomer PIs based on commercial photosensitive prepolymers.

The oligomer photoinitiators containing BAE were prepared by reacting the functional BAE-X with various oligomers. Their photocuring abilities were investigated in the formulations of photosensitive prepolymers and multifunctional acrylates.[9] These oligomer PIs, being similar to or having identical chemical structures with the photosensitive prepolymers, are expected to have better storage stability and higher photoreactivity in curing by attaining enhanced compatibility in the given formulations.

The functional benzoin alkyl ethers (BAE-X) in Fig. 6-2 with reactive substituents at the α-position were obtained by utilizing the reactive benzylic hydrogen of BAE photoinitiators. In addition to BAE-CO_2H and BAE-OH, α-(2-isocyanatoethyl)benzoin alkyl ethers (BAE-NCO) were newly prepared according to a procedure based on the Curtius rearrangement as shown in Fig. 6-4. Two BAE-NCO with methyl and isobutyl ether groups, α-(2-isocyanatoethyl)benzoin methyl ether (IEBME) and α-(2-isocyanatoethyl)benzoin isobutyl ether (IEBIBE), were subsequently used in the preparation of the oligomer PIs.

The oligomeric BAE photoinitiators were prepared by reacting BAE derivatives (BAE-X) with appropriate oligomers such as epoxy resin, PTMG-diisocyanate, and PTMG as shown in Figs. 6-5, 6-6 and 6-7 as follows:[9]

- Epoxy-bound BAE (Ep-BAE) by the reaction of BAE-CO_2H and epoxy resin (EpR) (Fig. 6-5)
- PTMG-bound BAE (PTMG-BAE) by the reaction of BAE-NCO and PTMG with mol. wt. of 1000 (Fig. 6-6)
- PTMG-urethane modified BAE (PTMG-U-BAE) via the reaction of BAE-OH and PTMG-diisocyanate (Fig. 6-7). PTMG-diisocyanate was prepared by the reaction of PTMG with two equivalents of diisocyanate, diphenylmethane diisocyanate (MDI) or toluene diisocyanate (TDI). The photosensitive prepolymer, PTMG-urethane diacrylate (PTMG-UA), was prepared by the reaction of PTMG-diisocyanate with 2-hydroxyethyl acrylate.

Three types of the oligomer PIs, Ep-BAE, PTMG-BAE, and PTMG-U-BAE

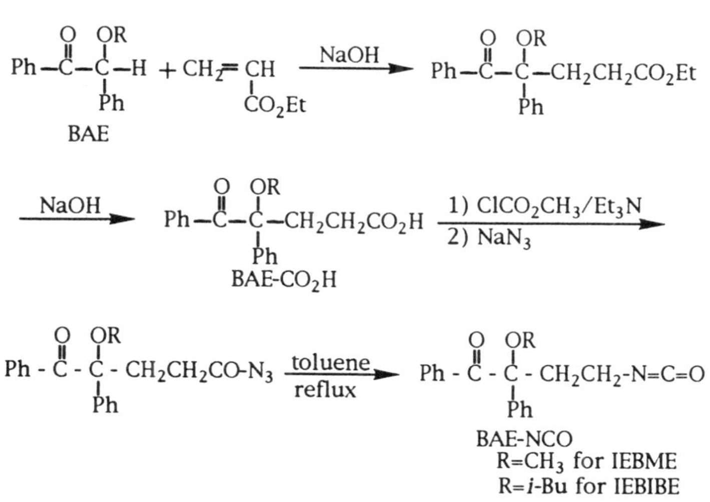

$$\underset{\text{BAE}}{Ph{-}C({=}O){-}C(OR)(Ph){-}H} + CH_2{=}CH{-}CO_2Et \xrightarrow{NaOH} Ph{-}C({=}O){-}C(OR)(Ph){-}CH_2CH_2CO_2Et$$

$$\xrightarrow{NaOH} \underset{\text{BAE-}CO_2H}{Ph{-}C({=}O){-}C(OR)(Ph){-}CH_2CH_2CO_2H} \xrightarrow[2)\ NaN_3]{1)\ ClCO_2CH_3/Et_3N}$$

$$Ph{-}C({=}O){-}C(OR)(Ph){-}CH_2CH_2CO{-}N_3 \xrightarrow[\text{reflux}]{\text{toluene}} \underset{\text{BAE-NCO}}{Ph{-}C({=}O){-}C(OR)(Ph){-}CH_2CH_2{-}N{=}C{=}O}$$

R=CH_3 for IEBME
R=*i*-Bu for IEBIBE

Figure 6-4. *Synthesis of α-(2-isocyanatoethyl)benzoin alkyl ether BAE-NCO.*

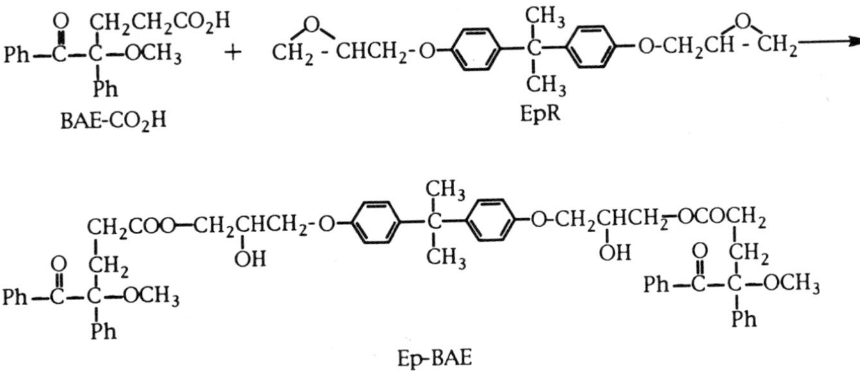

$$Ph-\overset{O}{\overset{\|}{C}}-\underset{Ph}{\overset{CH_2CH_2CO_2H}{C}}-OCH_3 + \overset{O}{CH_2-CHCH_2}-O-C_6H_4-\underset{CH_3}{\overset{CH_3}{C}}-C_6H_4-O-CH_2\overset{O}{CH-CH_2} \longrightarrow$$

BAE-CO_2H EpR

Ep-BAE

Figure 6-5. *Synthesis of an oligomer PI, Ep-BAE, bound to epoxy resin.*

were mixed with photosensitive epoxy acrylate (EpA) and PTMG-urethane acrylate (PTMG-UA) or reactive diluents such as trimethylolpropane triacrylate (TMPTA) and 1,6-hexanediol diacrylate (HDDA). The photoreactivity of these photocurable formulations was compared with the reactivity of the same formulations containing a low molecular weight BAE, usually benzoin isobutyl ether (BIBE) instead of oligomer BAEs.

The oligomer PIs generally exhibited a remarkable increase in the photoreactivity in these formulations. For example, in Fig. 6-8 an oligomer PI, PTMG-U-BAE, showed a higher photoreactivity in the formulation of PTMG-UA and HDDA in comparison with BIBE as the photoinitiator.[9] The degree of residual unsaturation (DRUS) measures the content of the unreacted acrylate double bonds or the extent of photocuring after UV exposure.

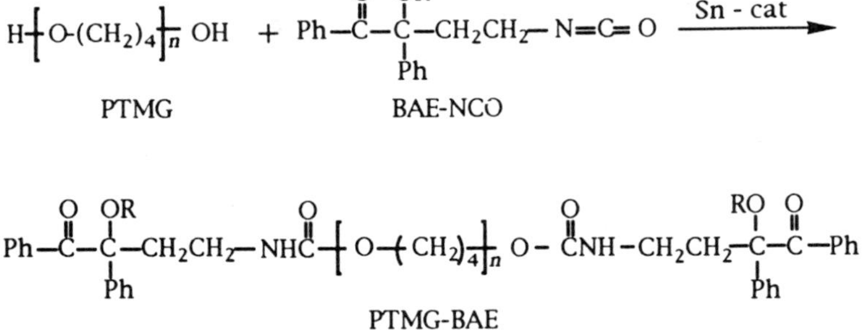

$$H\!\left[O-(CH_2)_4\right]_n OH + Ph-\overset{O}{\overset{\|}{C}}-\underset{Ph}{\overset{OR}{C}}-CH_2CH_2-N{=}C{=}O \xrightarrow{\text{Sn - cat}}$$

PTMG BAE-NCO

$$Ph-\overset{O}{\overset{\|}{C}}-\underset{Ph}{\overset{OR}{C}}-CH_2CH_2-NH\overset{O}{\overset{\|}{C}}\!\left[O\!\left(CH_2\right)_4\right]_n O-\overset{O}{\overset{\|}{C}}NH-CH_2CH_2-\underset{Ph}{\overset{RO}{C}}-\overset{O}{\overset{\|}{C}}-Ph$$

PTMG-BAE

Figure 6-6. *Synthesis of a modified polymeric photoinitiator, PTMG-BAE.*

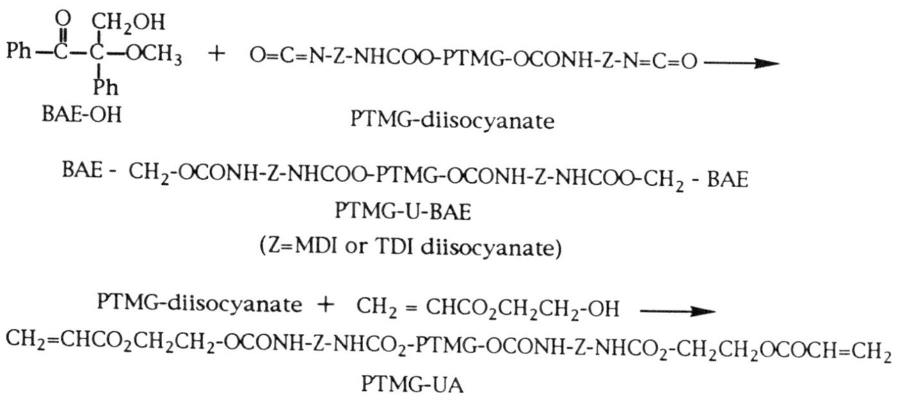

$$\mathrm{Ph{-}\overset{\overset{\displaystyle O}{\|}}{C}{-}\underset{\underset{\displaystyle Ph}{|}}{\overset{\overset{\displaystyle CH_2OH}{|}}{C}}{-}OCH_3} \quad + \quad \mathrm{O{=}C{=}N\text{-}Z\text{-}NHCOO\text{-}PTMG\text{-}OCONH\text{-}Z\text{-}N{=}C{=}O} \longrightarrow$$

BAE-OH PTMG-diisocyanate

$$\mathrm{BAE - CH_2\text{-}OCONH\text{-}Z\text{-}NHCOO\text{-}PTMG\text{-}OCONH\text{-}Z\text{-}NHCOO\text{-}CH_2 - BAE}$$

PTMG-U-BAE

(Z=MDI or TDI diisocyanate)

$$\text{PTMG-diisocyanate} + \mathrm{CH_2 = CHCO_2CH_2CH_2\text{-}OH} \longrightarrow$$

$$\mathrm{CH_2{=}CHCO_2CH_2CH_2\text{-}OCONH\text{-}Z\text{-}NHCO_2\text{-}PTMG\text{-}OCONH\text{-}Z\text{-}NHCO_2\text{-}CH_2CH_2OCOCH{=}CH_2}$$

PTMG-UA

Figure 6-7. *Synthesis of a urethane oligomer, PI, PTMG-U-BAE, and a urethane acrylate prepolymer, PTMG-UA.*

4. Benzoin Ether Photoinitiators Bound to Acrylated Prepolymers

A photoinitiator triggers the photopolymerization of acryloyl groups from both the multifunctional acrylate and the prepolymer in a photocurable formulation, thereby providing photocrosslinking of the components. Thus, when the photoinitiator and acryloyl groups are combined in one prepolymer molecule, they may bring enhanced compatibility and photoreactivity along with improved physical properties. Therefore, a new kind of acrylated prepolymer containing BAE photoinitiators was designed, and its photoreactivity in appropriate formulations is discussed in this chapter.

4.1. Materials and Synthesis

Benzoin methyl ether (BME) and benzoin isobutyl ether (BIBE) were purchased from Aldrich Chemical Co. and Wako Chemical Co., respectively. Poly(tetramethylene ether) glycol (PTMG) of mol. wt. 1000 (Quaker Oats) was purified by drying in vacuum for 3 hours at 60° C. Trimethylolpropane triacrylate (TMPTA), 1,6-hexanediol diacrylate (HDDA), and bisphenol A-epoxy diacrylate (EpA) were donated by Korea Chemical Co. Another EpA used was obtained from Shell Chemical Co. (Product No. Shell 828). Tetra-functional epoxy resin, tetraglycidyldiaminodiphenylmethane (TGAM) was purchased from Sumitomo Chemical Ind. (Product No. ELM-434). α-Methylolbenzoin alkyl ether (BAE-OH)[8] and α-2(-carboxyethyl)benzoin alkyl ether ($BAE\text{-}CO_2H$)[12]

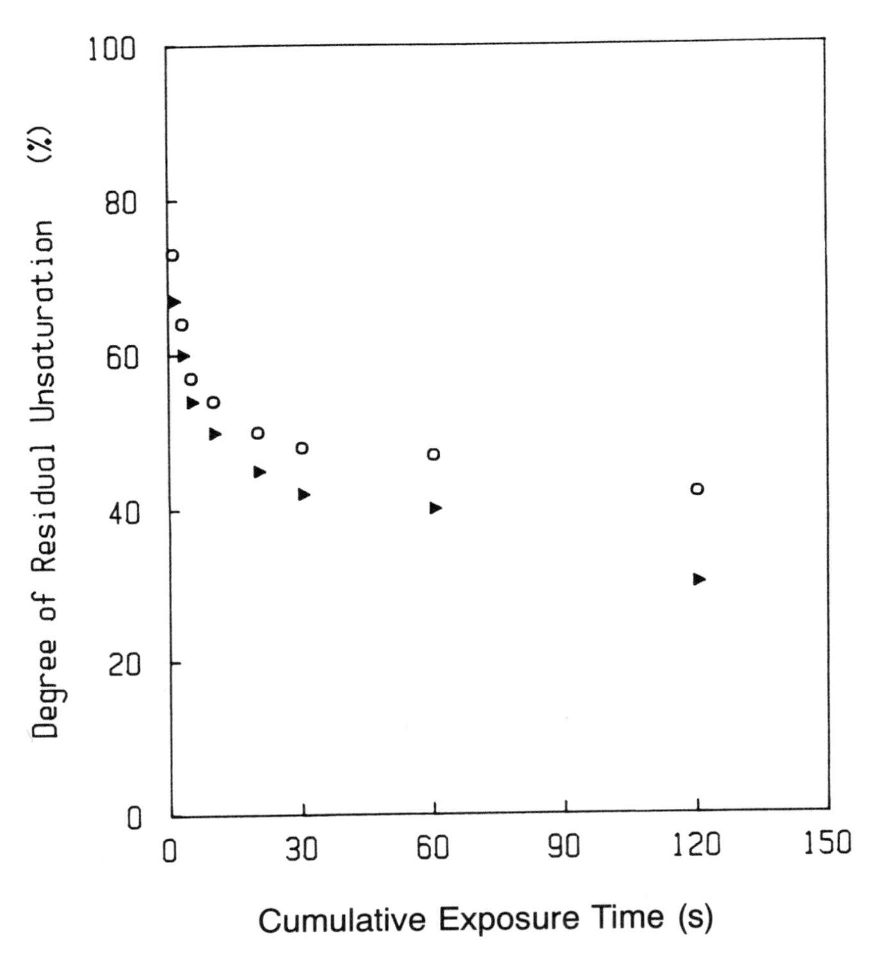

Figure 6-8. *The extent of photoreaction of PTMG-UA by two different photoinitiators (4 eq%): PTMG-UA and BIBE (○); PTMG-UA and PTMG-U-BAE (▶).*

were prepared according to known procedures. The preparation of PTMG-urethane diacrylate (PTMG-UA) was described elsewhere.[9]

α-(2-Isocyanatoethyl)benzoin alkyl ether (BAE-NCO)

Two kinds of BAE-NCO, α-(2-isocyanatoethyl)benzoin methyl ether (IEBME) and α-(2-isocyanatoethyl)benzoin isobutyl ether (IEBIBE) were prepared by a modified procedure based on the Curtius rearrangement (Fig. 6-4). The solution of 10.0 g (34 mmol) of α-(2-carboxyethyl)benzoin methyl ether in 200 ml of acetone was placed in a three-neck flask, and the mixture was cooled in an ice-water bath. 5.5 ml (38 mmol) of triethylamine was added to the mixture and stirred for 30 minutes, and then 3.6 g (39 mmol) of methyl chloroformate were added, dropwise, and reacted for 2 hours under cooling. The mixture was further reacted with 3.0 g (46 mmol) of sodium azide for 2 hours. From the cooled reaction mixture, precipitated triethylamine hydrochloride was filtered off, and the filtrate was concentrated by evaporation. The residue was extracted with ethyl acetate three times from water saturated with sodium chloride, and the ethyl acetate was dried over anhydrous magnesium sulfate. The viscous concentrate obtained by evaporation of ethyl acetate was dissolved in 150 ml of toluene, and the solution was refluxed for 4 hours under nitrogen. Toluene was evaporated in vacuo; the yellowish residue was confirmed to be the desired IEBME and it solidified in a refrigerator. Microdistillation in high vacuum (0.05 mm Hg) at 250° C gave high purity IEBME. The yield of IEBME was 60% based on the starting BAE-CO_2H. IR (liq.), 2250 (N=C=O), 1680 (benzoyl C=O), 1100 cm^{-1} (ether); NMR ($CDCl_3$), δ 7.8–8.3 (m, phenyl, 2H), 7.3–7.7 (m, phenyl, 8H), 3.9–4.3 (m, -CH_2-NCO, 2H), 3.3 (s, -OCH_3, 3H), 1.7–2.6 ppm (m, CH_2, 2H); Mass (m/e), 190, 147, 133, 105, 91, 77 (no molecular ion).

Waxy IEBIBE was prepared by the same procedure starting from α-(2-carboxyethyl)benzoin isobutyl ether with a yield of 63%. Its chemical structure was confirmed by NMR and IR spectral analyses.

PTMG-bound BAE photoinitiator (PTMG-BAE)

A mixture of 1.18 g (4.0 mmol) of IEBME, 2.0 g (2.0 mmol) of PTMG, and catalytic amounts of dibutyltin dilaurate without solvent was reacted at 50° C for 5 hours. Completion of the reaction was checked by the disappearance of isocyanate absorption at 2250 cm^{-1} in the IR spectra during the reaction.

Difunctional photoinitiator bound to acrylated epoxy prepolymer (EpA-BAE)

A solution of 16.7 g (50 mmol) of IEBIBE and catalytic amounts of dibutyltin dilaurate in 100 ml of chloroform was made, and then 12.1 g (25 mmol) of epoxy diacrylate (EpA) were added. The solution was maintained at 60° C for 10 hours. After completion of the reaction, the solution was filtered and the solvent was evaporated from the filtrate to obtain viscous modified epoxy acrylate-bound BAE photoinitiator, EpA-BAE.

Tetra-functional BAE photoinitiator bound to acrylated epoxy prepolymer (TGAM-A-BAE)

5.0 g (12 mmol) of TGAM and triethylbenzylammonium chloride (0.5 g) were dissolved in 50 ml of chloroform at room temperature. 3.6 g (50 mmol) of acrylic acid (AA) dissolved in 20 ml of chloroform were added, dropwise, to the solution, and the resulting solution was refluxed for 4 hours. The solvent was then stripped off, and the residue was extracted four times with ethyl acetate and saturated salt water. The ethyl acetate extract was dried over anhydrous magnesium sulfate, and the solvent was evaporated to obtain viscous tetra-acrylated TGAM, TGAM-A. Catalytic amounts of dibutyltin dilaurate were added to a solution of TGAM-A (5.0 g, 7 mmol) and IEBIBE (9.4 g, 28 mmol) in 50 ml of chloroform. The solution of the reactants was maintained overnight at 60° C. A highly viscous product, tetra-functional epoxy acrylate-bound BAE photoinitiator, TGAM-A-BAE, was obtained after evaporation of the volatiles in vacuo.

Photocurable Formulations and UV Irradiation

The photocurable formulations were made by mixing the pertinent photoinitiators and acrylated prepolymers and/or multifunctional acrylates in inert solvents or without solvents. The content of the initiators was employed in a range of 1 to 4 equivalent mol % of all reactive acrylate groups. The formulations were spin-coated on salt plates; the thickness of the coating was about 5 μm. The thin film on a salt plate was irradiated using a high pressure mercury lamp (500 W) with a light intensity of 45 mW/cm^2 measured at 365 nm using a power meter. The extent of photocuring was estimated in terms of DRUS by measuring the decrease in the intensity of IR absorption at 1405 cm^{-1} for a given exposure time.

4.2. Prepolymer Bound Photoinitiators in Formulations

A new kind of functional benzoin alkyl ether PI having a reactive isocyanate group, BAE-NCO, was synthesized with over 60% yield. α-(Isocyanatoethyl)benzoin methyl ether (IEBME) and α-(2-isocyanatoethyl)benzoin isobutyl ether (IEBIBE) were prepared according to the procedure[9] shown in Fig. 6-4. The isocyanate BAE PIs (BAE-NCO) were used in the preparation of an oligomer PI, PTMG-BAE (Fig. 6-6).

The hydroxyl groups of the acrylated epoxy prepolymers were advantageously utilized to introduce BAE moieties through urethane bonds by reacting with the isocyanate function of BAE-NCO. In this work two epoxy prepolymers of difunctional epoxy acrylate and tetra-functional epoxy resin were employed to make prepolymers containing BAE photoinitiators. One of the commercially important prepolymers in UV curing is an acrylated epoxy resin. Thus, epoxy diacrylate (EpA) was reacted with IEBIBE to obtain the desired epoxy-modified prepolymer PI as shown in Fig. 6-9. Formation of the acrylated epoxy prepolymer-bound PI, EpA-BAE, was checked by the disappearance of two IR absorption bands at 2250 cm^{-1} and 3600 cm^{-1} belonging to isocyanates and hydroxyl groups, respectively.

Another multifunctional epoxy resin chosen to obtain multi-substituted epoxy-BAE PIs was tetraglycidyldiaminodiphenylmethane (TGAM). The reaction scheme is shown in Fig. 6-10. First, TGAM was treated with acrylic acid (AA) to make the tetra-substituted acrylated epoxy resin, TGAM-A. Then the acrylate TGAM-A reacted with IEBIBE to produce the desired tetra-functional epoxy acrylate-bound BAE photoinitiator, TGAM-A-BAE, which also contains four acryloyl groups together in its molecule. The presence of both the acryloyl and BAE groups in TGAM-A-BAE was confirmed by IR and NMR spectra.

Both the acrylated prepolymer-bound PIs, EpA-BAE and TGAM-A-BAE, were mixed with the acrylated prepolymers, EpA, PTMG-UA, or other commercial urethane acrylates, and diluent acrylate monomers such as HDDA and TMPTA. The composition of the photocurable formulations studied in this work are described in Table 6-1. These formulations underwent UV exposure as films. The photoinitiating capability of different PIs was compared with the corresponding low molecular weight BAE analogues, usually BIBE in similar formulations.

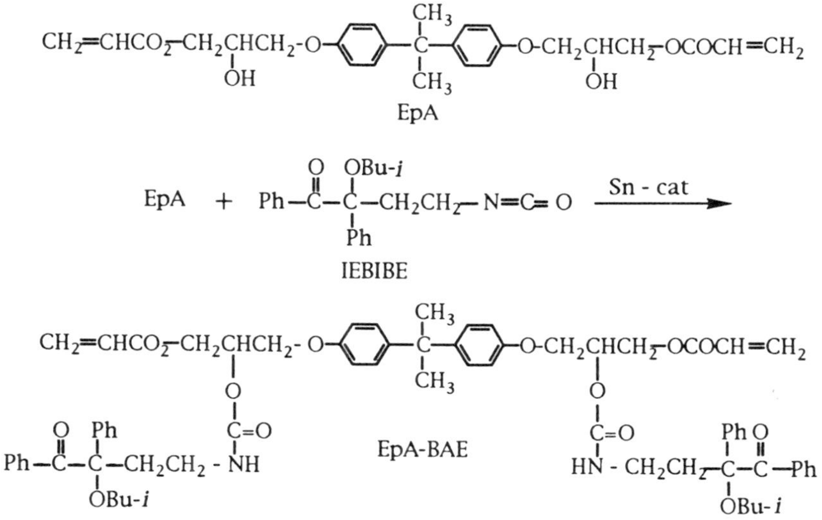

$CH_2{=}CHCO_2{-}CH_2CH(OH)CH_2{-}O{-}C_6H_4{-}C(CH_3)_2{-}C_6H_4{-}O{-}CH_2CH(OH)CH_2{-}OCOCH{=}CH_2$

EpA

EpA + $Ph{-}CO{-}C(Ph)(OBu\text{-}i){-}CH_2CH_2{-}N{=}C{=}O$ $\xrightarrow{\text{Sn - cat}}$

IEBIBE

$CH_2{=}CHCO_2{-}CH_2CH(O{-}CO{-}NH{-}CH_2CH_2{-}C(Ph)(OBu\text{-}i){-}CO{-}Ph)CH_2{-}O{-}C_6H_4{-}C(CH_3)_2{-}C_6H_4{-}O{-}CH_2CH(O{-}CO{-}NH{-}CH_2CH_2{-}C(Ph)(OBu\text{-}i){-}CO{-}Ph)CH_2{-}OCOCH{=}CH_2$

EpA-BAE

Figure 6-9. *Synthesis of an epoxy oligomer photoinitiator, EpA-BAE, bound to epoxy diacrylate prepolymer (EpA).*

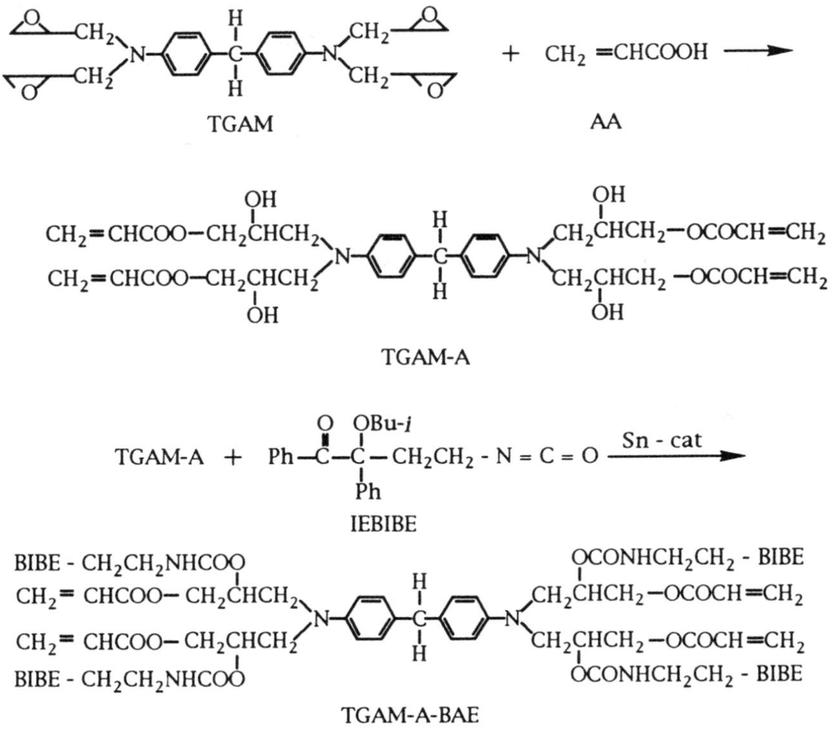

TGAM + $CH_2=CHCOOH$ (AA) ⟶

TGAM-A

TGAM-A + $Ph-C(=O)-C(OBu\text{-}i)(Ph)-CH_2CH_2-N=C=O$ (IEBIBE) $\xrightarrow{\text{Sn - cat}}$

TGAM-A-BAE

Figure 6-10. *Synthesis of an epoxy oligomer photoinitiator, TGAM-A-BAE, bound to TGAM-tetraacrylate prepolymer.*

Table 6-1. Composition of the Photocurable Formulations

PI	HDDA	TMPTA	PTMG-UA	EpA
BIBE	4 eq%	4 eq%	4 wt%	3 eq%
PTMG-BAE	4 eq%	4 eq%	4 wt%	—
EpA-BAE	—	—	—	3 eq%
TGAM-A-BAE	—	—	—	1, 3, 4 eq%

(eq% = percentage in equivalents)
PTMG-UA = urethane diacrylate oligomer based on PTMG and TDI
EpA = epoxy diacrylate of bisphenol A
HDDA = hexanediol diacrylate
TMPTA = trimethylolpropane triacrylate

The film cast on a salt plate was irradiated with UV light. The photosensitivities of the formulations are related to the relative disappearance rate of the acrylate functionalities over the given exposure time. The concentrations of acrylate groups were deduced by comparing the IR absorption intensities. The extent of photocuring (or photopolymerization) was evaluated by measuring the decrease in the intensity of IR absorption of the sample film at 1405 cm^{-1} corresponding to methylene in-plane bending mode of acrylate double bonds. Thus, the extent of photopolymerization was calculated in terms of the degree of residual unsaturation (DRUS). DRUS is the ratio of the concentration of acrylate groups in a formulation before and after UV irradiation for a given time. It is a measure of content of unreacted acrylate double bonds after irradiation.

The change in IR absorption spectra of a formulation containing prepolymer EpA and 4 eq% of the prepolymer-bound benzoin isobutyl ether TGAM-A-BAE is shown in Fig. 6-11. A remarkable reduction in the absorption intensities was observed at 1405 cm^{-1} due to the photopolymerization of acrylate groups. Other formulations also showed a similar IR spectral change depending on the curing time. From these IR spectral changes at 1405 cm^{-1}, DRUS values (percent

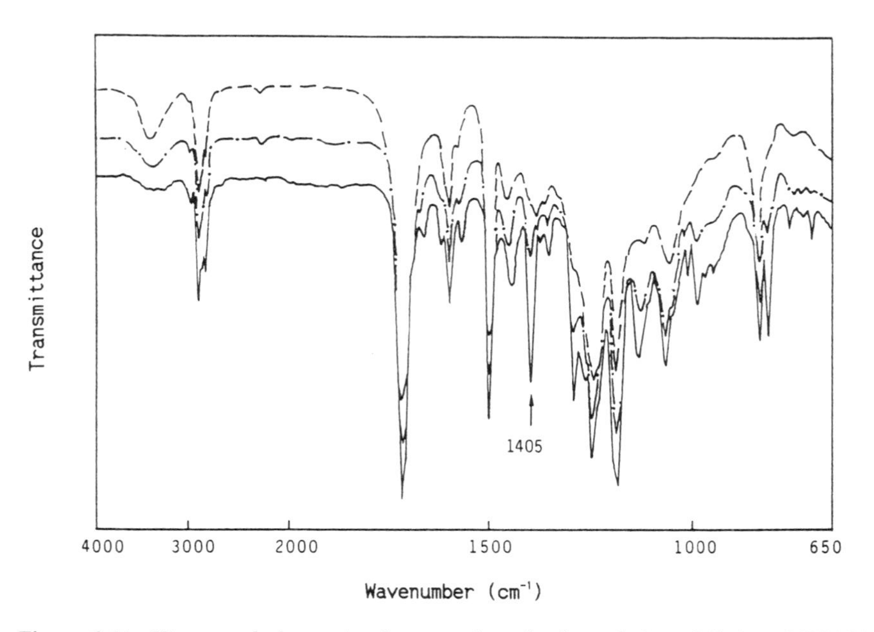

Figure 6-11. *IR spectral change in photoreaction of a formulation of EpA and TGAM-A-BAE in 4 eq%: uncured (——); cured 30 seconds (— · —); extensively cured, 15 minutes (– – –).*

equivalent) were calculated for comparison of the photopolymerization efficiencies of the prepolymer bound PIs.

For two representative examples, the extent of photoreaction was estimated in terms of DRUS for the formulations based on the acrylated epoxy prepolymer (EpA) as shown in Figs. 6-12 and 6-13. The commercially important prepolymer EpA was the main component and its photoreaction in the formulations was evaluated using EpA-BAE (Fig. 6-12) and TGAM-A-BAE (Fig. 6-13) as the

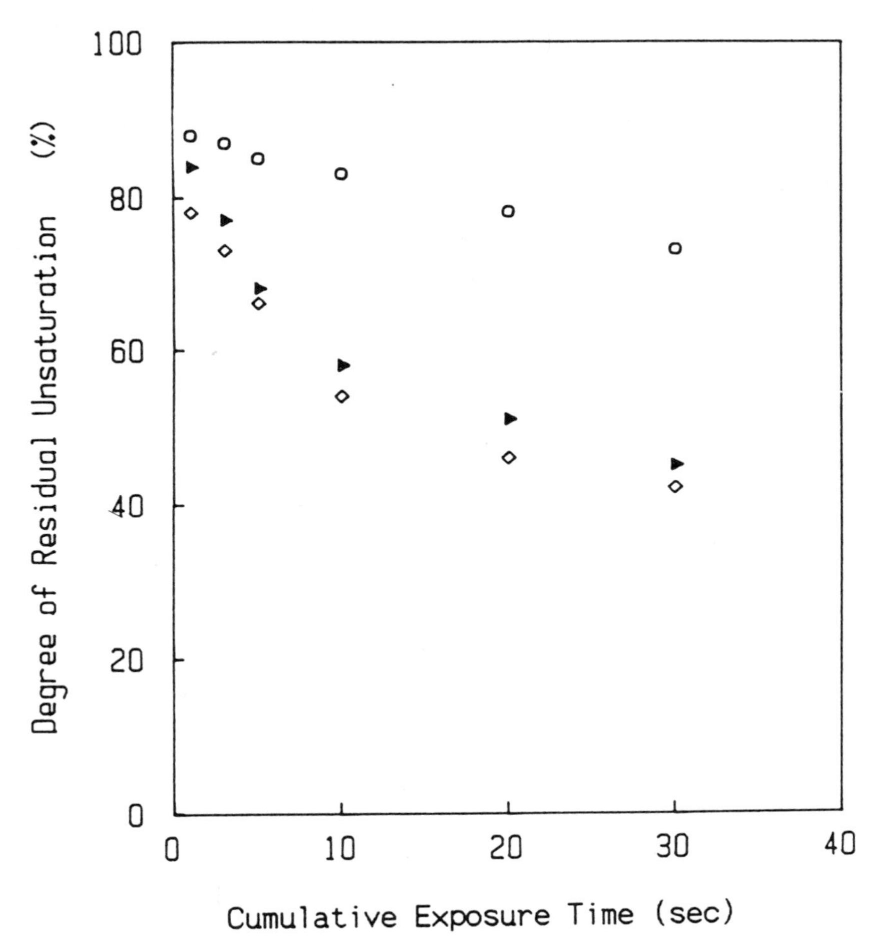

Figure 6-12. *The extent of photoreaction of formulations based on EpA by three kinds of BAE photoinitiators in 3 eq%: EpA and BIBE (○); EpA/TMPTA (6:4 by mol) and EpA-BAE (▸); EpA and EpA-BAE (◇).*

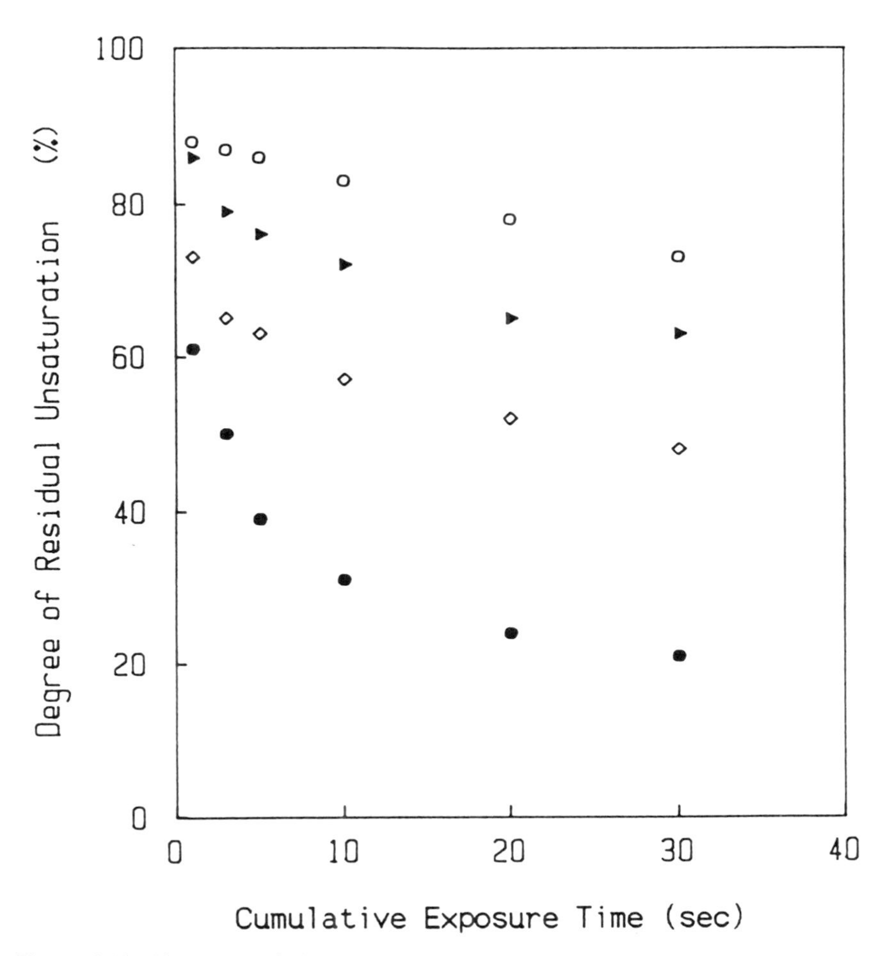

Figure 6-13. *The extent of photoreaction of formulations based on EpA by tetra-functional PI, TGAM-A-BAE: (○) for EpA and BIBE (3 eq%); EpA and TGAM-A-BAE, (▸) for 1 eq%, (◇) for 3 eq%, (●) for 4 eq%.*

photoinitiators. Both acrylated prepolymer-bound PIs appeared to be more effective in the photoreactions than the low molecular weight analogue BIBE. The formulations containing 3 eq% of EpA-BAE and TGAM-A-BAE revealed similar results in the photoreaction although EpA-BAE showed somewhat higher efficiency. With higher content of TGAM-A-BAE, the formulations render a remarkable increase in the photoreaction efficiency as shown in Fig. 6-13.

Because the two prepolymer PIs possess the same number of acrylate groups and BAE moieties in their molecules, the higher initiating efficiency should be possible at the moment of free radical generation from the BAE fragmentation by UV irradiation. Thus, whenever the free radicals from the BAE moiety are generated, the concomitant reactions with the acrylate groups in the same PI molecule of EpA-BAE or TGAM-A-BAE would follow. Then the propagation of the chain reactions among acrylate groups of the prepolymer EpA will be facilitated because of good compatibility of acrylated prepolymers and prepolymer-bound PIs in the formulation.

The higher efficiency observed in the acrylated prepolymer-bound PI system can be attributed to the previously mentioned "polymer effect," such as the reduced mobility of the free radicals (B), which prevents the radical coupling reactions, and leads to good compatibility and closer location of the BAE moiety and the acrylate group. In addition to improved photoreactivity, good mechanical properties are expected for cured products because of the photocrosslinkable acrylate groups in the same prepolymer-bound PIs. The complete formulation can be cured through the crosslinking of all the acrylate groups from the different acrylated components such as the acrylated prepolymer-bound PI, the acrylated prepolymer, and the acrylated diluent.

4.3. Conclusions

Two acrylated prepolymer-bound BAE photoinitiators, EpA-BAE and TGAM-A-BAE, were produced by reacting isocyanated BAE (BAE-NCO) with acrylated epoxy prepolymers, di-functional EpA and tetra-functional TGAM-A. EpA-BAE and TGAM-A-BAE with BAE and acrylate groups together exhibited higher photoreactivity in the formulations containing acrylated prepolymers and acrylated diluents than the corresponding low molecular weight BAE analogue. The observed high efficiency of the acrylated prepolymer PIs are ascribed to the polymer effect, good compatibility, and the closer vicinity of the acrylate and BAE groups in the PI molecule. The acrylated prepolymer-bound benzoin alkyl ether photoinitiators are very promising for improving storage stability, compatibility, and photoreactivity in formulation as well as for improving mechanical properties, odor (low volatility), and reducing the yellowing effect in the UV-cured products.

Acknowledgments

The author acknowledges the support of the project by the Korea Ministry of Science and Technology. The author also thanks Ms. Hyang-Sook Choi for her contribution to the experimental work.

References

1. K. Dietliker, ''Macromolecular photoinitiators'', in ''Photoinitiators for Free Radical and Cationic Polymerization, Chapters II-VIII, *Chemistry and Technology of UV & EB Formulation for Coatings, Inks & Paints*, P. K. T. Oldring, ed., Vol. 3 (SITA Technology Ltd., London, U.K., 1991) p. 204.
2. S. P. Pappas, ''Photoinitiation of Radical Polymerization'' in *UV Curing: Science and Technology*, S. P. Pappas, ed., Chapter 1 (Technology Marketing Corp., Norwalk, CT, 1983) p.1.
3. S. P. Pappas, *J. Radiat. Curing* **14**, 6 (July 1987).
4. J. P. Fouassier, D. J. Lougnot, G. Li Bassi, and C. Nicora, *Polymer Comm.* **30**, 245 (1989).
5. G. Li Bassi, L. Cadona, and F. Broggi, ''Radcure Europe '87'' Conference Proceedings (Munich, Germany, May 1987), SME; Dearborn, MI, U.S.A., 1987: p. 3-15.
6. C. Carlini, *Brit. Polym. J.* **18**, 236 (1986).
7. J. S. Shim, N. G. Park, U. Y. Kim, and K.-D. Ahn, *Polymer (Korea)* **8**, 34 (1984); *Chem. Abstr.* **100** (26), 210503a.
8. K.-D. Ahn, K. J. Ihn, and I.-C. Kwon, *J. Macromol. Sci. -Chem.* **A24**, 355 (1986).
9. K.-D. Ahn, I.-C. Kwon, and H.-S. Choi, *J. Photopolym. Sci. Technol.* **3**, 137 (1990).
10. L. Angiolini, C. Carlini, M. Tramontini, and A. Altomare, *Polymer* **31**, 212 (1990).
11. L. Angiolini and C. Carlini, *Chem. Ind. (Milan)* **72**, 124 (1990).
12. H.-G. Heine and H. Rudolph, *Liebiegs Ann. Chem.* **754**, 28 (1971).

sitive Liquid Crystalline Polymers

Carolyn Bowry

1. Introduction

Liquid crystal polymers (LCPs) are widely used in high tensile strength fiber materials such as Kevlar and Lydar, which can be found in products such as replacement asbestos, motorcycle helmets, and flame protection suits. However, there is growing research and development interest in using other functional properties of LCPs for optical and electro-optical uses.

LCPs are organic materials that combine liquid crystal molecules with a polymer backbone. The liquid crystal molecules can be part of the backbone itself to form a main chain LCP or they can be attached to the backbone as pendants to form side chain LCPs. Main chain LCPs are more viscous than side chain LCPs and therefore are less functional for electro-optic applications. However, for photosensitive applications, the main chain polymers are of equal importance because the activity can occur in the polymerization process itself.

Many effects that are observed in LCP materials are, in fact, photoactivated processes. The photoactivity can occur in the polymer backbone through polymerization processes or in the liquid crystal mesogen by isomerization processes, which can change the phase or alignment of the LC molecule. By adding other functional groups to the polymer backbone, it is possible to provide additional optical effects; for example, photochromic molecules can be added for color changes, organic dyes can be added for optical absorption, and nonlinear optical dyes can be added to induce electro-optic effects. These additional molecules add another dimension to the possible photosensitive properties of LCPs.

There are widespread potential uses of photosensitive LCPs, as demonstrated by the literature. Most uses are for optical applications: optical storage, holography, nonlinear optics, displays, and imaging. However, as liquid crystals have many anisotropic properties, it is not just the optics of the material that are altered by the light—the electrical and magnetic properties will also be affected. Such possibilities are only just beginning to be explored by researchers; it may be an interesting route to follow.

This chapter will describe the functionality that can be built into LCPs by making them photosensitive. Three main areas will be covered: photopolymerizing the LCP either from a reactive LC or from a reactive monomer to make the backbone; photoisomerizing the liquid crystal molecules or additional molecules, such as photochromics, in order to change the liquid crystal phase or alignment; and finally using the light to change the LC properties by absorption to thermally heat the material.

2. Liquid Crystal Polymers and Phases

When a liquid-crystalline molecule is attached to a polymer backbone, a new type of hybrid material is formed, a liquid crystal polymer (LCP). An LCP can have the same range of liquid crystal phases as a low molar mass (lmm) LC: nematic, cholesteric, smectic A, and ferroelectric.[1] A nematic liquid crystal phase is formed when anisometric-shaped molecules, such as rod-like or disc-like molecules, align to have orientational order but no positional order. The direction of orientational order is called the director n (Fig. 7-1(a)). A typical rod-shaped mesogen is the cyanobiphenyl. The cholesteric phase is a chiral nematic and, although there is no positional order, the director has a long range twist to form a helix (Fig. 7-1(b)). The smectic phase has a higher order than the nematic, and positional order is introduced by the molecules forming layers (Fig. 7-1(c)). For thermotropic liquid crystals, the phase changes occur by heating from the solid phase through the smectic, then nematic phases; finally, above the clearing point, the isotropic liquid phase is reached. In the solid phase, an LCP can be in two states: crystalline or glassy. The amorphous glass phase has a frozen random structure of the polymer backbone. The glass phase is important because the liquid crystal alignment is not disturbed on solidifying, therefore allowing textural changes to be frozen in.

In the smectic C phase, the director is tilted with respect to the layer axis so that the director is restricted to move on a cone, to produce a C2h symmetry. By making the molecules chiral, the symmetry is reduced further to C2 and the dipoles can only be in one unique direction for any given point on the cone. This gives the phase a spontaneous polarization $\mathbf{P}_s$. However, being chiral, the director twists between each layer to form a helielectric phase (Fig. 7-2). The phase only becomes ferroelectric when the helix is removed. In lmm LC, this is generally achieved by using very thin (2-μm) cells so that the surface forces

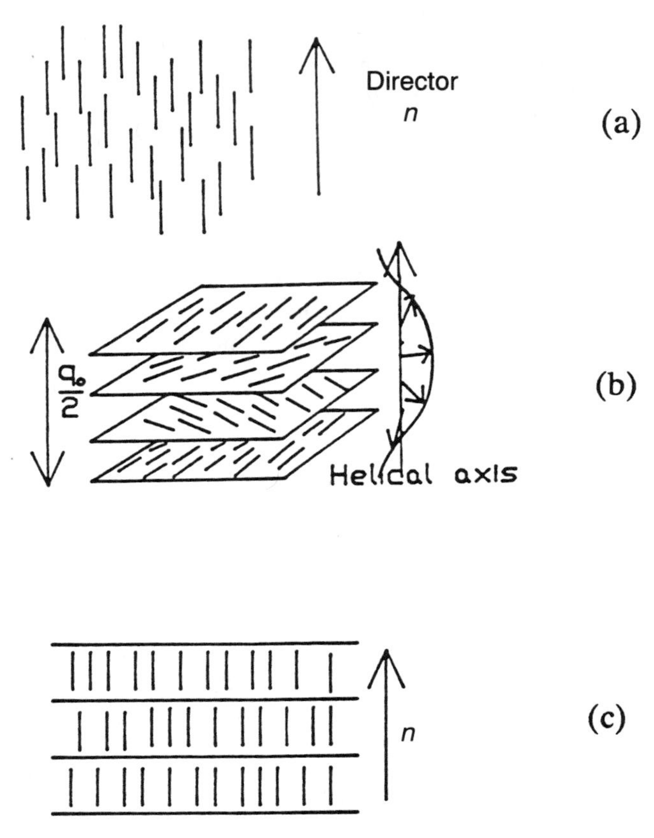

Figure 7-1. *Liquid crystal phases: (a) nematic, (b) cholesteric, (c) smectic A.*

are strong. The consequences of the LC being ferroelectric are that: it responds approximately a thousand times faster to an applied electric field than a nematic LC; it requires the polarity of the voltage to be reversed to switch in the opposite direction; it is bistable. Being ferroelectric, it is also piezoelectric and pyroelectric. There have recently been many ferroelectric LCP's synthesized, the first one of which was by Shibaev in 1984.[2]

The advantages of using an LC polymer over an lmm S_c^* are (a) that the polymer retains its alignment easily to produce a robust device, (b) the polymers can be processed into large areas and into many forms (e.g., a flexible film can be made using the polymer), and (c) the polarity of the ferroelectric phase can be frozen into the glassy phase of the polymer—this can be usefully exploited in nonlinear optics.

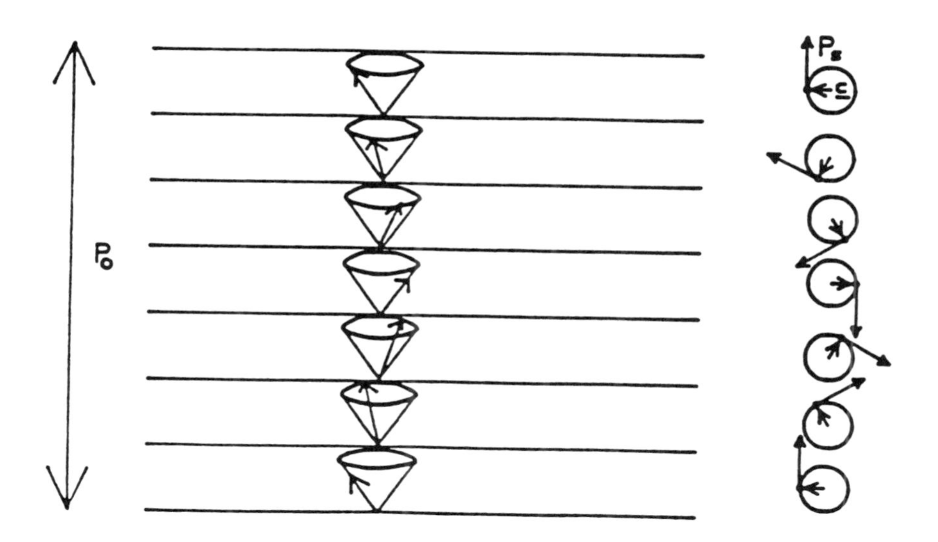

Position of director n on cone

Position of c vector and $\mathbf{P}_s$ dipole on cone (2π rotation).

Figure 7-2. *Ferroelectric helielectric phase.*

The functionality of the LCP is provided by both the liquid crystal (LC) properties and the rheology of the polymer. Some of the LC effects that can be exploited are: realignment of the director in an electric field or magnetic field; thermally induced texture changes or refractive index changes; and thermally induced pitch changes in cholesteric LCs. Such effects are well known in low molar mass (lmm) LCs. The polymer properties of the LCP enable standard polymer processing techniques to be used in sample preparation; e.g., solvent coating, lamination, and extrusion. This allows LCP devices to be made cheaply. LCPs can be combined with other plastics to produce multilayer films. They can also be extruded into fibers, or molded into complex shapes. Combining the LC properties with the flexibility in processing may open the door to novel uses of LCPs.

The first LCP to be discovered was by Bawden and Pirie[3] in 1937. They observed birefringence when tobacco mosaic viruses were put into solution. The viruses behaved as a main chain lyotropic polymer; the liquid crystalline behavior occurred within a certain concentration range only. The first synthetic

LCP was also a main chain lyotropic, a poly(γ-benzyl-L-glutamate), reported by Elliot and Ambrose in 1950.[4] This is a synthetic analogue of the naturally occurring polypeptide LCPs that are important in biological systems. The mesogens of a main chain LCP are joined together, via a flexible spacer group, to make a chain (Fig. 7-3(a)). The properties of such a polymer depend on the rigidity of the linkage group. The liquid crystalline nature of the mesogenic group is maintained if the spacer group is flexible enough to allow the mesogens to align with respect to each other.

In 1975, the E.I. Du Pont de Nemours & Co.[5] produced a para-aromatic polyamide, poly(*p*-phenylene teraphthalamide), called Kevlar. This could be spun from solution in the liquid crystal phase to form high modulus, high strength fibers that were very chemically resistant (they had to be spun from sulfuric acid) and were nonflammable. Kevlar can be made into cloths and composites to make many products such as replacement asbestos, motorcycle helmets, flame protection suits, and even car parts. The successful commercialization of Kevlar led to a tremendous growth in the field of LCPs, both in the underlying science and the applications. However, it was primarily directed at main chain polymers and the production of fibers.

The first thermotropic side chain LCPs (SCLCPs) were reported in 1979 by Shibaev et al.[6] in the USSR and by Finkelmann et al. in Germany.[7] The SCLCPs were produced by terminally attaching the mesogenic group to the polymer backbone in a pendant-like fashion (Fig. 7-3(b)). The side chain LC moiety is usually attached via a flexible spacer, which means that SCLCPs are free to behave like classical low molar mass liquid crystals (lmm LCs) in their optical, electrical, and magnetic characteristics. The side groups generally have molecular structures compatible with lmm LC systems.

The most commonly employed backbones are the polyacrylates, polymethacrylates, and polysiloxanes, each having a different degree of flexibility. The polysiloxanes are very flexible, which enables the LC moieties to align together to produce a high degree of order. These polymers therefore form smectic phases but have very low glass transition (T_g) temperatures (~0° C). The polyacrylates and polymethacrylates have a stiff backbone that restricts the movement of the mesogens so that they generally form nematic phases. The T_g of these polymers is also higher, generally around 40° C, making their properties at room temperature very different from those of the polysiloxanes.

3. Photopolymerization

There is a growing interest in producing oriented polymer structures because of their anisotropic properties. Such polymers can possess highly anisotropic optical, electrical, and thermomechanical properties that can be used in a variety

(a)

CH_3 CH_3

$\sim O-C_6H_4-C(CH_3)=N-N=C(CH_3)-C_6H_4-OCO(CH_2)_nCO\sim$

$n = 6;\ 8;\ 10$

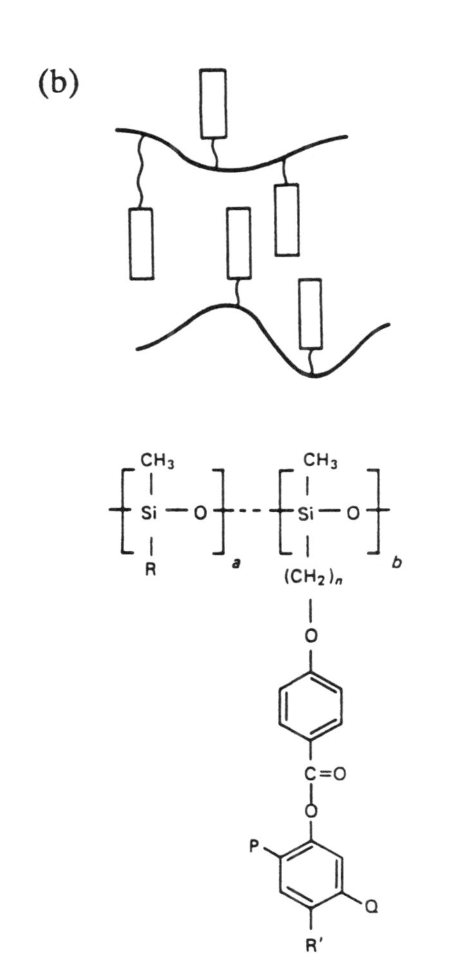

Figure 7-3. *Liquid crystal polymers: (a) main chain, (b) side chain.*

of applications, such as polarizers, optical filters, nonlinear optical switches, and optical storage.

Classical techniques to produce orientated systems include solid-state extrusion, and melt extrusion followed by drawing and aligning LCPs. However, these are difficult processes to control and are not suitable for large area oriented films or surfaces with complex geometries. A more controllable process is the in situ polymerization of reactive liquid crystal monomers to produce stable anisotropic polymers and networks. Liquid crystal monomers are far easier to align than LC polymers by the use of electric or magnetic fields or by the use of the surface forces from specially treated substrates. The alignment can then be frozen in by polymerization which can be thermally or optically induced. The liquid crystal phase is temperature dependent; therefore, a thermally induced polymerization is more difficult to control and may result in the wrong phase or a poorly aligned sample. For this reason, an isothermal photopolymerization process is preferred where the temperature is used to control the liquid crystal phase during the polymerization.

Photopolymerizing a liquid crystal monomer typically forms an anisotropic network. For example, Broer et al.[8] photopolymerized an oriented LC diacrylate to produce a three-dimensional network (polymer sandwich). By using crosslinking agents during or after the photopolymerization process, elastomers can be produced. A lightly crosslinked sample can be oriented by stretching, which changes the optical and electrical properties; therefore, piezoelectric or acousto-optic transducers may be possible. Such LC elastomers are interesting materials, but will not be discussed further here, even though polymerized LC monomers may well also be crosslinked.

A reactive liquid crystal monomer can be polymerized into a variety of states; a non-liquid crystalline polymer, a side chain LC polymer, a main chain LC polymer, an LC network, or a phase separated system. The first four resulting systems are generally used to produce aligned polymer structures, of which the type chosen will depend on the use, in particular whether it is a passive or an active application. The phase separated systems are generally produced from a reactive monomer with a nonreactive LC monomer and have very different properties and applications to the other systems. Therefore, they will be discussed separately in Section 4.

3.1. In situ Polymerization of Oriented LC Monomers

In situ polymerization is not new; in fact, it was used in the first reported syntheses of LC side chain polymers which were produced from solution.[9] The choice of monomer and polymerizable unit are important considerations in determining the success of in situ polymerization. The monomer must be liquid crystalline with either a rod shape, disc shape, or lyotropic structure.[10] The polymerizable unit must be chosen to minimize the release of heat so that the sample is not heated too much during the polymerization, which could disrupt the LC ordering. This effect is minimized by ensuring that the heat is released uniformly and that the monomer has an LC phase over a wide temperature range. In addition, the resulting polymer must be miscible with the monomer in order to prevent phase separation into a monomer-rich phase and a polymer-rich phase. This requires the polymerization to proceed at a faster rate than the kinetics of phase separation. Four commonly used polymerizable groups are LC diacrylates, diepoxides, divinylethers, and diacetylenic units.[10] Examples of diacrylate and monoacrylate LC reactive monomers are shown in Fig. 7-4.

In order to produce an oriented film, it is first necessary to align the LC monomer. Because the monomer has a low viscosity, it is possible to use standard LC techniques. The monomer can be placed between two substrates between 2 μm and 100 μm apart. If the substrates are coated with a rubbed polymer, such as nylon or polyimide, the LC will align along the rubbing direction of the polymer. This can be used to make uniform or twisted nematic structures, or even to produce tilted structures. Such techniques are also appropriate to align cholesteric and ferroelectric LC monomers in order to make aligned chiral LCPs. Alternatively, an electric field (~1V/μm) can be applied across the LC using electrodes on the substrates. If the LC has a positive dielectric anisotropy, it will align parallel to the field. For thick film samples (>50 μm) it is probably better to place the sample in a magnetic field (≥IT). As well

CB6

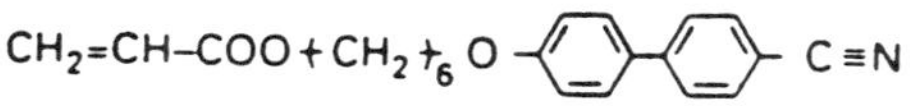

$CH_2{=}CH{-}COO{+}CH_2{+}_6\,O{-}C_6H_4{-}C_6H_4{-}C{\equiv}N$

Monoacrylate

C6M

CH_3

$CH_2{=}CH{-}COO{+}CH_2{+}_6\,O{-}C_6H_4{-}COO{-}C_6H_3(CH_3){-}OOC{-}C_6H_4{-}O{+}CH_2{+}_6\,OOC{-}CH{=}CH_2$

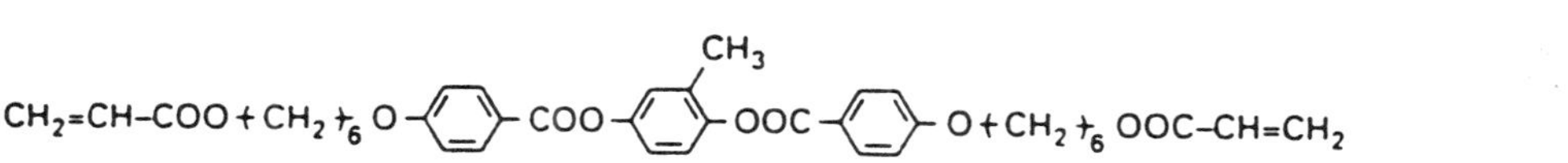

Diacrylate

Figure 7-4. *Examples of photoreactive liquid crystals.*

as structures with various tilts and twists, it is also possible to make patterned structures, either using the monomer alignment (e.g., patterned electrodes) or by using a mask during photopolymerization.

Once the LC is aligned into a monodomain, it is then heated to the required temperature, depending on the required phase, birefringence, or order parameter of the final polymer. The polymerization is then initiated using a UV light source so that the LC orientation is frozen in. The substrates can then be removed if necessary to produce a free-standing film.

The question as to whether the rate of polymerization is affected by the LC orientation has been investigated by several groups. Hoyle et al.[12] used a reactive acrylate monomer with a pendant cholesterol group and found that the polymerization rate was faster in the LC phase than in the crystalline or isotropic phases because of increased order and mobility. The LC phase changed from a cholesteric to a smectic during the initial stages of polymerization. In contrast, Hikmet et al.[11] and Doornkamp et al.[13] have investigated diacrylate monomers and concluded that the rate of polymerization was unaffected by the LC phase.

The resulting properties of the oriented film depend on the temperature at which the film was photopolymerized and the type of monomer used. Broer et al.[8] showed that, when photopolymerized, an aligned monoacrylate increased its birefringence and the temperature range of its smectic and nematic phases increased from 50° C to 90° C. The diacrylates they investigated would increase the birefringence if polymerized below 118° C and would decrease the birefringence above 118° C. This was attributed to the polymerization process decreasing the order parameter above 118° C. On heating the polymer, the birefringence changed very little up to about 240° C when thermal degradation started to occur. The thermal stability of an in situ polymerized film is significantly better than that of an LC monomer or aligned LCP.

The polymerized films also have anisotropic mechanical properties. The tensile modulus and strength in the direction of molecular orientation can be several times higher than in the lateral directions. However, both are lower than that found in linear main chain polymers. Thermal expansion of the polymers is also anisotropic. The film expands much less in the direction of molecular orientation, and at high temperatures it can even change to contraction.

3.2. Passive Uses

There are many passive uses of oriented polymer LC films. Such uses have the advantage that the film does not have to be liquid crystalline after polymerization, which makes it easier to find a suitable monomer.

Hikmet et al.[14] describe using an anisotropic network to produce a polarizer. A dichroic dye is added to the LC monomer, which is oriented along the director. When polymerized, the dye is frozen in place to produce a polarizer. There are two potential advantages with this type of polarizer: (a) it could be made with a very high T_g, thus making it more thermally stable than existing polarizers; (b) it could be made with different orientation of the dye molecules at different positions.

Hikmet et al.[14] also reported on the production of aligned cholesteric films by in situ polymerization. A cholesteric LC has the nematic director rotating in a helix which has the property of selectively reflecting circularly polarized light at the wavelength given by $\lambda = \bar{n}\rho$ where $\bar{n}$ is the average refractive index and ρ is the pitch of the helix. The pitch, and hence the position, of the selected wavelength of light is strongly dependent on temperature for LC monomers. LC polymers are less temperature dependent but are very difficult to align; a combination of shearing, heating, and alignment layers is required to align cholesteric LCPs. An easier approach is to polymerize an aligned cholesteric monomer to freeze in the helix, which results in an almost temperature-independent film. The pitch can be selected by polymerizing at different temperatures or by varying the monomer composition. These cholesteric films can be used as bandpass filters, notch filters (Fig. 7-5), circular polarizers, or birefringent compensators of LC displays. Belayev et al.[15] demonstrated a 70% polarized projection LC display using a cholesteric film as a polarizing mirror.

It is well known that to obtain a colorless, supertwisted nematic (STN) liquid crystal display, it is necessary to optically compensate for the birefringent twist in the display. This was initially achieved using an additional STN display with

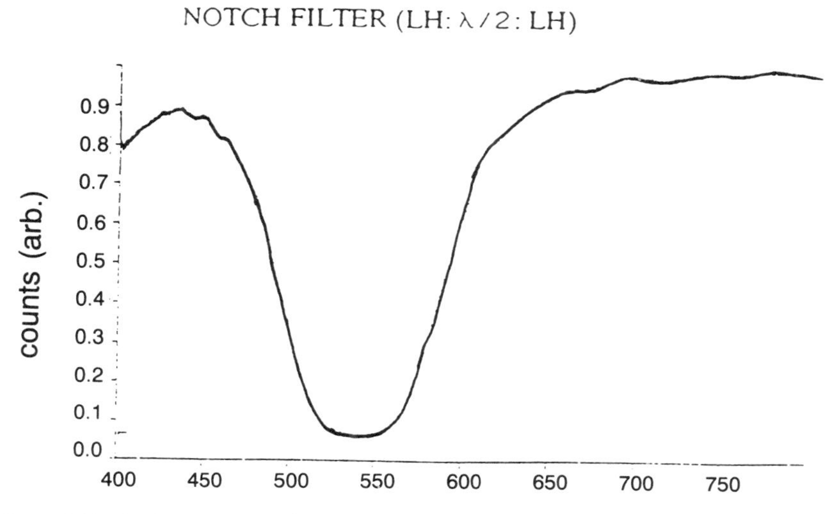

Figure 7-5. *Cholesteric LC polymer optical notch filter. (Abscissa is wavelength in nm.)*

similar optical properties but reverse twist. However, a cheaper and easier option is to use a birefringent compensating film even though the optical compensation is not as good. By using a cholesteric LC polymer film with matched twist, thickness, and birefringence, it is possible to produce a compensating film that can completely neutralize the wavelength dependence of the STN display, but with the advantages of using a polymer film.[7] It is critical that the optical retardation $d\Delta n$, where d is the thickness of the film, matches that of the STN display. The birefringence Δn can be controlled by the temperature at which the film is polymerized. The thickness decreases because of polymerization shrinkage in the direction perpendicular to the film, and hence is more difficult to control. Heynderickx et al.[16] showed that the shrinkage can be about 10% in the case of the film between substrates. This has to be taken into consideration when fabricating the compensating film.

By using photopolymerizable LCs to produce fiber-optic coatings, the thermal expansion of the resulting oriented polymer is more easily controlled than in conventional polymers and helps to prevent loss of transmitted data. The monomer is oriented around the fiber by melt flow, extrusion, or shearing, and is then UV cured. By using a helicoidal film the thermal expansion is minimized in both directions in the plane of the film. The uniaxial ordering is now averaged over all directions in the film plane, minimizing the thermal expansion; the expansion perpendicular to the plane is larger than in the case of a linear aligned film. Heynderickx et al.[16] showed that the volume expansion of the helicoidal film and the linear aligned film were the same. Such cholesteric films could be used for coating or encapsulation of electronic devices.

Another passive use of these oriented films is as controllable alignment layers for LC displays. Hikmet and de Witz[17] have demonstrated that by putting anisotropic gels over rubbed polyimide, it is possible to vary the tilt angle of the LC in contact with the gel. The pre-gel is an LC that is homogeneously aligned by the rubbed polyimide on one surface and, if sufficiently thick, it is aligned homeotropically at the air interface. As the pre-gel is made thinner, the LC becomes tilted at the air interface (Fig. 7-6). Polymerizing the pre-gel fixes the LC tilt angle, which can then be used to align the LC layer. A simpler method has been shown by Schadt et al.[18] Polarized light is used to polymerize the photoresist, PVMC, and the polymer orients along the direction of polarization. This can then be used to align the LC.

3.3. Active Uses

The main ''active'' use of oriented polymers is to create alignment of the liquid crystal molecules or of added dye molecules; either for nonlinear optical (NLO) applications or to create ferroelectric films. For NLO applications, a dye molecule is added to, and aligned by, the LC monomer before polymerization. After

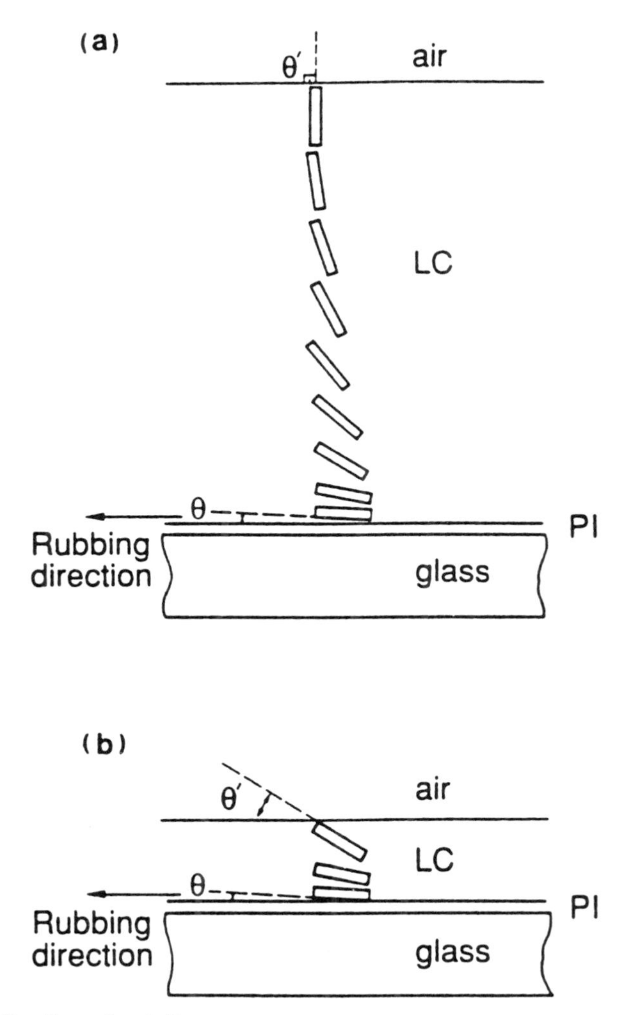

Figure 7-6. *Pretilt angle of director on rubbed polyimide layer and pre-gel LC mixture (Hikmet and de Witz, Ref. 17).*

polymerization the material need not be LC because the dye molecule remains active and does not have to physically move. For the ferroelectric films, the dipoles of the LC align and are made temperature stable by polymerizing. However, for active use of the films the LC molecules must move to switch the dipole; therefore, it is better to have the film remaining an LC after polymerization. This is more difficult to achieve.

To obtain NLO properties in organic molecules, it is necessary to have π-electron delocalization. Large third-order NLO responses can be obtained with unsaturated polymers such as polyacetylenes and polydiacetylenes. In order to get good third-order susceptibility values, χ^3, there must be macroscopic orientation of the polymer chain. Controlling the orientation of single crystals of polyacetylenes is difficult; therefore, using a liquid crystalline alignment to obtain the macroscopic orientation could be very advantageous. Such third-order NLO films have been produced by Le Moigne et al.,[19] although low χ^3 values were obtained. Photopolymerization of liquid crystal polydiacetylenes adds an additional processing advantage. Attard[10] used polymerizable diacetylene monomers consisting of disc-like molecules with a diacetylene unit in each of four alkyl chains radiating from the discotic core. Fibers were drawn from the discotic melt which, on cooling, formed a lamellar crystal structure based on the LC discotic alignment (Fig. 7-7). The fibers were then UV polymerized to form the poly(diacetylene) network which was well aligned and gave good χ^3 values.

Ferroelectric LCPs can potentially be made into aligned films far more easily if polymerized in situ from a monomer. To align a ferroelectric LC, the LC director and the layer structure have to be controlled. This can be achieved by using strong surface forces or by applying a large electric field (or both). Hikmet[20] has shown that by doping a smectic C photopolymerizable LC diacrylate

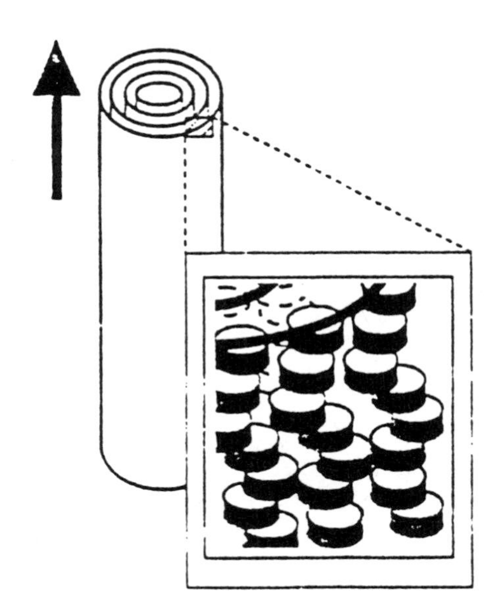

Figure 7-7. *Aligned discotic fiber with good third-order nonlinear optical properties (Attard, Ref. 10).*

with a chiral dopant, it is possible to produce an aligned ferroelectric LCP film. The LC chiral molecules were not chemically attached to the polymer backbone, which allowed the molecules more freedom of movement. The monomer could be polymerized in the ferroelectric S_c^* phase or the cholesteric N^* phase. By polymerizing while poling the S_c^* monomer, a network with piezoelectric properties was produced. The piezoelectric coefficient was dependent on the direction of the applied strain; perpendicular to the director, the value was 3.1 pC N^{-1}, parallel it was 1.4 pC N^{-1}. The large difference was due to the layer structure.

4. Phase Separation and PDLC

By photopolymerizing a monomer that is mixed with liquid crystal, a liquid crystal/polymer network is formed. The process used is polymerization induced phase separation (PIPS) to produce a polymer dispersed liquid crystal (PDLC). Mixtures containing a high proportion (e.g., >70%) of free LC molecules are referred to as gels, whereas others are called plasticized networks.

To make a PIPS PDLC, a reactive monomer (e.g., acrylate, epoxy, or vinyl-acrylate) is mixed with an LC to form a homogeneous solution (the monomer can be thermal or photoreactive). Such a mixture has its own distinct phase diagram including LC, isotropic, and phase separated (Fig. 7-8). The mixture is

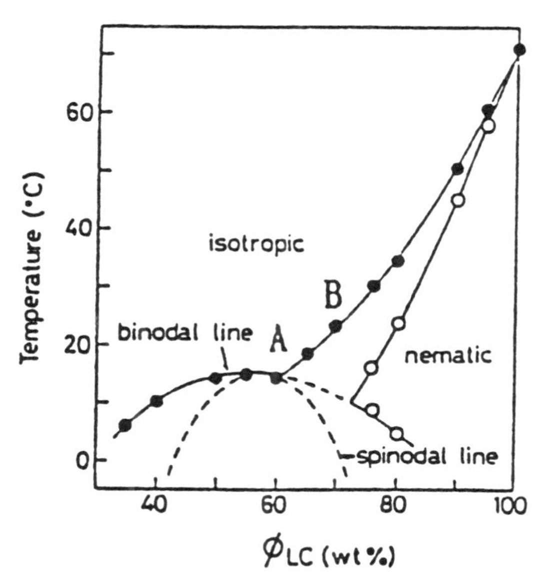

Figure 7-8. *Phase diagram of prepolymer mixture (Hirai, Ref. 23).*

spread into a thin film and UV polymerized while in the required phase. During polymerization the polymer is preferentially attracted to the reacting monomer and segregation occurs between the network and the LC solvent. This segregation occurs on a microscopic scale (De Gennes[21]) to form separate phases. As the mixture is curing it goes through several stages: homogeneous solution, immiscibility, droplet formation, gelation, and solidification.

The main application of PDLCs is as displays and electro-optic shutters; for example, high resolution projection displays and switchable windows. A PDLC device works on the principle of scattering the light when the LC refractive index mismatches the network refractive index, but is transparent when a field is applied and the LC reorients to match the refractive index of the network (Fig. 7-9). The reorientational forces of the device are controlled by the size and shape of the droplets and by the LC interaction with the polymer network. These are often weak and ill defined because the distribution of sizes and shapes can be large. In addition, the scattering is strongly dependent on the refractive index mismatch and the droplet size. Therefore, controlling the droplet formation is very important for device performance. This includes controlling the rate and uniformity of the phase separation process through careful selection of monomer, LC, temperature, and curing conditions. It has been shown by Hikmet[22] that the network dominates the behavior of the LC droplets, with up to 80% of the LC molecules being bound to the network in a 70% w/w LC: polymer PDLC.

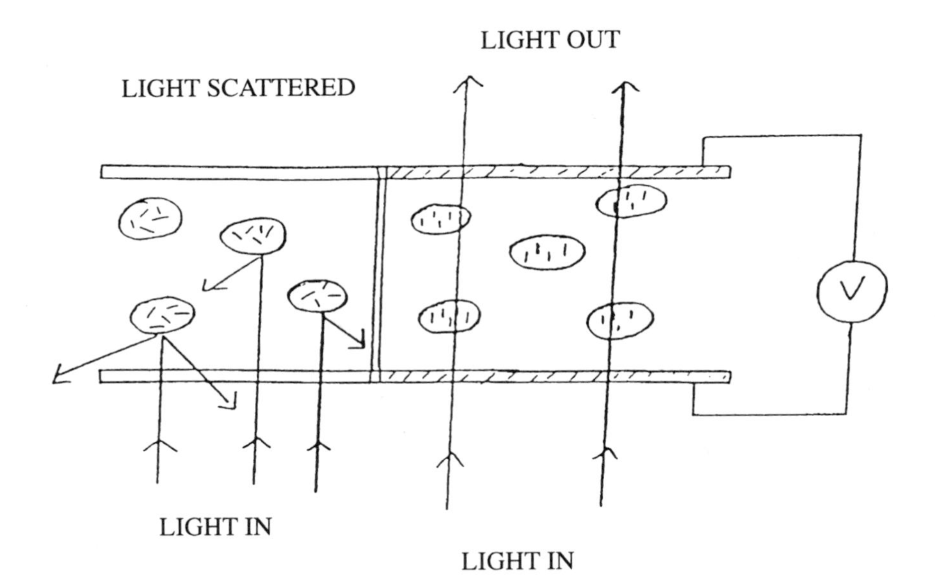

Figure 7-9. *Schematic representation of a PDLC shutter.*

Hirai et al.[23] reported that the key to obtaining a uniform structure of the network/LC droplets was to control the phase of the prepolymer mixture when curing. By using a monomer/oligomer combination as the network, the phase diagram can be modified so that optimum phase separation occurs. Hirai et al. state that the free energy of mixing has two important terms: the entropy term and the enthalpy term. Their results suggest that a monomer contributes to both terms, but an oligomer contributes only to the enthalpy term; hence they can control the solubility of the mixture by the selection of the prepolymers.

An alternative area of research has been to produce an anisotropic network by polymerizing the prepolymer while the solution is in the LC phase. Shimada and Uchida[24] demonstrated this by using a 95% LC sample which was nematic at 30° C. The mixture was aligned using rubbed polymer aligning layers and UV irradiated at 30° C to form the PDLC film. The network formed in the direction in which the LC was aligned (Fig. 7-10). They also showed that by

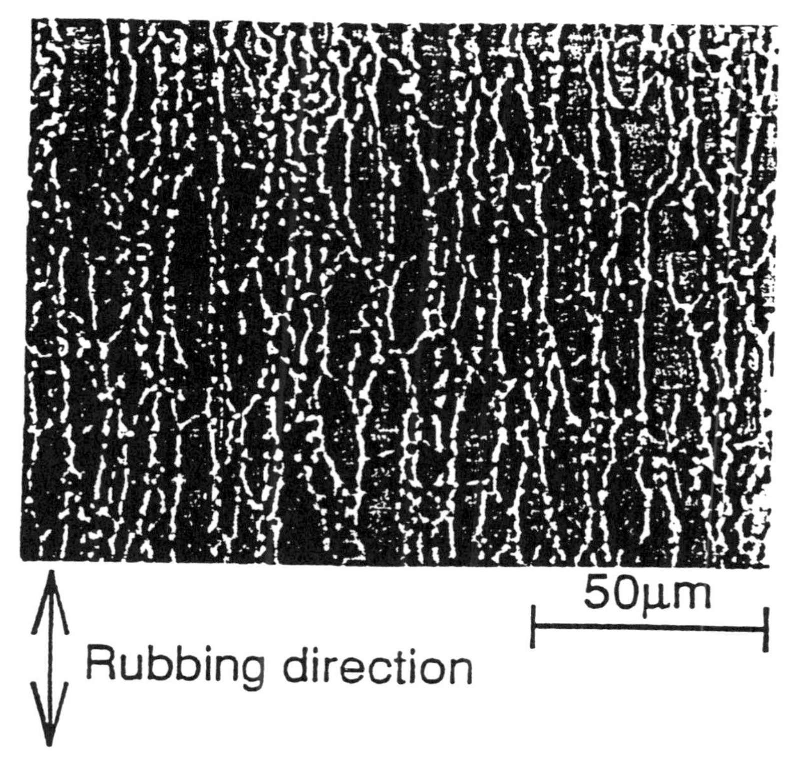

Figure 7-10. *Picture of an aligned PDLC network (Shimada and Uchida, Ref. 24).*

applying an electric field to the prepolymer mixture while curing, the network forms in the direction in which the LC is aligned; that is, parallel to the field. The LC droplets then align with the polymer network and alter the electro-optic performance of the film.

Braun et al.[25] demonstrated that a reactive LC monomer could be used to form the anisotropic network. A rigid mesogenic spacerless reactive diacrylate was photopolymerized in the presence of a nonreactive LC (20–80%), which was aligned using surface forces and a 1-tesla magnetic field. The result was an aligned network that controlled the orientation of the dispersed LC in the off-state, which provided a very good restoring force in the absence of the external electric field, therefore increasing the response time. When the LC was removed to form a dry network, it was found that the network was a continuous, regular channel structure with diameters of the order of 5 nm. Such dry networks may well be useful for membranes and absorbing agents.

The advantage of using LC reactive monomers is that higher concentrations of monomer can be used and the LC phase is still retained. By controlling the phase in which the mixture is polymerized, it is possible to produce more uniform droplet distributions and to produce anisotropic networks.

5. Photoisomerization

By attaching photoactive molecules as side chains to an LCP it is possible to change the LCP properties by exposure to light.[26] Generally the active moiety is a photochrome, which, on exposure to a specific wavelength, isomerizes and changes color. By putting photochromes into the LCP, exposure to light can either cause a color change or the isomers can disrupt the alignment of the LC. This can cause large changes in the Δn, scattering, viscosity, or dielectric properties, and these changes can be exploited in devices. The main area of interest for these materials is for optical storage and for optically controlling the alignment of LCs.

The spiropyran LCPs produce a color change from yellow to red when illuminated with λ ~365 nm which reverses when λ ~550 nm[27] (Fig. 7-11). By cooling the polymer film below T_g and irradiating with UV light, the film becomes blue. Three colors can therefore be obtained that could be used for new imaging applications by tailoring the polymer.

The azobenzene LCPs show very little color change when they trans-cis isomerize, but when illuminated with UV light (~370 nm), they change from a rod-shaped molecule to a bent shape that then disrupts the LC alignment (Fig. 7-12), and changes the optical density.[28] Using this technique, a nematic to isotropic phase transition can be induced by UV light that changes the birefringence of the system. This can be read out using crossed polarizers and can be

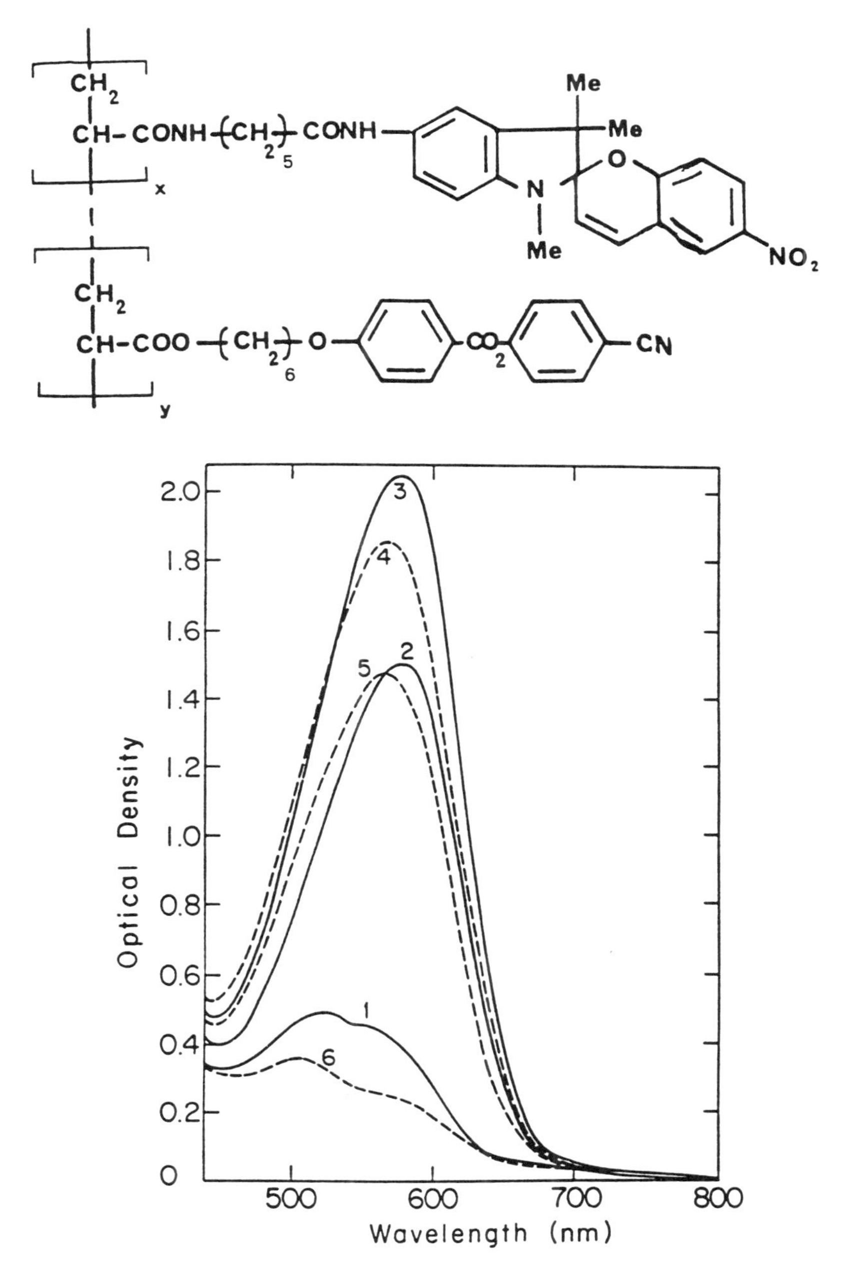

Figure 7-11. *Absorption spectra of spiropyran polyacrylate (Yitzchaik et al., Ref. 27). 1: before thermal color decay at 23° C; 2 and 3: after 12 and 24 min of successive UV irradiation; 4: after 17 hours of irradiation; 5 and 6: 3 min and 1 hour irradiation with visible light.*

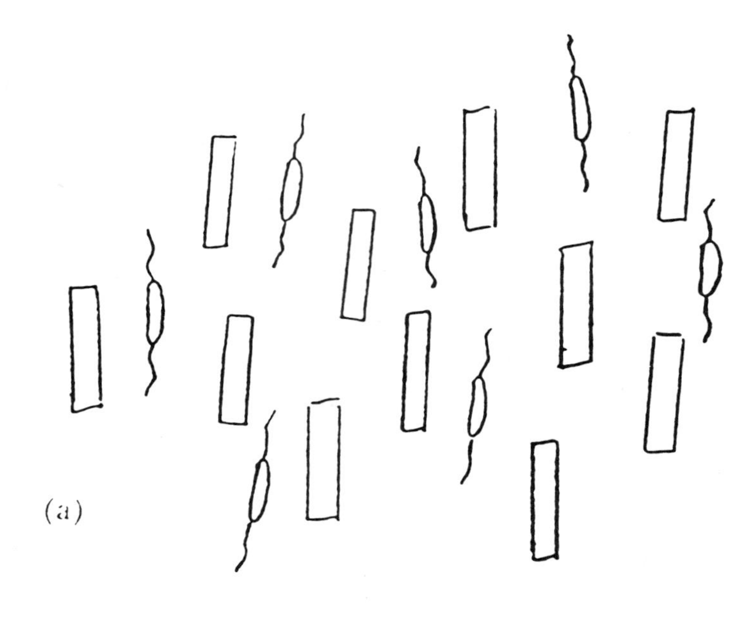

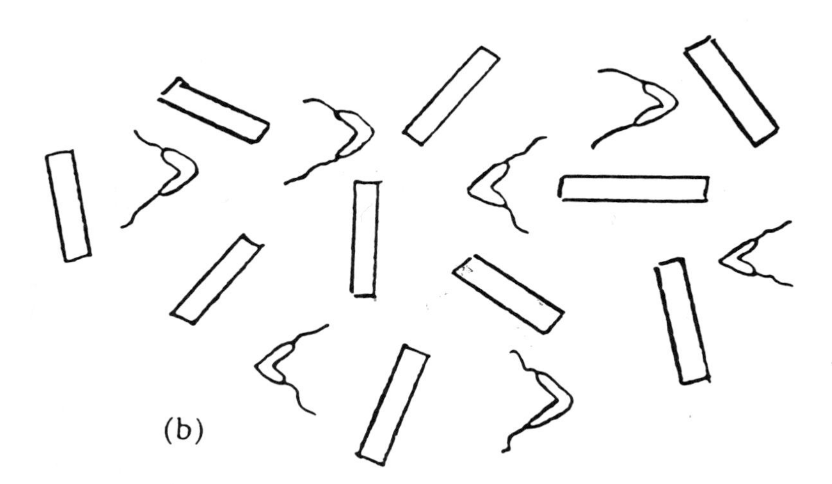

Figure 7-12. *Disruption of LC alignment by cis-trans isomerization of stilbene molecules: (a) trans molecular arrangement retains the LC alignment; (b) cis molecular arrangement disrupts the LC to form an isotropic phase; (c) an example of the stilbene isomerization.*

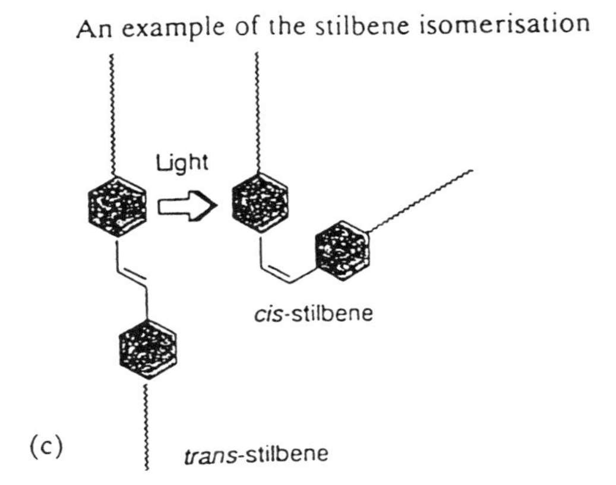

Figure 7-12. *Continued*

used for an optical storage device or imaging medium. The information can be stored by cooling the polymer into the glassy phase. Ikeda et al.[29] have reported 2–4 μm resolution using this technique. Similar effects have been shown in lmm smectic LCs by Ogura et al.[30] where the transition was from smectic to nematic. Contrast was obtained by shining light to change the alignment from scattering in the smectic phase to a clear, aligned state in the nematic phase. This was possible because the nematic phase could easily align on the surfaces but the smectic phase could not.

Other optical storage possibilities are to use the polarization sensitive effects that azo dyes show when they trans-cis isomerize.[31] Such effects have been used to produce polarization holograms. When linearly polarized UV light falls on the trans azo dye, it isomerizes to the cis form and then relaxes back into the trans form in a random direction. The dichroic absorption of the dye determines that the molecules aligned with the polarized light will isomerize more rapidly than those aligned perpendicular to the light. The molecules will then relax back to the perpendicular and the parallel in equal amounts. The perpendicular molecules will remain undisturbed longer than the parallel molecules so that eventually, more molecules align perpendicular to the light than parallel to it, and birefringence is induced. By rotating the polarization axis of the UV light, the position of the aligned azo dyes can be rotated and the system is rewritable. The reorientation time speeds up with subsequent cycles, probably as a result of reorientation effects in the photochromic material.

These materials have been used to demonstrate holography with diffraction efficiencies of 50% in thick phase gratings, with a sensitivity of 1 mW/cm^2. The resolution of the media was shown to be 0.2 μm (5000 lp/mm). The media is therefore also suited to digital optical storage. Eich and Wendorff[32] also demonstrated erasable holographic elements using LCP azo dyed materials by recording Fresnel zone plates to create a lens. This lens can be erased by heating above T_g and a new one can be rewritten as required.

Ortler et al.[33] have used cyclic cholesteric polysiloxanes doped with benzophenone (or carbon black) to produce a film that reflects in the visible. The reflection can be destroyed by exposure to laser light which destroys the helical activity of the cholesteric. The change can be irreversible or erasable depending on the dopant used. It may be possible to use such cholesteric LCPs to produce a colored optical storage/imaging medium.

Xerographic imaging using photochromic LCPs has also been demonstrated by Kimura et al.[34] A liquid crystalline azobenzene was incorporated into an ionically doped polymer binder. When UV illuminated, the azobenzene transforms from a nematic to an isotropic state which enhances the ionic conductivity by two orders of magnitude. The film is corona charged and the illuminated areas compensate the charge to produce a latent electrostatic image, which is developed using charged toner. The latent image is erased by illumination with visible light.

Azobenzenes can also be used to produce a switchable alignment layer for LC devices. Ichimura et al.[35] demonstrated that, by using an azobenzene surfactant as an alignment layer, a nematic can be switched from homeotropic to homogeneous by exposure to UV light. Such effects can be used in adaptive optics, optical memories, and displays, for example, to produce optically addressed SLMs that require no photoconductor, to produce high resolution switchable Fresnel zone plates without having to etch fine electrode structures, or, by exploiting the change in the capacitance, an optoelectronic transducer could be produced.

6. Photoabsorption and Optical Storage

Absorption of light by a liquid crystal polymer film can induce a phase change of the LCP through a thermo-optic effect. The light is absorbed by the LCP and converted into heat which heats up the film. By heating above the LC phase transition temperature and controlling the cooling conditions, it is possible to change the alignment state of the LC, which changes its optical properties. This is the basis of optical storage using LCP films.

The LCP itself is not an efficient absorber of light. The absorption spectrum of a typical polymer shows that above 350–400 nm, very little absorption occurs. Therefore, UV light is required for the pure LCP, which, for optical storage, would require an Ar^+ laser to achieve the required light intensity. Ar^+ lasers are large, expensive, and require water-cooling. For most thermo-optic applications,

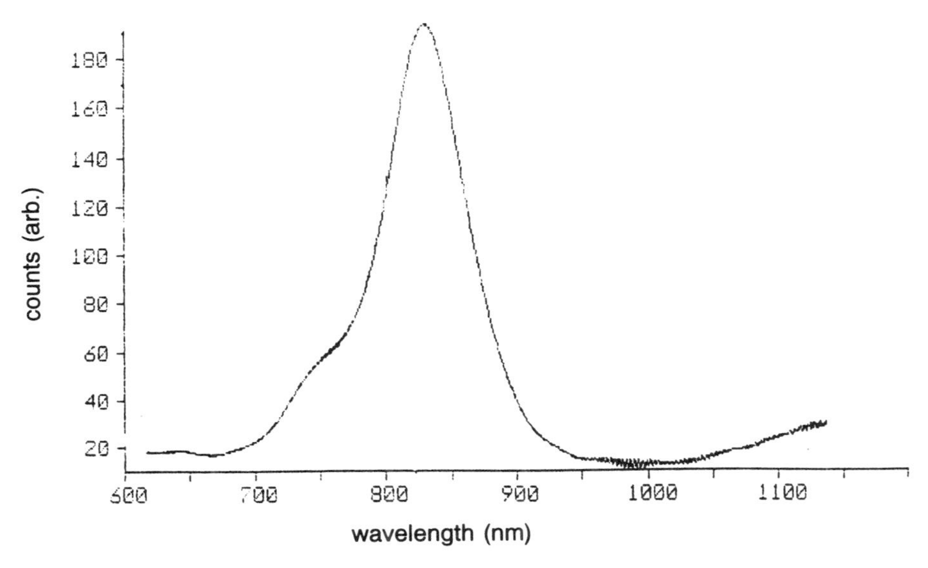

Figure 7-13. *Absorption spectrum of a typical LCP and squarilium dye.*

it is preferable to use a small, cheap, solid-state laser, which dictates a wavelength of 800–850 nm. In order to achieve sufficient absorption in the near infrared, a dye is added to the LCP. The function of the dye is to convert the laser light energy into thermal energy. In addition to having a high absorption coefficient, the dye must have good thermal and photo-chemical stability in the LCP. Bowry et al.[36] have developed appropriate dyes for optical storage in LCPs. Two chromophore types that were used were the squarilium dyes and the croconium dyes (Fig. 7-14). The squarilium dyes can have very intense absorption at 800 nm (ϵ_{max} = 200,000 $lmol^{-1}$ cm^{-1}) if powerful electron donor groups are present in the terminal groups. The croconium dyes are more bathochromic than the squarilium dyes, and thus simpler structures can give similar light absorbing properties. The dye shown in Fig. 7-14(b) has an absorption maximum at 815 nm (ϵ_{max} = 228,000 $lmol^{-1}$ cm^{-1}) in dichloromethane. The spectrum of a typical squarilium dye is shown in Figure 7-13. Both dye types absorb little in the visible region, and therefore appear almost colorless.

Using an LCP as the base material enables the dye to be chemically attached to the polymer backbone. This enables a high dye concentration to be achieved without the risk of separation of the dye and its migration to the film surface. Although chemically attaching the dye places larger stability requirements on the dye, attaching dyes to LCPs has been demonstrated.[36]

Producing optical storage film with dye has been described by Bowry et al.[37]

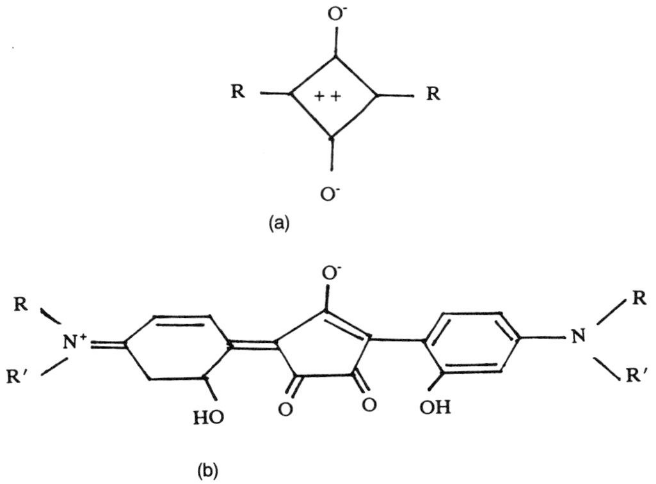

(a)

(b)

Figure 7-14. *Structure of IR dyes: (a) squarilium, (b) croconium.*

The simplest technique is to blade coat the LCP and dye mixture from solution onto a plastic substrate (e.g., PET). The solvent is then evaporated off and the polymer allowed to scatter in the LC phase. The optimum polymer layer is about 5 μm in order to maximize both the writing sensitivity and the scattering density. Alternatively, the LCP can be sandwiched between two substrates with ITO electrodes in order to enable electric field erasability.

The properties of the polymer used for thermo-optic effects greatly influences the alignment/phase changes that occur. By using a polymer with a glass transition temperature less than room temperature (e.g., <0° C) it is possible to realign the LCP using an electric field. Polysiloxanes are suitable LCPs and have been used to demonstrate selective erasure by changing the alignment of the LC using an electric field.[38] A schematic diagram of the process is shown in Fig. 7-15(a). The LCP is heated into the isotropic phase and cooled slowly with an electric field applied in order to align the LC director parallel to the field (homeotropic alignment). The film is irradiated using a focused laser beam to locally heat the LCP into the isotropic phase, and is allowed to cool without an applied field. The resulting state is that of scattering domains to produce a dark written point. If this point is then reheated using the laser and allowed to cool with the electric field applied, the point is erased. The electric field is chosen so that it is below the threshold for reorientation when the film is unheated.

If the polymer chosen has a high glass transition temperature (e.g., >30° C),

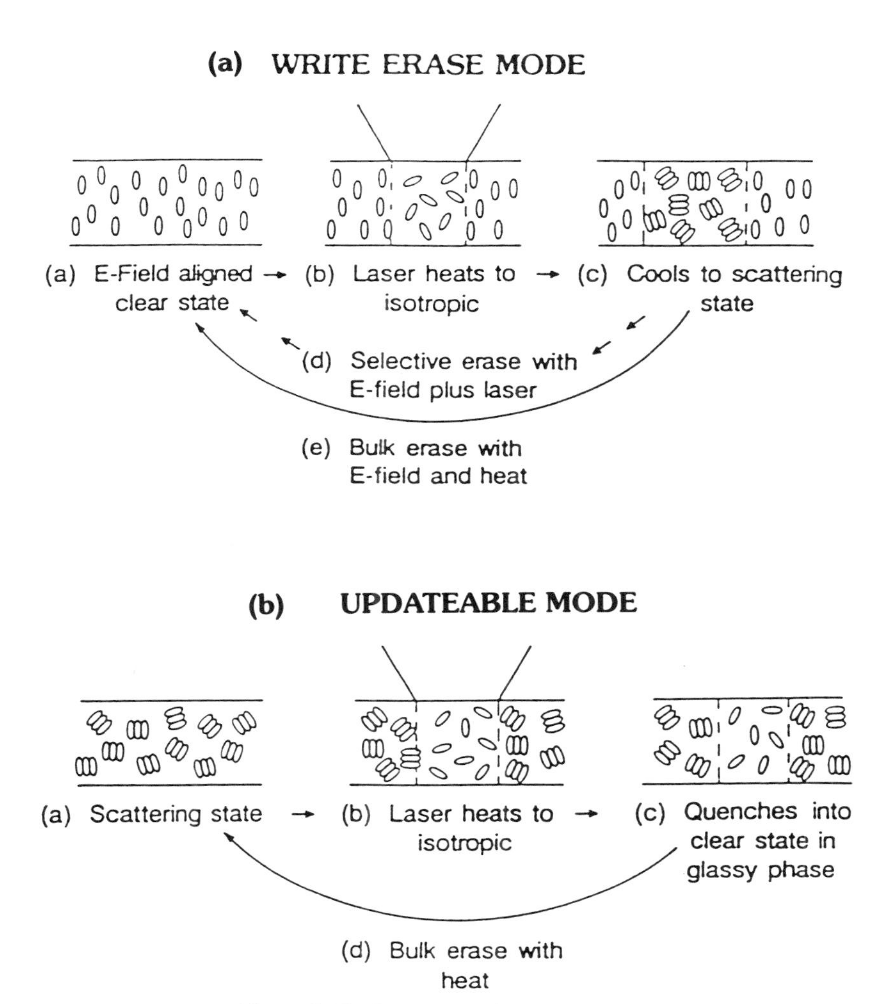

Figure 7-15. *Storage mechanisms in LCPs.*

then it becomes difficult to align the LCP in an electric field. A simpler approach is to make a write-once optical storage film. In this case (Fig. 7-15(b)) the film is heated into the isotropic phase and allowed to cool slowly without any field applied, so that a scattering texture is formed. When the laser irradiates the film, it is heated into the isotropic phase. On removing the laser, the heated point cools very quickly through the LC phase into the glass phase and is frozen as an isotropic texture. This phase change occurs because the LCP is not left in

Figure 7-16. *Photograph of image in LCP micrographic film.*

the LC phase long enough to realign. Modeling of the thermo-optic process has shown that cooling to the glass phase of the LCP after local laser heating can occur within 100 μs; the LCP requires seconds to minutes in order to reorient.

Both the above optical storage mechanisms have been demonstrated[36,37] and are useful for erasable and write-once storage in applications such as micrographics, optical discs, associative memories, and holographic storage. An example of a micrographics-type LCP film is shown in Fig. 7-16. The picture is approximately 16 mm × 70 mm and was written using an 820-nm, 40-mW laser diode. The focused laser beam had a 15-μm diameter and the energy density required to write on the film was 1 nJ/μm^2. The LCP was a polyester and the dye was a squarilium dye. The film had good mechanical robustness, and withstood extremes of cold (−25° C), humidity (90%) and heat (55° C).

7. Conclusion

This chapter has shown that there are many different photoinitiated effects in LCPs that can be used in a variety of applications. The spontaneous ordering of a liquid crystalline polymer enables alignment changes to be optically induced that would be extremely difficult to obtain with standard polymers. Photosensitive LCPs are therefore well suited to optical storage applications where large alignment changes can be thermally or sterically induced by a photosensitive dye, and these changes produce a large optical effect. Particularly, their use in micrographics, holography, or displays may well be seen commercially. Using photopolymerizable liquid crystals to form well-aligned polymeric layers could also have wide commercial implications. These layers could be used passively or actively, for example, as an aligning agent for nonlinear optical molecules or for liquid crystal displays to make high temperature polarizers or as optically compensating films for STN displays.

The fastest growing area for photosensitive LCPs, both in the market and in research, is as polymer dispersed liquid crystal displays. The need for high brightness projection displays and large area displays for such uses as road signs or travel indicators can be answered using PDLCs. These are already on the market, and new products are likely to appear.

References

1. G. W. Gray, *Polymer Liquid Crystals*, edited by A. Ciferri, W. Krigbaum and R. Meyer, eds. (Academic Press), 1982, Chapter 1, p. 1.
2. V. Shibaev, M. Kozlovsky, L. Beresnev, L. Blinov, and A. Platé, *Polymer Bull* **12**, 299 (1984).
3. F. C. Bawden and N. W. Pirie, *Proc. R. Soc. Ser. B* **123**, 1274 (1937).
4. A. Elliot and E. J. Ambrose, *Discuss. Faraday Soc.* **9**, 246 (1950).
5. A. Roviello and A. Sirigu, *J. Poly. Sci., Polym. Lett.* **13**, 455 (1975).
6. V. Shibaev, N. Platé, and Y. Freidzon, *Polym. Sci. Chem.* **17**, 1655 (1979).
7. H. Finkelmann, H. Ringsdorf, M. Happ, and M. Portugall, *Makromol. Chem.* **179**, 2541 (1978).
8. D. J. Broer, R. G. Gossink, and R. A. M. Hikmet, *Die Angewandte Makromolekulare Chemie* **183**, 45-66 (1990).
9. Y. Bouligand, P. E. Cladis, L. Liebert, and L. Strzelecki, *Mol. Cryst. Liq. Cryst.* **25**, 233 (1974).
10. G. S. Attard, 1990 *Mat. Res. Soc. Symp. Proc.* Vol. **175**, 239 (1990).
11. R. A. M. Hikmet, J. Lub, P. Maassen van der Brink, *Macromolecules* **25**, 4194 (1992).
12. C. E. Hoyle, C. P. Chawla, and A. C. Griffin, *Polymer* Vol. **60**, 1909 (1989).
13. A. T. Doornkamp, G. O. R. Alberda van Ekenstein, and Y. Y. Tan, *Polymer* Vol. **33**, No. 13, 2863 (1992).
14. R. A. M. Hikmet, J. Lub, and D. J. Broer, *Adv.Mater* 3, No.**7/8**, 392 (1991).
15. S. V. Belayev, M. Schadt, M. Barnik, J. Funfschilling, N. V. Malimoneko, and K. Schmit, *Jap. J. Appl. Phys.* **29**(4), L634 (1990).
16. I. Heynderickx and D. J. Broer, *Mol. Cryst. Liq. Cryst.* Vol. **203**, 113 (1991).
17. R. A. M. Hikmet and C. de Witz, *J. Appl. Phys.* **70**, (3), 1265 (1991).
18. M. Schadt, K. Schmitt, V. Kozinkov, and V. Chigrinov, *Jpn. J. Appl. Phys.* **31**, 2155 (1992).
19. J. Le Moigne, B. Francois, D. Guillon, A. Hilbere, A. Skoulios, A. Soldera, and F. Kajzar, *Inst. Phys. Conf.*, Ser. No. 103: Section 2.4, 209 (1989).

20. R. A. M. Hikmet, *Macromolecules* **25**, 5759 (1992).
21. P. G. De Genne, Chapter V, The Physics of Liquid Crystals. 1993, Oxford University Press.
22. R. A. M. Hikmet, *Liquid Crystals* Vol. 9, No. 3, 405 (1991).
23. Y. Hirai, S. Niiyama, H. Kumai, and T. Gunjima, *Spie Proc.* **1257**, 2, 285 (1990).
24. E. Shimada and T. Uchida, *Jpn. J. Appl. Phys.* Vol. **31**, P L352, Part 2, No. 3B. (1992).
25. D. Braun, G. Frick, M. Grell, M. Klimes, and J. H. Wendorff, *Liquid Crystals* Vol. 11, No. 6, 929 (1992).
26. C. B. McArdle, Applied Photochromic Polymer Systems, *Blackie*, ed., Chapter 4 V. Krongauz, p. 121 (1992).
27. S. Yitzchaik, I. Cabrera, F. Buchholtz, and V. Krongauz, *Macromolecules* **23**, 707 (1990).
28. J. H. Wendorff and M. Eich, *Mol. Cryst. Liq. Cryst.* **169**, 133 (1988).
29. T. Ikeda, S. Horiuchi, D. B. Karanjit, S. Kurihara, and S. Tazuoke, *Macromolecules* **23**, 36 (1990).
30. K. Ogura, A. Hirabayashi, A. Ueijima, and K. Nakamura, *Jpn. J. Appl. Phys.* **21**, 969 (1982).
31. M. Eich and J. H. Wendorff, *J. Opt. Soc. Am.B.* **7**(8), 1428 (1990).
32. M. Eich and J. H. Wendorff, *Makromol. Chem. Rapid. Commun.* **8**, 467 (1987).
33. R. Ortler, C. Brauchle, A. Miller, and G. Riepl, *Makromol. Chem. Rapid Comm.* **10**, 189 (1989).
34. K. Kimura, Y. Suzuki, and M. Yokoyama, *J. Chem. Soc. Chem. Commun.* 1570 (1989).
35. K. Ichimura, Y. Suzuki, T. Seki, Y. Kawanishi, and A. Aoki, *Makromol. Chem. Rapid Commun.* **10**, 5 (1989).
36. C. Bowry, P. Bonnett, M. G. Clark, G. Mohlmann, C. Wreesman, P. de Erdhuisen, L. W. Jenneskens, G. W. Gray, A. M. McRoberts, R. Denham, D. Lacey, R. M. Scrowston, J. Griffiths, and S. Tailor, *Proc. Eurodisplay '90*, p. 158 (1990).
37. C. Bowry and P. Bonnett, *Opt. Comp. & Proc. 1* No.1, 13 (1991).
38. C. B. McArdle, M. G. Clark, and C. M. Haws, *Proc. Eurodisplay '87*, p. 160 (1987).

III

Photoreactive Polymers in Advanced Applications

1

Holographic Recording Materials

Roger A. Lessard, Rupak Changkakoti, and Gurusamy Manivannan

1. Introduction

The invention of holography by Denis Gabor[1] in 1948 initiated a new direction in information recording. However, because of limitations such as unavailability of coherent light sources, holography lay dormant for almost a decade. The discovery of the off-axis holographic technique by Leith and Upatneiks[2] in 1962 and the development of highly coherent laser sources created waves of excitement and a resurgence in the field of holography. Simultaneously, Denisyuk's[3] and Van Heerdeen's[4] proposal of Lippman holography and deep (volume) holograms at the same period gave a new life to the domain of holography. The reconstructed three-dimensional images seen through a hologram struck the imagination of scientists all over the world and there was a burst of research activity to find new recording materials and techniques capable of recording the complete optical information carried by object wavefronts.[5–13] The demand for better materials with stringent characteristics such as high resolution, better energy sensitivity, broad wavelength sensitivity, simpler processing procedure, and erasability was severely felt and still exists today.

The most readily available and widely used recording material viz., silver halide photographic emulsion (SHPE) was extensively investigated for the purpose of recording holograms. Detailed accounts of the use of silver halide emulsions and their holographic characteristics are well chronicled by various authors.[5–9,14–24] The limitations of these media, the high grain noise resulting in low signal-to-noise ratio, were soon realized and led investigators to search for grainless recording media with other characteristics comparable to those of SHPE.

The development of new materials is important because of the extensive use of holography in optics and photonics. Various applications of holography have been proposed, such as the fabrication of holographic optical elements, holographic displays, scanners and optical disc systems, optical computing, and integrated optics. One of the distinct advantages of holography is the large capacity of storage using volume holographic techniques. A comparison of the storage capacity of various commercially available memory devices is presented in Fig. 1-1; it is shown that holographic memory devices exhibit the highest storage capacity. As this chapter is limited to polymer-based materials for holographic recording, only biopolymer- and synthetic polymer-based recording materials will be discussed.

To better comprehend the characteristics of various materials, a brief discussion of the different type of holograms based on recording geometry and material characteristics follows.

2. Holographic Principle

Denis Gabor described holography as a technique of wavefront reconstruction by recording of the complete information (both phase and amplitude) emanating

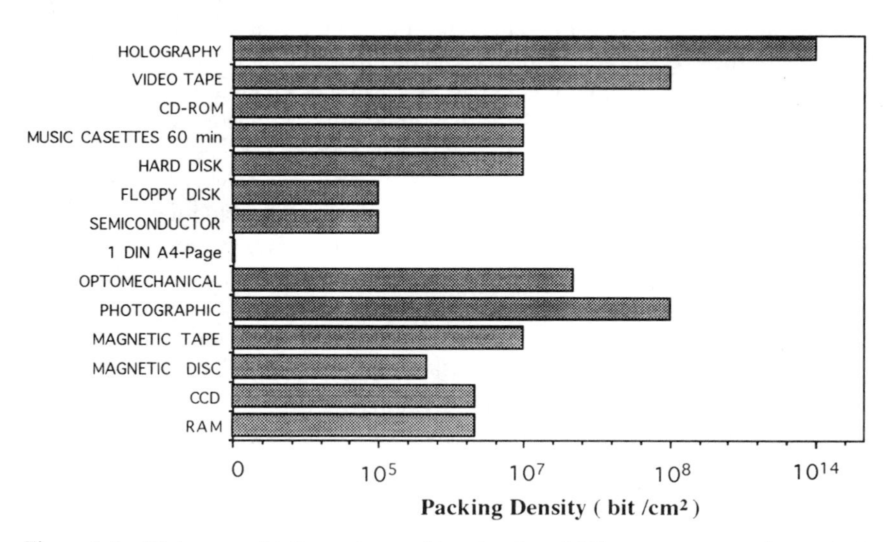

Figure 1-1. *Histogram of information packing density of different commercially available memory devices.*

from an object wave field as it intercepts the recording plane.[1] Because most recording media are square law detectors, the problem consisted of recording the phase of the wave field. Gabor solved this problem by providing a background wave, called reference wave, which converted the phase differences between two points to intensity differences that could be easily recorded on any detector. This record of information was named a hologram (*complete record* in Greek). The record is essentially a pattern of interference fringes, which, when illuminated with an appropriate beam, exactly regenerates the object wave field. The reconstructed wave field propagates the same way as it did during recording.

A simple classification of the various hologram types (without indulging in mathematical details) based on the recording geometry, the type of modulation imposed on the illuminating wave, and the thickness of the recording medium is given below. The first hologram recorded by Gabor was made by two interfering beams traveling in almost the same direction. This type of interference pattern is called a *Gabor hologram* or an *in-line hologram.* When a record is made by the interference of two beams arriving at the recording plane from substantially different directions, the pattern is called a *Leith-Upatneiks hologram* or an *off-axis hologram.*[2] And when a hologram is recorded by two interfering beams traveling in essentially opposite directions, it is called a *Lippman or Denisyuk hologram (reflection hologram).*[3]

Holograms can be recorded as either *thick or thin holograms*, depending on the thickness of the recording medium and the recorded fringe width. A *thin (or plane)* hologram is one where the thickness of the recording medium is small compared to the fringe spacing (*d*). A *thick* (or volume) hologram is one where the thickness of the medium is of the order of or greater than the fringe spacing. These two holograms can be easily distinguished by the *Q*-parameter defined by:

$$Q = \frac{2\pi\lambda d}{n\Lambda^2} \tag{1}$$

where λ is the illuminating wavelength, d is the thickness of the medium, n is the refractive index, and Λ is the fringe spacing. Generally, a hologram is considered thick when $Q \gg 10$ and thin otherwise.[25]

Holograms are also classified according to the modulation they impose on an illuminating wavefront dependent on the change incurred on the medium while recording the hologram. In the case of an *amplitude hologram* (e.g., holograms on photographic emulsions), the interference pattern is recorded by a density variation of the recorded medium, and thus the amplitude of the illuminating wave is modulated. In the case of a *phase hologram*, the fringe pattern is recorded by virtue of thickness or refractive index change and accordingly the phase of the illuminating wave is modulated. Thus a hologram can be either of the thick/thin amplitude or phase type.

2.1. Holographic Efficiency or Diffraction Efficiency

Before going into a detailed description of the materials, various fundamental characteristics of the recorded holograms such as diffraction efficiency and spatial frequency will be defined. The real-time diffraction efficiency of the recorded hologram is defined as:

$$\eta = \frac{I_1}{I_0} \tag{2}$$

where I_1 is the intensity of the first-order beam and I_0 is the intensity of the transmitted incident beam through the unexposed holographic recording film. This method of diffraction efficiency evaluation compensates for the losses in the incident beam due to inherent absorption of the film and the reflections at two surfaces of the substrate on which the film is coated.

The spatial frequency of a recorded grating having unslanted fringes is evaluated as being the inverse of the spatial grating period (Δ):

$$f = \frac{1}{\Lambda} = \frac{2 \sin \theta}{\lambda} \tag{3}$$

where θ is the half-angle between two writing beams and λ is the wavelength of recording beam.

3. Recording Materials

Many state-of-the-art reviews on holographic recording materials have been documented in detail over the past two decades.[5-24,26-30] As mentioned earlier, we will discuss mainly polymer-based recording materials used for hologram recording. Accordingly, the holographic materials that will be described in this chapter are classified into three broad categories, viz., *(1) photopolymerizable systems, (2) photocrosslinking systems, and (3) doped polymer systems,* based on the mechanism of photochemical change they undergo on photoirradiation or on the composition. The classification is laid out in the flow chart shown in Fig. 1-2. As in descriptions of most holographic recording materials, significant characteristics and mechanisms of recording will be discussed in all their ramifications.

Most of these volume phase recording materials are in the developmental stage except for a few systems that are commercially available. The general mechanism of recording involves either *photopolymerization* or *photocrosslinking* reactions, which creates a refractive index or thickness (surface relief) modulation after exposure to a suitable light source. The advantage of these materials

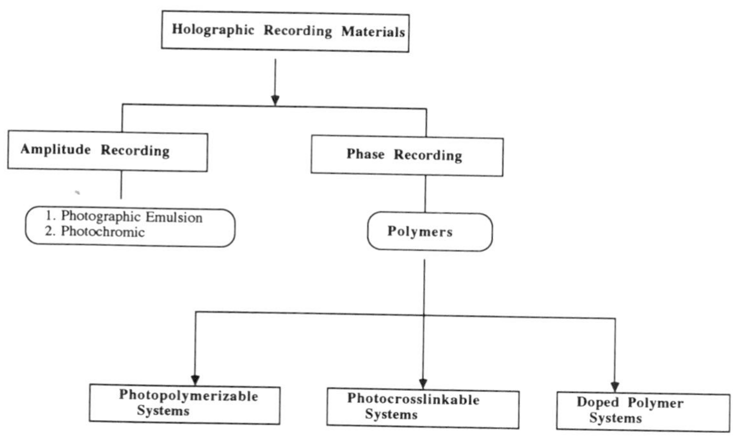

Figure 1-2. *Flow chart describing the various categories of holographic recording materials outlined in this chapter.*

is that very thick layers (μm to mm) of recording medium can be made; thus they serve as true volume holograms, giving very high diffraction efficiency with good angular selectivity and high storage capacity. However, with very thick layers, the scattering increases and consequently the signal-to-noise (SNR) ratio is lower.

A distinct feature of these materials is that most of them are either self-developing during the recording process in contrast to latent image forming media or require a simple, dry post-development process such as an overall light and/or thermal exposure. This can be greatly advantageous for real-time applications such as interferometry and information storage, if the development time is very short. On the contrary, if the hologram develops significantly during exposure it can also lead to interaction between the recording beams and the recorded hologram, resulting in degeneration of the hologram quality. It can also limit the recording of multiplexed holograms in a specified area, which is one of the great advantages offered by volume type materials.

3.1. Photopolymerizable Systems

Photopolymerizable systems are generally compositions consisting of a photopolymerizable monomer, a photoinitiator, and a sensitizer in a polymer matrix (a binder). These systems can be used either as *liquid compositions* in a cavity or *dry films* casted in a mold or on a substrate. In principle, they can achieve quite high efficiency because a single photon can initiate a chain polymerization reaction involving a large number of reactive monomer molecules. The various characteristics and mechanisms of these photopolymerizable materials are grouped under two categories: (1) photopolymerizable liquid compositions and (2) photopolymerizable dry films.

3.1.1. Photopolymerizable Liquid Compositions

(a) Acrylate and acrylamide monomer systems. Close et al.[31] in 1969 were the first to report the use of photopolymers for volume hologram recording. A liquid free radical polymerizable *metal monomer* developed by Hughes Research Laboratories [later commercially marketed by Newport Research Corporation (NRC)] was used. The monomers were polymerized using photocatalysts such as methylene blue, *p*-toluene sulfinic acid sodium salt. The mixed solution has a short lifetime; thus, it must be exposed immediately after mixing. The photosensitive dye, methylene blue, is sensitive to red light and therefore holograms were made using a ruby laser at 694 nm. Energy sensitivities were ≈300 mJ/cm^2 and a diffraction efficiency of 45% was obtained. The holograms were fixed with a 15- to 30-second exposure to a 200-W mercury arc lamp, which bleaches the unreacted methylene blue, rendering a clear hologram. Relatively low SNR due to the presence of light scattering polymer particles, low sensitivity, and short shelf life were the disadvantages of this medium. Jenney[32–35] carried out detailed experiments using the same Hughes photopolymer system and studied the influence of various parameters on the hologram formation. The mechanism of hologram formation was also investigated in detail using scanning electron microscopy and refractive index matching.

The holograms stored in these photopolymers were the result of both the thickness and refractive index modulation. Holograms with a very high spatial frequency of ≈3000 cycles/mm could be recorded, but between 2000-2500 cycles/mm, the efficiency fell to a very low level. These photopolymers were sensitized to both a green and red spectral range by different dye sensitizers. The highest energy sensitivity of 0.6 mJ/cm^2 was reported for low spatial frequency gratings giving 10–20% diffraction efficiency at 633 nm. The slow response of the materials was due to the induction period caused by the presence of *p*-toluene sulfinic acid. As earlier reported by Close et al.,[31] even though these photopolymers were self-developing, fixing was necessary to preserve the recorded holograms. Long-term fixing could be obtained by leaving the hologram in dark for 1–2 days. During this time, the catalyst becomes deactivated by a slow reaction with the monomer. Rapid fixing methods such as exposing the recorded hologram to a flash lamp or thermal treatment that stabilized the holograms within a few hours were proposed by Jenney.

The development of photopolymerizable liquid compositions and their utilization for holographic recording carried out in the Commonwealth of Independent States (former USSR) has been discussed in detail by Mikaelian and

Barachevsky.[36] The photopolymerizable compositions (PPC) were based mainly on a class of acrylate oligomers such as oligourethaneacrylate (OUA), oligoetheracrylate (OEA), and oligocarbonateacrylate (OCA) with a diacetyl photoinitiator. From the various experiments conducted, it was observed that the highest sensitivity was found in PPC based on low-molecular weight OUA with acrylic unsaturated groups. It was also observed that the diffraction efficiency maximum increased as the intensity of the laser beam was increased. When the holograms were made in PPC with exposure energy less than the optimum (energy for maximum diffraction efficiency), the diffraction efficiency of the recorded holograms declined in the process of dark storage. Because of light scattering by the polymerized matrix, values of diffraction efficiency and resolution >60% and >2500 lines/mm, respectively, could not be obtained. The source of light scattering was associated with microheterogeneities (formation centers of polymeric network) at the stage of gelation. Therefore, to overcome the problem of high light scattering and thus upgrade the hologram quality, the PPC was passed through a preliminary photopolymerization process with incoherent light, where the exposure was limited to the point of passage of the gelation point. Using this technique holograms with higher sensitivity, diffraction efficiency (80%), and resolution (6000 lines/mm) were recorded.

Boiko et al.[37–39] investigated a similar type of liquid composition based on OUA as the monomer, diacetyl as the initiator, and triethanolamine and camphorquinone as the sensitizers for recording transmission, reflection, and relief gratings. Gyulnazarov et al.[40,41] improved these systems by introducing a chemically inert component such as bromonaphthalene, which does not participate in the photopolymerization reaction but has a higher refractive index in the visible region compared to the oligomer and monomer. The chemical composition consisted of an α,ω-acryl-bis(propylene glycol-2,4-toluene diurethane) oligomer, a dimethylacrylate ethyleneglycol monomer, a photoinitiator such as naphthoquinone, and Michler's ketone, as well as bromonaphthalene. The above composition was called FPK-488. The main contribution of the inert component was the photochemical stability of the recorded holograms even during the process of their recording. This phenomenon was explained by microsynergesis (pushing the neutral component out of the polymeric network formed) of the monomer and oligomer polymerization in the region of irradiation and their diffusion into this region. FPK-488 has a sensitivity of 20 mJ/cm^2 over the 300–500 nm spectral range with a resolution of >6000 lines/mm, making it possible to record holograms[40] with diffraction efficiencies of up to 100%.

The mechanism of hologram recording on these photopolymerizable compositions (FPK-488) involves at least three simultaneous processes that determine the refractive index modulation: (1) density modulation leading to shrinkage of the material, (2) polarizability modulation, differences in polarizability caused by different degrees of polymerization at the dark and bright regions of recorded fringe pattern, and (3) concentration (chemical composition) modulation[41] caused by the interdiffusion of the ingredients of the PPC.

In Japan, Sukegawa et al.[42,43] and Sugawara et al.[44] used a similar system sensitive to red wavelengths containing an *acrylamide monomer*, *N,N*-methylene bisacrylamide as a crosslinking agent, methylene blue as a sensitizer, hydroquinone as a transparency maintaining reagent, and triethanolamine (TEA) or acetylacetone (ACA) as an initiator. The system was used in a liquid phase placed in cells of 50-μm thickness. Among the various studies conducted using TEA and ACA, TEA was found to be more efficient with regard to diffraction efficiency and sensitivity. Efficiency up to 65% was achieved with exposures of about 50 mJ/cm^2 at 633 nm.

Another system proposed by the same research group consisted of the same monomer and crosslinking agent but with ferric ammonium citrate as the sensitizer and *t*-butyl hydrogen peroxide (tBH) as the initiator. A very high diffraction efficiency of 80% was achieved with an exposure energy of 20 mJ/cm^2, in which the holograms were recorded with 457.9 nm line of an argon ion laser and had a resolution of 1500 lines/mm. The holograms fixed with a tungsten lamp exposure were found to be stable for more than six months.

The *mechanism of hologram formation* is due to the diffusion of monomer from the low intensity to the higher intensity regions, until the higher intensity regions become hard from photopolymerization and the diffusion of monomers is no longer possible. The polymerized regions have a higher density compared to the lower intensity regions, and this creates a refractive index difference.

Two-photon holographic recording with continuous wave (cw) near infrared (IR) lasers on liquid photopolymerizable layers was reported by Carré et al.[45] The liquid photopolymerizable layer consists of a mixture of a liquid monomer, dimethylacrylamide used as the solvent, a solid monomer, acrylamide used to obtain a higher viscosity, and a difunctional reactive compound, methylene bisacrylamide that ensures fast gelation through a curing process. An IR dye sensitizer viz., tricarbocyanine and a singlet oxygen trap, diphenyl isobenzofuran (DPBF), are used in the photopolymerizable composition. The oxygenated mixture is enclosed between two glass plates separated by a 50-μm Mylar spacer. The recording process works only when irradiation of the entrapped mixture is done with both the UV and IR sources. Holograms recorded at 715 nm with a CW ring dye laser and monitored with a 633 nm He-Ne laser exhibited a maximum diffraction efficiency of 2.9%.

The general mechanism of hologram formation in these materials was explained[46] as a multistep process in which the first step generates singlet oxygen in the bright regions when the sample is illuminated by the holographic pattern. This involves the excitation of a photosensitizer to its triplet state; the energy transfer from this transient level to molecular oxygen, which is dissolved in the reactive mixture, then produces the first excited singlet state. The second step involves the thermal addition of singlet oxygen to a selected diene that finally converts to a carbonyl compound. The local concentration of this new compound, synthesized in situ, parallels the power density of the interference pattern. In the third step, a noncoherent UV source excites this carbonyl compound to its' triplet state, thus giving rise to radicals capable of initiating the polymerization of a mixture of acrylic monomers. This uniform intensity UV exposure, which superimposes on the holographic pattern, causes part of the monomers to polymerize with the amount of polymerization a function of the intensity of the UV illumination and of the local concentration of the initiator. Monomer concentration gradients, which reflect local variations in the amount of polymerization, then give rise to the diffusion of the monomer molecules from regions of higher concentration to regions of lower concentration.

Other photopolymerizable liquid compositions based on acrylate monomers were also proposed[47–55] and the reactive mixture contained typically three components: a multifunctional acrylate monomer, pentaerythritol triacrylate (PETA); an amine cosynergist, methylethanolamine; and a sensitizer dye, a xanthene dye (eosin Y, erythrosin B, rose bengal). Some specific additives or agents enhancing the initiating activity were occasionally used such as 1,2-propanedione-1-phenyl-2-[*o*-(ethoxycarbonyl)oxime] or tetrabromomethane. Because these materials have sensitivity in the region of 450–550 nm, they were exposed with 514-nm beams, and the diffraction efficiency development was monitored using a 632.8-nm beam from a He-Ne laser. Various experiments to optimize the different parameters contributing to the diffraction efficiency of the holograms were conducted. Fig. 1-3 presents the diffraction efficiency of this system where diffraction efficiency as a function of exposure energy is plotted for samples having different concentrations of hydrogen donor (methylethanolamine). The highest diffraction efficiency achieved was 80%.[49] Spatial frequency analysis of this medium exhibited a decline of efficiency at frequencies greater than 2500 lines/mm.[50] The contribution of the various components of the medium and the photopolymerization kinetics has been described in detail by Lougnot et al.[47] Real-time holographic interferometry, recording of image and computer generated holograms, and chopped light recording were some of the applications demonstrated.[48,53,54]

(b) Methacrylate monomer system. A dye-sensitized photopolymerizable recording medium prepared with 2-hydroxyethyl methacrylate for fabrication of water immersible holograms was suggested by Yacoubian et al.[56] The coating solution consists of the monomer sensitized with the visible wavelength sensitive camphorquinone dye and the oxidizing agent benzoyl peroxide as the polymerization initiator. The recording monomer solution sandwiched between two glass plates with a 120-μm spacer is semi-polymerized before exposure to make it suitable for holographic recording. The diffraction efficiency of the holograms

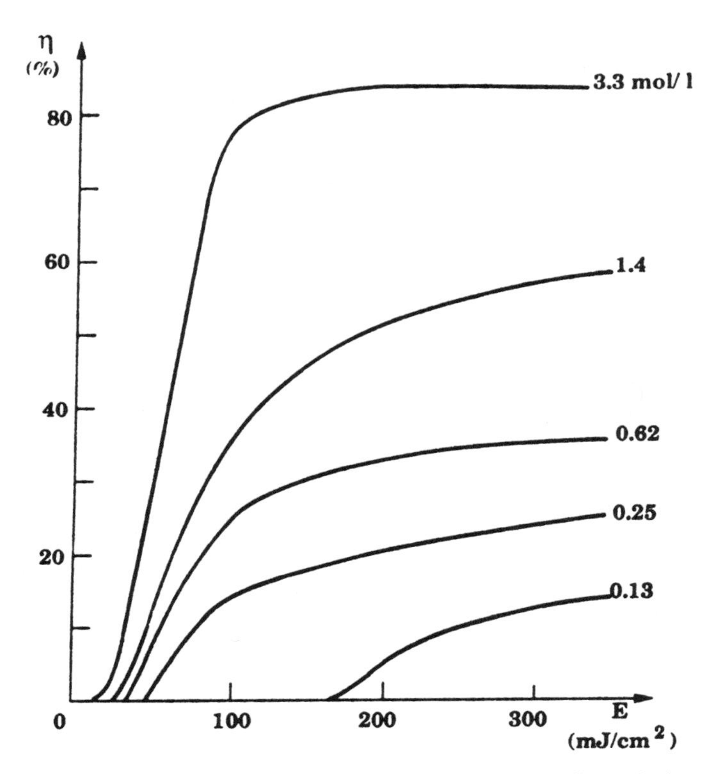

Figure 1-3. *Diffraction efficiency as a function of exposure for different hydrogen donor concentrations in the sensitive mixture. Fringe visibility: 1, exposure power density: 6.7 mW/cm², fringe spacing: 1.2 μm, emulsion thickness: 55 μm) (Ref. 49.)*

recorded with a 457-nm beam is low (≈0.5%) after recording, but shows an increase when swelled in water (≈14%). Differential swelling of the high and low index areas (bright and dark fringe areas) causes an increase in the index modulation and hence an increase in the diffraction efficiency.

3.1.2. Photopolymerizable Dry Films

(a) Acrylate monomer systems. *(i) Du Pont's photopolymer system.* A high diffraction efficiency photopolymer was developed for commercial use by E. I. du Pont de Nemours & Co. Booth[57,58] has investigated this system in detail and found it to produce an index modulation in excess of 10^{-2} on irradiation. The system, as many of the aforementioned materials, consists of an *acrylate* type photopolymerizable monomer, initiator system, and a cellulose polymer binder. The initiator is sensitive to both UV and visible light in the blue-green region of the optical spectrum from 350–550 nm with the addition of a dye sensitizer. The specific composition, being proprietary, is not known. The photopolymer coated on a glass substrate with thickness ranging from 1.25–200 μm, has a shelf life of 6–8 months when stored in a closed container kept at 5–10° C. A refractive index modulation of ≈1% can be produced and efficiencies of transmission gratings are up to 88–90%. The sensitivity of these photopolymer films is inhibited by atmospheric oxygen. Thus Booth found that 10–40 mJ/cm^2 was the typical exposure energy in air, whereas the exposure energy was 1–4 mJ/cm^2 in a nitrogen or oxygen-free atmosphere for achieving a maximum diffraction efficiency. The spatial frequency range of the material was relatively flat up to 3000 lines/mm, which is quite sufficient for transmission type holograms, but frequencies beyond that might be limited because of the presence of dye and other scattering centers. Although the material is self developing, diffraction efficiency was considerably enhanced by a uniform post-exposure with a fluorescent lamp for about 2–6 min. This post-exposure illumination also bleaches the unreacted dye giving a clear hologram. Holograms stored in these media were not significantly affected when exposed to severe temperatures of −60° C or 100° C or when placed under water.

The mechanism of hologram formation proposed by Colburn and Haines[59,60] and Wopschall et al.[61] was based on material diffusion into the partially polymerized regions. However, Booth differed from this proposed hypothesis on the basis that, if the formation of the hologram is due to the diffusion, then the temperature during exposure should significantly affect the material response. On the contrary, an unexpected result was obtained where the final diffraction efficiency of different exposure temperatures ranging from 10–61° C did not show a significantly different trend except a faster polymerization and consequently a faster increase in diffraction efficiency.

However, Kurtzner and Haines[60] based their argument on experimentally derived results and proposed that hologram formation in photopolymers is a complicated process involving both polymerization and monomer diffusion. First, when a normal exposure is made with the interference pattern to be recorded, a part of the monomer polymerizes with the rate being a function of the intensity of illumination. Monomer concentration gradients cause variations in the amount of polymerization, then give rise to the diffusion of the relatively small monomer molecules from regions of higher concentration to regions of lower concentration. With the completion of the diffusion step the photopolymer is exposed to light of uniform intensity until the remaining monomer is polymerized. As the surface relief in regions of polymerization was not experimentally observed, Kurtzner and Haines deduced that the mass transport during diffusion must create density changes and thus refractive index variations that cause diffraction of light by the hologram.

(ii) Du Pont's Omnidex photopolymer system. One of the most attractive photopolymers with many outstanding features has been developed by Monroe et al.[10,62,63] at E.I. du Pont de Nemours & Co. The improved photopolymer for holographic recording comprising liquid aromatic or aliphatic monomers such as 2-phenoxyethyl acrylate (POEA) and polymeric binders such as cellulose acetate butyrate (CAB) can record holograms of ≈100% diffraction efficiency with comparatively low exposure energy ≈50–100 mJ/cm^2.

The coating compositions usually consist of the following general formula: a photosensitizing dye, an initiator, a chain transfer agent, a plasticizer, an acrylic monomer, and a polymer binder.[64] Various aliphatic and aromatic acrylic monomers in different polymeric binders such as CAB, poly(methyl methacrylate) (PMMA), polystyrene (PS), poly(styrene-methyl methacrylate) (PS-MMA), and poly(styrene-acrylonitrile) (PS-AN) were investigated. The visible light absorbing dye 2,5-{bis[4-(diethylamine)-phenyl] methylene cyclopentanone} (DEAW) (λ_{max} at 480 nm), an ultraviolet light absorbing initiator hexaarylbiimidazole (HABI), and a chain transfer agent, 2-mercaptobenzoxazole (MBO), used in the photopolymer system were synthesized in Du Pont laboratories.

The photosensitive solution is coated onto 0.01-cm thick clear polyethylene-terephthalate film at a speed of 2–4 cm/s using a Talboy coater equipped with a doctor's knife, a drier set at 40–50° C, and a laminator station. Doctor's knives with different gaps were used to obtain coatings of varying thicknesses. A cover sheet of 23-μm-thick clear polyethylene-terephthalate film was laminated on each of the coatings as it emerged from the drier, and coated samples were stored at room temperature in black polyethylene bags. These films have a long shelf life. The Du Pont polymers are sensitized from the UV to red spectral region and the recorded holograms need a UV curing and heat processing to achieve a maximum refractive index modulation. Diffraction efficiency as a function of grating thickness was measured for compositions containing triethyleneglycol diacrylate (TEGDA), 2-phenoxyethyl acrylate (POEA), and POEA mixed with 2-naphthyl acrylate (see Fig. 1-4). Good agreement with theoretically calculated values was obtained and several applications of Du Pont's Omnidex polymers are discussed.[64,65] Highly efficient transmission as well as reflection holograms have been recorded in their HRF series of films and diffraction efficiencies of >99% have been reported in both cases.[65] A maximum refractive index modulation of 0.073 has been achieved. Curtis and Psaltis[66] have also shown the possibility of recording 10 multiplexed holograms.

The recording mechanism in these photopolymer systems is similar to that of Du Pont's photopolymer involving a diffusion mechanism. The interference pattern can be described as a combination of light-struck and non-light-struck regions. When polymerization is initiated in the light-struck regions, monomer is converted to polymer. Additional monomer diffuses to these regions from the non-light-struck regions. Formation of the photopolymer network inhibits the

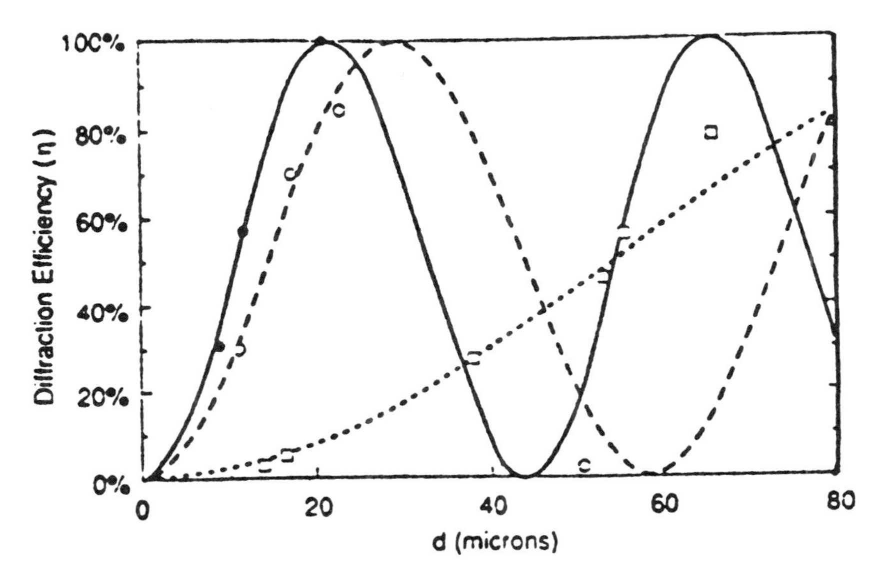

Figure 1-4. *Fit of experimental data to the coupled wave theory.* □ *TEGDA;* ○ *POEA;* ● *80% POEA-20% 2-naphthyl acrylate. (Ref. 62.)*

diffusion of large molecules. In addition, as the polymer is being formed, the solubility of the binder in monomer may also decrease so that initially at least, there may be some net migration of binder away from these regions. The polymerizing regions expand as monomer diffuses in and is polymerized. This expansion increases the concentration of monomer-derived polymer and decreases the concentration of binder in these regions. Because the binder in the non-light-struck regions is no longer swollen by the dissolved-in monomer, it contracts as monomer diffuses away. Contraction increases the concentration of binder in these regions. This process continues until there is no more monomer to migrate or until the hologram is fixed with an overall exposure that polymerizes the remaining monomer.

A partial segregation of monomer and binder between the light-struck and non-light-struck regions is produced. After the final overall exposure, the light-struck and non-light-struck regions have different compositions. These light-generated composition differences are responsible for the measured refractive index modulation.

(iii) DMP-128 system. A photopolymer system was also developed by Ingwall et al.[67–71] at the Polaroid Corporation and marketed as Polaroid DMP-128. These films, sensitive from 442- to 647-nm wavelengths have a shelf life of more than nine months when stored under dry conditions. The holographic recording is

based on dye-sensitized photopolymerization of a mixture of lithium acrylate, acrylic acid, and methylenebisacrylamide incorporated into a film forming a poly-*N*-vinyl-pyrrolidone matrix.[68] Films of 1- and 30-μm thickness can readily be coated onto glass or flexible plastic substrates. The dry films are inactive and must be incubated for a few minutes in an environment of 50% relative humidity before exposure. In Fig. 1-5 the square root of diffraction efficiency of transmission holograms is plotted against exposure energy for different values of visibility of the fringes. Transmission holograms with 80–95% diffraction efficiency were recorded with an exposure energy of 5 mJ/cm^2 whereas bright reflection holograms were obtained with 30 mJ/cm^2. The sequential processing steps include: (1) a uniform white light illumination for 0.5 to 2 min, (2) 2–5 min incubation in a specially formulated developer/fixer bath, (3) a rinse to re-

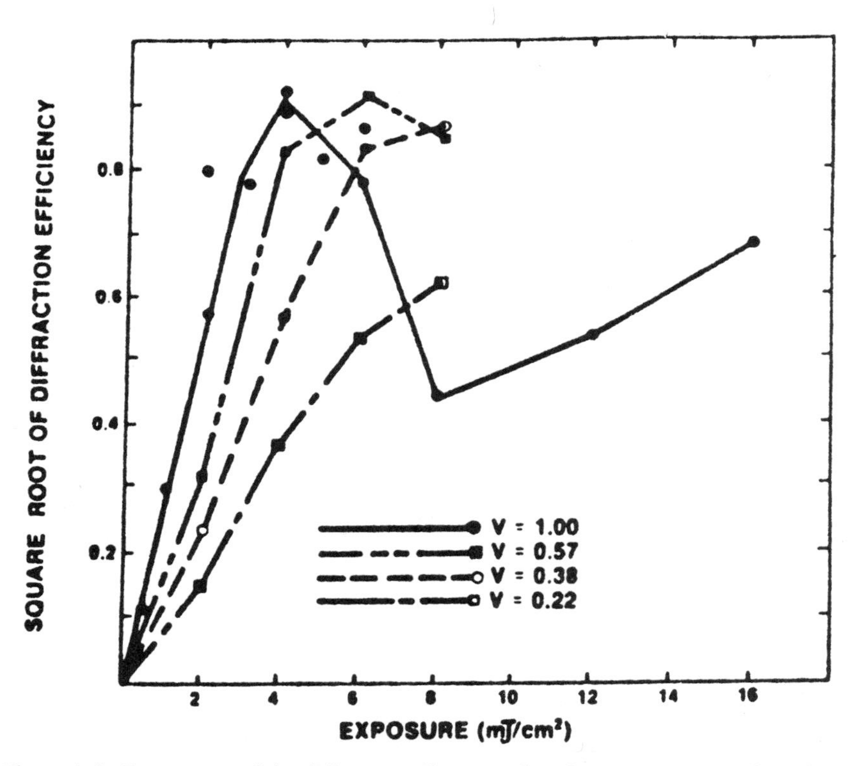

Figure 1-5. *Square root of the diffraction efficiency plotted against exposure for selected beam visibilities. The visibility* V *was calculated from the measured beam irradiance ratio* $K = I_{ref}/I_{obj}$ *according to the equation* $V = 2\sqrt{K}/(1 + K)$. *Exposure was with 488-nm light. (Ref. 67.)*

move the processing chemicals followed by (4) careful drying. Holograms recorded in DMP-128 exhibited good stability under high humidity environments. The stability at 95% relative humidity at room temperature was evaluated both for transmission and reflection holograms.[67] In the case of transmission holograms, the diffraction efficiency does not diminish even after 260 days of incubation, whereas, for reflection holograms, although the wavelength of maximum diffraction efficiency shifts by 60 nm to the blue after one day and essentially remains constant thereafter, there is no visible loss of diffraction efficiency.

The mechanism of hologram formation is well elucidated,[69–71] using scanning electron microscopy (SEM) and refractive index matching. Laser recording initiates photopolymerization reaction and creates concentration gradients that drive monomer from dark regions of the fringe pattern to light regions. In DMP-128 the hologram that is formed after exposure and before chemical processing is weak and the diffraction efficiency rarely exceeds 0.1%. Bright high diffraction efficiency holograms appear after development in selected processing fluids. The processing fluids remove soluble components and produce differential swelling and shrinkage in the films. SEM pictures show light solid layers alternating with layers containing numerous dark spots that appear to be holes or pores. The spacing between adjacent layers is commensurate with the exposure fringe pattern.

Immersion of DMP-128 holograms in liquids profoundly affects their optical properties. The diffraction efficiency decreases as the refractive index of the immersion liquid approaches the refractive index of the material; these effects are completely reversible as the original hologram is restored upon removal of the immersion liquid. The refractive index modulation is reduced upon immersion as air in the voids is replaced with a liquid. In the case of immersion in trifluoroethylene, the lowest refractive index liquid used, the reduction in index modulation improves efficiency to slightly over 85%. Similar results have also been obtained by Hay and Guenther.[72]

Special features of the DMP-128 polymer were exploited by Whitney and Ingwall[73] to form composites comprising a supporting matrix derived from the photopolymer with a dispersed second phase. Composites prepared by filling the pores in a patterned matrix by a nematic liquid crystal were used for holography and optical waveguides. The diffraction efficiency of holograms containing the liquid crystal can be rapidly changed by applying an electric field. High diffraction efficiencies of $\approx$100% could be easily achieved by switching the electric field, and the change in diffraction efficiency for a given composite was dependent on the field strength and the polarization of light used to measure the diffraction efficiency.

(iv) Other acrylate dry photopolymerizable films. A photopolymer material based on an acrylate monomer was also proposed by Zhang et al.[74] where the photosensitive coating solution consisted of 0.2 g diethylene glycol diacrylate,

14 g of chlorobenzene, 1.2 g of methacrylic acid benzyl ester, 12 g of polyvinyl acetate solution (25% polyvinyl acetate, 75% methanol), 0.6 g of *N*-vinyl carbazole, 0.1 g of 3-mercapto-4-methyl-4*H*-1,2,4-triazole, 0.2 g of 2,2′-bis(*o*-chlorophenyl)-4,4′,5,5′-tetraphenyl-1,2-biimidazole, and 0.005 g of sensitizer dye (synthesized by the authors). The films were coated on glass substrates and used for recording reflection holograms using the 488-nm line of an argon ion laser. A heat treatment was used to fix the holograms, but this resulted in a blue shift of the reconstruction wavelength. Therefore, in order to fix the peak reflection wavelength and enhance the refractive index modulation, rephotopolymerizing or rethermopolymerizing methods were employed by coating the heated holograms with a monomer and photosensitizer or thermoinitiator. The monomer coated holograms were either exposed to a uniform UV radiation or heated in an oven to get the desired peak reconstruction wavelength. Using this technique trichromatic filters and pseudocolor holograms were fabricated. Bright reflection holograms comparable to dichromated gelatin were obtained.

The mechanism of hologram recording involved in this system is based on the dye-sensitized photopolymerization and monomer diffusion. As the photosensitive layer is irradiated, the light absorbing photosensitizing dye is activated to an excited state, then transfers its energy to the initiator. The excited initiator molecule dissociates into two imidazole radicals that oxidize the chain transfer agent, 3-mercapto-4-methyl-4*H*-1,2,4-triazole, to generate the free radical for initiating the polymerization of monomers. The crosslinking monomers were added to improve the image quality and physical properties of the photopolymer. On recording the holographic pattern, as in systems discussed earlier, the monomer is first polymerized in the high intensity regions of the pattern and the simultaneous diffusion of monomer from the dark regions creates a monomer concentration and density gradient. The resulting nonuniform polymer distribution yields a refractive index variation. This is later enhanced by a uniform UV exposure, heat treatment, and repolymerization. A unique feature of these holograms is that they are totally waterproof.

(b) Methyl methacrylate (MMA) monomer. Another widely investigated material used as a photopolymer has been poly(methyl methacrylate) (PMMA), derived from methyl methacrylate (MMA) monomer, because of its high intrinsic susceptibility for refractive index change when exposed to UV ($\approx$325 nm) radiation. Tomlinson et al.[75] reported a refractive change of 3×10^{-3} and resolutions of $\approx$5000 lines/mm in PMMA films prepared with MMA as the starting material. Polymerization was initiated by irradiation from a mercury (Hg) lamp (253 nm) in the presence of an initiator, azobisisobutyronitrile (AIBN). An essential step in the preparation consisted of oxidation of the monomer that was assumed to produce peroxides. Various oxidation levels were obtained by normal autooxidation in the dark or (to produce higher oxidation levels) by irradiation with a 5-W low pressure Hg lamp for 48 hours in an oxygen atmosphere.[75,76]

Three-dimensional dielectric gratings were recorded by two interfering UV laser beams of 325 nm in these 2-mm thick PMMA samples. Gratings with a fringe spacing of up to 0.22 μm have been recorded. A diffraction efficiency as large as 70% was observed.[75] An interesting application of this material was demonstrated by the fabrication of an optical waveguide having a loss of <1 dB. Moran and Kaminow[76] carried out detailed experiments with PMMA films of the same composition and observed a peak refractive index change of 2.8×10^{-3}. They characterized the index change in relation to its sensitivity, temperature dependence, and development time. Diffraction efficiency values of up to 96% were obtained in gratings with a spatial frequency of 1260 lines/mm in 2-mm-thick samples. However, a development time of several days was required before the refractive index reached a constant value. Angular sensitivity characteristics of these gratings indicated that they exhibited wavelength independence and that the effective thickness of the gratings was less than the physical thickness of the films because of nonuniformity of the index variation in the material. The most significant problem encountered in this media was the irreproducibility of results, which was explained as a result of variations of residual monomer content in different samples.

The mechanism of hologram formation in PMMA was suggested by Tomlinson et al.[75] to be due to the photoinduced refractive index change caused by the density change. This density change is caused by the crosslinking of adjacent chains by free radicals produced during the irradiation of oxidation products. However no evidence was offered in support of this mechanism. Based on their various experimental findings, Bowden et al.[77] concluded that the refractive index increase induced by laser irradiation at 325 nm of peroxidized PMMA is due to an increase in density in the irradiated region, where the increase in density results not from the crosslinking of polymer chains but instead from a photoinduced polymerization of unreacted monomer (≈1–2%) within the 2-mm PMMA films. In addition to an increase in density, exposure also causes a slight reduction in the thickness of the irradiated region. The peroxides (oxidized products) of both polymer and monomer act as photoinitiators. The sensitivity of the layer is related to the concentration of photoinitiator, but the absolute value of refractive index change (Δn) will depend on the amount of unreacted monomer. Because this variable generally cannot be controlled, it was believed that irreproducibility will always present a problem in this process.

Sensitization of PMMA with *p*-benzoquinone (abbreviated as Q) for recording at 488 nm was proposed by Laming.[78] Sensitized PMMA films were made by solvent evaporation on flat glass plates by pouring a solution of PMMA and Q in chloroform. Self-developing holographic diffraction gratings that can be fixed with a UV light have been produced in this system with diffraction efficiencies greater than 70% for writing beam intensities of 2.5 J/cm^2. Diffraction efficiencies of these holographic gratings as a function of thickness and angular sensitivity were also studied. The experimental results varied from the theoretical prediction in the case of very thick films (>300 μm) mainly because of a discrepancy in the effective and measured thickness of the gratings. The mechanism of recording was assumed to be due to photodegradation rather than crosslinking based on the fact that the refractive index decreased in contrast to the other investigated systems. These films when exposed to UV radiation were found to have drastically reduced sensitivity to 488-nm light.

Recently, thick PMMA blocks ($0.5 \leq$ thickness ≤ 3 mm) containing residual monomer have been investigated for holographic recording at $\lambda = 514$ nm by Marotz[79] and Lückemeyer and Franke.[80] The PMMA blocks of size 10 mm × 10 mm × d ($0.5 < d < 3$mm) were made using MMA monomer prepolymerized at 50° C using the initiator AIBN. The polymer blocks exhibited a monomer content of ≈10%.[81] Titanium biscyclopentadienyl dichloride (titanocene chloride) was used as the photoinitiator for a 514.5-nm wavelength and was added at the stage of preparation of the PMMA blocks. Fig. 1-6 depicts the characteristic development of diffraction efficiency as a function of irradiation time with a recording beam intensity of 130 mW/cm^2. The diffraction efficiency developed

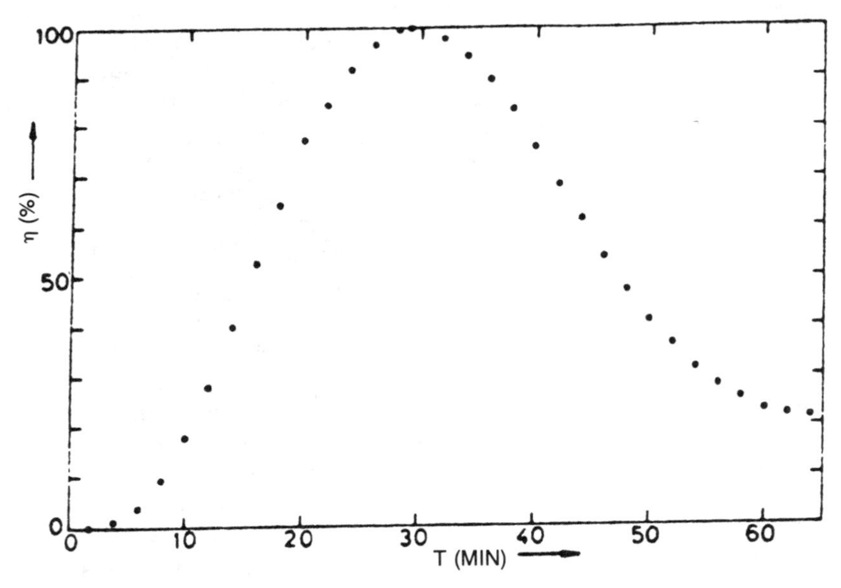

Figure 1-6. η-*Development vs. illumination time* (d = *2.5 mm,* I = *130 mW/cm^2. (Ref. 79.)*

to a value of $\approx 100\%$ at an exposure time of 30 mins. Thermal fixing of the exposed samples at 70° C for a few hours increased the diffraction efficiency as well as the refractive index amplitude significantly. Without thermal fixing, the development takes several weeks. A characteristic light scattering as noted by Moran et al.[76] was also observed in thick samples ($d > 1$mm). The marked self-developing nature of this media demonstrates that refractive index change is due to an increasing density caused by light-induced polymerization of the residual monomer (in contrast to crosslinking). However, the low sensitivity of the material and the delay time before development offer the possibility of superposing a large number of holograms in the same volume without mutually disturbing the other holograms during writing.

Lückemeyer et al.[80] have demonstrated methods of measuring the photosensitivity profiles in volume phase gratings made of titanocene chloride sensitized PMMA blocks. Refractive index profiles in the depth of the thick films were studied by slicing the blocks using a microtome. The influence of parameters such as intensity of the recording beams, storage time, aging, and the diffusion of oxygen, nitrogen, and monomer on the refractive index profile was also evaluated. The mechanism of increase in refractive index in the photosensitive PMMA samples is mainly due to the photopolymerization of residual monomer. From the various experimental results, it was concluded that the profiles of the volume phase gratings recorded in aged samples were primarily due to the monomer distribution. Exposure to oxygen/nitrogen atmosphere does not significantly affect the refractive index change. But the strong influence of monomer in-diffusion on regaining the photosensitivity gives strong evidence of monomer concentration domination in the photosensitivity profiles.

(c) Polyester host systems. Bloom et al.[82–85] of RCA laboratories developed various photopolymer materials composed of polyesters doped with photosensitive α-diketones. They investigated a number of α-diketones such as 2,3-pentanedione, 2,3-heptanedione, 3,4-heptanedione, and camphorquinone dissolved in various combinations of unsaturated polyester hosts. The most suitable system was obtained by dissolving camphorquinone in a polyester host; the various characteristics of this medium are described in detail in Ref. 83. The photosensitive resin consisted of an unsaturated polyester (whose components are propylene glycol, maleic acid, and phthalic acid) dissolved in styrene containing a small amount of MMA monomer. The camphorquinone and methyl ethyl ketone peroxide catalyst were added to the casting resin. After degassing, the mixture was poured into a Teflon mold and allowed to cure at room temperature for several days. The obtained rigid samples were than cut and polished. These films had an absorption band centered around 460 nm and recording of holograms could be carried out at 514.5 nm. Final diffraction efficiencies of >70% have been obtained for exposure energies of 240 J/cm^2. It was observed that inclusion of larger concentrations of camphorquinone increases the sensitivity of the material, but, for obtaining uniform holograms throughout the thickness of emulsion, 5% was chosen as the preferred concentration. Post-exposure heat treatment of the holograms resulted in higher sensitivity of the holograms as well as an increase in the final diffraction efficiency. Characteristic curves of diffraction efficiency as a function of exposure energy (represented as log E) are shown in Fig. 1-7. It can be observed that films with different concentrations of camphorquinone reach the peak diffraction efficiency value of 70% at different exposure energies.

A detailed study of the effect of different polymer hosts on the performance of volume phase holograms was also carried out.[84] Various influencing criteria for the performance of the resins were considered such as sensitivity, diffraction efficiency, exposure, and Δn. In a few of these materials, information can be stored only for a few days, whereas others can be used for permanent storage. Because of the large depth available and high angular sensitivity, multiple holograms, up to 100, have been recorded in a 2-mm-thick sample.[85]

To elucidate the mechanism of hologram formation, several experiments using electron paramagnetic resonance (EPR), absorption, and emission spectroscopy were carried out.[82] It was observed that permanent holograms can be recorded

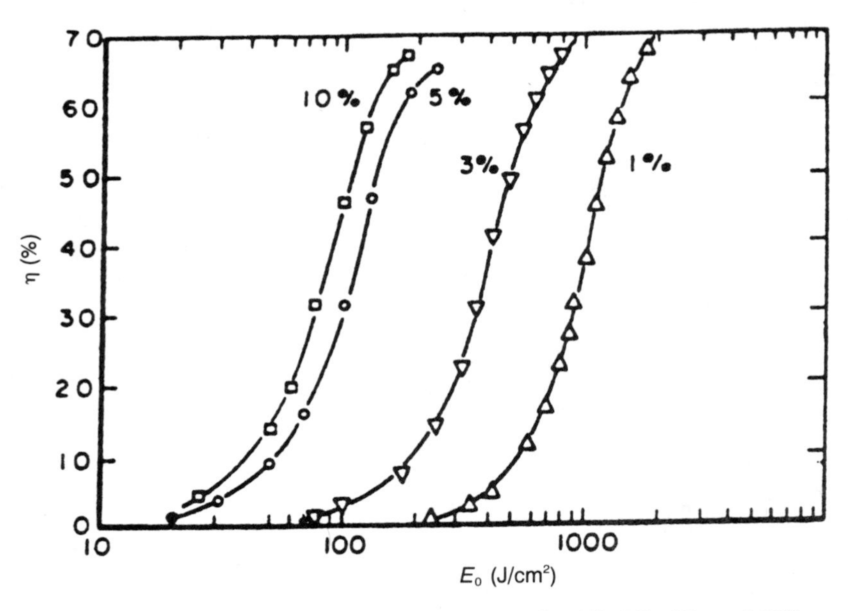

Figure 1-7. *Diffraction efficiency* η *vs. exposure* (E_0) *for 1%, 3%, 5%, and 10% camphorquinone. (Ref. 83.)*

when the guest (α-diketone) contains no intramolecular abstractable hydrogen, whereas transient holograms may be recorded using those guests that may intramolecularly abstract hydrogen. The mechanism of permanent hologram formation in the cured, unsaturated polyester resins involves addition of residual monomer to the growing free radical chains. The reaction is initiated by excited state camphorquinone abstracting hydrogen atoms from the polymer backbone. The addition of monomer to the free radical site leads to an increase in density and the concomitant observed increase in refractive index. Free radical diffusion in the solid host is the rate-limiting step in the kinetics leading to the refractive index change.

(d) Multicomponent monomer system. A novel photopolymer system for the fabrication of high resolution volume phase holograms and grating devices was reported by Tomlinson et al.[86] of Bell Laboratories. In contrast to the previously investigated single monomer polymerizable systems, they thought that it would be advantageous to use a mixture of components with different reactivities and polarizabilities resulting in a composition modulation. Various combinations of high and low reactivity monomers were combined with an initiator and used in closed cavities. On recording volume holograms on these samples with exposure energies of 0.75–3.0 J/cm^2 at 325 nm, peak-to-peak refractive index modulations of up to 7.8×10^{-3} were obtained. A uniform UV preexposure was used to eliminate the dissolved oxygen and other polymerization inhibitors; post-exposure was used to fix the holograms by polymerizing the remaining unreacted monomer.

The mechanism of hologram formation in these multicomponent systems involves a greater extent of polymerization in the areas of higher intensity illumination than that of the lower intensity areas. The polymer formed preferentially from the high reactivity monomer produces a concentration gradient that will cause additional high reactivity monomer to diffuse into the high intensity areas. Because the remaining monomer mixture has been enriched with the low reactivity monomer, the low intensity areas will then contain an excess of low reactivity monomer, and the high intensity areas will contain a polymer with an excess of the high reactivity monomer. In this way, a modulation of the chemical composition of the final fully polymerized material is achieved. The refractive index modulation arising from this modulation of chemical composition is much larger than that resulting from the density effect. Thus a two-way diffusion results in the refractive index modulation.

A unique system among all these photopolymer systems is the latent imaging photopolymer developed by Chandross et al.[87] The material consists of a photosensitive polymerization initiator chemisorbed on the surfaces of a pourous glass matrix (Vycor 7930, Corning Glass) with an average pore diameter of 4 nm. The latent image is recorded by selectively destroying the initiator by exposure to an incident light pattern. At this stage, there is a very little change in the optical properties of the sensitized glass. The image is then developed by filling the glass with one or more polymerizable monomers and using a uniform overall exposure to initiate polymerization. The development exposure activates the remaining photoinitiator, which is still bound to the surface of the glass and initiates polymerization of the monomer mixture. As the concentration of the remaining initiator has been spatially modulated by the image exposure, the rate of initiation of polymerization is similarly modulated and results in a modulation of the refractive index of the final polymer. Different combinations of monomers such as cyclohexyl methacrylate, 1-chloronaphthalene, dimethyl suberate, dibutyl adipate, and ethylene glycol dimethacrylate were used in conjunction with benzoin as a photoinitiator, and the best peak-to-peak refractive index change of up to 3.2×10^{-4} was achieved.

(e) Acrylamide system. A novel improvement of *acrylamide* photoinduced polymerization, first proposed by Sugawara et al.,[44] was reported by Jeudy and Robillard[88] involving a reversible photochrome as the sensitizer. The material coated on optically flat glass substrates consisted of acrylamide monomer, TEA as promotor, indolinospiropyran (P 265) as sensitizer, and PVA as binder. The photochrome is fully transparent in the visible region and can be switched with UV light. The UV irradiation shifts the absorption of the dye to the red region with the peak absorption at 600 nm. Thus very high diffraction efficiency ($\approx$80%) holograms using 633-nm light from a He-Ne laser could be recorded at a spatial frequency of 3000 lines/mm with an exposure energy of 100 mJ/cm^2. After the photochrome reversal, absolutely transparent holograms were obtained. Transparency of the emulsion over the substrate was better than 92% at 633 nm. In spite of the material's capability of recording transparent holograms with a high diffraction efficiency, the stability of the emulsion after its preparation is only 4 to 5 days (before recording).

Variations of the acrylamide system were investigated by Van Renesse.[89] Tests concerning the resolving power and photosensitivity of the acrylamide/methylenebisacrylamide system were carried out, and a pronounced reciprocity failure in the material and a photosensitivity of $\approx$5 mJ/cm^2 were observed. Sadlej and Smolinska[90] also continued investigations on photopolymer layers proposed by Jenney.[32] They modified the system with an essential difference; the photoreactive components were dissolved in a protective polymer that produced stable photosensitive layers, as photopolymers used by Jenney were not very stable for a longer period. Polymers such as poly(vinyl acetate) (PVAc), methyl cellulose (MeC), PVA, and gelatin were tested as protective polymers. The diffraction efficiency and the optical quality of the recorded holograms remained unchanged over a period of several months ($\approx$2 years). The protective layers made of PVAc and MeC were found to be inhomogeneous with respect to transparency and photosensitivity. Therefore layers with the required properties were obtained using PVA and gelatin. Further, buffered solutions of PVA and gelatin were used to get greater diffraction efficiencies than those that result from dissolving these polymers in water. The basis of this behavior of polymers is not very clearly understood. Films of 15–50-μm thickness were made of acrylamide as the photosensitive monomer, methylene blue as the sensitizer, *p*-toluene sulfinic acid as reducing agent, and *N,N*-methylenebisacrylamide as the crosslinker in a PVA or gelatin protective polymer by spin coating on glass substrates. On recording holograms, both surface relief and refractive index modulation were detected on the media. A special feature of these layers was their very good frequency response. Spatial frequencies were found to be flat over the range of about 3000 lines/mm, and gratings with up to 4700 lines/mm were recorded. The layers had a relatively high sensitivity of 10 m/Jcm^2, and transmission gratings with a diffraction efficiency of 0.5% for layers without thickness modulation and 4% for layers having surface deformation were recorded.

Dry acrylamide-poly(vinyl alcohol) (AA-PVA) films were used in holographic storage by Calixto.[91] The transmittance of the films and the diffraction efficiency of the recorded holograms as a function of exposure and the beam ratio were studied. Diffraction efficiencies of ~10% could be obtained with a combined reading beam power of ~230 μW/cm^2. The polymeric mixture presented a self-developing process and so it could be ready just after the exposure time or it could be fixed by illuminating with a UV source.

The mixture consisted of acrylamide monomer, TEA as promotor, and methylene blue (MB) as sensitizer, with PVA as binder. The composition of the mixture was 15 ml of PVA made by mixing 7.5 g of PVA in a 3:2 mixture of water and alcohol, 0.3 g acrylamide, 0.1 ml of TEA, and 4.0 mg of MB. A He-Ne laser at 632.8 nm was used to record the information in the plates. The action of light on the photosensitive material is two-fold; it bleaches the dye and simultaneously promotes polymerization of acrylamide. The result is a medium with spatial changes in refractive index and absorption. A fixing technique employed the placement of the recorded holograms in close proximity to a Hg lamp for a period of 20 min so that the dye was bleached and the polymerization was terminated. Recorded holograms were stored in a light tight box for ~3 months after fixing and no recovery of the dye was noticed.

The chief characteristics of this holographic recording material are the ease of fabrication and attainment of diffraction efficiencies of ~10% with exposures of ~94 mJ/cm^2. The use of lasers having output power of ≈100 mW was suggested because it was found that higher diffraction efficiencies can be attained if the writing power density is high. Recently, computer generated infrared radiation focusing elements were fabricated involving these materials.[91b]

The salient characteristics of the different liquid and dry film photopolymerizable materials are summarized in Table 1-1.

Table 1-1. Photopolymerizable Systems

Material	Type of Modulation	Preparation Method	Thickness (μm)	Recording Wavelength (nm)	Sensitivity (mJ/cm^2)	Resolution (lines/mm)	Diffraction Efficiency %	Refs.
A. *Liquid photopolymerizable compositions*								
Metal acrylate	Thickness and refractive index	Liquid between two glass plates	10–20	694	0.3	3000	45	31
Oligourethane (FPK-488)	Density, polarizability, concentration	Liquid between two glass plates	20	300–500	0.02	>6000	>80	40
Acrylamide	Refractive index	Liquid between two glass plates	50	300–500	0.02	1500	80	42–44
Multifunctional acrylate	Refractive index	Liquid between two glass plates	55	450–550	0.1–0.2	>2500	80	47, 52
Methacrylate	Refractive index	Liquid between two glass plates	120	457	0.5–0.7	—	14	57

Table 1-1. *Continued*

Material	Type of Modulation	Preparation Method	Thickness (μm)	Recording Wavelength (nm)	Sensitivity (mJ/cm^2)	Resolution (lines/mm)	Diffraction Efficiency %	Refs.
B. *Photopolymerizable dry films*								
Du Pont's photo-polymer	Refractive index	Coated on glass substrate	1.25–200	350–550	0.1–0.4	3000	88–90	58, 59
Du Pont's Omnidex	Refractive index	Coated on plastic sheet	6–78	450–650	0.01–0.1	6000	>99	63–66
DMP-128	Voids and solid	Coated on flexible plastic	1–30	442–647	0.005–0.03	5000	80–95	68–73
Methyl methacrylate	Refractive index	PMMA samples	2000	325	50–150	5000	70	75–77
Q-PMMA	Refractive index	Coated on glass	10–900	488	2.5	>1000	70	78
Titanocene chloride PMMA	Refractive index	PMMA blocks	500–3000	514	3.9	—	≈100	79–81
Polyester with α-diketones	Refractive index	Molded samples	2000	514	100–300	—	>70	82–85
Acrylamide	Refractive index	Coated on glass	100	633	0.1	3000	80	88
Acrylamide	Refractive index	Coated on glass	—	633	0.094	—	10	91

3.2. Photocrosslinking Polymers

The photocrosslinking polymers are another major class of recording materials that have been employed by many research groups. Different sensitizers such as metal ions and dyes have been employed to sensitize the polymer matrix. On exposure to an appropriate light source the sensitizer undergoes photochemical reaction which in turn induces a crosslinking in the polymer matrix. The changes induced by the photochemical reactions are responsible for the changes in the refractive index modulation and hologram formation. Most of these materials developed to date are used in the form of dry films. The characteristic features of these materials will be discussed below.

3.2.1. Dichromated poly(vinyl alcohol) (DCPVA)

Dichromate sensitized poly(vinyl alcohol) (DCPVA) has been used for many industrial applications such as printing plates, stencils, lithographic plate making, and color television picture tubes.[92–95] Ziping et al.[96] used DCPVA as a holographic recording material for the first time and later there was a considerable interest in employing this material for many optical applications. They improved the final diffraction efficiency up to ≈32% using thermal development of the exposed holograms.

Recently, real-time recording and reading of polarization volume transmission holograms were performed with DCPVA films by Lelièvre and Couture.[97] They used two parallel and cross polarized blue recording beams of 488 nm from an argon ion laser, and the growth of diffraction efficiency was monitored in real time by the reading beam at 632.8 nm from a 5-mW He-Ne laser. All experiments were carried out with an interbeam angle $2\theta_B = 30°$ (1061 cycles/mm) and a beam intensity ratio of 1. They found that saturation of diffraction efficiency was observed for a total exposure energy of ≥400 mJ/cm^2. The maximum diffraction efficiency achieved was 18% for a dichromate concentration of 1.4% in the films. Introducing the corrections for reflections and absorption, the diffraction efficiency was evaluated to be 32%. For holograms recorded with two beams having linear crossed polarizations, a diffraction efficiency (without considering the corrections) of about 0.8% was observed, a value fifty times smaller than that obtained for parallel polarization.

The dependence of η as a function of exposure energy for varying power of the writing beams was also investigated, and Fig. 1-8 illustrates the energy sensitivity of the material. Each point denoted by E in the figure indicates the end of exposure period and after these points the curves illustrate the temporal grating growth. It indicates that the photocrosslinking process continued even after the laser irradiation was stopped. Similarly, the beam intensity ratio K was taken as a parameter for the investigation involving linear parallel polarizations.

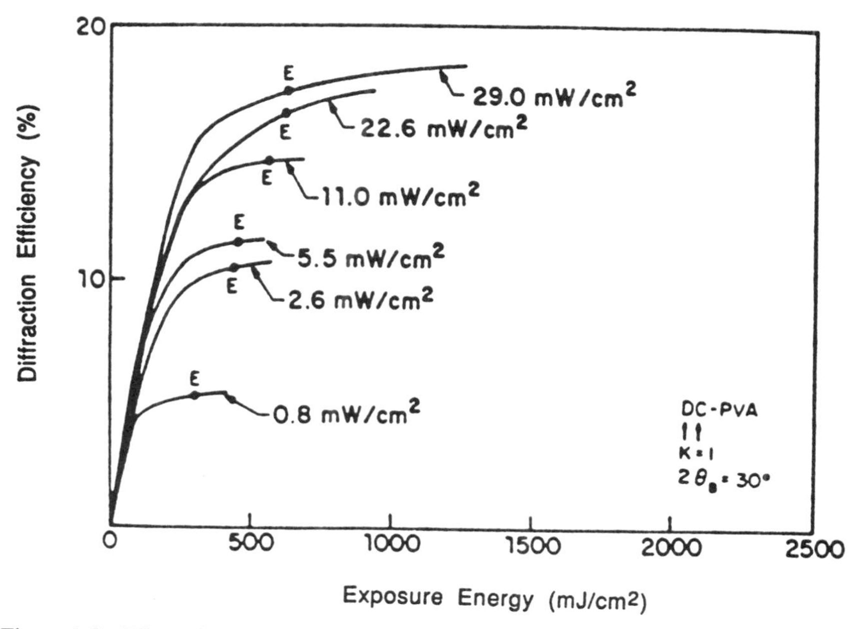

Figure 1-8. *Effect of varying power of the linear parallel polarization writing beams on the diffraction efficiency of DCPVA films (Ref. 98.)*

Maximum diffraction efficiency was achieved for a K value of 1 (Fig. 1-9) and the value of η changes rapidly with an increase in K. Similar results were obtained for crossed linear polarization.[98]

The spatial frequency response of DCPVA films was determined for parallel and cross polarization recording with beams of equal power. In both cases, Modulation Transfer Function (MTF) curves for the DCPVA films are linear between 1000 cycles/mm up to 3500 cycles/mm. The spatial frequency response is explained by the short length of photocrosslinked polymer chains. When the spatial frequency exceeds 3000 cycles/mm, the photocrosslinking mechanism is less efficient because it implies that very short chain lengths are required to record a fringe period Λ that is <0.25 μm. On the other hand, whenever the spatial frequency is <500 cycles/mm, the chain formation is more important near the first surface of the DCPVA films; consequently the recorded fringes do not penetrate the full depth of the recording films.[97]

The angular selectivity response of the recorded holograms in DCPVA films was investigated and Fig. 1-10 represents the same for natural and thermally fixed (200° C, 1 min) holograms.[98] The symmetrical angular selectivity response curve indicates that unslanted holographic fringes stay in their recorded positions

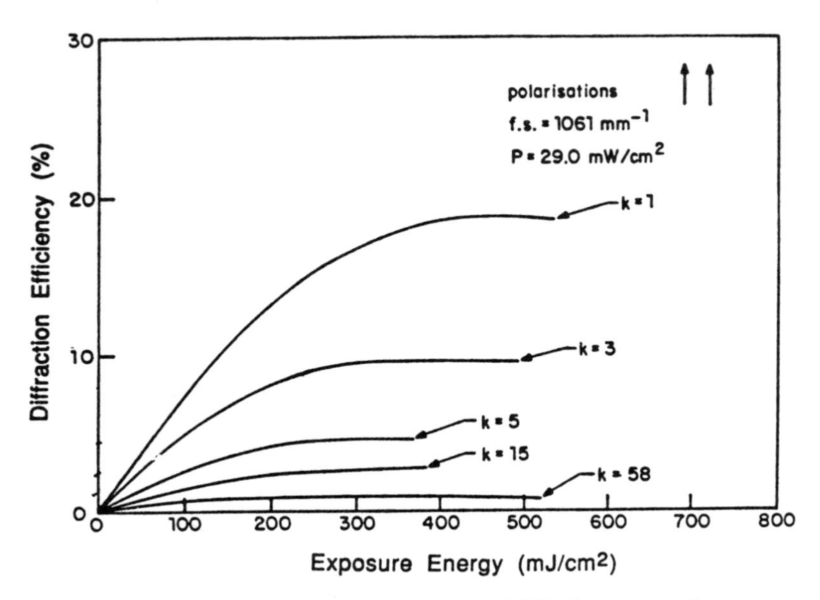

Figure 1-9. *Diffraction efficiency dependence of DCPVA films on total exposure with different intensity beam ratios for parallel polarizations. (Ref. 98.)*

without any significant movement. The effective thickness of 25.3 μm and 24.7 μm was determined for the natural DCPVA film and for the thermally treated film respectively. It indicates that the recorded fringes penetrate almost the entire depth of the film. The phase modulation parameter was calculated to be 1.36×10^{-3} for natural DCPVA film.

In a continuing effort to improve the real-time diffraction efficiency of DCPVA holograms, Lessard et al.[99] have performed more experiments to understand the effect of various influencing parameters, namely, the pH of the coating solution and the addition of an external electron donor, dimethyl formamide.

In the photoredox reaction involving Cr(VI), the pH of the medium was found to play an important role. Similarly, the pH variation of the coating solution from which the polymer films were fabricated was found to play a significant role in the diffraction efficiency of the films. DCPVA films of varying pH ($2.8 < pH < 10.3$) were prepared by the suitable addition of either acetic acid or aqueous ammonia. The shape and evolution of diffraction efficiency profiles remain the same for all the DCPVA films made with varying pH. Fig. 1-11 depicts the maximum diffraction efficiency achieved with respect to different films of varying pH. It clearly indicates an increasing trend, corresponding to

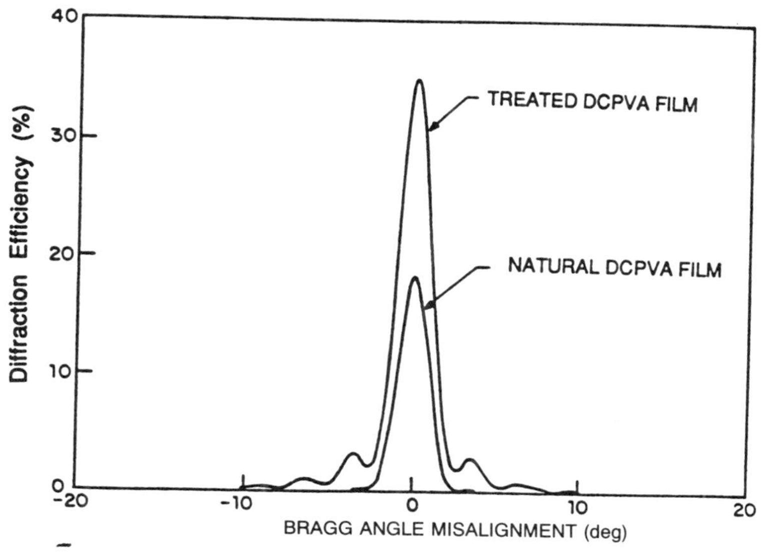

Figure 1-10. *Angular selectivity response of natural and thermally treated DCPVA films. (Ref. 98.)*

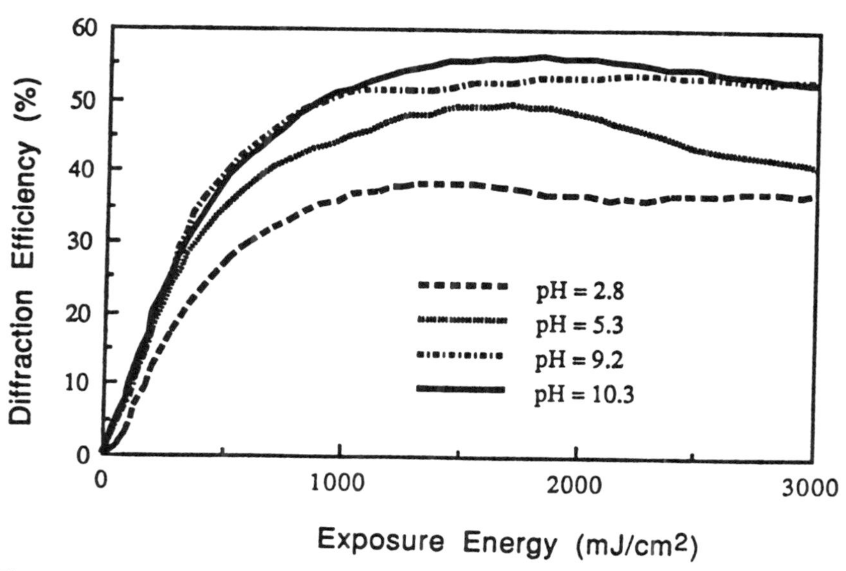

Figure 1-11. *Effect of pH of the coating solution on the diffraction efficiency of DCPVA system. [Ref. 99(c).]*

an increase in pH (going from an acidic to basic pH). This behavior will be explained in the mechanism of hologram recording.

The photoreduction of dyes[100] and holographic performance of methylene blue dichromated gelatin[101–103] were found to be greatly influenced by the presence of electron donors such as aliphatic and aromatic amines, and dimethyl formamide. As observed in earlier studies, incorporation of dimethyl formamide (DMF) in DCPVA films resulted in a good positive response on the improvement of diffraction efficiency. An optimum concentration of DMF (0.062 M) resulted in peak real-time diffraction efficiency of ≈65% as compared to 49% diffraction efficiency for holograms without DMF (Fig. 1-12).

Ziping et al. developed the recorded holograms with a thermal treatment and obtained a maximum diffraction efficiency of 32%.[96] Even though a high real-time diffraction efficiency was achieved for DCPVA films with DMF and optimized pH, Lessard et al. tried two chemical developing methods for improving the final diffraction efficiency, and Table 1-2 illustrates the involved simple processing steps.[99(b)] When the exposed holograms were developed by the first method, a 60% diffraction efficiency for an exposure of 500 mJ/cm^2 was achieved. The second method was found to give higher energy sensitivity and a better recording rate, as λ_{max} was 68% for an exposure level of 200 mJ/cm^2 (Fig. 1-13).

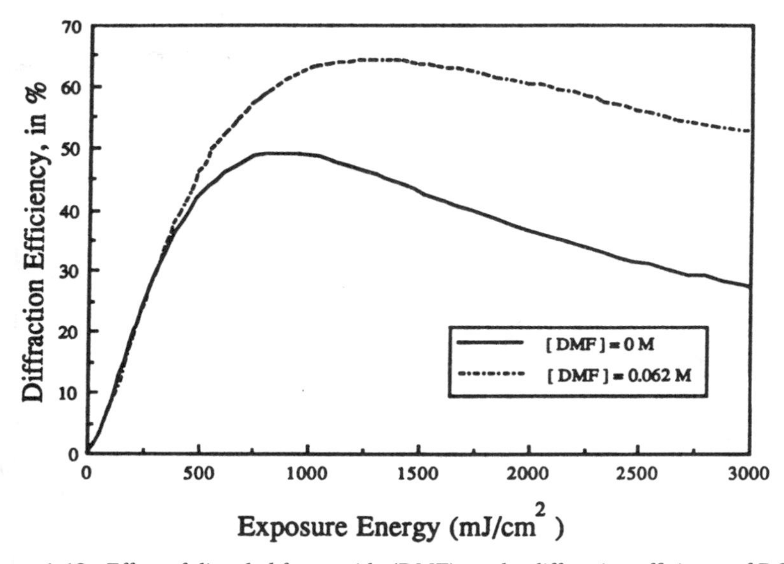

Figure 1-12. *Effect of dimethyl formamide (DMF) on the diffraction efficiency of DCPVA holograms. [Ref. 99(a).]*

Table 1-2. Development Procedure for DCPVA Holograms [Ref. 99(b)]

Method I

(a) Develop in 1:9 water-ethanol mixture for 10 mins with agitation.
(b) Soak in 100% ethanol for 3 mins.
(c) Dry the developed holograms with a blower.

Method II

(a) Develop the holograms in 100% ethanol heated to 70° C for 10 mins.
(b) Dry the developed holograms with a blower.

Similar experiments have been performed by Mazakova and Pantcheva[104] on DCPVA for holographic recording, and they have achieved a diffraction efficiency of 40% for a recording energy of 200 mJ/cm^2. They have also found that diffraction efficiency increases with an increase in the molecular weight of the PVA up to 50,000 and then starts to diminish. Moreover, an attempt to increase the sensitivity of this method was made by incorporating the salts of dicarboxylic

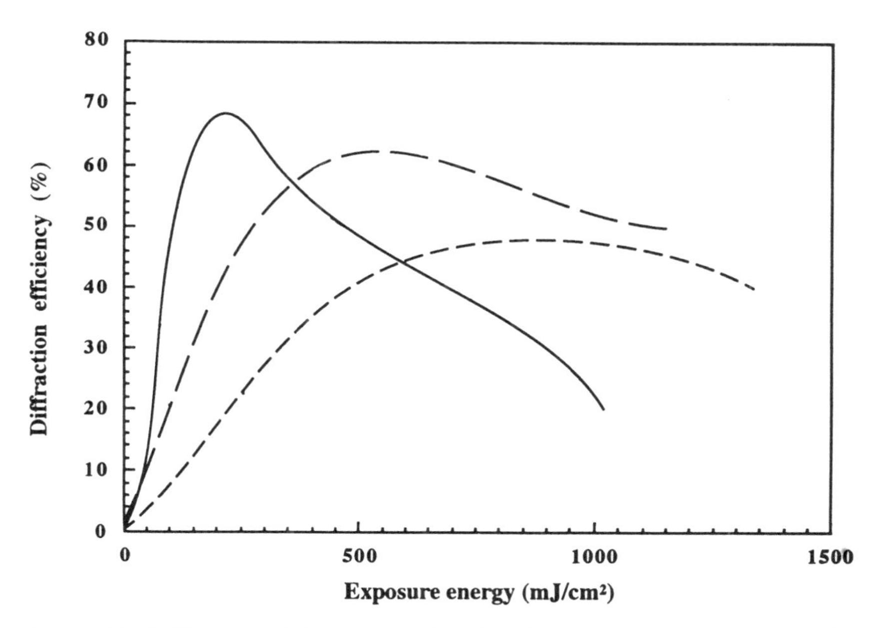

Figure 1-13. *Diffraction efficiency of developed DCPVA holograms as a function of exposure energy.----Undeveloped holograms;——Holograms developed by Method 1;—— Holograms developed by Method 2. [Ref. 99(b).]*

acid such as ammonium succinate, and an increase in diffraction efficiency from 20% to 40% at a recording energy of 200 mJ/cm^2 was observed for a concentration of 0.3 wt % of ammonium succinate.

Reflection holograms have also been recorded in DCPVA films using the Lippman configuration with an argon ion laser at 488 nm. A peak diffraction efficiency value of ≈38% was observed for an exposure energy of 500 mJ/cm^2. The angular selectivity response was also determined and the angular bandwidth was found to be 4.5 mins. Holographic characterization of DCPVA as a recording material for reflection holograms has been carried out for the first time.

The basic chemical process responsible for the mechanism of hologram formation is the photochemical reduction of Cr(VI) doped in the polymer matrix (PVA). According to Kosar,[12] the light-sensitive nature of dichromate coupled with organic matter was discovered as far back as 1830. The photochemistry of Cr(VI) has been investigated over decades, and it has been generally accepted that the photoreduction of Cr(VI) leads to Cr(III) in the presence of organic reducing substances like secondary alcohols.[105] The photoreduction of dichromate in PVA films has been studied by Duncalf and Dunn,[106] and they suggested that the insolubility of DCPVA on illumination with light was caused by the complexation of PVA and Cr(III) [the photoproduct of Cr(VI)].

In aqueous solution, chromium (VI) exists in different forms, and three species were present in the polymer films: $HCrO_4^-$ and $Cr_2O_7^{2-}$ in the acidic medium, and CrO_4^{2-} in the basic medium[107–110] as:

$$HCrO_4^{2-} \rightarrow CrO_4^{2-} + H^+ \qquad pK = 6.49 \tag{4}$$

$$2HCrO_4^- \rightarrow CrO_7^{2-} + H_2O \qquad k_d = 48\ mol^{-1},L \tag{5}$$

$HCrO_4^-$ and CrO_4^2 have been considered as the active chromium (VI) species at pH < 5.0 and > 8.0, respectively.[111]

The primary photoprocesses of the interaction of Cr(VI) with PVA have been followed through real-time electron spin resonance (ESR) spectroscopy, to understand the mechanism of hologram formation.[112] Thermal and photochemical evolution of Cr(V) have been identified from its characteristic ESR signal of $g = 1.9789$ at normal room temperature. The photoproduct, Cr(III) was also identified in the polymer films, and spin-trapping experiments with *N-t*-butyl-α-phenyl nitrone (PBN) gave evidence of a polymer radical. From the holographic and ESR spectroscopic measurements on polymer films with varying pH it has been concluded that CrO_4^{2-} also functions as a better active species than $HCrO_4^-$. Further, it is clear that Cr(V) is the primary photospecies formed upon irradiation from Cr(VI) leading to the hologram formation through an electron transfer mechanism involving the polymer matrix. In other words, Cr(VI) removes the needed three electrons for the photoconversion from the polymer matrix, leading to the crosslinking in the polymer (scheme shown in Fig. 1-14).

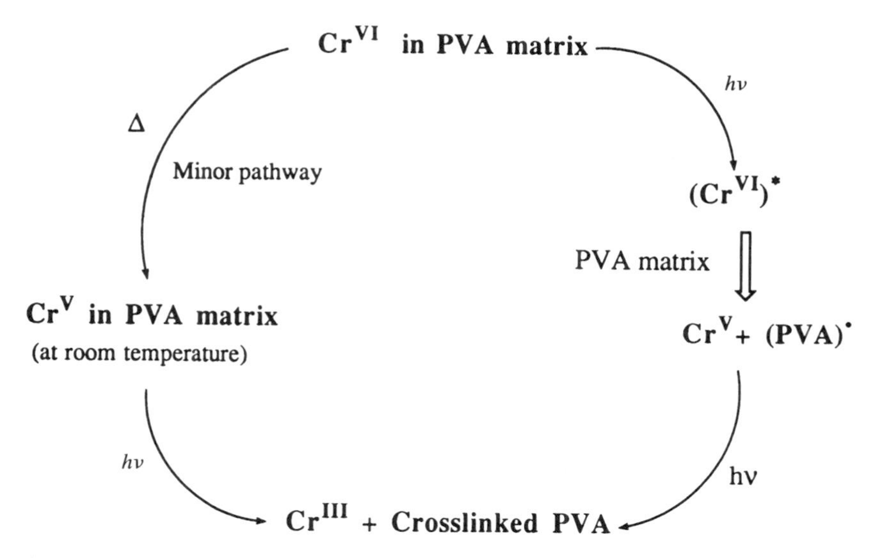

Figure 1-14. *Scheme representing the photochemical mechanism of hologram formation in DCPVA system. (Ref. 112.)*

Similarly, when an electron donor is added to the polymer matrix, the source of available electrons for the conversion of Cr(VI) to Cr(III) is doubled; thus the percentage and rate of conversion of Cr(VI) to Cr(III) are increased.[112] This observation suggests that the holographic efficiency of these materials can be suitably modified at will by the appropriate incorporation of electron donors. Upon irradiation, orange-colored DCPVA films became brown/brownish blue in color. The characteristic physical parameters such as solubility, refractive index, and absorption constant also changed as a result of blue light exposure.[113]

3.2.2. Dichromated Poly(acrylic acid) (DCPAA)

As a continuation of the ongoing search for an efficient real-time holographic recording material, dichromated poly(acrylic acid) (DCPAA) ($\overline{M}_w$ = 90,000, 25% weight percentage in water) was used to record volume transmission holograms.

The effect of weight percentage of poly(acrylic acid) (PAA) on the diffraction efficiency of the holograms was investigated, and for the polymer films with a higher concentration of PAA a maximum diffraction efficiency of 28.4% for an exposure energy of 4 J/cm^2 was observed. Because the energy requirement to achieve a better diffraction efficiency is very high compared to that of DCPVA, more experiments were performed to improve the energy sensitivity of the medium. As in DCPVA, electron donors have been introduced into the DCPAA films during fabrication to check their influence on the diffraction efficiency of the holograms. Dimethyl formamide (DMF) shows a remarkable effect on the diffraction efficiency of the material as well as on the energy sensitivity of the medium. The real-time diffraction efficiency shows more than a twofold increase in the presence of DMF (2.22 *M*) and 68% efficiency was obtained at an exposure energy of ≈200 mJ/cm^2 as shown in Fig. 1-15. This interesting observation also suggests that an electron transfer process is more efficient in the DCPAA system. A similar holographic recording mechanism as in DCPVA can be envisaged in the DCPAA system. The orange-yellow-colored films undergo a color change to greyish green on irradiation.

3.2.3. Ferric Chloride Doped Poly(vinyl alcohol) (FePVA)

Budkevich et al. have studied the photo transformation in Fe(III) (2%) doped PVA at low temperature in thin films of 10–15 μm thickness.[114] The holograms were recorded using an He-Cd laser of 441.6 nm and reconstructed with a 632.8-nm laser. The spatial frequency was found to be 80 mm^{-1} because of the geo-

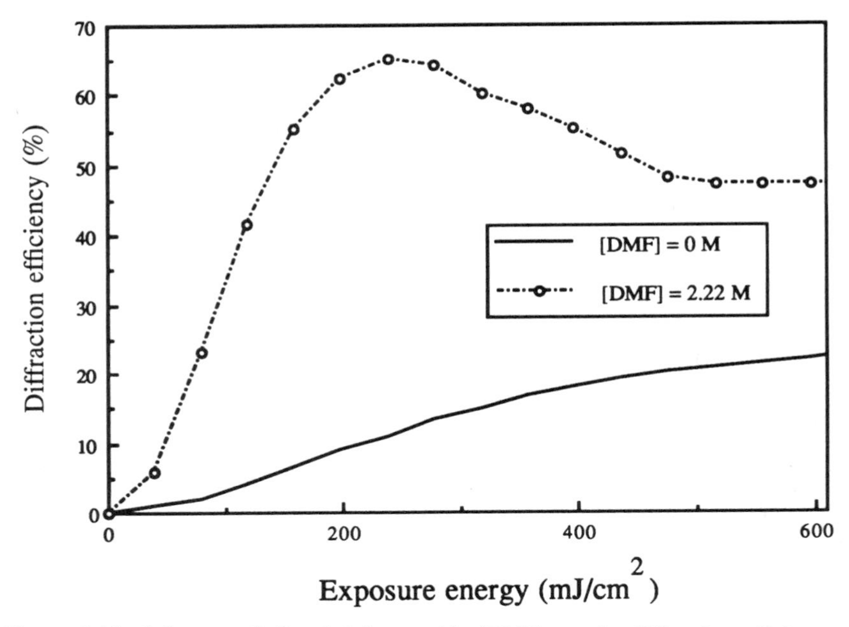

Figure 1-15. *Influence of dimethyl formamide (DMF) on the diffraction efficiency of DCPAA holograms [Ref. 99(c).]*

metrical limitations. Amplitude and phase components of the diffraction efficiency of recorded holograms were studied and a diffraction efficiency of ≈16% was reported. We have employed[99(a)–(c),115] ferric chloride doped PVA (FePVA) films for the fabrication of volume transmission holograms, and different experimental parameters have been optimized.

The effect of molecular weight of PVA used on the holographic efficiency was investigated. A maximum real-time diffraction efficiency of 45% was obtained for films with average molecular weight of 11,000–31,000 and 14,000. Similarly, the effect of $FeCl_3$ concentration on the diffraction efficiency η was investigated, and it was found that films containing 7 wt % of $FeCl_3$ exhibited a better energy response with maximum η of 72% for an exposure energy of 12 J/cm^2. Fig. 1-16 depicts the observed effect of concentration of Fe(III) on η. Moreover, the power of recording beams was varied in the range of 13 mW to 60 mW and it was observed that the speed of recording and peak diffraction efficiency increases with an increase in power of recording. A maximum η of ≈80% was achieved for a total beam power of 40 mW at 17 J/cm^2 (shown in Fig. 1-17).

As observed in DCPVA and DCPAA, the holographic efficiency of FePVA films was also found to be influenced by the addition of various electron donors such as triethanolamine, DMF, and an initiator *t*-butyl peroxide (tBP). Addition

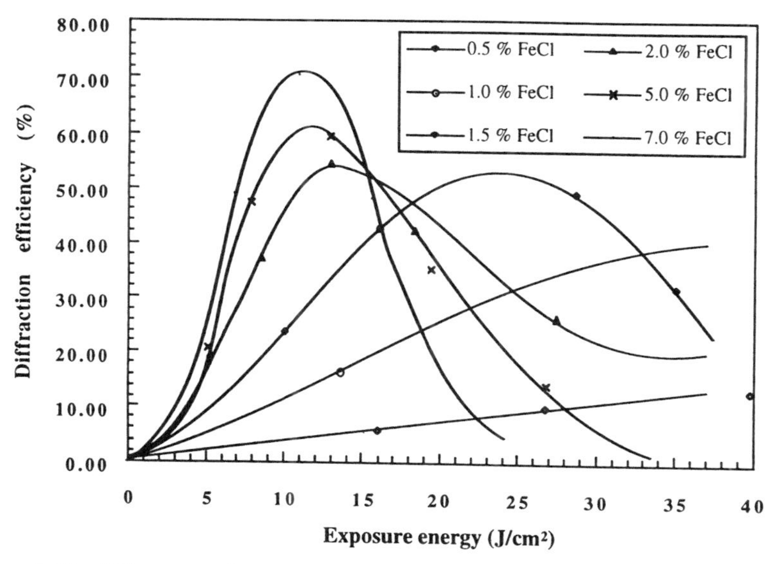

Figure 1-16. *Diffraction efficiency of FePVA holograms for different concentration of $FeCl_3$. [Ref. 99(a).]*

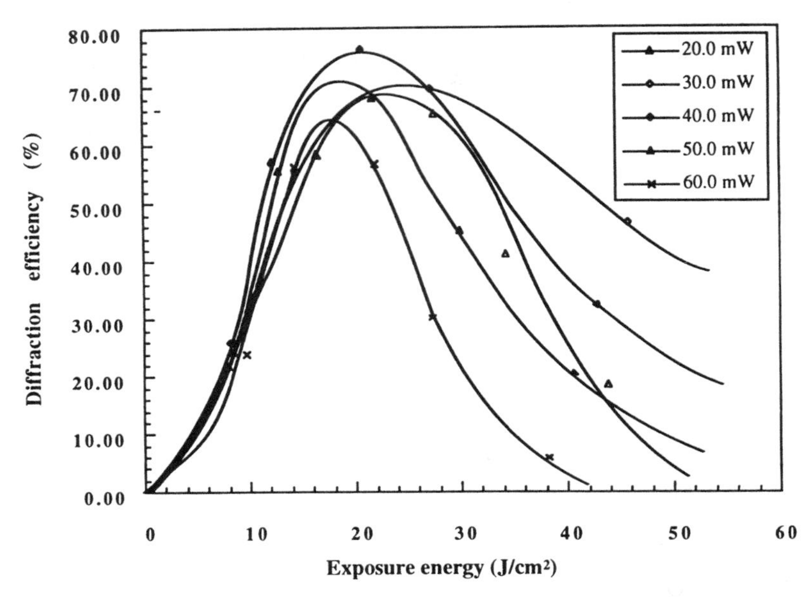

Figure 1-17. *Diffraction efficiency of FePVA holograms as a function of exposure energy for different power of the recording beams [Refs. 99(a), 115.]*

of the initiator tBP showed promising results. The films without tBP exhibited a diffraction efficiency of ≈60% at an exposure energy of 25 J/cm^2, whereas films with 0.2 *M* tBP gave an η value of 80% for the same exposure energy.

The physical characteristics such as angular selectivity of the holograms and refractive index change in the exposed films were also determined. The spatial frequency response of this material was also determined by varying the inter-beam angle between 14° to 94° to get the frequency of recording from 500 to 3000 cycles/mm. Dependence of the diffraction efficiency on the spatial frequency was found to be almost linear with a frequency bandwidth of 2500 cycles/mm. An angular bandwidth of 40 mins was observed from the angular selectivity curve, and using this value the effective thickness of the grating was found to be 55 μm for films of 60-μm thickness using the method suggested by Couture and Lessard[116] The refractive index changes of the exposed films was evaluated at 632.8 nm for different exposure energies using an adaptation of the prism-film coupler method,[113] and the Δn was found to be 1.9×10^{-2}.

Photoreduction of ferric salts by UV light has been known for more than a century. Oster et al.[100] found that Fe(III) can be photoreduced to Fe(II) by visible light in the presence of an electron donor. Behavior of the PVA matrix as an electron donor has been suggested in the aforementioned work on the DCPVA system.[112] Budkevich et al.[114] suggested the removal of chlorine radical from ferric chloride which reacts with the PVA matrix to give HCl and a polymer radical. The EPR[117] data on the photolysis of this system indicated that only 5–10% of the Cl˙ radical leaves the intermediate at 77 K and reacts with alcohol, with hydrogen atom detachment from the α-position. Tests also showed that HCl accumulates in the system.[118]

The mechanism of hologram formation in FePVA also can be considered from the understanding of photochemical reactions in DCPVA and DCPAA films and well-known photochemical processes. The possible reactions are:

$$\text{Fe(III) in PVA matrix} \xrightarrow{h\nu} \text{Fe(III)*} \tag{6}$$

$$\text{Fe(III)*} \xrightarrow{\text{PVA}} \text{Fe(II)} + \text{crosslinked polymer} \tag{7}$$

3.2.4. *Poly-(N-vinyl carbazole) (PVCz) System*

Poly-(N-vinyl carbazole) (PVCz) has been used as a volume phase holographic recording material by two different research groups of Japan.[119,120] The holographic material consists of PVCz as a base polymer, camphorquinone as an initiator, and thioflavine T as a sensitizer.[119] After recording a latent image of a fringe pattern by exposure to argon ion laser light, a hologram was developed by swelling and shrinking of the film with two kinds of solvents. The thickness of the hologram could be reduced to 2.5 μm, because the PVCz hologram has a large amplitude of the refractive index modulation related to the crystallinity modulation. Advantages of this material are a higher refractive index modulation and higher stability under high humidity conditions.

The holograms were recorded on spin-coated polymer films of thickness 1.4 μm, and dried in a nitrogen atmosphere at room temperature. An argon ion laser of 488 nm was used to record the holograms, and the film was developed after exposure by the following procedure: (1) pretreatment with ethyl alcohol to remove the unreacted camphorquinone and thioflavine T; (2) swelling of the PVCz film with the hologram by toluene (a good solvent); and (3) shrinking of the PVCz film by dipping in pentane (a poor solvent).

The diffraction efficiency was monitored by a He-Ne laser beam and the crystallinity was evaluated by the difference in X-ray diffraction pattern. The durability of the uncovered hologram was evaluated from the change in η under typical environmental conditions of 40° C and 95% relative humidity (RH). It was discovered that the holograms maintained a high η for more than 1000 hours. An exposure of about 500 mJ/cm^2 was required to achieve an efficiency of 80% with a spatial frequency of 800–2500 lines/mm. A positive change in refractive index of 0.32 was found between exposed and unexposed areas.

Another research group from the Canon Research Center[120] successfully installed a holographic display element fabricated from PVCz film in a commercialized 8mm movie camera. The letters "END" are displayed on an imaging plane of the finder system by using an image plane hologram. This group has also made a holographic lens for which aberration is well corrected at laser diode wavelengths. Here, the PVCz is sensitized with carbon tetraiodide (Cl_4) that is exposed to an ar ion laser of 488 nm. The CI_4 generates radicals on exposure leading to crosslinking reaction of the polymers. The interference pattern is transferred as the pattern of crosslinking. The group also employed a solvent treatment involving the removal of unreacted material followed by shrinking the swelled sensitive material. Finally, the recorded hologram was covered with a glass plate. The spectral sensitivity span was from 400 to 560 nm. A maximum η of 96% was obtained by the exposure of 50 mJ/cm^2 with a resolution of >3500 lines/mm; moreover, it has been discovered that further exposure diminishes the efficiency of the hologram. The durability characteristics (at 70° C and 95% RH) of the holograms with protective glass plates were found to be better than the naked hologram.

The same material has also been used to record thick reflection holograms[121] involving an argon ion laser at 488-nm wavelength. A maximum η of 91% at 515-nm wavelength was observed for an exposure of 75 mJ/cm^2. The spectral bandwidth was 17 nm for a film thickness of 8.8 μm (effective thickness of 7.1 μm). The effective refractive index (n) and index modulation (Δn) were found to be 1.72 and 0.042, respectively.

3.2.5. Poly(methyl methacrylate) (PMMA) System

Thick photodielectric holograms were recorded by Freilich et al.[122] on PMMA or poly(methyl-α-cyanoacrylate) with *p*-benzoquinone (PBQ) as the sensitizer using a 488-nm argon ion laser line. As the spectral response is limited to blue-green, recording was done at 488 nm, and the read out wavelength was 632.8 nm from a He-Ne laser. Fig. 1-18 presents the diffraction efficiency as a function of exposure energy for films of different thicknesses.[123] A diffraction efficiency of 100% was achieved by Friesem et al.[123] for a film thickness of 130 μm with an exposure energy of 7 J/cm^2. The spatial frequency response was 100–2000 lines/mm. The holograms could be fixed with a uniform post-exposure by an illumination beam of 100 mW/cm^2 at 488 nm for about 1 hour. The excellent angular discrimination capabilities of the thick photodielectric materials were illustrated by recording 20 superimposed, but independent, holograms.

Simple reflection holograms were also recorded with the materials. The hologram formation mechanism involves a change in index of refraction due to the local increase of density. This is further corroborated by viscosity measurements demonstrating that the photodielectric materials crosslinked when irradiated with actinic iradiation. Involved photocrosslinking processes are: absorption of actinic radiation by PBQ and production of free radicals that crosslink ad-

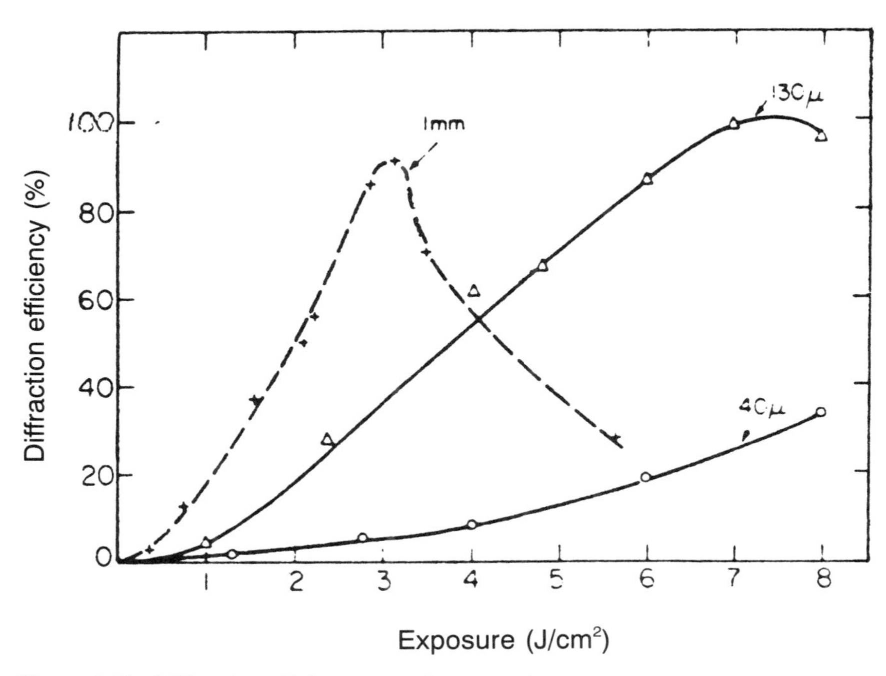

Figure 1-18. *Diffraction efficiency as a function of exposure for different thicknesses. (Ref. 123.)*

jacent polymer chains. The holographic recording mechanism was explained based on an increase in refractive index as a result of the above-mentioned photocrosslinking process. The first reaction of the triplet PBQ is probably the hydrogen abstraction from the polymer chain resulting in the combination of two radicals to give $PBQH_2$ attached to the polymer backbone or abstraction of another H by the PBQH radical to yield a free $PBQH_2$ and another macro radical. Note that low exposure sensitivity and orthochromatic sensitivity are some shortcomings of these materials. Similarly, metal dithizonates such as diphenyl thiocarbazone, H_2D_z doped polystyrene, PMMA, polyvinyl acetate, and epoxy resin were employed.[124] The sensitivity of these materials was 300–800 mJ/cm^2 with a spatial fequency reaching 2800 lines/mm and an initial diffraction efficiency of about 10%.

3.2.6. Dichromated Gelatin (DCG)

Ever since the first proposal of dichromated gelatin's (DCG) use as a holographic recording material by Shankoff,[125] it has been one of the most widely investigated materials for recording holographic optical elements. As the name implies, DCG consists of gelatin sensitized with a dichromate (from ammonium dichromate). The recording mechanism results from the photoinduced crosslinking of the biopolymer (gelatin) by Cr(III) derived from the photoreduction of the dichromate [Cr(VI)]. DCG possesses many desirable properties to qualify as an ideal hologram recording medium. When processed under carefully controlled conditions (e.g., temperature, relative humidity, and chemical environment) it forms clear holograms that exhibit very little absorption and scattering. Also, a large refractive index modulation (≈0.08) can be achieved in the interior of the gelatin layer,[126] thus making it possible for the realization of transmission/reflection volume phase holograms possessing near-ideal diffraction efficiencies (≈100%). The large refractive index modulation capacity of DCG enables its use for superimposing and multiplexing a large number of holograms into the depth of the material.[127] Further, DCG is endowed with desirable properties such as high spatial resolution (as high as 5000 lines/mm) and uniform spatial frequency response over a broad range of spatial frequencies (e.g., from 100 to 5000 lines/mm). A unique feature of DCG is its capability for reprocessing.[126] For example, the properties of a hologram such as diffraction efficiency and signal-to-noise ratio (SNR) can deteriorate because of the influence of environmental factors, but can be regained by reprocessing.[128,129] Another important characteristic of DCG, which is very useful in the fabrication of wavelength selective tunable optical elements, is its amenability for thickness control through pre-processing or post-processing.[126]

DCG films can be prepared by one method from gelatin films derived from conventional photographic emulsions such as Kodak 649F[130] or Agfa 8E75HD[131] by desensitizing them in a photographic fixer containing a hardener and then washing them in water followed by methyl alcohol.[127] Another method consists of coating one's own gelatin films using off-the-shelf gelatin and using gravity settling, doctor blading, dip coating, spin coating, or spraying techniques.[132] Better control of film parameters can be achieved using the second method because precise knowledge on the film characteristics of company-manufactured photographic emulsions is often hard to obtain. Also, many of the film parameters are not the same from batch to batch.

An important criterion in the fabrication of DCG films is the precise control of the prehardness of the films before exposure. It appears that an optimum prehardness, known as bias hardness, is required for efficient hologram recording.[126] Bias hardness is most commonly achieved by chemical and thermal hardening techniques. For example, Brandes et al.[132] prehardened the films by introducing trace of ammonium dichromate and subsequently baking the dried films at 150° C. The influence of organic and inorganic chemical prehardeners (e.g., formaldehyde and chrome alum) on the diffraction efficiency of the holograms was studied by Samoilovich et al.[133] Prehardening with noble metals such as gold (Au) and silver (Ag) etc., has been tried by Mazakova et al.[134] Their results indicated that the energy sensitivity of DCG films is much higher if the amino groups of the gelatin chain are involved for bias hardness, leaving the carboxyl groups free for photochemical hardening during exposure of the holograms. Films derived from standard photographic emulsions are already prehardened during the manufacturing process. The fine-tuning of prehardening (required for recording holograms) is achieved during the desensitization step by incorporating a chemical hardener (e.g., potash alum, chrome alum, or formaldehyde) along with the photographic fixer.[127] The prehardening achieved depends on the concentration of the hardener added to the fixer bath, and in this regard the study of Keinonen[135] revealed that the best diffraction efficiency for DCG holograms results when the concentration of the hardener is about 15% of the fixer solution. Sjölinder found a relation between the pH of the gelatin solution used and the diffraction efficiency of the holograms.[136] An optimum diffraction efficiency value was obtained when the pH of the gelatin solution was about 8–9. Monitoring of the bias hardness can be done in several ways such as by determining the swelling, melting point, scratch resistance, the number of crosslinks, or the resistance to boiling. The most commonly used technique has been the determination of the swelling of the layers.[133,137] The best films are obtained when they are dried at normal room temperature (25–27° C) with a relative humidity of 40%.[138–140] The drying time also influences the quality of the films, and a typical drying time is of the order of 24 hrs.[132, 141]

Photosensitization of gelatin films is achieved by incorporating ammonium dichromate by absorption through a solution bath, and such films are sensitive in the wavelength region of 350–514 nm. Solano et al.[140] found that pure DCG is also sensitive to red light (633 nm), and the energy required to record gratings having a diffraction efficiency of about 70% in 30-μm-thick films is 300–400 J/cm^2. However, it is known that DCG films can be made sensitive to red light by adding a suitable dye such as methylene blue (MB).[101–103,142–157] The main criterion in preparing the dye-sensitized DCG films is the compatibility of the dye with the dichromate. Though some of the dyes have high solubility in the solution, when introduced in the films they precipitate if they are not compatible. Important parameters found to influence the quality of methylene blue-sensitized DCG (MBDCG) films are the pH of the coating solution and the controlled film drying atmosphere.[146] The problem of low solubility of the dye in MBDCG films was overcome by a technique suggested by Solano et al.[142] where the plain methylene blue-sensitized gelatin films were exposed and later processed in dichromate solution.

The energy sensitivity of MBDCG was found to be much lower than that of DCG. Based on the findings of Oster and Oster,[100] experiments to improve the diffraction efficiency of MBDCG holograms using electron donors were carried out.[101,144–146] A detailed investigation was carried out by Changkakoti et al.[101] on this aspect, and they found the contribution of three electron donors viz., dimethyl formamide (DMF), ethylene diamine tetraacetic acid (EDTA), and triethanolamine (TEA) to the diffraction efficiency of the holograms. However, improvement of the energy sensitivity of the electron donor-doped MBDCG was not significant. The search for better electron donors by Capolla et al.[156–157] and Blyth[102] led to the discovery of a class of amines such as dimethylaminoethylmethylaminoethanol, dimethylethanolamine, and tetramethylguanidine. These amines improved the energy sensitivity tenfold; high efficiency holograms (≈80%) were obtained with an exposure energy of ≈50 mJ/cm^2. Similar observations were made by Mazakova et al.[158] when DCG films with dimethyl formamide showed higher diffraction efficiency than those without it.

The chemical development of the exposed films amplifies and fixes the recorded information in the gelatin base in an irreversible fashion. In this step the removal of the untransformed dichromate and dye from the film takes place. The most commonly used process for development of DCG/MBDCG holograms involves a water wash followed by dehydration in a dehydrant liquid such as isopropyl alcohol. The dehydrated plates are then dried in a flow of hot air. Plates developed by this process show a milky white opacity.[130,138] Various developing techniques have been proposed over the past two decades to improve the final quality of the processed holograms. Lin[130] suggested the use of an ammonium dichromate bath before the water wash of the plates to overcome the problem of milkiness in DCG made from a photographic emulsion. Kubota et al.[145] introduced a hot water wash (43–47° C) between the water wash and dehydration steps. Close and Graube[30] developed the exposed plates in an aqueous solution of triethanolamine. Sosnowski and Kogelnik[159] modified the dehydration process by using ethanol as a dehydrant. Other influencing parameters during development of the holograms are pH and temperature. Changkakoti and Pappu[150] found that the pH of the first developing solution contributes to the final efficiency of the MBDCG holograms. Larger refractive index modulation has been achieved[30] for plates developed at 50° C and baked[160] at a temperature of 71° C. Also, the energy sensitivity is seen to improve if a time delay is introduced between exposure and development.[126,148] A similar behavior is noticed when the MBDCG plates are heated[148] at 60° C. Similarly, heat treatment of recorded but undeveloped holograms leads to higher holographic sensitivity.[158]

After the exposed DCG film is chemically processed, it must be dried in a flow of dried air or in an oven. This drying process is potentially destructive if the medium still contains a significant amount of water. Thermal drying is usually associated with a blue shift of the reconstruction wavelength; when the distance between Bragg planes is reduced and the Bragg wavelength lies in the violet or beyond, the holograms lose applicability. To overcome this problem, Rebordâo and Andrade[161] proposed a technique in which the developed holograms were dried in a microwave oven. This considerably increased the resistance to heating, and no wavelength shift was observed for wide band holograms. Jeong et al.[162] and Georgekutty and Liu[160] have suggested other techniques to decrease the time of processing and development to obtain high-efficiency holograms in DCG films made from a Kodak 649F photographic emulsion.

A fair amount of research has been done with regard to the characterization and the use of DCG holograms in several application areas such as holographic optical elements (HOEs), optical processing, head-up displays (HUDs), laser scanners, fiber-optic couplers, multiplexers and demultiplexers, filters, relief holograms, and optical interconnects.[163–194]

For the realization of high-diffraction efficiency holograms in DCG, it is strongly believed that adequate refractive index modulation should manifest itself in the bulk of the recording medium. It is also believed that this refractive index modulation comes about by the photoinduced transformation of the Cr(VI) to Cr(III) under light which subsequently crosslinks the gelatin matrix. The mechanism of the transformation of this crosslinked gelatin to the final refractive index modulation in the hologram is still an unresolved issue. Various mechanisms have been proposed to explain this phenomenon.[126,133,141,195]

The various salient features of photocrosslinkable polymers discussed earlier are grouped in Table 1-3.

3.3. Doped Polymer Systems

3.3.1. Dye Doped Polymers

Dye doped PVA materials are used for *polarization holographic recording*. They satisfy the needed characteristics of polarization-sensitive materials, namely, high sensitivity, quick and simple chemical development, good spatial frequency response, reversibility with reasonable diffraction efficiency, and high signal-to-noise ratio. Quite interesting is the possibility of obtaining reusable recording films giving many thousands of write/read/erase (WRE) cycles without apparent fatigue. Kakichashvili[196] was the first to introduce polarization holography applications obtained with photochromic trimethylspirane-benzopyran. Later Todorov et al.[197–202] pointed out that methyl orange (MO) and methyl red (MR) layers introduced in a PVA and PMMA, matrix respectively, may record polarization holograms by trans-cis photoisomerization.

Polarization holographic recording was accomplished by two plane waves with mutually orthogonal polarization at the 488-nm wavelength of an argon ion laser. The induced optical anisotropy (dichroism or birefringence) is spatially modulated in accordance with the polarization modulation of the recording light field; i.e., a polarization holographic grating is recorded. Simultaneous dichroism

Table 1-3. Photocrosslinking Systems

Material	Type of Modulation	Preparation Method	Thickness (μm)	Recording Wavelength (nm)	Sensitivity (mJ/cm^2)	Resolution (lines/mm)	Diffraction Efficiency %	Refs.
DCPVA	Refractive index	Coated on glass	30–60	488	0.5	3000	68	96–99
DCPAA	Refractive index	Coated on glass	>60	488	0.2	3000	68	99 (c)
FePVA	Refractive index	Coated on glass	60	488	17	3000	80	115
PVCz	Refractive index	Spin coated	1.4–7.1	488	0.05, 0.5	>3500	96	119–121
PMMA	Refractive index	Coated on glass	130	488	7.0	2000	≈100	123
DCG	Refractive index	Coated on glass	15–20	488–514	0.01–0.1	>5000	≈100	125–127
MBDCG	Refractive index	Coated on glass	15–20	633–650	0.05–0.5	>5000	80–90	101–103, 144–157

and birefringence were found to be induced in the investigated layers. Readout was done with a He-Ne laser beam at Bragg's angle and the diffraction efficiency of +1 order was measured. After about 20 sec, diffraction efficiency reaches a saturation level (20%) and remains constant as long as the recording continues. When the recording light is turned off, initially the diffraction efficiency falls to around one-third and remains practically unchanged for more than 24 hours. Such memory effect is observed only in thermally pretreated samples.

If the recording medium has an intrinsic birefringence before the recording due to preliminary mechanical stretching, the phase difference between the two recording waves and their polarizations will vary with the depth of the sample, as will the polarization interference pattern. The diffraction efficiency depends on the polarization direction of the linearly polarized reconstructing wave and reaches a maximum when it coincides with that of the corresponding recording wave (polarization wave).[200] Sensitivity of the diffraction efficiency to the polarization of the recording wave gives additional possibilities such as using polarization holographic recording for raising the information capacity of the recording and solving problems in the field of optical image processing. Possible applications of polarization holography are recording of pairs of superimposed holograms with one spatial frequency, logical operations on two data arrays recorded simultaneously in the material, and double-exposure holographic interferometry.[201]

A theoretical generalization of the method of polarizational holographic recording for photoanisotropic media with a residual scalar response is considered. The so-called polarization correspondence curve is derived, which characterizes the photoanisotropic constants of the recording medium. Experimental results are presented on polarizational recording of holograms; these agree with the theoretical approach developed. Curves of the polarization correspondence of three photoanisotropic media are provided: (1) bichromic layer of gelatin with introduced malachite dye, (2) chlorinated film of silver on a glass substrate, and (3) inorganic photochromic glass.[203] The diffraction efficiency of the materials as a function of the relative orientation of the reference and object waves was measured, which is less than 1.0%.

The photochromic properties of three dyes (fluorescein, methyl red, and methyl orange) introduced in different matrices were investigated and it was found that the photochromic changes in the dyes are accompanied by inducing optical anisotropy. These materials can be used for transient holographic recording of the conventional (scalar) as well as the polarization type. Quite attractive is the feasibility of controlling the kinetics of the transient photochromism and photoinduced anisotropy by selecting appropriate matrices. Retardation of the reverse thermal reactions makes it possible to use continuous lasers with relatively low intensities.[203]

Moreover, rigid solutions of two types of materials are used: (1) xanthene dyes in orthoboric acid, in which photoanisotropy is caused by selective excitation of optical transitions from a singlet to a triplet state, and (2) azo dyes in polymers, in which photoanisotropy is caused by reversible photoisomerization of the dye molecules. Three different dye/polymer systems, namely, fluorescein (F)/orthoboric acid (OBA), methyl red (MR)/PMMA, and methyl orange (MO)/ PVA were used to investigate the kinetics of photoinduced anisotropy. All of the three systems allow real-time recording, and the photochromic transformations are entirely reversible for more than 10^4 cycles without noticeable fatigue. The system F/ OBA possesses a faster response time (about 0.1 s) than the other two systems. The phase character of the recording in the system MO/PVA explains a high-diffraction efficiency up to 35% for polarization holographic recording.[203]

The holographic recording in an F/OBA system has an amplitude character and hence the diffraction efficiency is low. In MR/PMMA and MO/PVA systems, the holographic gratings have a phase character because of the matrix involvement in the recording. During the photoprocess, under the pressure of molecules of an isomerizing dye, a spatial displacement and ordering of the polymer molecules takes place. In addition, substantial birefringence is induced in the layers and the polarization holographic recording has a high diffraction efficiency (up to 35%).

Diffraction efficiency and selectivity of polarization holographic recording on photoinduced optical anisotropy in the recording material are discussed for two types of polarization modulation: recording with two orthogonal linear waves and recording with circular polarized waves. It is shown that during polarization recording in materials with an intrinsic birefringence, there is a gradual transition between two types of polarization modulation. In these combined types of recording both the diffraction efficiency of the holograms and the polarization of the diffracted wave depend on the polarization of the reconstructing wave.[204] Interest in this type of holographic recording is aroused by the improved SNR and the possibilities it offers for image processing, such as subtraction and selective erasure.

Modulation transfer function (MTF) curves of MO/PVA and acidified methyl red (AMR)/PVA for the 500–4000 cycles/mm spatial frequency domain were reported.[205] Refs. 206 and 207 present the indicator action mechanism of acidified methyl red molecules and the basic mechanism of photodichroism of the azo dyes. First, the recording process can be seen as induced by an $n \rightarrow \pi^*$ transition occurring under light illumination (polarized). After that, a rotation over the σ bond is achieved because the π bond is strongly disturbed by one electron of the π^* state changing a trans isomer into a cis isomer.[208,209]

The MO/PVA films showed nearly uniform diffraction efficiency for many thousands of write/erase cycles. The erasure cycles were observed and for circular orthogonal polarizations, the kinetic erasure process giving the highest diffraction energy was observed with a longer time of erasure. Moreover, some preliminary results using AMR/PVA are also presented such as MTF curves. The diffraction efficiency was found to be the same for the frequency range 500–3300 cycles/mm. In general, it can be concluded that the spatial frequency response of transmission volume holograms is good until 4000 cycles/mm and depends on the polarization states of writing and reading beams.

Moreover, an erasure process is achieved when azo samples are heated or placed in the dark; then the *cis* form disappears and the most stable *trans* form occurs. The two isomer species existing in equilibrium depend on the temperature and specific polarized light action. The steric arrangement, local viscosity, and a good choice of solvent promote this *trans*↔*cis* photoisomerization. Further, the addition of acetic acid to an aqueous azo dye/PVA solution gives a bathochromic shift resulting in a strong absorption in the 450–580-nm range and a good transmittance at 632.8 nm. Macroscopically, when the recording process is completed azo dye/PVA layers demonstrate an increased transmittance at 488 nm under light action of the reconstruction beam, which has the same polarization as the reference beam used in the recording process. Hence the kinetic erasure process depends on the polarization states of the recording beams. This photodichroic behavior is in agreement with the basic *trans*↔*cis* photoisomerization mechanism. The kinetic erasure times and spatial frequency responses change according to the polarization mode used during the recording process. A photostationary state of the recording layers was found to attain for an exposure time of 60 s.

3.3.2. Azobenzene Side Chain Polymers

Recently a novel series of azo polymers, copolymers, and polymer blends for reversible optical storage devices were also described.[210] Various azobenzenes can be bound in the side chain or within the main chain of long chain polymers such as polyester, polystyrene, and polymethacrylates with very short or no spacers to form amorphous polymers. These polymers, as a film or deposited on a transparent substrate, can be used to record optical information using a linearly polarized laser beam, which induces optical anisotropy in the film.

The information or image is written by exposing the film to linearly polarized light, which causes a reorientation and alignment of the molecules in the film. The materials have relatively high glass transition temperatures, making them less thermally sensitive and enhancing long-term storage. The writing procedure orients polarizable groups, as done in electrical field poling to obtain nonlinear optical properties of a polymerizable material.

Reading of information was achieved by monitoring the light transmitted through a crossed polarizer set-up where three stages of polarization can be separated: (1) linearly polarized horizontal or vertical, (2) linearly polarized at $\pm$ 45°, and (3) no polarization. For a thin film of about 500 nm on a glass substrate the optical writing and erase times are 1 or 2 seconds to achieve 90% saturation when using an argon laser at 514 nm at 5-mW power. It is possible to write a grating on the film with line separations of one μm, and the resolution limit of the recording is established by the optics and not by the recording medium. The stored information can be erased thermally by heating the films above their glass transition temperature, or optically by overwriting with light polarized in a predefined zero direction or with circularly polarized light which also permits local, selective erasure.

3.3.3. Bacteriorhodopsin Doped Polymers

Bacteriorhodopsin (BR) is the light transducing protein in the purple membrane of *Halobacterium halobium.* The purple membrane, which contains the protein in a lipid matrix (3:1 protein to lipid), is synthesized by the bacterium when the oxygen content of its surroundings is too low to sustain aerobic respiration. The important photochemical intermediates in the bacteriorhodopsin photocycle are referred to as bR and M. The initial state, bR, is characterized by a large absorption maximum in the yellow region of the visible spectrum with a maximum at 568 nm. Upon the absorption of light energy, bR passes through a series of photochemical intermediates to the blue light absorbing M state, with a maximum at 413 nm. Because M has a strongly blue shifted absorption maximum relative to bR, this intermediate plays an important role in most optical applications in which a large shift in absorption maxima and a corresponding change in refractive index are required for optimal function. Under standard biological conditions, M thermally reverts to the ground state with a time constant of ± 10 ms. By suspending bacteriorhodopsin in a polymer matrix in the presence of certain chemical agents, the lifetime of M can be dramatically increased to 15–30 minutes. This property, coupled with the high quantum yields of formation in the forward and reverse direction ($\phi = 0.64$), high photocyclicity ($>10^6$ write/read/erase cycles), and excellent thermal stability (<90° C), give bacteriorhodopsin photochromic qualities unavailable in most other organic photochromics.[211]

Regulation of the lifetime of the M intermediate is accomplished through the addition of various chemicals such as guanidine hydrochloride and arginine. Polyvalent metal salts (e.g., $LaCl_3$, $AlCl_3$, $ZnCl_2$) and several diaminoalkane derivatives also have a profound effect on the lifetime of the M intermediate.

A thin film was made from a solution containing PVA, guanidine hydrochloride, 1,4-diaminobutane, and purple membrane. The pH of the solution was adjusted to 10.8 and this mixture was degassed for three minutes under high vacuum. Films were cast on glass substrate by the pouring method and dried under low humidity for several days. Resulting films had a thickness of ~150 μm and the film was then exposed to a brief flux of yellow light to initiate a

50% photoconversion from bR → M and the absorbance was measured a second time.

Use of bacteriorhodopsin in oriented PVA films as a medium for dynamic holograms and random access memory has been discussed.[212–216] The diffraction efficiency and nonlinear transmission properties of chemically enhanced thin films of bacteriorhodopsin are analyzed by using absorption spectroscopy, the Kramers-Kronig transformation, coupled-wave theory, and a simplified kinetic model of the bacteriorhodopsin photocycle. A maximum diffraction efficiency of 11% for a 2.5 OD, 150-μm-thick film occurs at read-out wavelengths between 620–700 nm. These films also exhibited significant nonlinearity in transmissivity at low laser intensities and could find potential use in spatial filtering applications.

As there is significant difficulty encountered in forming uniform, thermally stable, and robust BR films, another method has also been employed.[217] An oriented BR-PVA film was made from PVA with a molecular weight of 40,000 dissolved in a 50-mM *N*-2-hydroxyethylpiperazine-*N'*-2-ethanesulfonic acid (HEPES) buffer heating to 98° C. A volume of 1 ml of a 0.15-mM BR solution dissolved in the same buffer was mixed with 9 ml of the PVA solution after it had cooled down to room temperature. The BR-PVA solution was then degassed by spinning the solution at 5000 rpm in a centrifuge to remove any residual bubbles. The substrate was treated with Chromerge at ~80° C. A volume of 4 ml of the BR-PVA solution was then spread on a leveled glass substrate and filtered air was used to uniformly purge the film surface during the drying process.

Oriented samples of purple membrane have been obtained in a number of ways such as incorporation into positively charged bimolecular lipid membranes or in polyacrylamide gel and adsorption to cationic surfaces. In addition to these methods, dry films of oriented BR can also be made by electrophoresis of suspensions of purple membrane. Nondestructive reading was achieved with second harmonic (SH) microscopy. A problem with all photochromic materials in optical memories is that erasure of the stored information is unavoidable when using a beam for readout that is in resonance with one of the states. However, the large and different nonlinear properties of BR and M allow SH microscopy to be effectively applied to read an image without erasure. Chen et al. have succeeded in producing oriented BR films with high optical quality in which images can be impressed and erased over $>10^6$ cycles.[117]

During the last decade, attempts have been made to generate artificial derivatives of BR to meet the demands of holographic recording processes.[218] One possibility is chemical modifications of the chromophoric group.[219] However, these variants cannot be produced in substantial amounts by conventional biochemical methods. Alternatively, the amino acid sequence of BR can be changed by conventional mutagenesis of wild type bacteria and a sophisticated isolation procedure. The variant BR_{D96N} has been chosen, and this material has a strong retardation of the proton-dependent thermal relaxation of M-B. Therefore these films have an approximately 50% higher recording sensitivity and a twofold higher diffraction efficiency compared to wild type films.[220,221] Some interesting applications, such as dynamic optical filtering with spatial light modulators, and optical pattern recognition, can be visualized with these films.

3.3.4. Spectral Hole Burning and Data Storage in Dye Doped Polymers

The spectral hole burning technique to store the data in the form of a hologram has been employed extensively by the research groups of Wild[222–226] and others.[227] Spectral hole burning in combination with holography has not only generated interest as an experimental technique in fundamental research, but is also a promising approach in high-density optical storage.[225] Hole burning is a method that allows selective detection of the sharp homogeneously broadened line profile within an inhomogeneously broadened absorption band. The molecules that absorb the monochromatic light (on irradiation with a spectrally narrow laser beam) can undergo a photoreaction; and the photoproducts, which generally absorb at different spectral positions, will leave a dip or hole in the absorption band at the spectral position of the irradiation.[222] The dip in the absorption spectrum can be associated with a stored bit. Repeating the hole burning process at different spectral positions offers the possibility of high capacity information storage.[228]

It has also been demonstrated that by using an electric field, an additional dimension in data storage is possible.[222,229] In this regard, it has been shown that holography is a sensitive zero-background technique that can be advantageously employed in high-resolution spectroscopic studies of organic molecules in condensed phase.[230] For a detailed description of the phenomenon and the experimental set-up, interested readers are referred to ref. 225. The photochemical hole burning mechanism involved here is quite well understood.[231] A time- and space-domain holographic method that provides the storage and reproduction of the time dependence of optical signals with a duration of 10^{-8}–10^{-13}s, utilizing inhomogeneously broadened media has also been reported by Saari et al.[232]

Wild et al. used chlorin I (2,3-dihydroporphyrin) or oxazine 4 in a polyvinylbutyral (PVB) film at 4.2 K. Several holograms were stored in the same sample area using different electric field strengths and laser frequencies.[223] Holograms were recorded by simultaneously exposing the sample to object and reference beams and by adjusting the laser frequency and electric field applied to the sample to specific values, thus associating each hologram with specific values. A specific image was addressed by adjusting the electric field strength and laser frequency to the values used during recording. It was retrieved by illuminating the sample with an attenuated reference beam, and the diffracted light was detected by a TV camera system. Typical hologram efficiencies were of the order of 10^{-4}–10^{-3}. The different images were well separated by distances of 20 kV/cm in the electric field dimension corresponding to voltage stops of 200 V. It is interesting to note that the 2-D addressing in the electric field/optical frequency space allows storage of at least 50 holograms in the experimental range of 200 kV/cm $\times$ 30 GHz. Similar to the investigated systems, dye doped PVA films were used.[226–228]

3.3.5. *Other Materials—Sol-gel System*

A two-dimensional permanent transmission grating was formed on a novel polymer-gel composite film by ultrashort (~0.5 ps) and visible (~602 nm) pulsed laser radiation. With an arrangement of three non-coplanar coherent laser beams, two approaches were used to produce direct formation of a two-dimensional grating on the film. One approach was to expose the sample twice to different combinations of two beams; the other was to expose the sample to three laser beams simultaneously. The diffraction patterns and the relative intensity distributions for different order diffraction of the two-dimensional gratings were formed on the conjugated polymer, poly-*p*-phenylene vinylene (PPV)/sol-gel processed V_2O_5 oxide films. They were analyzed for the different two-beam combinations and relative orientations among the three laser beams.[233] It has been found that the PPV/V_2O_5-gel film and the PPV/silica film are good candidates for grating fabrication. The thickness of the film was ~1 μm. The relative absorption spectral distribution was 300 to 800 nm for a ~1-μm-thick PPV/V_2O_5-gel film sample. The measured threshold value was ~35 mJ/cm^2 for the pulse energy density or ~70 GW/cm^2 for the pulse power density. The best exposure time was 1–2 s for the PPV/V_2O_5-gel film and 3–5 s for the PPV/silica films. The diffraction pattern of the grating could be recorded with a vidiocon detector. The total diffraction efficiency for the incident probe laser beam into all the non-zero-order diffraction beams reached 48%.

An optical grating system was used for the following purposes: (1) as a dispersion element with high spectral resolution for spectroscopic purpose, (2) as a pulse compressor for generating ultrashort laser pulses, (3) as a key resonator element for distributed feedback and/or tunable laser systems, (4) as an input-output coupler in optical waveguide systems, and also (5) as a multiple beam splitter for a multichannel laser system or optical communication system. The formation of a two-dimensional grating results from the laser radiation without the involvement of any developing or chemical etching process. The instant absorption may lead to two possible manifestations. One is a permanent refractive index change within the film due to photorefractive, photochemical, or photothermal processes. The other is a surface-relief effect from photoablation (ther-

Table 1-4. Doped Polymer Systems

Material	Type of Modulation	Preparation Method	Thickness (μm)	Recording Wavelength (nm)	Sensitivity (mJ/cm^2)	Resolution (lines/mm)	Diffraction Efficiency %	Refs.
Azo dyes PVA	Polarization grating	Coated on glass	15–30	488	—	500–4000	0.3	203
MO/PVA	Polarization grating	Coated on glass	—	488	0.6	—	35	200
Poly-*p*-phenylene vinylene/V_2O_5	—	Sol-gel film	1	UV-602	—	—	48	233
Oxazine 4 in PVB	Spectral hole burning	Films	100	—	—	—	10^{-4}–10^{-3}	223
Acidified MR/PVA	Polarization grating	Coated films	30	488	—	500–4000	1	205
Azo side chain	Polarization holography	Coated films	0.5	514	—	1000	—	210
Bacteriorhodopsin		Coated films	200–500	350–620	—	>5000	10	211

mal vaporization) and/or photoionization processes. The optical microscopic examination shows that, for a longer exposure time, the surface-relief effect is observed. If the exposure time is short enough, no obvious surface-relief effect can be observed with ordinary microscope illumination.[233]

Table 1-4 summarizes the different properties and salient features of the discussed doped polymer systems.

4. Conclusion

A review of all the materials reported in the literature with all their ramifications has been presented. From the discussion it can be observed that dry film photopolymerizable compositions have an advantage over other existing materials with regard to energy and spectral sensitivity. Photocrosslinkable materials can produce high-efficiency holograms, but the energy sensitivity is still a limitation. Doped polymer materials can be very useful in applications of holographic interferometry, but still the diffraction efficiencies of recorded holograms in these media are extremely low. From all these considerations it can be concluded that an ideal material for holography is still to come.

Acknowledgments

The authors wish to express their gratitude to the authors of all the papers cited in this chapter and to SPIE, OSA, SPSE, and Springer-Verlag for permitting us to reproduce some of the figures. The authors wish to apologize to those whose articles inadvertently might not have been cited.

References

1. D. Gabor, *Nature* **161**, 777 (1948).
2. E. N. Leith and J. Upatneiks, *J. Opt. Soc. Am.* **52**, 1123 (1962).
3. Y. N. Denisyuk, *Soviet Phys-Doklady* **7**, 543 (1963).
4. P. J. Van Heerden, *Appl. Opt.* **7**, 393 (1963).
5. P. Hariharan, *Optical Holography: Principles, Technology and Applications* (Cambridge University Press, Cambridge, 1984), pp. 88–115.
6. L. Solymar and D. J. Cooke, *Volume Holography and Volume Gratings* (Academic Press, New York, 1981), pp. 254–304.
7. W. Gladden and R. D. Leighty, ''Recording Media,'' in *Handbook of Optical Holography*, H. J. Caulfield, ed. (Academic Press, New York, 1979), pp. 277–298.
8. H. M. Smith, *Holographic Recording Materials* (Springer-Verlag, Berlin, 1977).
9. R. J. Collier, C. B. Burckhardt, and L. H. Lin, *Optical Holography* (Academic Press, New York, 1971) pp. 265–336.
10. B. M. Monroe and W. K. Smothers, ''Photopolymers for Holography and Wave-guide Appli-

cations,'' in *Polymers for Lightwave and Integrated Optics:* Technology and Applications, L. A. Hornak, ed. (Marcel Dekker, New York, 1992) pp. 145–169.

11. R. A. Lessard, J. J. A. Couture, and P. Galarneau, ''Application of Third Order Non-linearities of Dyed PVA to Real Time Holography,'' in *Nonlinear Optical effects in Organic Polymers*, J. Messier, F. Kajzar, P. Prasad, and D. Ulrich, eds. (Kluwer Academic Publishers, Dordrecht, Boston, 1989), **162**, pp. 343–349.
12. J. Kosar, *Light Sensitive Systems* (John Wiley and Sons, New York, 1965).
13. S. I. Peredereeva, V. M. Kozenkov, and P. P. Kisilitsa, *Photopolymers for Holography* (Moscow, 1978), p. 51.
14. H. Thiry, *J. Photog. Sci.* **35**, 150 (1987).
15. S. V. Pappu, *Int. J. Optoelectronics. (UK)* **5**, 251 (1990).
16. P. Hariharan, *Opt. Eng.* **19**, 636 (1980).
17. R. A. Bartolini, *Proc. SPIE* **123**, 2 (1977).
18. R. A. Bartolini, H. A. Weaklien, and B. F. Williams, *Opt. Eng.* **15**, 99 (1976).
19. R. L. Kurtz and R. B. Owen, *Opt. Eng.* **14**, 393 (1975).
20. K. Biedermann, *Opt. Acta* **22**, 103 (1975).
21. J. Bordogna, S. A. Keneman, J. J. Amodei, *RCA Rev.* **33**, 227 (1972).
22. E. G. Ramberg, *RCA Rev.* **33**, 5 (1972).
23. J. C. Urbach, *Proc. SPIE* **25**, 17 (1971).
24. J. C. Urbach and R. W. Meier, *Appl. Opt.* **8**, 2269 (1969).
25. H. Kogelnik, *Bell Sys. Techn. J.* **48**, 2909 (1969).
26. G. A. Delzenne, ''Organic Photochemical Imaging Systems,'' in *Advances in Photochemistry*, J. N. Pitts, Jr., G. S. Hammond, and K. Gollnick, eds. (Wiley-Interscience, New York, 1980), **11**, pp. 1–103.
27. W. J. Tomlinson and E. A. Chandross, ''Organic Photochemical Refractive-Index Image Recording Systems,'' in *Advances in Photochemistry*, J. N. Pitts, Jr., G. S. Hammond, and K. Gollnick, eds. (Wiley-Interscience, New York, 1980), **12**, pp. 201–281.
28. C. M. Verber, R. E. Schwerzel, P. J. Perry, and R. A. Craig, *Holographic Recording Materials Development*, N.T.I.S. Rep. N76-23544 (1976).
29. W. S. Colburn, R. G. Zech, and L. M. Ralston, *Holographic Optical Elements*, Tech. Report AFAL, TR-72-409 (1973).
30. D. H. Close and A. Graube, *Materials for Holographic Optical Elements*, Tech. Report AFAL, TR-73-267 (1973).
31. D. H. Close, A. D. Jacobson, J. D. Margerum, R. G. Brault, and F. J. McClung, *Appl. Phys. Lett.* **14**, 159 (1969).
32. J. A. Jenney, *J. Opt. Soc. Am.* **60**, 1155 (1970).
33. J. A. Jenney, *Appl. Opt.* **11**, 1371 (1972).
34. J. A. Jenney, *J. Opt. Soc. Am.* **61**, 1116 (1971).
35. J. A. Jenney, ''Recent Developments in Photopolymer Holography,'' *Proc. SPIE* **25**, 105 (1971).
36. A. L. Mikaelian and V. A. Barachevsky, *Proc. SPIE* **1559**, *Photopolymer Device Physics, Chemistry and Applications II*, 246 (1991).

37. Y. B. Boiko and E. A. Tikhonov, *Sov. J. Quantum Eletron.* **11**, 492 (1981).

38. Y. B. Boiko, V. M. Granchak, I. I. Dilung, V. S. Solovjev, I. N. Sisakian, and V. A. Sojfer, *Proc. SPIE* **1238**, 253 (1990).

39. Y. B. Boiko, V. M. Granchak, I. I. Dilung, and V. Y. Mironchenko, *Proc. SPIE* **1238**, 258 (1990).

40. E. S. Gyulnazarov, V. V. Obukhovskii, and T. N. Smirnova, *Opt. Spectrosc.* **69**, 109 (1990).

41. E. S. Gyulnazarov, T. N. Smirnova, and E. A. Tikhonov, *Opt. Spectrosc.* **67**, 99 (1989).

42. K. Sukegawa, S. Sugawara, and K. Murase, *Electron. Commun. Jap.* **58-C**, 132 (1975).

43. K. Sukegawa, S. Sugawara, and K. Murase, *Rev. Elect. Comm. Labs.* **25**, 580 (1977).

44. S. Sugawara, K. Murase, and T. Kitayama, *Appl. Opt.* **14**, 378 (1975).

45. C. Carré, D. Ritzenthaler, D. J. Lougnot, and J. P. Fouassier, *Opt. Lett.* **12**, 646 (1987).

46. (a) D. J. Lougnot, D. Ritzenthaler, C. Carré, and J. P. Fouassier, *J. Appl. Phys.* **63**, 4841 (1988).
(b) J. P. Fouassier, C. Carré, and D. J. Lougnot, *Proc. SPIE* **1213**, *Photopolymer Device Physics, Chemistry and Applications*, 201 (1990).
(c) D. J. Lougnot, J. P. Fouassier, C. Carré, and P. Van de Walle, *Proc. SPIE* **1026**, 22 (1988).

47. D. J. Lougnot and C. Turck, *Pure Appl. Opt.* **1**, 251 (1992).

48. D. J. Lougnot and C. Turck, *Pure Appl. Opt.* **1**, 269 (1992).

49. C. Carré and D. J. Lougnot, *J. Optics* **21**, 147 (1990).

50. C. Carré, D. J. Lougnot, Y. Renotte, P. Leclère, and Y. Lion, *J. Optics* **23**, 73 (1992).

51. D. J. Lougnot, "Photopolymer Materials for Holographic Recording," *OPTO '92*, Paris, France 14-16 April, 99 (1992).

52. C. Careé and D. J. Lougnot, "Photosensitive Material for Holographic Recording in the Region of 500 nm Applications," *OPTO '91*, ESI publications, 317 (1991).

53. C. Carré, C. Maissiat, and P. Ambs, *Proc. OPTO '92*, Paris, 165 (1992).

54. C. Carré and D. J. Lougnot, "Developments of Recording Materials Based on the Photopolymerization of Monomers and Oligomers," *OPTO '90*, ESI publication, Paris, 541 (1990).

55. N. Noiret-Roumier, D. J. Lougnot, and I. Petitbon, "Photopolymer Materials for Holographic Recording in the Red and Near Infra-Red," *OPTO '92*, ESI publications, Paris, 104 (1992).

56. A. Yacoubian, G. Sawant, and T. M. Aye, *Proc. SPIE* **1559**, *Photopolymer Device Physics, Chemistry and Applications* II, 403 (1991).

57. B. L. Booth, *Appl. Opt.* **11**, 2994 (1972).

58. B. L. Booth, *Appl. Opt.* **14**, 593 (1975).

59. W. S. Colburn and K. A. Haines, *Appl. Opt.* **10**, 1636 (1971).

60. E. T. Kurtzner and K. A. Haines, *Appl. Opt.* **10**, 2194 (1971).

61. R. H. Wopschall and T. R. Pampalone, *Appl. Opt.* **11**, 2096 (1972).

62. B. M. Monroe, W. K. Smothers, D. E. Keys, R. R. Krebs, D. J. Mickish, A. F. Harrington, S. R. Schicker, M. K. Armstrong, D. M. T. Chan, and C. I. Weathers, *J. Imaging Sci.* **35**, 19 (1991).

63. B. M. Monroe, *J. Imaging Sci.* **35**, 25 (1991).

64. W. K. Smothers, B. M. Monroe, A. M. Weber, and D. E. Key, *Proc. SPIE* **1212**, Practical Holography IV, 20 (1990).

65. A. M. Weber, W. K. Smothers, T. J. Trout, and D. J. Mickish, *Proc. SPIE* **1212**, Practical Holography IV, 30 (1990).
66. K. Curtis and D. Psaltis, *Appl. Opt.* **31**, 7425 (1992).
67. R. T. Ingwall and H. L. Fielding, *Opt. Eng.* **24**, 808 (1985).
68. H. L. Fielding and R. T. Ingwall, U.S. Patent 4,588,664 (13 May 1986).
69. R. T. Ingwall and M. Troll, *Opt. Eng.* **28**, 586 (1989).
70. R. T. Ingwall and M. Troll, *Proc. SPIE* **883**, Holographic Optics: Design and Applications, 94 (1988).
71. R. T. Ingwall, M. Troll, and W. T. Vetterling, *Proc. SPIE* **747**, Practical Holography II, 67 (1987).
72. W. C. Hay and B. D. Guenther, *Proc. SPIE* **883**, Holographic Optics: Design and Applications, 102 (1988).
73. D. H. Whitney and R. T. Ingwall, *Proc. SPIE* **1213**, Photopolymer Device Physics, Chemistry and Applications, 18 (1990).
74. C. Zhang, M. Yu, Y. Yang, and S. Feng, *J. Photopolymer Sci. Tech.* **4**, 139 (1991).
75. W. J. Tomlinson, I. P. Kaminow, E. A. Chandross, R. L. Fork, and W. T. Silfvast, *Appl. Phys. Lett.* **16**, 486 (1970).
76. T. M. Moran and I. P. Kaminow, *Appl. Opt.* **12**, 1964 (1973).
77. M. J. Bowden, E. A. Chandross, and L. P. Kaminow, *Appl. Opt.* **13**, 112 (1974).
78. F. P. Laming, *Polym. Eng. & Sci.* **11**, 421 (1971).
79. J. Marotz, *Appl. Phys.* **B37**, 181 (1985).
80. T. Lückemeyer and H. Franke, *Appl. Phys.* **B46**, 157 (1988).
81. M. Kopietz, M. D. Lechner, and D. G. Steinmeier, *Eur. Polym. J.* **20**, 667 (1984).
82. A. Bloom, R. A. Bartolini, and H. A. Weakliem, *Opt. Eng.* **17**, 446 (1978).
83. R. A. Bartolini, A. Bloom, and H. A. Weakliem, *Appl. Opt.* **15**, 1261 (1976).
84. A. Bloom, R. A. Bartolini, and P. L. K. Hung, *Polym. Eng. Sci.* **17**, 356 (1977).
85. A. Bloom, R. A. Bartolini, and D. L. Ross, *Appl. Phy. Lett.* **24**, 612 (1974).
86. W. J. Tomlinson, E. A. Chandross, H. P. Weber, and G. D. Aumiller, *Appl. Opt.* **15**, 534 (1976).
87. E. A. Chandross, W. J. Tomlinson, and G. D. Aumiller, *Appl. Opt.* **17**, 566 (1978).
88. M. J. Jeudy and J. J. Robillard, *Opt. Comm.* **13**, 25 (1975).
89. R. L. Van Renesse, *Opt. Laser Tech.* **4,** 24 (1972).
90. N. Sadlej and B. Smolinska, *Opt. Laser Tech.* **7**, 175 (1975).
91. (a) S. Calixto, *Appl. Opt.* **26**, 3904 (1987).
    (b) Y. B. Boiko, V. S. Solovjev, S. Calixto and D. J. Lougnot, *Appl. Opt.*, **33**, 787 (1994).
92. M. Bravar, V. Rek, and R. Kostelac-Biffl, *J. Polym. Sci., Polym. Symp.* **40**, 19 (1973).
93. M. P. Ritt and L. M. Saulnier-Elbert, U.S. Patent 4,561,931A, RCA Corporation.
94. K. Schalaepfer, *Advan. Print. Sci. Technol.* **6**, 1 (1971).
95. A. M. Morrell, H. B. Law, E. G. Rambert, and E. W. Herold, *Color Television Picture Tubes* (Academic Press, New York, 1974).
96. F. Ziping, Z. Jugin, and H. Dahsiung, Guangxue Xuebao, *Optica Acta Sinica* **4**, 1101 (1984).

97. S. Lelièvre and J. J. A. Couture, *Appl. Opt.* 2**9**, 4384 (1990).

98. S. Lelièvre, "Holographie de polarisation au moyen de films de PVA bichromate", MS dissertation, Université Laval, Québec, Canada (1989).

99. (a) R. A. Lessard, R. Changkakoti, and G. Manivannan, *Proc. SPIE* **1559**, Photopolymer Device Physics, Chemistry and Applications II, 438 (1991).
(b) R. A. Lessard, R. Changkakoti, and G. Manivannan, *Optical Memory and Neural Networks* **1**, 75 (1992).
(c) G. Manivannan, R. Changkakoti, and R. A. Lessard, *Opt. Eng.* **32**, 671 (1993).
(d) R. A. Lessard, C. Malouin, R. Changkakoti, and G. Manivannan, *Opt. Eng.* **32**, 665 (1993).
(e) G. Manivannan, R. Changkakoti and R. A. Lessard, *Polym. for Adv. Technologies*, **4**, 569 (1993).

100. (a) G. K. Oster and G. Oster, *J. Am. Chem. Soc.* **81**, 5543 (1959).
(b) R. H. Kayser and R. H. Young, *Photochem. Photobiol.* **24**, 395 (1976).

101. R. Changkakoti, S. S. C. Babu, and S. V. Pappu, *Appl. Opt.* **27**, 324 (1988).

102. J. Blyth, *Appl. Opt.* **30**, 1598 (1991).

103. R. A. Lessard, N. Capolla, R. Changkakoti, and G. Manivannan, *Proc. SPIE* **1731**, Holography and Optical Information Processing, 99 (1991).

104. M. Mazakova and M. Pantcheva, *J. Inf. Rec. Mater.* **18**, 191 (1990).

105. F. H. Westheimer, *Chem. Rev.* **45**, 419 (1949).

106. B. Duncalf and A. S. Dunn, *J. Appl. Polym. Sci.* **8**, 1763 (1964).

107. M. Pourbaix, *Atlas d'équilibres électrochimiques à 25° C*, (Gauthier-Villars, Paris, 1963), p. 258.

108. *Handbook of Chemistry and Physics*, 64th Edition (CRC, Cleveland 1983-84), p. D169.

109. H. G. Linge, A. L. Jones, *Aust. J. Chem.* **21**, 2189 (1968).

110. J. Y. P. Tong, E. L. King, *J. Am. Chem. Soc.* **75**, 6180 (1953).

111. P. Fageol, M. Bolte, and J. Lemaire, *J. Phys. Chem.* **92**, 239 (1988).

112. (a) G. Manivannan, R. Changkakoti, R. A. Lessard, G. Mailhot, and M. Bolte, *Proc. SPIE* **1774**, Nonconducting Photopolymers and Applications, 24 (1992).
(b) G. Manivannan, R. Changkakoti, R. A. Lessard, G. Mailhot, and M. Bolte, *J. Phys. Chem.* **97**, 7228 (1993).

113. F. Trépanier, G. Manivannan, R. Changkakoti, and R. A. Lessard, *Can. J. Phys.*, **71**, 423 (1993).

114. B. A. Budkevich, A. M. Polikanin, V. A. Pilipovich, and N. Ya. Petrochenko, *Optical Spectroscopy* (translated from Zhurnal Prikladnoi Spectroskopii in Russian), **50**, 621 (1989).

115. R. Changkakoti, G. Manivannan, A. Singh, and R. A. Lessard, *Opt. Eng.* **32**, 2240 (1993).

116. J. J. A. Couture and R. A. Lessard, *Can. J. Phys.* **64**, 553 (1986).

117. V. F. Plyusnin and N. M. Bazhin, *Zh. Nauchn. Prikl Fotogr. Kinematogr.* **25**, 90 (1980).

118. O. F. Syrets, V. V. Sviridov, V. G. Guslev et al., *Zh. Nauchn. Prikl Fotogr. Kinematogr.* **30**, 408 (1985).

119. Y. Yamagishi, T. Ishizuka, T. Yagishita, K. Ikegami, and H. Okuyama, *Proc. SPIE* **600**, Progress in Holographic Applications, **14** (1985).

120. K. Matsumoto, T. Kuwayama, M. Matsumoto, and N. Taniguchi, *Proc. SPIE* **600**, Progress in Holographic Applications, **9** (1985).

121. T. Kuwayama, N. Taniguchi, Y. Kuwae, and N. Kushibiki, *Appl. Opt.* **28**, 2455 (1989).
122. Y. L. Freilich, M. Levy, and S. Reich, *J. Polym. Sci., Chem. Ed.* **15**, 1811 (1977).
123. A. A. Friesem, Z. Rav-Noy, and S. Reich, *Appl. Opt.* **16**, 427 (1977).
124. T. Ganko, N. Sadlej, and B. Smolinska, *Appl. Opt.* **13**, 2770 (1974).
125. T. A. Shankoff, *Appl. Opt.* **7**, 2101 (1968).
126. B. J. Chang and C. D. Leonard, *Appl. Opt.* **18**, 2407 (1979).
127. D. Meyerhofer, *RCA Review* **33**, 110 (1972).
128. B. J. Chang, *Optics Commun.* **17**, 270 (1976).
129. R. Changkakoti and S. V. Pappu, *Appl. Opt.* **28**, 340 (1989).
130. L. H. Lin., *Appl. Opt.* **8**, 963 (1969).
131. J. Oliva, P. G. Boj, and M. Pardo, *Appl. Opt.* **23**, 196 (1984).
132. R. G. Brandes, E. E. Francois, and T. A. Shankoff, *Appl. Opt.* **8**, 2346 (1969).
133. D. M. Samoilovich, A. Zeichner, and A. A. Freisem, *Photographic Sci. and Eng.* **24**, 161 (1980).
134. (a) M. Mazakova, M. Pancheva, P. Kandilarov, and P. Sharlandjiev, *Opt. Quant. Elect.* **14**, 311 (1982).
(b) M. Mazakova, M. Pancheva, P. Kandilarov, and P. Sharlandjiev, *Opt. Quant. Electr.* **14**, 317 (1982).
135. T. Kienonen, ''On dichromated gelatin as a recording material,'' Ph.D. dissertation, University of Joensuu, Finland (1983).
136. S. Sjölinder, *J. Imag. Sci.* **30**, 151 (1986).
137. S. Sjölinder, *Photographic Sci. and Eng.* **28**, 180 (1984).
138. M. Chang, *Appl. Opt.* **10**, 2550 (1971).
139. S. P. McGrew, *Proc. SPIE* **215**, Recent Advances in Holography, 24 (1980).
140. C. Solano, R. A. Lessard, and P. C. Roberge, *Appl. Opt.* **2**4, 1189 (1985).
141. J. R. Magarinos and D. J. Coleman, *Opt. Eng.* **24**, 769 (1985).
142. C. Solano, R. A. Lessard, and P. C. Roberge, *Appl. Opt.* **26**, 1989 (1987).
143. A. Graube, *Opt. Comm.* **8**, 251 (1973).
144. M. Akagi, *Photographic Sci. and Eng.* **18**, 248 (1974).
145. T. Kubota, T. Ose, M. Sasaki, and K. Honda, *Appl. Opt.* **15**, 556 (1976).
146. A. Graube, *Photographic Sci. and Eng.* **22**, 39 (1978).
147. T. Kubota and T. Ose, *Opt. Letts.* **4**, 289 (1979).
148. T. Kubota and T. Ose, *Appl. Opt.* **18**, 2538 (1979).
149. C. Solano and R. A. Lessard, *Appl. Opt.* **24**, 1776 (1985).
150. R. Changkakoti and S. V. Pappu, *Appl. Opt.* **25**, 798 (1986).
151. N. Capolla and R. A. Lessard, *Appl. Opt.* **27**, 3008 (1988).
152. R. Changkakoti and S. V. Pappu, *Opt. Laser Tech.* **21**, 259 (1989).
153. C. Solano, *Appl. Opt.* **28**, 3524 (1989).
154. T. Mizuno, T. Goto, M. Goto, K. Matsui, and T. Kubota, *Appl. Opt.* **29**, 4757 (1990).
155. S. V. Pappu and R. Changkakoti, *Proc. SPIE* **1213**, Photopolymer Device Physics, Chemistry and Applications, 39 (1990).

156. N. Capolla and R. A. Lessard, *Proc. SPIE* **1389**, Microelectronic Interconnects and Packages: Optical and Electrical Technologies, 612 (1990).
157. N. Capolla, R. A. Lessard, C. Carré, and D. J. Lougnot, *Appl. Phys.* **B52**, 326 (1991).
158. M. Mazakova, P. Sharlanjiev, M. Pancheva, and G. Spassov, *Appl. Opt.* **24**, 2156 (1985).
159. T. P. Sosnowski and H. Kogelnik, *Appl. Opt.* **9**, 2186 (1970).
160. T. G. Georgekutty and H. K. Liu, *Appl. Opt.* **26**, 372 (1987).
161. J. M. Rebardâo and A. A. Andrade, *Appl. Opt.* **28**, 4393 (1989).
162. M. H. Jeong, J. B. Song, and I. W. Lee, *Appl. Opt.* **30**, 4172 (1991).
163. B. J. Chang, *Opt. Eng.* **19**, 642 (1980).
164. R. K. Howard, *The Marconi Review*, Third Quarter, 179 (1979).
165. M. H. Jeong and J. B. Song, *Appl. Opt.* **31**, 161 (1991).
166. D. G. McCaulay, C. E. Simpson, and W. J. Murbach, *Appl. Opt.* **12**, 232 (1973).
167. T. Keinonen and O. Salminen, *Appl. Opt.* **27**, 2573 (1988).
168. T. Kubota, *Appl. Opt.* **25**, 4141 (1986).
169. A. M. Kursakova, M. N. Meshalkina, V. V. Smirnov, M. K. Topunova, and T. V. Shedrunova, *Opt. Spectr.* **69**, 539 (1990).
170. Y. E. Kuzilin, Y. B. Melnichenko, and V. V. Shiliov, *Opt. Spectr.* **69**, 106 (1990).
171. B. O. Maier, D. I. Staselko, and L. A. Yurlova, *Opt. Spectr.* **56**, 676 (1984).
172. V. N. Krylov, V. N. Sizov, G. A. Sobolev, S. B. Soboleva, D. I. Staselko, and M. K. Shevtsov, *Opt. Spectr.* **69**, 115 (1990).
173. T. Y. Kalnitskaya, A. M. Kursakova, V. V. Smirnov, M. K. Topunova, and T. V. Shchedrunova, *Opt. Spectr.* **67**, 121 (1989).
174. P. G. Boj, J. Crespo, and J. A. Quintana, *Appl. Opt.* **31**, 3302 (1992).
175. O. Salminen and T. Keinonen, *Optica Acta* **29**, 531 (1982).
176. L. T. Blair and L. Solymar, *Appl. Opt.* **30**, 775 (1991).
177. S. Calixto and R. A. Lessard, *Appl. Opt.* **23**, 1989 (1984).
178. Y. Ishii and K. Murata, *Appl. Opt.* **23**, 1999 (1984).
179. J. L. Horner and J. E. Ludman, *Proc. SPIE* **215**, Recent Advances in Holography, 46 (1980).
180. D. J. McCartney, D. B. Payne, and S. S. Duncan, *Opt. Lett.* **10**, 303 (1985).
181. S. S. Duncan, J. A. McQuoid, and D. J. McCartney, *Opt. Eng.* **24**, 781 (1985).
182. D. Meyerhofer, *Appl. Opt.* **10**, 416 (1970).
183. Y. Z. Liang, D. Zhao, and H. K. Liu, *Appl. Opt.* **22**, 3451 (1983).
184. Y. G. Jiang, *Appl. Opt.* **21**, 3138 (1982).
185. J. Oliva, A. Fimia, and J. A. Quintan, *Appl. Opt.* **21**, 2891 (1982).
186. Y. Z. Liang and H. K. Liu, *Opt. Lett.* **9**, 627 (1984).
187. B. Robertson, M. R. Taghizadeh, J. Turunen, and A. Vasara, *Appl. Opt.* **29**, 1134 (1990).
188. B. Robertson, E. J. Restall, M. R. Taghizadeh, and A. C. Walker, *Appl. Opt.* **30**, 2368 (1991).
189. K. H. Brenner and F. Sauer, *Appl. Opt.* **27**, 4251 (1988).
190. F. Sauer, *Appl. Opt.* **28**, 386 (1989).
191. R. K. Kostuk, M. Kato, and Y. T. Huang, *Appl. Opt.* **29**, 3848 (1990).

192. J. B. McManus, R. S. Putnam, and H. J. Caulfield, *Appl. Opt.* **27**, 4244 (1988).
193. F. Lin, E. M. Strzelecki, C. Nguyen, and T. Jannson, *Opt. Lett.* **16**, 183 (1991).
194. M. R. Wang, G. J. Sonek, R. T. Chen, and T. Jannson, *Appl. Opt.* **31**, 236 (1992).
195. R. K. Curran and T. A. Shankoff, *Appl. Opt.* **9**, 1651 (1970).
196. Sh. D. Kakichashvili, *Opt. Spektrosk.* **33**, 324 (1972) [*Opt. Spectrosc.* **33**, 171 (1972)].
197. L. Nikolova and T. Todorov, *Opt. Acta* **24**, 1179 (1977).
198. T. Todorov, L. Nikolova, N. Tomova, and V. Dragostinova, *Opt. Quantum Electron.* **13**, 203 (1981).
199. T. Todorov, N. Tomova, and L. Nikolova, *Opt. Commun.* **47**, 123 (1983).
200. T. Todorov, L. Nikolova, and N. Tomova, *Appl. Opt.* **23**, 4309 (1984).
201. T. Todorov, L. Nikolova, and N. Tomova, *Appl. Opt.* **23**, 4588 (1984).
202. T. Todorov, L. Nikolova, K. Stoyanova, and N. Tomova, *Appl. Opt.* **24**, 785 (1985).
203. T. Todorov, L. Nikolova, N. Tomova, and V. Dragostinova, *IEEE J. Quantum Electron.* QE-**22**, 1262 (1986).
204. L. Nikolova and T. Todorov, *Optica Acta* **31**, 579 (1984).
205. J. J. A. Couture and R. A. Lessard, *Appl. Opt.* **27**, 3368 (1988).
206. C. R. Noller, *Chemistry of Organic Compounds* (Saunders, Philadelphia, 1957), p. 699.
207. A. Streitweiser and C. H. Heathcock, *Introduction to Organic Chemistry* (Macmillan, New York, 1985), pp. 751–52.
208. J. March, *Advanced Organic Chemistry* (Wiley-Interscience, New York, 1985), pp. 110 and 215.
209. C. R. Noller, *Chemistry of Organic Compounds* (Saunders, Philadelphia, 1957), p. 663.
210. A. L. Natansohn, P. L. Rochon, S. Zie, Queen's University, Kingston, Canada, U.S. Patent 5,173,381 (22 Dec., 1992).
211. R. B. Gross, K. Can Izgi, and R. R. Birge, *Proc. SPIE* **1662**, Image storage and retrieval systems, 1 (1992).
212. R. R. Birge, C. F. Zhang, and A. L. Lawrence, ''Optimal Random Access Memory Based on Bacteriorhodopsin,'' *Proceedings of the Fine Particle Society*, Santa Clara, CA, 1988.
213. G. Rayfields, *Phys. Bull.* **34**, 483 (1989).
214. N. N. Vsebolodov and G. R. Ivanitsky, *Biofizika* **30,** 883 (1985).
215. N. N. Vsevolodov and V. A. Poltoratski, *Sov. Phys. Tech. Phys.* **30**, 1235 (1985).
216. A. Lewis and V. del Priore, *Phys. Today* **41**, 38 (1988).
217. Z. Chen, A. Lewis, H. Takei, and I. Nebenzahl, *Appl. Opt.* **30**, 5188 (1991).
218. Ch. Brauchle and N. Hampp, *Makromol. Chem., Macromol. Symp.* **50**, 97 (1991).
219. E. Kölling, W. Gärtner, D. Oesterhelt, and E. Ludger, *Agnew. Chemie* (Int. Ed. Engl.), **23**, 81 (1984).
220. N. Hampp, C. Brauchle, and D. Oesterhelt, *Biophys. J.* **58,** 83 (1990).
221. N. Hampp and C. Brauchle, *Photochromism*, H. Dürr and H. Bouas-Laurent, eds., (Elsevier, Amsterdam, 1990), p. 954.
222. U. P. Wild, S. E. Bucher, and F. A. Burkhalter, *Appl. Opt.* **24**, 1526 (1985).
223. A. Renn and U. P. Wild, *Appl. Opt.* **26**, 4040 (1987).

224. U. P. Wild, A. Renn, C. De Caro, and S. Bernet, *Appl. Opt.* **29**, 4329 (1990).

225. U. P. Wild and A. Renn, *J. Mol. Electronics* **7**, 1 (1991).

226. K. Holliday, C. Wei, A. J. Meixner, and U. P. Wild, *J. Luminescence* **48** and **49**, 329 (1991).

227. (a) Y. Kanematsu, R. Shiraishi, S. Saikan, and T. Kushida, *Chem. Phys. Lett.* **147**, 53 (1988).
(b) Y. Kanematsu, R. Shiraishi, S. Saikan, and T. Kushida, *IQEC '88 Tokyo Technical Digest*, 406 (1988).
(c) Y. Kanematsu, R. Shiraishi, A. Imaoka, S. Saikan, and T. Kushida, *J. Chem. Phys.* **91**, 6579 (1989).

228. G. Gastro, D. Haarer, R. M. Macfarlane, and H.-P. Trommsdorff, U.S. Patent 4,101,976 (1976).

229. U. Bogner, K. Beck, and M. Maier, *Appl. Phys. Lett.* **46**, 534 (1985).

230. A. J. Meixner, A. Renn, S. E. Bucher, and U. P. Wild, *J. Phys. Chem.* **90**, 6777 (1986).

231. S. Völker and R. M. Macfarlane, *IBM J. Res. Devel.* **23**, 547 (1979).

232. P. Saari, R. Kaarli, and A. Rebane, *J. Opt. Soc. Am.* **B3**, 527 (1986).

233. G. S. He, C. J. Wung, G. C. Xu, and P. N. Prasad, *Appl. Opt.* **30**, 3810 (1991).

2

Photoresists and Their Development

Nigel P. Hacker

1. Abstract

A review of the chemistry associated with photoresists based on photochemical generation of carboxylic acids from diazoketones and strong Brönsted acids from aromatic onium salts is presented. The mechanisms for changes in dissolution properties for these photopolymers, as well as the underlying photochemistry that initiates these changes, are discussed. The photoinitiator, or photoactive compounds, and the polymers used in photoresists have traditionally been considered as separate entities. Thus the photoinitiator absorbs light and undergoes a chemical reaction that results in a change in dissolution of the photopolymer formulation. This chapter will describe the role of the polymer matrix for the dissolution changes of diazoketone resists and the role of the polymer in the photochemical generation of Brönsted acids from onium salts. It will be shown that the photochemistry and photophysics of both the photoinitiator and the matrix polymer have profound effects on the performance of photoresists.

2. Introduction

The invention of resists for microlithography was derived from earlier chemistries primarily designed for use in the printing industry.[1] The demands of the electronics industry drove the development of new resists for the manufacture of circuit boards and silicon chips. The latter require very stringent feature sizes for larger circuit density and have pushed microlithography into the sub-micron regime. A schematic of a photoresist process is shown in Fig. 2-1. The resist is

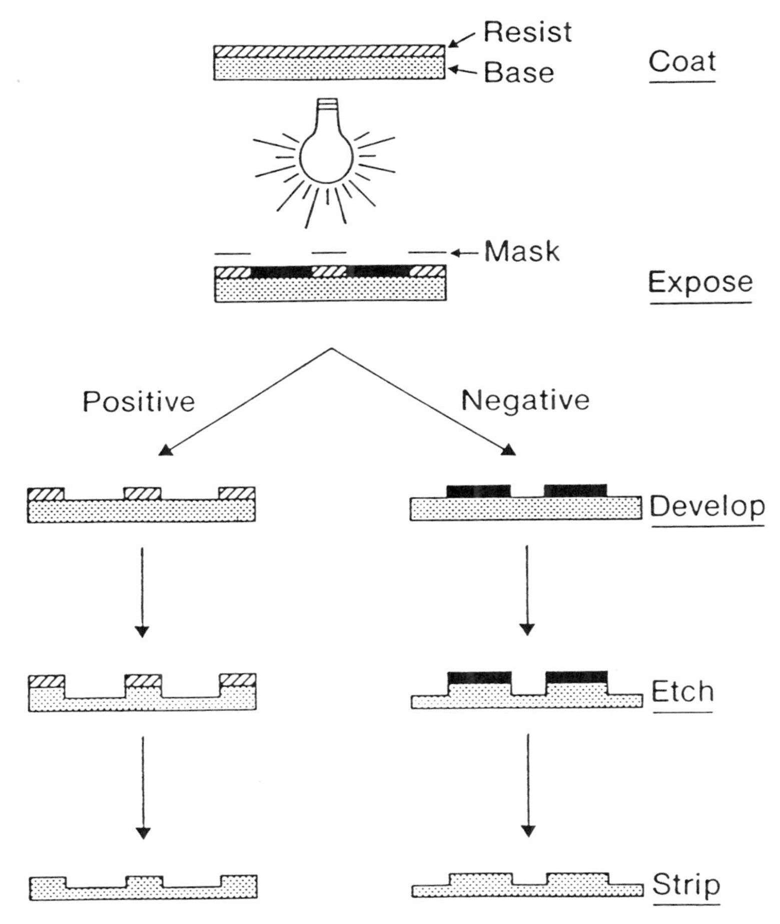

Figure 2-1. *The photolithographic process for positive and negative photoresists. The metallization process is not shown (reproduced from Ref. 2).*

coated onto a substrate and exposed to light through a mask. The incident light causes a chemical reaction to occur that renders the exposed areas more soluble (positive resist), or less soluble (negative resist), than the unexposed areas upon development with a suitable solvent. The exposed substrate is then subjected to etching or metallization processes and finally the resist is stripped. The product from this process is a selectively etched, or metallized, substrate that is an imaged reproduction of the mask artwork in the positive or negative mode. Willson has described a hierarchy for resists as a positive or negative tone, and a one component or two component process.[2] Today the resist hierarchy is much more complex. Most resists are multicomponent formulations, requiring top coats to act as contamination barriers, bottom coats to prevent light reflection from the substrate into the resist, and adhesion promoters. All of these components help meet the demands for smaller feature sizes (currently less than 0.2 μM in research laboratories).

Rather than review resist chemistry as a whole, this chapter will describe, from a mechanistic standpoint, the design of resists based on the more traditional processes derived from formation of carboxylic acids using diazoketone chemistry, as well as the more recent developments of acid catalyzed "chemically amplified" resists. These quite different chemical approaches to making photopolymers represent the current and future technologies of photoresists.

3. Diazoketones

3.1. Photochemistry

Resists based on diazonaphthoquinones have been "the workhorses" for resist processing and currently account for the bulk of the 2300 tons/year resist market.[3] This type of resist represents the classical photochemical concept, a one photon-one chemical event process. The 2-diazonaphthoquinone (2-DNQ) is formulated with Novolac resin, spin coated onto a silicon wafer, baked, and exposed to light, and the resist is developed with an aqueous base that selectively dissolves the exposed regions. The photolysis of DNQ was originally studied by Sus who reported that the reaction proceeded by loss of nitrogen followed by a Wolff rearrangement to give an indene ketene that reacted in the presence of water to give the 1-indene carboxylic acid (1 of Fig. 2-2).[4] It is now known that the product is the isomeric 3-indenecarboxylic acid (3 of Fig. 2-2).[5] The 3-indenecarboxylic acid could form by isomerization of 1-indenecarboxylic acid or by direct reaction with water.[5,6] Quantum yields for the photolysis of DNQ are reported to be 0.15–0.30. In addition to the formation of 3-indenecarboxylates, photolysis of 2-DNQ also gives 1-hydroxynaphthalene and 2-alkoxy-1-hydroxynaphthalenes as minor products in alcoholic solvents.

There is strong evidence for formation of the indene ketene intermediate (2) from DNQ photolysis from both product and spectroscopic studies. The observation that photolysis of both 2-DNQ or 1-DNQ give identical indene carboxylic acid derivatives, in water or alcohol solvents, suggests a common intermediate from two isomeric precursors.[5] Photolysis of 2-DNQ and 1-DNQ at 77 K in hydrocarbon glass gives an intermediate with UV absorption bands at 310 nm and 250 nm ($\epsilon = 5.4 \times 10^3\ M^{-1}\ cm^{-1}$ and $1.6 \times 10^4\ M^{-1}\ cm^{-1}$, respectively) that disappear upon warming the glass to room temperature.[7] Infrared spectros-

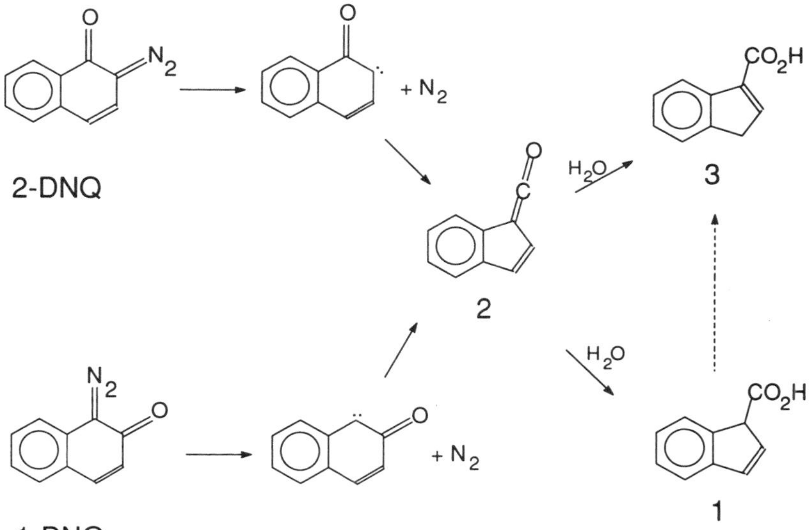

Figure 2-2. *Photochemical Wolff rearrangement of 2-diazo- and 1-diazonaphthoquinones.*

copy detects a common intermediate when 2-DNQ and 1-DNQ are photolyzed at 77 K.[7] Also substituted DNQs give intermediates with ketene stretching frequencies.[7–10] It has been reported that photolysis of a thin film of 2-DNQ at 77 K results in a loss of bands at 2151 cm^{-1} and 2122 cm^{-1} (C = N = N stretch) and 1615 cm^{-1} (C = O stretch) accompanied by the formation of new bands at 2130 cm^{-1}, 2115 cm^{-1}, (C = C = O stretch) 1450 cm^{-1}, 770 cm^{-1}, 745 cm^{-1}, and 705 cm^{-1}.[7] The latter series of peaks is also formed from photolysis of 1-DNQ. Later experiments in argon matrices at 20 K report similar IR and UV spectra for the ketene intermediate, and a number of sulfonate substituted indene ketenes have also been detected by IR spectroscopy at low temperatures.[8–10]

Flash photolysis experiments with sulfonated DNQs in aqueous solution have given a transient that absorbs at <330 nm with a lifetime of 2 ms and a second transient at 350 nm with a lifetime of 40 ms.[11,12] The absorption of the latter transient increases with the addition of water. While both of these absorbances are red-shifted from those reported for the parent indene ketene, it should be noted that the ketene observed from photolysis of 2-DNQ-5-sulfonylchloride has red-shifted UV absorbances at 317 nm and 264 nm.[7] The transients both decayed with pseudo first-order kinetics in quenching experiments that were run with water at concentrations from 10^{-6} to 1 M. A plot of the pseudo first-order rate constant versus the square of the water concentration for the <330-nm transient was linear. The formation of a ketene complex (4a or 4b) with water dimer was proposed to account for the second-order reaction and thus the 350-nm species was assigned to this complex (Fig. 2-3).[11] A second group studied the laser flash photolysis of similar DNQs in water and assigned the two UV transients to an oxirane and ketene respectively.[12] The assignment of the shorter lived transient as an oxirane is unlikely according to results obtained from isotopic labeling experiments (vide infra).

Thus, while product analysis, flash photolysis, and low temperature spectroscopy present strong evidence for the generation of ketene from photolysis of DNQs, there is less spectroscopic evidence concerning the events prior to ketene

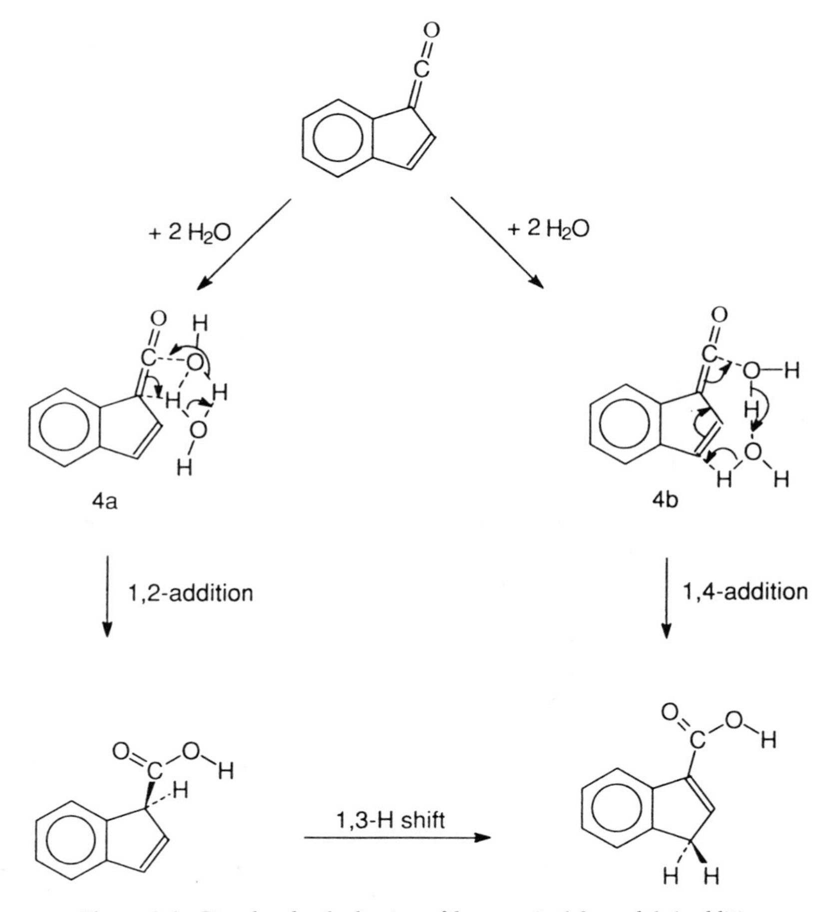

Figure 2-3. *Bimolecular hydration of ketone via 1,2- and 1,4-addition.*

formation. The detection of 3-indenecarboxylic acid from photoproduct studies is rationalized as a product of Wolff rearrangement from the singlet excited state of the DNQ. This reaction may be concerted from $[DNQ]^1$ or proceed via a ketocarbene.[3–5] Evidence for ketocarbene (5) comes from the detection of 1-hydroxy-2-alkoxynaphthalenes (1,2-HAN) from photolysis or thermolysis reactions of 2-DNQ in methanol, ethanol, 2-propanol, and *tert*-butanol (Fig. 2-4).[4] 1,2-HANs are formed from reaction of ketocarbene (5) with alcohol through intermediate yields or by a concerted mechanism through a three-membered ring transition state. Also, photolysis of 2-DNQ gives 1-hydroxynaphthalene, the formation of which is quenched in the presence of added oxygen and thus may be formed from the triplet manifold. The triplet intermediate is either a triplet ketocarbene or a triplet diradical; the latter species is expected to abstract hydrogen from the solvent. The yield of 1-hydroxynaphthalene increases with H donor ability of the solvent; i.e., 2-propanol > ethanol > methanol which follows the

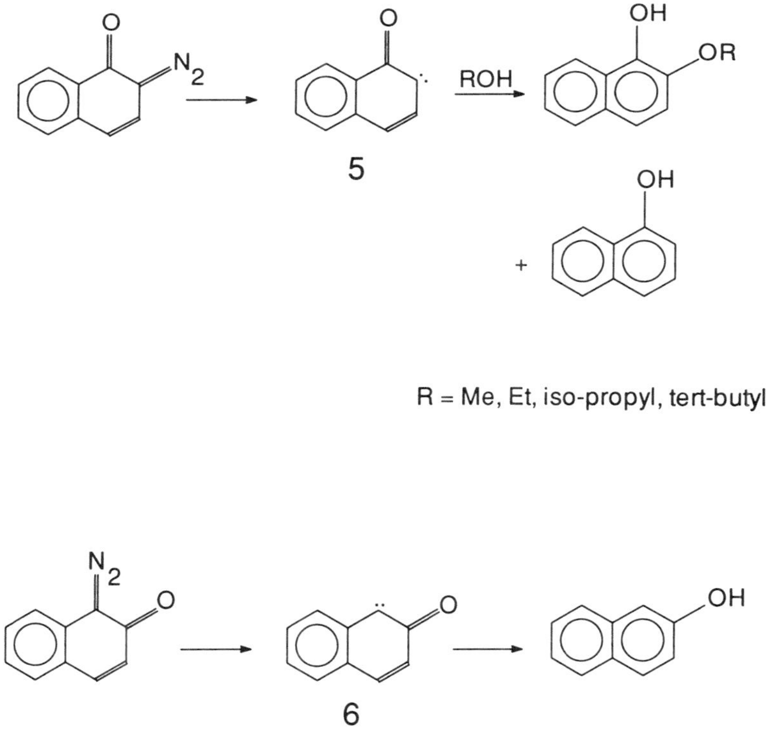

Figure 2-4. *Reaction of ketocarbene isomers with alcohols. 1- or 2-Hydroxynaphthalenes are not formed in* tert-*butanol (see text).*

trend for C—H homolysis, $E_{C—H}$: 3.8, 4.0, 4.1 eV respectively. Also 1-hydroxynaphthalene is not formed in solvents that are poor hydrogen donors, e.g., water or *tert*-butanol ($E_{O—H}$: 4.8 eV). Photolysis of 1-DNQ shows similar trends in these solvents except that 2-hydroxynaphthalene, formed from the reaction of ketocarbene (6) with solvent, is now the triplet product and 2,1-HANs are not formed, probably because of steric crowding at the 1-position of the naphthalene ring.

Although photoproduct studies present evidence for ketocarbene intermediates (5 or 6) from photolysis of DNQs at room temperature, these intermediates have eluded detection by low temperature spectroscopy or flash photolysis. It has been reported that oxiranes have been detected from the photolysis of certain diazoketones.[13] However the assignment of either the <330-nm or 350-nm transients from the laser flash photolysis studies to the oxirane of naphthalene is questionable because of the previously reported isotopic labeling experiments.[14,15] 1-DNQ and 2-DNQ, both labeled with ^{13}C in the 1-position, were synthesized and photolysis of these compounds gave indenecarboxylic acids labeled in different positions (Fig. 2-5). Thus the 2-DNQ compound gives 3-indenecarboxylic acid with the label exclusively on the carboxyl group, whereas the 1-DNQ puts the label in the indene ring at the 3-position. If oxirane (7) were an intermediate, it would be expected that there would be some scrambling of the ^{13}C label in these two positions from each precursor.

Finally, a peculiarity of aromatic esters of 2-DNQ-4-sulfonic acid is that in addition to rearrangement to the 3-indenecarboxylic acid derivatives, the ester group is also cleaved to give the free sulfonic acid upon photolysis.[16] It is proposed that the reaction proceeds via a sulfene intermediate (8), which adds water to give the sulfonic acid (Fig. 2-6). The ester cleavage reaction does not occur with DNQ-5-sulfonic acid derivatives.

3.2. Resist Chemistry

A typical resist based on diazoketone chemistry is usually formulated with a 2-DNQ-sulfonic acid derivative at loadings of up to 20% in a Novolac resin. Novolac is a general term for a resin formed from condensation of formaldehyde with derivatives of phenol. A Novolac resin used in resist systems is usually an acid-catalyzed condensation product from *meta*- and *para*-cresol and formaldehyde with a number average molecular weight M_n of 1000–3000 which is about 8–20 repeat units (Fig. 82-7). Phenol is not used for resists because it has three reactive sites that can crosslink the resin and *para*-cresol is not usually used by itself because it is unreactive in the condensation reaction. Dammel has described in detail the art of preparing Novolacs for resist applications.[3] Thus chemically, Novolac is an oligomeric crosslinked phenolic resin optimized for

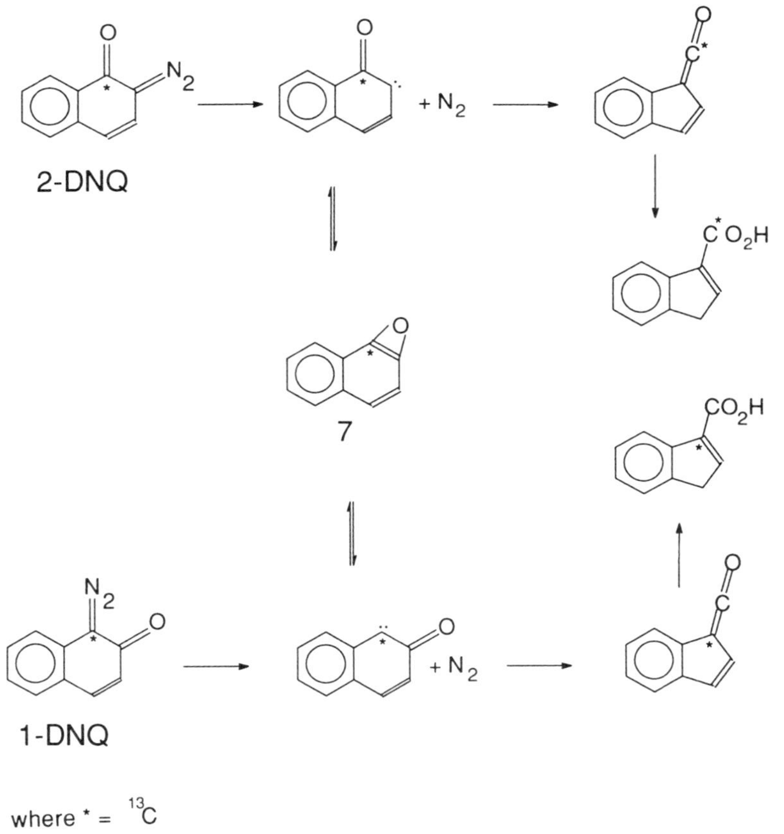

2-DNQ

+ N_2

$\overset{*}{C}O_2H$

CO_2H

7

1-DNQ

+ N_2

where * = ^{13}C

Figure 2-5. *Photochemistry of ^{13}C-labeled diazonaphthoquinones.*

dissolution properties, compatibility with DNQs, T_g, adhesion, spin-coating characteristics, and UV transmission.

In addition to the basic photochemical reactions described above, there are a number of side reactions that occur with DNQ in the resist process. Novolac is a phenolic medium and thus the intermediates produced from photolysis of DNQs may react with the polymer. For example, the ketocarbene intermediate could become covalently attached to Novolac by a reaction analogous to the formation of 1,2-HANs from the photolysis of 2-DNQ in alcoholic solvents, and the ketene intermediate can react with the O—H bond in Novolac to give a phenolate ester (Fig. 2-8). There are other functionalities in Novolac besides the O—H that may react with the ketocarbene. For example the bridging methylene

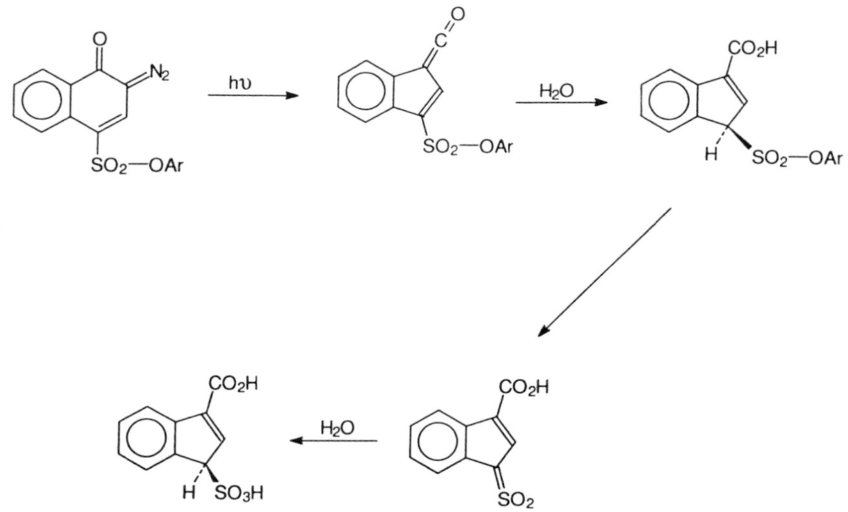

Figure 2-6. *Photochemical generation of sulfonic acid from 2-diazonaphthoquinone-4-sulfonates.*

groups of Novolac, or the pendant methyl groups derived from the cresol monomer, can undergo hydrogen abstraction with the ketocarbene to give 1-hydroxynaphthalene and a diphenylmethyl, or a benzyl radical, respectively. These radicals may result in crosslinking of the resin. Other side reactions include formation of methylene dioxole (9), which can be visualized as an insertion product between ketocarbene and ketene, the formation of azodyes, and decarboxylation to give indene, or indene dimer (Fig. 2-8). All of these reactions can

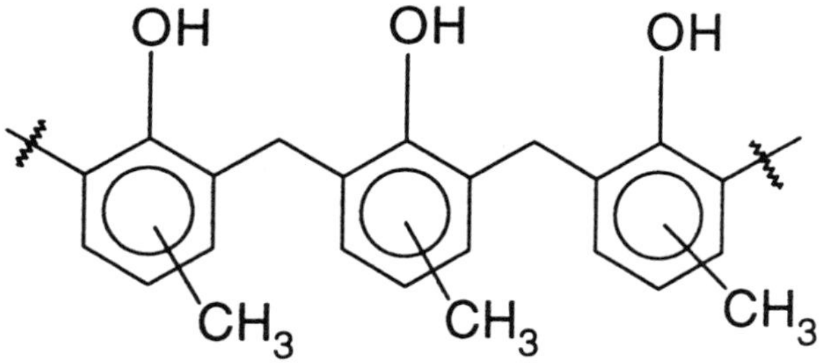

Figure 2-7. *Structure of Novolac resin.*

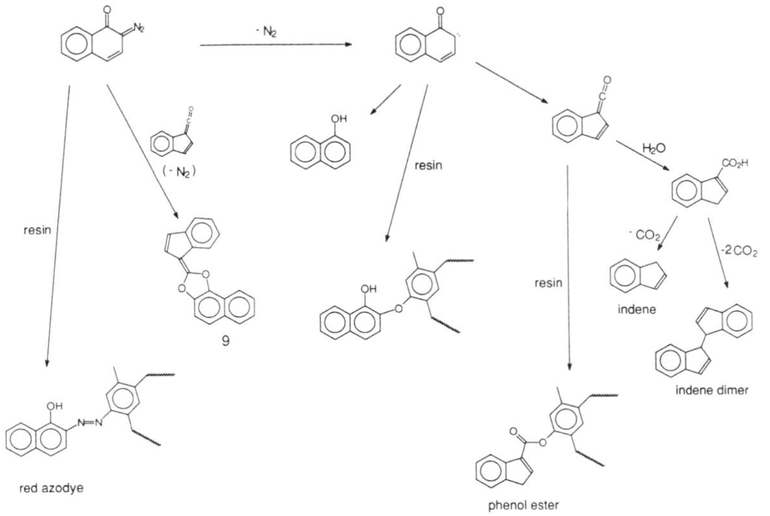

Figure 2-8. *Side reactions from photolysis of 2-diazonaphthoquinone that may lead to image-reversal.*

contribute to different types of behavior of DNQ/Novolac resists during processing.

The DNQs used in these resists are dissolution inhibitors and photolysis will, in the presence of water, give 1-indenecarboxylic acid, a species that may be expected to act as a dissolution promoter. Similarly the DNQ-4-sulfonates form 1-indenecarboxylic-4-sulfonic acid derivatives that are also expected to be dissolution promoters. Thus after the photolysis of DNQ/Novolac the resin becomes more soluble and selectively dissolves making a positive resist. However, some of the side reactions outlined above would lead to a different behavior for the resist. First the photochemical formation of sulfonic acid, described earlier for DNQ-4-sulfonates, can lead to acid-catalyzed crosslinking of the Novolac resin, and this results in image reversal; i.e., the DNQ/Novolac formulation becomes a negative resist. The latter reaction does not occur with DNQ-5-sulfonates, which do not give sulfonic acids upon photolysis.[17] Similarly, insertion of ketocarbene, or ketene, into the Novolac resin can result in dissolution inhibition because the hydrophilic phenolic O—H sites are replaced by ethers or esters. The latter reaction has been claimed as the basis for a negative resist simply by baking the resist prior to exposure, a process that removes water and suppresses indenecarboxylic acid formation.[18] The decarboxylation of indene carboxylic acid can be catalyzed by base and this has been used to make a negative resist by adding imidazoles to the DNQ/Novolac formulation.[19] Finally, the formation of azodyes, an insertion reaction, can be used to render the Novolac resin more insoluble during processing.[3]

The formation of indene carboxylic acid is an oversimplification of the mechanism for the DNQ/Novolac system behaving as a positive resist. Deprotonation of Novolac gives pK_a values of 4–6, whereas the pK_a for monomeric phenols is in the range of 7–9.[20] This hyperacidity of Novolacs indicates that Novolac is more acidic than the 1-indenecarboxylic acid derivatives that are the expected photoproducts from DNQ photolysis. Therefore if Novolac is more readily deprotonated by base than the DNQ photoproducts, it is unlikely that DNQ photoproducts are dissolution promoters; i.e., the dissolution characteristics of the resist are determined by dissolution inhibition properties of DNQ before exposure to light. Infrared spectroscopy shows a 30 cm^{-1} shift to shorter wavelengths for the C = O vibration of 2-DNQs in Novolac, whereas the C = N = N vibration is not shifted.[21] In addition the O—H stretching vibration of the phenolic resin is red-shifted by 200–500 cm^{-1}. These results have been interpreted as stabilization of Novolac monoanions to give DNQ selectively co-ordinated in the resin matrix, similar to a "host-guest" complex.[3] An explanation for the dissolution inhibition of Novolac is that DNQ selectively occupies, or blocks, O—H sites in the Novolac resin. Unfortunately the parent DNQ is not a very good dissolution inhibitor. It has been shown that the sulfonyl group imparts the dissolution inhibition properties to DNQ-sulfonates and that simple aromatic sulfonate esters are better dissolution inhibitors than diazoketones in Novolac resins.[22] Thus an overall view of the resist process is explained by the fact that the DNQ sulfonates selectively occupy sites co-ordinated to multiple phenolic O—H groups in the resin, and that the sulfonate function inhibits dissolution at these base soluble sites. Photolysis results in formation of 1-indenecarboxylic acid sulfonates which do not selectively co-ordinate to multiple O—H groups and thus free these sites for deprotonation by developer. Although 1-indenecarboxylic acid is less acidic than the resin, it will still dissolve in the developer solution, but at a slower rate than the Novolac resin. Indeed it is a fortuitous combination of the ability of the resin to selectively co-ordinate with the photoactive compound, the photosensitivity of diazoketones, and the dissolution inhibition properties of aromatic sulfonates at those specific resin sites, that makes the DNQ/Novolac resists so efficient. As will be discussed later, it is not easy to transfer all of these properties to other resin/diazoketone combinations to give resists of comparable performance to the DNQ/Novolac photoresists.

3.3. Deep UV Photoresists Using Diazoketone Chemistry

The main driving force for the development of deep UV resists is the need for resolution of smaller features (<1 μM) in lithography. For projection printing the resolution (R, minimum feature size) is determined from the process constant (k), the numerical aperture of the lens (NA) and the wavelength of light (λ) used in the process, by the following relationship:

$$R = k\frac{\lambda}{NA}$$

However if the numerical aperture of the lens is increased, the depth of focus (D_f) decreases:

$$D_f = \frac{\lambda}{(NA)^2}$$

Most photoresists for microlithography require a minimum thickness of 0.5–1.0 μM for processing and thus reduction of λ or k is the best way to improve resolution while maintaining depth of focus. The disadvantage with using Novolac/DNQ as a deep UV resist is that both the resin and the diazoketone have strong absorbances at wavelengths below 300 nm. Also DNQs do not bleach in the deep UV upon photolysis. Consequently approaches to making deep UV resists based on diazoketone chemistry have involved the formulation of resins and diazoketones with lower absorbances in the deep UV.

One approach involves the use of aliphatic diazoketo-derivatives, which bleach at wavelengths below 300 nm upon photolysis, in Novolac resin. A number of diazo-Meldrum's acid (10) derivatives were evaluated in Novolac resin and features could be resolved with exposure doses of 50 mJ cm^{-2}. The drawback with these resists was the lack of contrast and the observation that many of the diazo-Meldrum's acid derivatives decomposed or simply sublimed from the resist during the post-apply bake.[23] The dissolution mechanisms that were proposed included carboxylic acid formation, analogous to DNQ, and also photo-void formation, which could enhance diffusion of the developer because it was previously reported that diazo-Meldrum's acid formed nitrogen, carbon monoxide, and acetone upon direct photolysis (Fig. 2-9).[24] The poor contrast of these resists was probably due to weaker dissolution inhibition to the developer solution by diazo-Meldrum's acid, which lacks the important sulfonate ester functional groups that are present in the photoactive compounds used in DNQ-based resists. The dissolution rate for Novolac film containing 2-DNQ-4-phenylsulfonate in aqueous base is reported as 0.11 μM min^{-1}, whereas the rate for diazo-Meldrum's acid under similar conditions is 2.40 μM min^{-1}.[22] Another system, which had the advantage that the photoactive compound did not sublime from the resist during post-apply bake, used diazotetramic acid derivatives.[25] These resists showed no loss of diazo-function by IR spectroscopy even after baking at 105° C for 30 minutes. Again the contrast for these systems was not as good as that of the DNQ-sulfonate/Novolac resists.

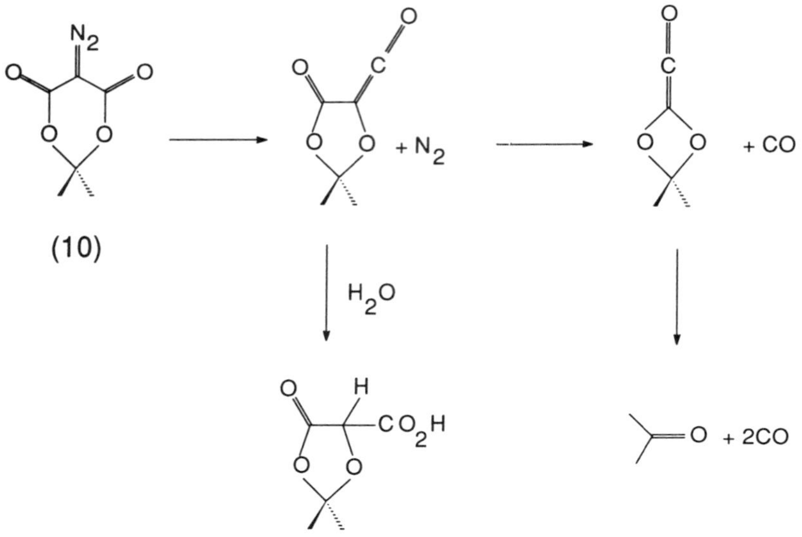

Figure 2-9. *Photoproducts from diazo-Meldrum's acid in aqueous and anhydrous solvents.*

More recent studies have used poly(4-hydroxystyrene) (poly-HOST) as the resin for the resist. Formulations of aliphatic diazoketones and diazoesters in poly-HOST were found to be positive-acting deep UV resists.[26] These resists suffered from the same problems associated with sublimation of the photoactive compound and poor dissolution inhibition properties of both the diazo compounds and the poly-HOST resin. A modification of this approach was to use a blend of Novolac and poly-HOST as the resin, which resulted in lowering the deep UV absorbance yet retained the desirable dissolution properties associated with Novolac.[27] DNQ sulfonates that had good dissolution inhibition properties were employed in this formulation. However these compounds have strong absorptions and do not photobleach in the deep UV.

3.4. Chemical Amplification

The processes described so far involve diazoketone-based photoresists that undergo the classic one photon-one chemical reaction process. The photospeed (the dose required to satisfactorily image the polymer) associated with this chemistry is typically 100–1000 mJ cm^{-2}. These photospeeds are too slow for processing chips. To increase throughput, tools have been designed to accommodate larger wafers. Most processes involve exposing only part of the wafer, to give the best optical quality for the light beam, and then moving the beam, or substrate, until the whole wafer is imaged. Thus the number of exposures per wafer has increased and therefore the exposure time per site must decrease to maintain the improved throughput. Also, the drive for finer features required the development of new photoresists that could be imaged at shorter wavelengths, the so-called deep UV region between 200 and 300 nm. If the emission from a mercury-xenon lamp at wavelengths below 300 nm is to be used, processing will be even slower because the emission from this lamp at 248 nm is orders of magnitude weaker than at 365 nm or 405 nm. Even if excimer lasers are used as a source of 248-nm light, a second problem arises: Novolacs strongly absorb at this wavelength. A general rule of thumb is that the total absorbance of the photoresist, polymer, and photoactive component should be less than 0.35 for maximum light transmission to the bottom of the film.[28] Thus new photoresists that absorbed less in the deep UV and had photospeeds orders of magnitude faster than diazoketone resists, were required. The latter cannot be achieved by a one photon-one chemical event process because DNQs already have high quantum yields. The maximum improvement in photospeed would be two- to threefold for a process with a photochemical quantum yield of 1. Consequently photoresist chemists started thinking about catalytic processes as a way to compensate for the limitation of photochemical quantum yield. Indeed acid-catalyzed crosslinking reactions that could be photochemically initiated were first reported by Smith and Crivello.[29(b),30] Later a catalytic deprotection reaction of a protected poly-HOST, using photogenerated acid, was described by Ito and Willson.[31] Common to all of these reactions was the use of onium salts to photochemically produce acid.

4. Photochemistry of Onium Salts

4.1. Direct Photolysis

Triphenylsulfonium and diphenyliodonium salts are the most actively used photoacid generators. These salts are involatile, thermally stable, have good solubility in most organic solvents, and most importantly, they efficiently generate acid upon photolysis. This section will review the photochemistry of triphenylsulfonium salts. Other onium salts exhibit similar reactivity and a detailed account of their chemistry has recently been reported.[32] Early studies on the photochemistry of triphenylsulfonium salts (TPS) detected diphenylsulfide, benzene, biphenyl, anisole, and acid as photoproducts when methanol was the solvent.[33] To account for these products McEwen proposed that both homolytic and heterolytic cleavage of the carbon-sulfur bond could occur from the TPS excited state. Heterolysis generated diphenylsulfide and phenyl cation intermediates which could react with methanol to give anisole and acid. Homolysis gave phenyl radical and diphenylsulfinyl radical cation intermediates which reacted with the solvent to give benzene, diphenyl sulfide, and acid. Biphenyl was a minor photoproduct which could be formed by dimerization of phenyl radical or was a secondary photoproduct. Later studies supported the homolysis[34,35] or heterolysis[36] cleavage mechanisms. No matter which mechanism is occurring there should be one mole of acid produced for every mole of diphenylsulfide formed. Yet an anomalous radio of 2–3:1 for acid to diphenylsulfide was consistently found from photolysis of TPS salts, and an additional mechanism was proposed to account for the excess acid.[34]

More recently the photolysis of TPS was re-examined and it was discovered that 2-, 3-, and 4-phenylthiobiphenyl (PTB) were formed in addition to the previously detected photoproducts (Fig. 2-10).[37] The PTBs were detected by capillary gas-liquid chromatography (GLC), identified by GLC-mass spectral

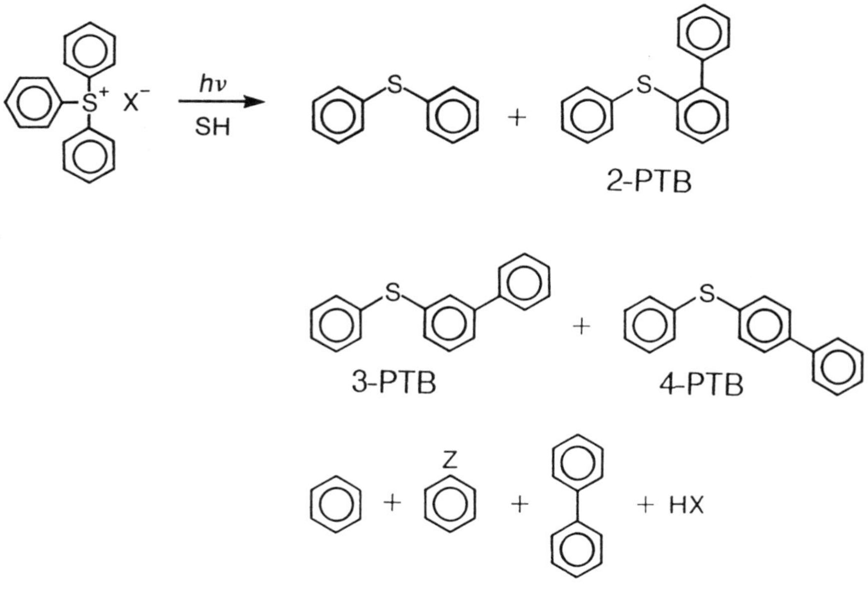

SH = CH_3CN, Z = $-NHCOCH_3$

SH = CH_3OH, Z = $-OCH_3$

SH = C_2H_5OH, Z = $-OC_2H_5$

Figure 2-10. *New photoproducts from photolysis of triphenylsulfonium salts (reproduced from Ref. 37).*

(MS) analysis, and confirmed by comparison of GLC retention times with those from authentic samples. In addition consumption of TPS was measured by both high performance liquid chromatography (HPLC) and titration via the TPS-cobalt thiocyanate complex. Finally acid production was measured by photometrically using sodium nitrophenoxide as an indicator. Within experimental error the consumption of TPS equaled the formation of acid and also equaled the sum of the PTBs plus diphenylsulfide formation. The PTBs accounted for 60–70% of the sulfur-containing photoproducts and thus there was no excess acid formed relative to total sulfide, PTBs, and diphenylsulfide.

While the formation of PTBs solved the "excess acid" problem, there were still questions about how the PTBs were formed and also whether the excited state of TPS cleaves by homolysis or heterolysis of the C—S bond. Experiments to probe the effects of solvent viscosity and polarity, experiments to trap possible intermediates, and cross-over experiments were run to explain how the PTBs were formed. PTB formation could be visualized as an initial fragmentation followed by diffusion of these fragments through the solvent to react with other fragments by an escape reaction mechanism, as an intramolecular rearrangement or as an in-cage fragmentation-recombination reaction. Although the escape reaction mechanism is unlikely because the highly reactive phenyl fragments are more likely to react with solvent than a sulfur-containing moiety at the low TPS concentrations (0.01M) used, a cross-over experiment was run to eliminate this mechanism. A mixture of tris(4-tolyl)sulfonium salt (TTS) and TPS was irradiated and the products analyzed by GLC and GLC-MS. Fifty-nine peaks were observed and 12 peaks accounted for 93% of the peak area. These 12 peaks corresponded to the photoproducts of TPS and TTS when each salt was irradiated separately. Thus the lack of cross-over products makes the escape mechanism for PTB formation unlikely. Photolysis of TPS in a variety of solvents revealed that the ratio of PTB to diphenylsulfide increased with increasing viscosity. For example the ratio of PTB:DPS was 1.28:1 in methanol and 2.56:1 in 2-propanol. Also the relative quantum yield decreased by an order of magnitude from low viscosity solvents such as dichloromethane (ρ = 0.41 cP) to more viscous solvents like glycerol (ρ = 1500 cP). These results support the premise that PTBs are formed by an in-cage reaction. The nature of the in-cage reaction, concerted or fragmentation-recombination, was determined from the isomeric mixture of PTBs and by trapping experiments. The ratio of 2-PTB:3-PTB:4-PTB was 5:1:2 in acetonitrile. It is possible that 2- and 4-PTB are formed by a concerted reaction; however, it is very unlikely that 3-PTB is formed by a concerted rearrangement. Further support for the cleavage reaction is that addition of bromide ion traps the phenyl fragment to give bromobenzene, and the yield of PTB decreases concomitant with an increase in diphenylsulfide. A concerted rearrangement would be unaffected by the addition of bromide.

The above results give strong support for the in-cage fragmentation-recombination mechanism for PTB formation, but also help understand how the C—S bond cleaves from photolysis of TPS. The phenyl group escape fragments support both homolysis and heterolysis mechanisms. It is well known that the phenyl radical reacts with solvent to form benzene, a product always observed from TPS photolysis. Also anisole, phenetole, and acetanilide are observed from TPS photolysis in methanol, ethanol, and acetonitrile, respectively, which suggests that the phenyl cation is an intermediate. The observation that the bromide anion traps a phenyl fragment to give bromobenzene also implies that the phenyl cation is an intermediate. The selectivity of PTB formation 2-PTB >> 4-PTB supports a selective reaction typical of the phenyl radical-diphenylsulfinyl radical cation pair from homolytic C—S cleavage. However, 3-PTB is always detected from TPS photolysis, and this suggests the intermediacy of a nonselective very reactive species such as the phenyl cation. Although there is strong evidence for both homolytic and heterolytic cleavage of the C—S bond, it is unlikely that the singlet excited state of TPS will cleave by two different pathways. The two sets of intermediates, phenyl cation-diphenylsulfide pair and phenyl radical-diphenylsulfinyl radical cation pair, are related by an electron transfer reaction:

$$Ph^{+} + Ph_2S \rightarrow Ph^{\cdot} + Ph_2S^{+\cdot}$$

The oxidation potential for phenyl cation is 2.1 V (estimated from the gas phase ionization potential) and the oxidation potential for diphenylsulfide is 1.41 V (vs. SCE in CH_3CN, Pt electrode). Therefore the above reaction is exothermic by 0.69 V (16 kcal mol^{-1}), making the phenyl radical-diphenylsulfinyl radical cation pair more stable. Thus it is proposed that the excited state of TPS cleaves initially by heterolysis to the phenyl cation pair of intermediates, which undergo a rapid electron transfer to give the more stable phenyl radical pair of intermediates. This mechanism (Fig. 2-11) gives a satisfactory explanation for all of the observed photoproducts.

4.2. Triplet Sensitized Photolysis

Photolysis of TPS in the presence of triplet sensitizers with $E_T > 74$ kcal mol^{-1} gives diphenylsulfide, benzene, and acid in methanol.[37,38] The absence of anisole suggests that phenyl cation is not an intermediate, and the detection of benzene suggests that phenyl radical is an intermediate. To explain the absence of PTBs it was proposed that the triplet excited state of TPS cleaved by heterolysis to give the phenyl radical-diphenylsulfinyl radical cation *triplet* radical pair which reacted with solvent rather than undergoing spin inversion to the singlet radical pair and recombining (Fig. 2-12). Further evidence for this mechanism is that a transient with absorbances at 340 and 750 nm, assigned to diphenylsulfinyl radical cation, is detected by flash photolysis of TPS salts with triplet sensitizers.

$$Ph_3S^+\ X^- \xrightarrow{h\nu} \underline{[Ph_3S^+\ X^-]^\bullet} \qquad (1)$$

$$[Ph_3S^+\ X^-]^\bullet \longrightarrow \underline{Ph_2S \quad Ph^+ \quad X^-} \qquad (2)$$

$$\overline{Ph_2S \quad Ph^+ \quad X^-} \longrightarrow Ph_2S^{+\bullet} \quad Ph^\bullet \quad X^- \qquad (3)$$

$$\overline{Ph_2S \quad Ph^+ \quad X^-} \longrightarrow \text{2-PTB} + \text{3-PTB} + \text{4-PTB} + H^+ \qquad (4)$$

$$\overline{Ph_2S^{+\bullet} \quad Ph^\bullet \quad X^-} \longrightarrow \text{2-PTB} + \text{4-PTB} + H^+ \qquad (5)$$

$$\overline{Ph_2S \quad Ph^+ \quad X^-} \longrightarrow Ph_2S + Ph^+ + X^- \qquad (6)$$

$$\overline{Ph_2S^{+\bullet} \quad Ph^\bullet \quad X^-} \longrightarrow Ph_2S^{+\bullet} + Ph^\bullet + X^- \qquad (7)$$

$$Ph^+ + RH \longrightarrow PhR + H^+ \qquad (8)$$

$$Ph_2S^{+\bullet} + RH \longrightarrow Ph_2S^+{-}H + R^\bullet \qquad (9)$$

$$Ph_2S^+{-}H \longrightarrow Ph_2S + H^+ \qquad (10)$$

$$Ph^\bullet + RH \longrightarrow PhH + R^\bullet \qquad (11)$$

$$Ph^\bullet + R^\bullet \longrightarrow Ph{-}R \qquad (12)$$

$$Ph^\bullet + Ph^\bullet \longrightarrow Ph{-}Ph \qquad (13)$$

$$R^\bullet + R^\bullet \longrightarrow R{-}R \qquad (14)$$

Figure 2-11. *Mechanism for product formation from direct photolysis of triphenylsulfonium salts (reproduced from Ref. 37).*

Also, an enhanced signal for the benzene peak is detected by photo-CIDNP from triplet sensitized photolysis of TPS, which confirms that phenyl radical reacts with the solvent by a cage-escape from a triplet radical pair.[39] In contrast direct photolysis of TPS gives an emission for benzene in photo-CIDNP experiments.

4.3. Sensitization by Photoinduced Electron Transfer

Photolysis of TPS in the presence of aromatic hydrocarbons, e.g., anthracene, gave diphenylsulfide, phenylanthracenes, and acid.[40,41] Again no phenylthiobiphenyls or anisole were detected in methanol solvent, which suggests that the reaction is different from direct photolysis and that phenyl cation is not an intermediate. However this sensitization reaction does not involve triplet energy transfer because the triplet energy of anthracene (E_T = 42 kcal mol^{-1}) is too low

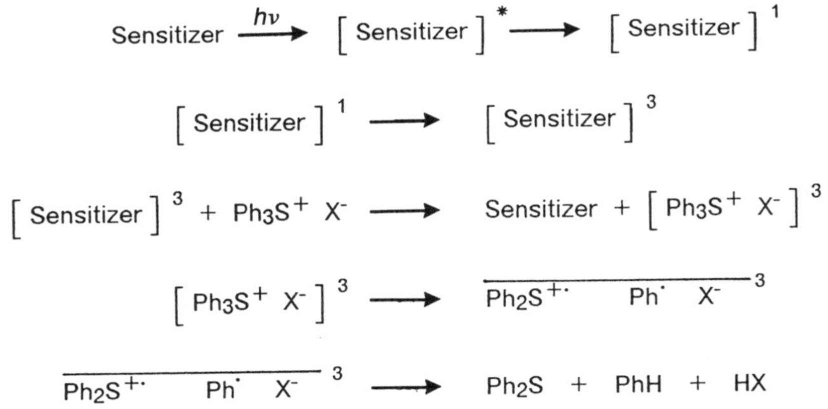

$$\text{Sensitizer} \xrightarrow{h\nu} [\text{Sensitizer}]^{*} \longrightarrow [\text{Sensitizer}]^{1}$$

$$[\text{Sensitizer}]^{1} \longrightarrow [\text{Sensitizer}]^{3}$$

$$[\text{Sensitizer}]^{3} + Ph_3S^{+}\,X^{-} \longrightarrow \text{Sensitizer} + [Ph_3S^{+}\,X^{-}]^{3}$$

$$[Ph_3S^{+}\,X^{-}]^{3} \longrightarrow \overline{Ph_2S^{+\cdot} \quad Ph^{\cdot} \quad X^{-}}^{\,3}$$

$$\overline{Ph_2S^{+\cdot} \quad Ph^{\cdot} \quad X^{-}}^{\,3} \longrightarrow Ph_2S + PhH + HX$$

Figure 2-12. *Mechanism for product formation from triplet sensitized photolysis of triphenylsulfonium salts (after Ref. 38).*

to sensitize TPS (E_T = 74 kcal mol^{-1}). This sensitization reaction also results in phenylation of the sensitizer, a reaction not observed with triplet energy transfer. However TPS is a good electron acceptor (E_{red} = -1.2 V), and the excited state of an aromatic hydrocarbon is a good donor, making an electron transfer reaction feasible. The excited state oxidation potential (E^{*}_{ox}) of a molecule can be determined from its excited state energy and ground state oxidation potential (E_{ox}):

$$(E^{*}_{ox}) = S_1\ (T_1) - E_{ox}$$

where S_1 and T_1 are the energies of the first singlet and triplet excited state energies, respectively.

For anthracene the ground state oxidation potential is 1.09 V (vs. SCE Pt electrode in CH_3CN), the first singlet excited state energy is 76.3 kcal mol^{-1} (3.31 eV), and the triplet excited state energy is 42.0 kcal mol^{-1} (1.82 eV). Thus the excited state potentials for anthracene are 2.22 eV (51.2 kcal mol^{-1}) and 0.73 eV (16.9 kcal mol^{-1}) for the singlet and triplet excited states, respectively. For photoinduced electron transfer to occur, the reaction must be energetically favorable according to the Rehm-Weller equation:[42]

$$\Delta G = -(E^{*}_{ox} + E_{red})$$

Using the values from above, electron transfer from the singlet excited state of anthracene is energetically favorable, ΔG = -1.02 V (-23.5 kcal mol^{-1}), whereas from the triplet excited state the reaction is endothermic, ΔG = + 0.47

V (+ 10.8 kcal mol^{-1}). The mechanism for the reaction is outlined in Fig. 2-13. The singlet excited state of anthracene transfers an electron to the ground state of TPS to generate triphenylsulfur radical and anthracene radical cation in a solvent cage. Triphenylsulfur radical rapidly dissociates to diphenylsulfide and phenyl radical; the latter species can react with anthracene inside the solvent cage to give phenyl anthracenes and acid.[40,41] Alternatively, phenyl radical can react with solvent to give benzene, and anthracene radical cation becomes the source of acid.[34]

A further new in-cage reaction can occur from photolysis of TPS with aromatic hydrocarbons.[41] Naphthalene sensitized photolysis of TPS proceeds by an electron transfer reaction to give naphthalene radical cation and triphenylsulfur radical, which dissociates diphenylsulfide and phenyl radical. In addition to the in-cage phenylation of naphthalene radical cation, electron transfer from diphenylsulfide to naphthalene radical cation gives diphenylsulfinyl radical cation. The later species reacts with phenyl radical in the solvent cage to give 2- and 4-PTB and acid (Fig. 2-14). Diphenylsulfide has an oxidation potential of 1.31 V, naphthalene 1.54 V, and anthracene 1.09 V (all vs. SCE Pt electrode in CH_3CN). Thus the radical cation of naphthalene can oxidize diphenylsulfide, whereas anthracene radical cation cannot. These results also confirm the premise that reaction of phenyl cation with diphenylsulfide is responsible for 3-PTB formation upon direct photolysis of TPS. Naphthalene sensitized photolysis of TPS shows no evidence for phenyl cation escape products, e.g., acetanilide or phenyl cation in-cage reaction to form 3-PTB.

Photo-CIDNP experiments of anthracene and naphthalene sensitized photolysis of TPS show strong emissive signals for benzene indicating that phenyl radical escapes from a singlet radical pair and reacts with solvent. The absorbances at 340 and 750 nm, assigned to diphenylsulfinyl radical cation, are de-

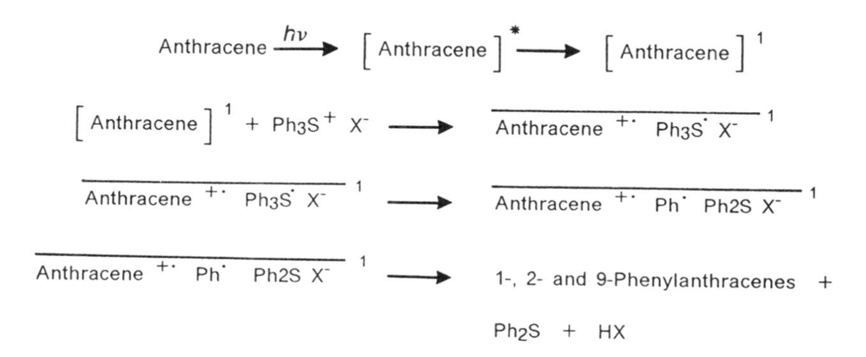

Figure 2-13. *Mechanism for product formation by electron transfer sensitized photolysis of triphenylsulfonium salts from the singlet excited state of anthracene.*

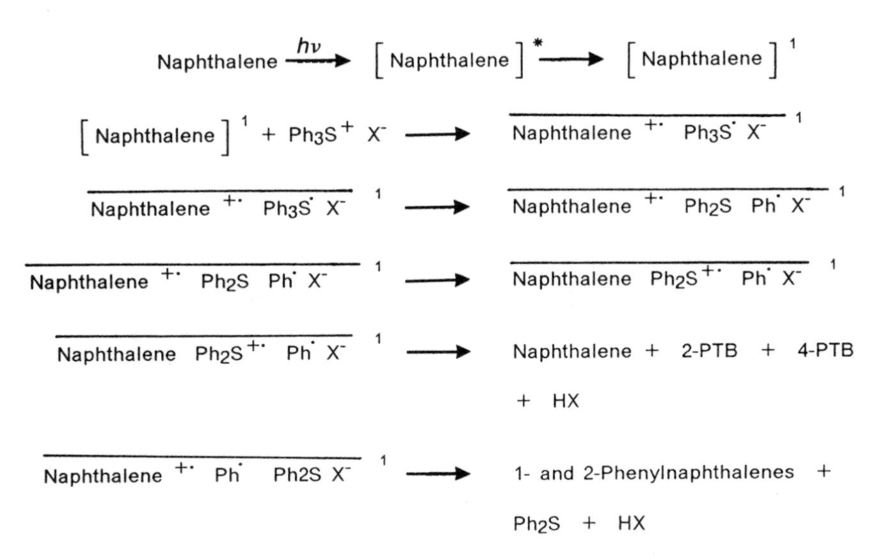

Figure 2-14. *Mechanism for product formation by electron transfer sensitized photolysis of triphenylsulfonium salts from the singlet excited state of naphthalene.*

tected by flash photolysis of TPS salts with naphthalene sensitizer, but not with anthracene, which supports the above mechanism.[39]

5. Resist Chemistry using Onium Salt Photoinitiators

The resist chemistry employing onium salts is diverse and fascinating. The use of iodonium salts as photoinitiators for an acid-catalyzed reaction was reported by Smith at 3M in 1975.[29(b)] The formulation of 6 wt % diphenyliodonium hexafluorophosphate in bisphenol A diglycidyl ether was cured in 25 seconds upon exposure to a 275-W sunlamp. Crivello later reported his studies on a number of different crosslinking reactions of epoxy monomers using a variety of substituted diaryliodonium salts.[30] The reaction for both of these processes is photochemical generation of a Brönsted acid that protonates the epoxy ring and is subsequently attacked by nonprotonated epoxides catalytically to give a crosslinked polymer (Fig. 2-15). This reaction has been used as the basis for polymerizing many different types of cyclic ethers and vinylethers.[30,35,43]

It became apparent to those working in microlithography that other catalytic processes using photogenerated acid could be applied to create sensitive deep UV resists. In 1973 Smith reported that a Novolac resin formulation that contained a tetrahydropyranyl (THP) protected phenol and a nonionic photo-acid

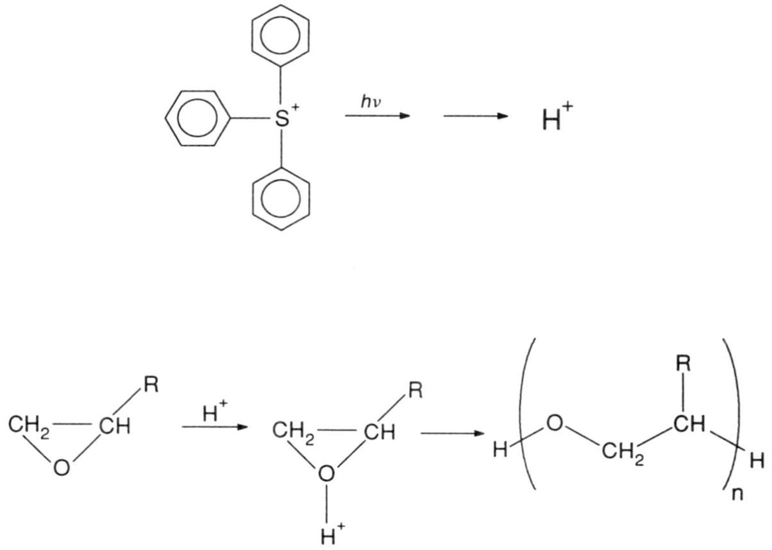

Figure 2-15. *Acid catalyzed crosslinking of an epoxide initiated by triphenylsulfonium salt photolysis.*

generator, could be used as a positive acting resist for printing plates.[29(a)] The THP group is catalytically cleaved by the photolytically generated acid to give a substituted phenol that is soluble in aqueous base. The THP protected phenol is insoluble in base developer and so the unirradiated areas do not dissolve. As discussed earlier, Novolacs absorb considerably in the deep UV and it became necessary to synthesize new polymers with better transmission at 250 nm. This was realized when Ito and Willson first reported the ''TBOC Resist'' in 1983.[44] For many years peptide chemists have used the *tert*-butyl and *tert*-butoxycarbonyloxy (TBOC) groups for the protection of alcohols, carboxylic acids, and amines.[45] These groups are readily cleaved catalytically by Brönsted acids under mild conditions. Cleavage of the *tert*-butyl group generates the *tert*-butyl cation intermediate which gives isobutylene and acid. Thus for the TBOC-resist, poly-HOST protected with TBOC groups is formulated with an onium salt to make a deep UV resist. The polymer has good transmittance in the deep UV and photolysis generates acid from the onium salt which cleaves the TBOC group to give the free phenolic OH group. The acid consumed during this reaction is regenerated by the formation of isobutylene (Fig. 2-16). The chain length for the catalytic process is estimated to be 1000 and so an initiation reaction with a quantum yield of 0.1 for acid formation can give an effective quantum yield

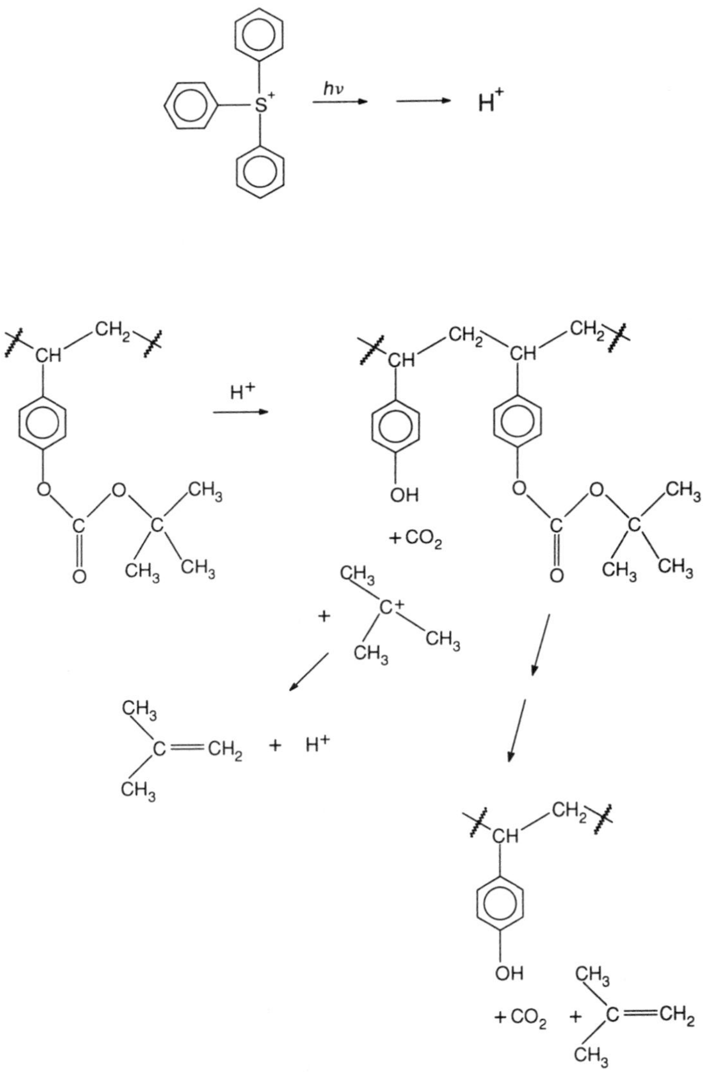

Figure 2-16. *Photochemically initiated acid-catalyzed deprotection of TBOC resist.*

of 100 for TBOC cleavage.[46] The term ''chemical amplification'' has been coined for these catalytic reactions.[44,47] The deprotection reaction generates poly(4-hydroxystyrene) in the exposed areas and thus the change in polarity from the protected to deprotected areas gives a positive- or negative-acting resist after development with polar or nonpolar solvents, respectively. A similar type of resist in which poly-HOST is protected with THP groups that are cleaved by photogenerated acid has been reported.[48]

Photolysis of a mixture of onium salt and poly(tert-butyl-4-vinylbenzoate) gives poly(4-vinylbenzoic acid), which can be developed with aqueous base.[49] Two approaches in which the photoinitiating species undergoes catalytic deprotection have recently been described. A TBOC-protected phenolic version of triphenylsulfonium salt (11) photochemically generates acid that diffuses through the resin to cleave the TBOC groups on any unreacted onium salt (Fig. 2-17).[50] Similarly *tert*-butyl-protected naphthalene carboxylic acid derivatives (e.g., 12) can undergo a photoinduced electron transfer reaction with onium salts and generate acid.[51] The acid cleaves the *tert*-butyl groups to give the free naphthalene carboxylic acid which can be developed with aqueous base (Fig. 2-17).

A self-developing resist based on a formulation of poly(phthalaldehyde) with onium salt has been reported.[31] Poly(phthalaldehyde) has a low ceiling temperature and must be end-capped to be stable above room temperature. The depolymerization of poly(phthalaldehyde) is catalyzed by acid, and thus photolysis of the formulation generates acid which cleaves the poly acetal backbone (Fig. 2-18). Simply heating the irradiated polymer results in depolymerization. Unfortunately, the unreacted onium salt leaves residues that can interfere with subsequent processing.

Numerous examples of negative photoresists involving the use of photogenerated acid to crosslink epoxies, and catalyze pinacol rearrangements, Claisen rearrangements, or electrophilic aromatic substitution of substituted poly(styrene) derivatives have been reported.[30,35,43,52–55]

The need for making photoresists sensitive in the deep UV, 200 nm to 300 nm, was discussed earlier. Future technologies will require photosensitive chemically amplified resists that are imageable at quite different wavelengths. As features <0.25 μm will be required, there is a drive to make photoresists that can be imaged at even shorter wavelengths, <200 nm. Today there are commercial Kr-F excimer laser tools that operate at 248 nm, and it is quite reasonable to expect that Ar-F tools will be available in the future for 193-nm exposure. The requirements for the polymer are complicated by the fact that most polymers that are resistant to reactive ion etching (RIE), the required method for the image-transfer process in chip manufacturing, are aromatic and absorb strongly at 193 nm. Aliphatic polymers that have good transmittance at 193 nm generally have poor RIE etch resistance. One approach has been to synthesize acrylate and methacrylate polymers that have pendant *tert*-butyl groups for sensitivity to photogenerated acid.[56] Features down to 0.1 μm in a 0.5-μm film

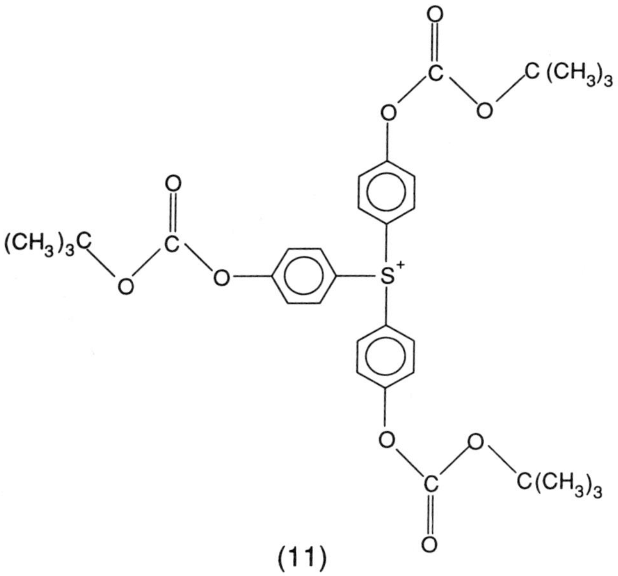

(11)

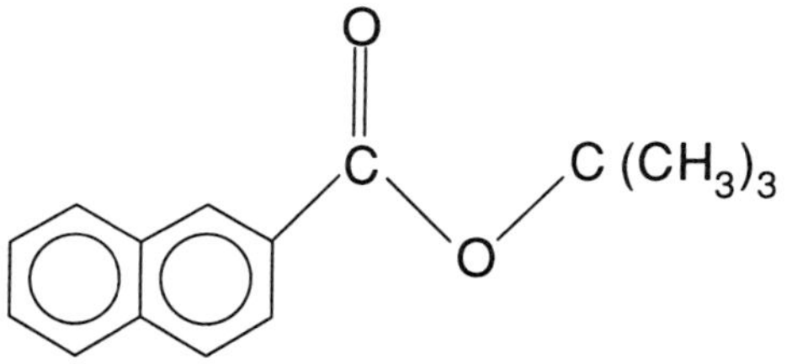

(12)

Figure 2-17. *Chemically amplified photoinitiator (11) and photosensitizer (12).*

have been resolved using a phaseshift mask. To impart etch resistance, pendant isobornyl and adamantyl groups are incorporated in the polymer. These groups are transparent at 193 nm, but should impart etch resistance to the polymer because they have a high carbon content.

The second wavelength regime, in the visible region, is also driven by laser imaging technology. New tools that rapidly scan an argon ion laser beam serially across a substrate are potentially useful for manufacturing printed circuit boards (PCBs). This approach to lithography has the advantage that it eliminates the

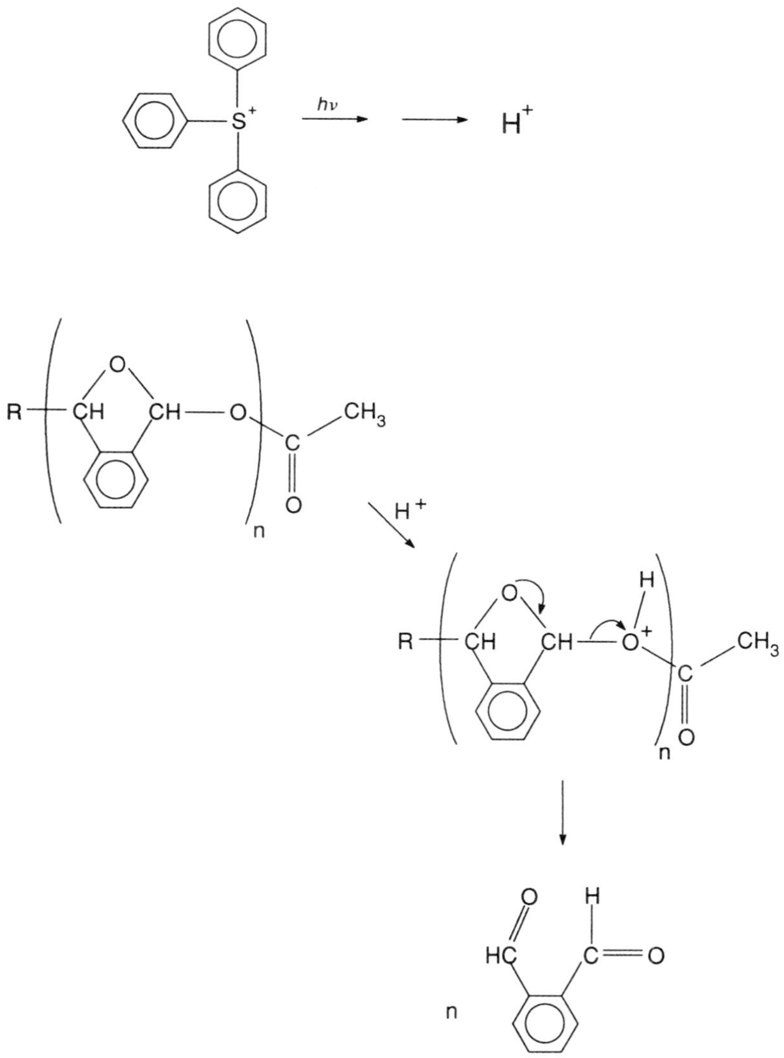

Figure 2-18. *Photochemically initiated acid-catalyzed depolymerization of poly (phthaladehyde).*

need for artwork and mask fabrication tools. This process is computer controlled and is ideal for making low volume custom PCBs and also for cutting cycle times for PCB design. Thus an engineer can design, test, and modify a PCB without having to prepare a new mask for each modification. Although the minimum feature size requirements for PCBs, 25 μM, are not very demanding, there are two additional features the resist needs for optimum processing: sensitivity to the output of the argon ion laser (488 nm and 514 nm); and a photospeed of better than 15 mJ cm^{-1} for optimum throughput of the tool. Thus a chemically amplified resist is needed. The material that satisfies these requirements is a terpolymer of methylmethacrylate (MMA), *tert*-butylmethacrylate (TBMA), and methacrylic acid (MA).[57] The MMA and MA impart desirable mechanical, solubility, and adhesion properties to the terpolymer, whereas the TBMA has a group that can be cleaved by photochemically generated acid. Onium salts do not absorb at 488 or 514 nm and have triplet energies too high for triplet energy transfer. However onium salts are good electron acceptors, and sensitization by photoinduced electron transfer was chosen to make the photoresist sensitive to the visible laser. A diphenyliodonium salt was used as the photoinitiator of acid because it has a lower oxidation potential than that of triphenylsulfonium salts.[40] Several substituted anthracene derivatives were examined as sensitizers, and it was found that methoxy-substituted 9,10-bis(phenylethynyl)anthracenes had strong absorptions that matched the output of the argon ion laser and were efficient excited state electron donors.[57]

6. Interaction between Polymer and Photoinitiators in Chemically Amplified Resists

Conventional wisdom has treated the onium salt photoinitiator and polymer as two separate entities. The onium salt may have some dissolution inhibition properties to be utilized; there have also been concerns about phase separation because onium salts are ionic materials, whereas the polymer matrix is nonionic. Indeed, recent papers on resist chemistry show mechanisms where the onium salt absorbs the incident light, produces acid, and then the catalytic chemistry resulting from the acid produced is discussed. Even though the polymers frequently absorb >50% of the incident light it has been assumed that this results in energy wasting processes. However recent studies on TPS in TBOC polymer films and in solution reveal that this is not the case.[58] Photolysis of TPS in polymer films that do not absorb the incident light results in ratios of PTB:TPS (C/E ratio) with values of 2.8–3.5:1. In nonviscous solution the C/E ratio is approximately 1.1:1, and for viscous solutions C/E values up to 5.6:1 have been reported.[37] Thus the polymer matrix is a viscous environment and this favors in-cage recombination reactions to form PTBs rather than the cage-escape reactions that produce diphenylsulfide. However photolysis of TPS in TBOC polymer films gives C/E values of 1.7–2:1 at 254 nm and 0.78–0.86:1 at 300 nm, depending on the loading of TPS in the film. These C/E values are low for a viscous polymeric environment, and it seemed possible that a sensitization reaction may occur. Sensitization by both triplet energy transfer and electron transfer are known to favor diphenylsulfide formation. It is unlikely that the unusual C/E values are due to aggregation phenomena because it is known that the C/E ratio is 4.1:1 for the photolysis of TPS.SbF_6, the onium salt used in the polymer films, as a crystalline solid.[59] It would be expected that aggregated TPS in polymer films would behave more like the crystalline solid and result in high C/E values. To eliminate the viscosity effects of TBOC polymer on TPS, the films were dissolved in acetonitrile and irradiated. Under these conditions that C/E ratios were even lower, 0.70–0.79:1 at 254 nm and 0.04 at 300 nm, depending on the TPS loading. At 300 nm TBOC polymer absorbs >90% of the incident light for 1–10 wt % loadings of TPS, which suggests that sensitization by TBOC polymer is the major mechanism for the photodecomposition in solution at this wavelength.

Sensitization by triplet energy transfer and electron transfer are both feasible processes. Although the photophysical and electrochemical data for TBOC polymer are not available, anisole serves as a good model. Anisole has an oxidation potential of 1.35 V (vs. SCE Pt electrode in CH_3CN), a singlet excited state energy of 4.47 eV (103 kcal mol^{-1}), and a triplet excited state energy of 3.50 eV (80.8 kcal mol^{-1}). Thus triplet energy transfer to TPS (E_T = 74 kcal mol^{-1}) is energetically favorable. Also, using the Rehm-Weller equation, electron transfer from the singlet and triplet excited states of anisole are both exothermic, −44 and −22 kcal mol^{-1}, respectively. UV emission studies were used to differentiate between triplet energy transfer and electron transfer from the singlet excited state of TBOC polymer to TPS. Fig. 2-19 shows the emission spectra for the polymers of 4-hydroxystyrene (poly-HOST), 4-methoxystyrene (poly-MOST), and TBOC (poly-TBOC) in acetonitrile. All three polymers show a main emission peak that is broadened and tails off to the red. This tailing emission is also observed in the solid state and is not a feature of model monomeric compounds, which show only narrow fluorescent peaks. Thus the tailing emission is a feature only of the polymers and is attributed to partial ordering of the pendant phenyl groups in the polymer, somewhat akin to excimer fluorescence.

In solution the fluorescence of each of the three polymers is efficiently quenched by the addition of TPS. The Stern-Volmer plots for dynamic quenching of the fluorescence of the polymers by triphenylsulfurium salts give a good linear relationship for I_0/I vs. TPS concentration in all three polymers (Fig. 2-20). From the gradients of these plots, the lifetimes of the polymers are esti-

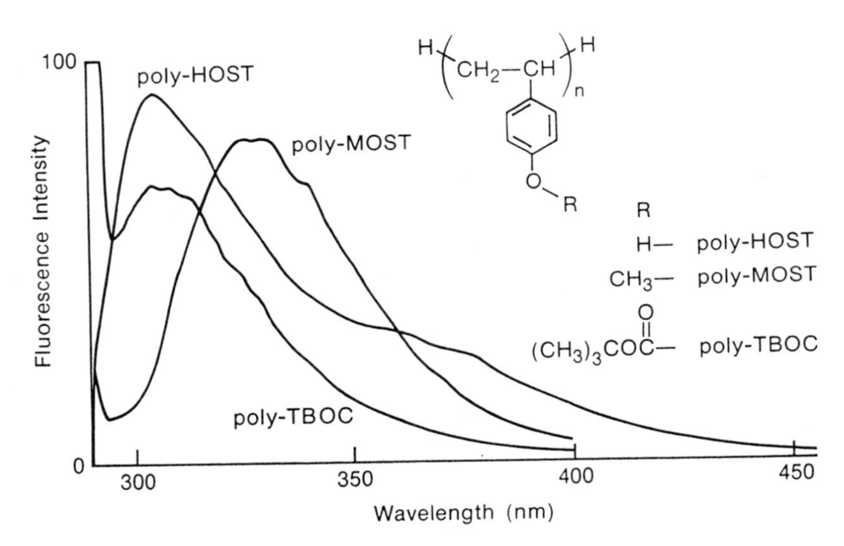

Figure 2-19. *Fluorescence spectra of 4-oxystyrene polymers in acetonitrile (reproduced from Ref. 58).*

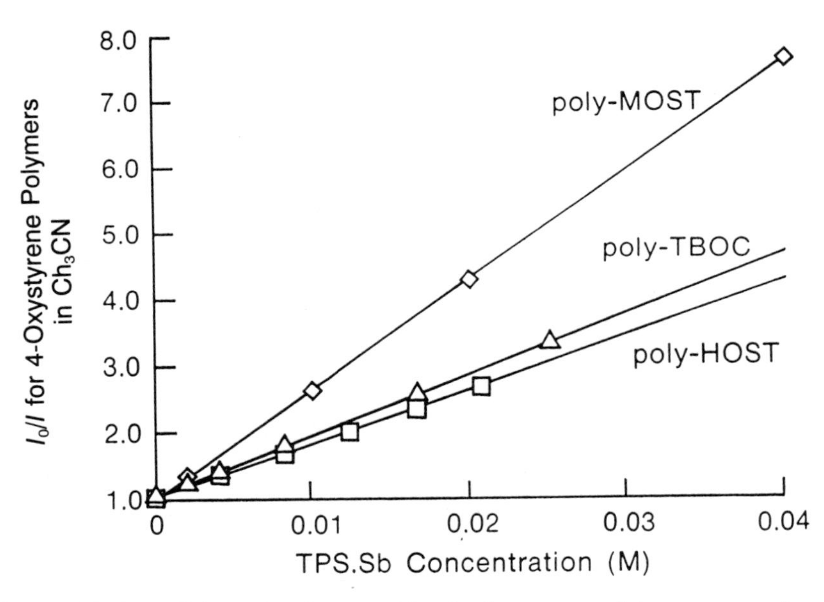

Figure 2-20. *Stern-Volmer plots for quenching the fluorescence of 4-oxystyrene polymers by triphenylsulfonium salts in acetonitrile (reproduced from Ref. 58).*

mated to be 8.3, 4.5, and 4.0 ns for poly-MOST, poly-TBOC, and poly-MOST, respectively. Based on the model monomers anisole (τ = 8.3 ns) and phenol (τ = 2.1 − 7.4 ns), the quenching reaction occurs at the diffusion-controlled rate. These results and the observation that the emission from each of the polymers is not significantly quenched by oxygen suggest that there is sensitization of TPS photodecomposition by electron transfer of the singlet excited state of the polymer.

The quenching of the poly-TBOC by TPS as films was also studied. At 7.0 wt % loading of TPS in the film approximately 68% of the excited states were quenched; however, plots of I_0/I vs. TPS concentration were nonlinear for the films. The nonlinearity suggests that the fluorescence quenching may be static rather than dynamic. Fig. 2-21 shows good linearity for the Perrin plot of $\ln(I_0/I)$ versus TPS concentration, the model for a static quenching mechanism. From the gradient of this plot the radius of the active quenching sphere for poly-TBOC was estimated to be 16°. Similar plots for poly-HOST gave 19 Å for the

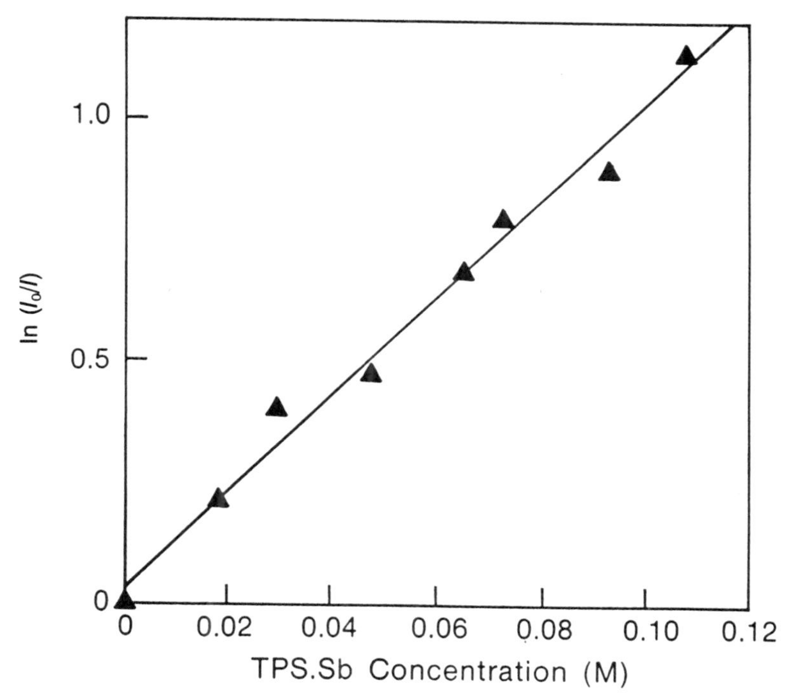

Figure 2-21. *Perrin plot for quenching the fluorescence of TBOC polymer films by triphenylsulfonium salts (reproduced from Ref. 58).*

active radius. These values are too large for normal static quenching (8 Å) and may reflect some additional dynamic component where the photon, or exciton, can migrate through 1–3 units of the polymer. This premise is also supported by the red-shifted tailing emissions detected from the polymers described earlier.

From the photoproduct and the fluorescence quenching studies there is strong evidence for photoinitiation by electron transfer from the singlet excited state of the TBOC polymer to the TPS photoinitiator. However this is not the only photoinitiation mechanism because the polymer does not absorb 100% of the incident light. This is supported by the observation that PTB isomers, products from direct photolysis of TPS, are also detected from photolysis of the polymer films. It can be concluded from these results that a dual photoinitiation mechanism (DPM) occurs where the excited state of the polymer initiates the photogeneration of acid by an electron transfer reaction and TPS can also initiate acid formation by direct photodecomposition (Fig. 2-22). The relative amount of each component of the DPM is determined by the relative absorbances of the polymer and photoinitiator, the donor capability of the polymer excited state and the rate of back electron transfer, and the quantum yield for direct photolysis of the photoinitiator. Thus in these systems the polymer is an important component in the photogeneration of acid.

A crosslinking reaction of substituted poly(styrenes) has recently been reported that takes advantage of the photoinitiated electron transfer reaction of these polymers with onium salts.[60]

New types of photoinitiators for acid generation have recently been reported. The pyrogallol tris-(methanesulfonate) (PTM) and succinimidoyl triflates (SIT) have both been reported to generate acid upon photolysis in polymer films.[61,62] Both of these photoacid generators have only weak UV absorbances and so the

$$P + Ph_3S^+X^- \xrightarrow{h\nu} [P]^* + Ph_3S^+X^- + [Ph_3S^+X^-]^* + P$$

$$[Ph_3S^+X^-]^* \longrightarrow PhPhSPh + Ph_2S + HX$$

$$[P]^* + Ph_3S^+X^- \longrightarrow P^{+\cdot} + Ph_3S^\cdot + X^-$$

$$P^{+\cdot} + Ph_3S^\cdot + X^- \longrightarrow P^{+\cdot} + Ph^\cdot + Ph_2S + X^-$$

$$P^{+\cdot} + Ph^\cdot + X^- \longrightarrow P\text{-}Ph + HX$$

where P = poly[4-[(tert-butoxycarbonyl)oxy]styrene]

Figure 2-22. *Dual photoinitiation mechanism (DPM) for TBOC photoresist (reproduced from Ref. 58).*

majority of the incident light is absorbed by the polymer. From the above discussion on TPS interactions with polymers, one simple set of experiments will show how these compounds generate acid. Fig. 2-23 shows the Stern-Volmer plots for quenching poly-HOST fluorescence by TPS, PTM, and SIT in acetonitrile.[63] It can be seen that both TPS and PTM efficiently quench poly-HOST fluorescence at the diffusion controlled rate, whereas SIT quenches poly-HOST fluorescence but at about 30% efficiency of the other two photoinitiators. Indeed, recent matrix isolation ESR studies have reported a dissociative electron transfer for SIT in the presence of sodium atoms.[64] A similar reaction likely occurs in the polymer film except that the singlet excited state of the polymer is the electron donor. The need for the excited state of the polymer to be an electron

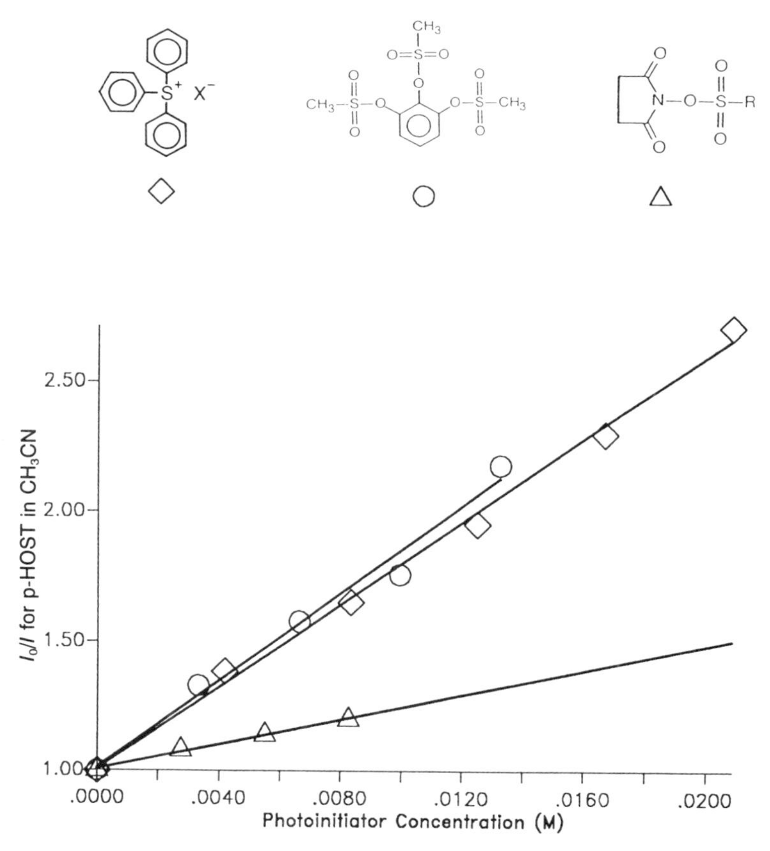

Figure 2-23. *Stern-Volmer plots for quenching the fluorescence of 4-hydroxystyrene polymers by various photogenerators of Brönsted acid in acetonitrile.*

donor to generate acid from PTM and SIT is demonstrated in the 193-nm laser resist.[56] The acrylate-type polymers used in the resist are poor excited state donors, and formulations with PTM or SIT derivatives were markedly inefficient for the acid-catalyzed deprotection reaction compared with onium salts.

7. Conclusions

Although resist chemistry is usually thought of in a quite simplistic manner as a photoinduced change in hydrophilicity or hydrophobicity in a polymer film, it is apparent that the underlying mechanisms for a simple photoinitiated polarity change are very complex for the polymers that have been discussed. Even though the photochemistry of diazoketones is well understood, the mechanism for the change in dissolution properties of the Novolac/diazonaphthoquinone resist involves a number of complex processes that are still under investigation. Most of the work on the chemically amplified resist systems has been done on resist design and formulation, yet it is clear that the underlying chemistry is quite complex. From the mechanistic studies described in this chapter it is concluded that in addition to designing resists with optimized chemistry and mechanical properties, the photophysical and photochemical properties of both the polymer and photoinitiator ultimately determine the success of a photoresist formulation.

Acknowledgment

The American Chemical Society is gratefully acknowledged for permission to reproduce Figures 2, 10–12, 20–22.

References

1. A. Reiser, *Photoreactive Polymers—The Science and Technology of Resists* (J. Wiley and Sons, New York, 1989).
2. C. G. Willson, *Introduction to Microlithography*, ACS Symposium Series, No. 219, Eds. L. F. Thompson, C. G. Willson and M. J. Bowden, (eds. American Chemical Society, Washington, D.C., 1984), p. 87.
3. R. Dammel, *Diazonaphthoquinone-based Resists*, Tutorial Texts in Optical Engineering, Volume TT **11**, SPIE, Bellingham, Washington 1993.
4. O. Sus, *Annalen* **556**, 65 (1944).
5. R. M. Ponomareva, A. M. Komagorov, and N. A. Andronova, *J. Org. Chem., USSR* **16**, 140 (1980).
6. M. Yagihara, Y. Kitihara, and T. Asao, *Chem. Lett.* 1015 (1974).
7. N. P. Hacker, and N. J. Turro, *Tetrahedron Lett.* 1771 (1982).

8. G. N. Rodionova, Yu. G. Tuchin, N. P. Protsenko, and R. D. Erlikh, *Zh. Vses. Khim. Obshchest* **18**, 355 (1973). (*Chem. Abstr.* **79**, 151556t [1973]).

9. J. Pacansky, and D. Johnson, *J. Electrochem. Soc.* **124**, 862 (1977).

10. G. A. Bell, and I. R. Dunkin, *J. Chem. Soc., Faraday Trans.* **2**, 81, 725 (1985).

11. J. A. Delaire, J. Faure, F. Hassine-Renou, and M. Soreau, *New J. Chem.* **11**, 15 (1987).

12. (a) K. Tanigaki, and T. W. Ebbeson, *J. Am. Chem. Soc.* **109**, 5883 (1987); (b) K. Tanigaki, and T. W. Ebbeson, *J. Phys. Chem.* **93**, 4531 (1989).

13. H. Meier, and K.-P. Zeller, *Angew. Chem.* **14**, 32 (1975).

14. K.-P. Zeller, *Chem. Ber.* **108**, 3566 (1975).

15. A. M. Komagorov, R. M. Ponomareva, I. S. Isaev, and V. A. Koptyug, *Bull. Acad. Sci. USSR, Div. Chem. Sci.* **4**, 925 (1976).

16. F. A. Vollenbroek, C. M. J. Mutsaers, and W. P. M. Nijssen, *Polym. Mater. Sci. Eng.* **61**, 283 (1989).

17. M. Spak, D. Mammato, S. Jain, and D. Durham, Proc. 7th. Tech. Conf. Photopolymers, Ellenville, NY. 1985, 247 (see reference 3, page 24).

18. C. M. J. Mutsaers, F. A. Vollenbroek, W. P. M. Nijssen, and R. J. Visser, *Microelectron. Eng.* **11**, 497 (1990).

19. S. A. MacDonald, H. Ito, and C. G. Willson, *Microelectron. Eng.* **1**, 269 (1983).

20. G. R. Sprengling, *J. Am. Chem. Soc.* **76**, 1190 (1954).

21. M. Koshiba, M. Murata, M. Matsui, and Y. Harita, *Proc. SPIE, ''Advances in Resist Technology and Processing V''* **920**, 364 (1988).

22. M. Murata, M. Koshiba, and Y. Harita, *Proc. SPIE, ''Advances in Resist Technology and Processing VI''* **1086**, 48 (1989).

23. B. D. Grant, N. J. Clecak, R. J. Twieg, and C. G. Willson, *IEEE Trans. Electron Devices*, ED-**28**, 1300 (1981).

24. (a) Jones Jr., W. Ando, M. E. Hendrick, A. Kulczycki, Jr., P. M. Howley, K. I. Hummel, and D. S. Malamant, *J. Am. Chem. Soc.* **94**, 7469 (1972).
(b) S. L. Kammula, H. L. Tracer, P. B. Shevlin, and M. Jones, Jr. *Org Chem.* **42**, 2931 (1977).
(c) T. Livinghouse, and R. V. Stevens, *J. Am. Chem. Soc.* **100**, 6479 (1978).
(d) R. V. Stevens, G. S. Bisacchi, L. Goldsmith, and C. E. Strouse, *Org Chem.* **45**, 2708 (1980).

25. C. G. Willson, R. D. Miller, D. R. McKean, L. A. Pederson, and M. Regitz, *Proc. SPIE, ''Advances in Resist Technology and Processing IV''* **771**, 2 (1987).

26. G. Schwartzkopf, K. B. Gabriel, and J. B. Covington, *Proc. SPIE, ''Advances in Resist Technology and Processing VII''* **1262**, 456 (1990).

27. M. J. Hanrahan, K. S. Hollis, and M. Regitz, *Proc. SPIE, ''Advances in Resist Technology and Processing IV''* **771**, 128 (1987).

28. A. R. Gutierrez, and R. J. Cox, *Polym. Photochem.* **7**, 517 (1986).

29. (a) G. H. Smith, and J. A. Bonham, *U.S. Patent* 3,779,778 (1973), (equivalent to *Ger. Offen* 2,306,248. *Chem. Abstr.*, 1974, **80**, 114851k).
(b) G. H. Smith, *Fr. Demande* 2,270,269, 1975 (*Chem. Abstr.*, 1976, **85**, 22871s).

30. J. V. Crivello, and J. H. W. Lam, *J. Polym. Sci. Symposium* **56**, 383 (1976).

31. H. Ito, and C. G. Willson, in *Polymers in Electronics*, ACS Symposium series No. 242 T. Davidson, ed. (American Chemical Society, Washington, D.C. 1984), p. 11.

32. N. P. Hacker, *Radiation Curing in Polymer Science and Technology Volume II*, J. P. Fouassier, and J. F. Rabek, eds. (Elsevier, London, U.K., 1993, p 473).

33. J. W. Knapzyck, and W. E. McEwen, *J. Am. Chem. Soc.* **91**, 145 (1969).
(b) J. W. Knapzyck, and W. E. McEwen, *J. Org. Chem.* **35**, 2539 (1970).

34. (a) S. P. Pappas, *Prog. Org. Coat.* **13**, 35 (1985).
(b) S. P. Pappas *J. Imag. Tech.* **11**, 146 (1985).

35. J. V. Crivello, *Adv. Polym. Sci.* **62**, 1 (1984).

36. R. S. Davidson and J. W. Goodin, *Eur. Polym. J.* **18**, 589 (1982).

37. (a) J. L. Dektar and N. P. Hacker, *J. Chem. Soc., Chem. Commun.* 1591 (1987).
(b) J. L. Dektar and N. P. Hacker, *J. Am. Chem. Soc.* **112**, 6004 (1990).

38. J. L. Dektar, and N. P. Hacker, *J. Org. Chem.* **53**, 1833 (1988).

39. K. M. Welsh, J. L. Dektar, M. A. Garcia-Garibaya, N. P. Hacker, and N. J. Turro, *J. Org. Chem.* **57**, 4179 (1992).

40. R. J. Devoe, M. R. V. Sahyun, E. Schmidt, N. Serpone, and D. K. Sharma, *Can. J. Chem.* **66**, 319 (1988).

41. J. L. Dektar and N. P. Hacker, *J. Photochem. Photobiol., A. Chem.* **46**, 233 (1989).

42. (a) D. Rehm and A. Weller, *Ber. Bunsenges, Phys. Chem.* **73**, 834 (1969).
(b) D. Rehm and A. Weller, *Isr. J. Chem.* **8**, 259 (1970).

43. (a) J. V. Crivello, *UV Curing: Science and Technology*, S. P. Pappas, ed. (Technology Marketing Corporation, Stamford, 1978), p. 23.
(b) J. V. Crivello, *CHEMTECH* **10**, 624 (1980).
(c) J. V. Crivello, *Makromol. Chem., Macromol. Symp.* **13/14**, 145 (1988).

44. H. Ito and C. G. Willson, *Polym. Eng. Sci.* **23**, 1012 (1983).

45. (a) J. F. W. McOmie, *Protective Groups in Organic Chemistry* (Plenum, London, 1973).
(b) T. W. Greene, *Protective Groups in Organic Synthesis* (Wiley-Interscience, New York, 1981).

46. D. R. McKean, U. Schaedeli, and S. A. MacDonald, *J. Polym. Sci.*, Chem. Ed. **27**, 3927 (1989).

47. G. A. Delzenne, *Adv. Photochem.* **11**, 1 (1974).

48. S. A. M. Hesp, N. Hayashi, and T. Ueno, *J. Appl. Polym. Sci.* **42**, 877 (1991).

49. H. Ito, C. G. Willson, and J. M. J. Frechet, *Proc. SPIE, ''Advances in Resist Technology and Processing IV''* **771**, 24 (1987).

50. R. Schwalm, R. Bug, G-S. Dai, P. M. Fritz, M. Reinhardt, S. Schneider, and W. Schnabel, *J. Chem. Soc., Perkin I* 1803 (1991).

51. M. J. O'Brien, *Polym. Eng. Sci.* **29**, 846 (1989).

52. K. J. Stewart, M. Hatzakis, J. M. Shaw, D. E. Seeger, and E. Neumann, *J. Vac. Sci. Technol.* B7 1734 (1989).

53. R. Sooriyakumaran, H. Ito, and E. Mash, *Proc. SPIE, ''Advances in Resist Technology and Processing VIII''* **1466**, 419 (1991).

54. H. Stover, S. Matuszczak, R. Chin, K. Shimizu, C. G. Willson, and J. M. J. Frechet, *Polym. Mater. Sci. Eng.* **61**, 412 (1989).

55. J. M. J. Frechet, S. Matuszczak, B. Reck, and C. G. Willson, *Polym. Mater. Sci. Eng.* **60**, 147 (1989).

56. G. Wallraff, R. Allen, W. Hinsberg, C. Larson, R. Johnson, R. DiPietro, G. Breyta, N. Hacker, and R. R. Kunz, *J. Vac. Sci. Technol. B.*, **11**, 2783 (1993).

57. G. M. Wallraff, R. D. Allen, W. D. Hinsberg, L. L. Simpson, and R. R. Kunz, *CHEMTECH* **23**, 22 (1993).

58. (a) N. P. Hacker and K. M. Welsh, *Macromolecules* **24**, 2137 (1991).
(b) N. P. Hacker and K. M. Welsh, *Structure-Property Relations in Polymers: Spectroscopy and Performance*, ACS Advances in Chemistry Series No. 236, M. W. Urban, and C. D. Claver, eds. (American Chemical Society, Washington, D.C. 1993), p. 557.

59. N. P. Hacker, D. V. Leff, and J. L. Dektar, *Mol. Cryst. Liq. Cryst.* **183**, 505 (1990).

60. J. V. Crivello, *J. Electrochem. Soc.* **136**, 1453 (1989).

61. L. Schlegel, T. Ueno, H. Shirashi, N. Hayashi, and T. Iwayanagi, *Chem. Mater.* **2**, 299 (1990).

62. C. A. Renner, *U.S. Patent* 4,371,605 (1983), (*Chem. Abstr.*, 1983, **98**, 152865x).

63. N. P. Hacker, unpublished results.

64. P. H. Kasai, *J. Am. Chem. Soc.* **114**, 2875 (1992).

Index*

acidified methyl red (AMR)/PVA system, 352, 353
acrylamide monomer systems, 312, 314–15, 328–29
acrylamide-poly(vinyl alcohol) (AA-PVA) films, 329
acrylated prepolymers, 267–76
acrylate monomers
  alkoxy caked acrylate monomers, 38
  photopolymerizable dry films, 316–22
  photopolymerizable liquid compositions, 312, 314–15
  radiation curing chemistry, 47–49
acrylate resin, **42**
acryloyl groups, 161
acylphosphine oxides, 40–41
aligned PDLC network, **293**
aluminum complex-silanol systems, 155–58
ammonium borates, 147
amplitude hologram, 309
anisotropic network, 293–94
anisotropic photopolymerization
  basic principles of, 185–90
  fluorescent monomer in high optical density film, 190–200
  kinetic models of in films, 200–10
  modeling of and experiments on, 210–53
aryldiazonium salts, 148
aryliodonium salts, 150
asphaltene, 12–14, **15**
azobenzene LCPs, 294, 297, 298
azobenzene side chain polymers, 353–54
azo dyes, 297

bacteriorhodopsin doped polymers, 354–56
BAE-NCO, 268–69

*__Boldface__ numbers indicate illustrations.

benzoin ether photoinitiators
  acrylated prepolymers bound to, 267–76
  description of, 261–64
  oligomer photoinitiators bound to prepolymers, 264–66
benzophenone tetraperester (BTTB), 72–73, **74**, 123–29
bias hardness, holographic films, 346–47
biimidazolols, 108
biphotonic process, 144
bis(pentafluorophenyl)titanocene, 154
bitumen
  absorption spectrum of film, **17**
  history of photography, 7–10, 11–30
  principle of image formation with, **16**
  varnish and absorption of light, **19**, **20**
bromonaphthalene, 313
bulk monomers, 83–87

camera obscura, 4, 21–22
cationic photopolymerization, 162
cellulose acetate butyrate (CAB), 222, 227
chemical amplification, of sensitivity in photopolymers, 162
chloromethyl substituted triazine, 42–43
cholesteric LC polymer films, 287, 288
contact printing, 21
coumarin dyes, 123–29, **130**
croconium dyes, 299
cyanine dye borates, 146–48

Daguerre, Louis Jacques Mandé, 10, 21, 30
daguerreotype, invention of, 3
deep UV photoresists, 378–80
deep (volume) holograms, 307
degree of residual unsaturation (DRUS), 266
Denisyuk hologram, 309
diacrylates, **285**
diazoketones, 370–81

diazo-Meldrum's acid, 379, **380**
diazonaphthoquinones, 370–78
diazonium salts, 63–64, 148
dichromated gelatin (DCG), 345–49
dichromated poly(acrylic acid) (DCPAA), 339–40
dichromated poly(vinyl alcohol) (DCPVA), 332–39
diffraction efficiency, of holograms, 310
digital optical storage, 298
dimethyl formamide (DMF), 336, 340
diphenyliodonium salts, 150, 381
DMP-128 system, 319–21
doped polymer systems, 349, 351–59
dry films, photopolymerizable, 316–29
dual photoinitiation mechanism (DPM), 398
dye doped polymers, 349, 351–53, 356–57

electron transfer
  hexaarylbiimidazole and visible sensitizer, 170–83
  photoinduced bond cleavage via, 68–71
  photoresists and sensitization by photoinduced, 385–88
  visible light photoinitiation and, 123–42, **143**, 146–48
electro-optic shutters, 292
energy transfer
  laser spectroscopy of excited state processes, 67–68, **69**
  singlet oxygen sensitization and, 143–46
eosin/amine/ketone systems, 77–83
epoxy acrylate-bound BAE photoinitiator (EpA-BAE), 269, 271, 274–75, 276
epoxy-silicone monomers, 49–50
etching, history of photography, 10
excited state processes, in laser photosensitive systems, 64–83

ferric chloride dope poly(vinyl alcohol) (FePVA), 340–43
ferroelectric films, 288–91
fiber-optic coatings, 288
flexographic printing, 187
fluorescein (F)/orthoboric acid (OBA) system, 352
fluorescence intensity, anisotropic photopolymerization, 207–208
fluorescence photobleaching recovery kinetics, 191
fluorinated diaryltitanocene, 41–42
4-hydroxystyrene polymers, **399**
4-oxystyrene polymers, **396**
4-phenylthiobiphenyl (PTB), 382–84
FPK-488, 313

Gabor, Denis, 307, 308–309
Gaïac (tree), resin of, 6

Halobacterium halobium, 354
heliography, 3–4, 10–11, **27**, **31**, **32**
hexaarylbiimidazole (HABI) molecule
  diffusivity of as initiator, 203–204
  photodissociation and recombination of, 248, 250–53
  sensitivity of to oxygen, 205–207
hexaarylbiimidazolyl
  advances in chemistry and technology, 107–109
  chromotropism of with isomerization, 98–101
  discovery of, 90–91
  kinetic study of photochromism in solution, 94–98
  photochromic behavior of, 91–94
  photochromic color change of photodimers at low temperatures, 101–106
high optical density photopolymer film, 190–200
high resolution projection displays, 292
holographic recording
  applications of, 308
  development of, 307–308
  diffusion in polymer matrix and, 187, 189
  invention of, 307
  liquid and dry film photopolymerizable materials, 330–31
  principles of, 308–10
  recording materials, 310–59
  sensitivity of photopolymers, 162
hydrogen transfer, NPG/TXD system, 130–42, **143**

image resolution, swelling and, 236–39
immobile initiator, 243
induced transient grating relaxation methods, 191
in-line hologram, 309
Inokuti-Hirayama (IH) theory, 179, 183
in situ polymerization, 284–86

iodine, bitumen images on silver, 28–30
iodonium salts, 79, 148–49
Irgacure 369, 39, 45
Irgacure 784, 41–42
iron-arene salts, 64–65

JAW molecule, 170–83

ketocarbene, 373, 374
ketocoumarins, 62–63, 75–77
ketone
  bimolecular hydration of, **372**
  ketone/amine/bromocompound, 83
  sensitizers, 62
ketone/amine/bromocompound, 83
Kevlar, 282
kinetic analysis, 34, 188–89

lasers
  applications of in polymer photochemistry, 59–60
  excited state processes in visible light photosensitive systems, 64–83
  induction of polymerization reaction by, 60–64
lavender oil, 7, 13
Leith-Upatneiks hologram, 309
letterpress plate, 111
light attenuation, 200
limestone, bitumen images on, 22–23, **24**
line drawings, photoreproduction of, 7–8, 25–28
Lippman holography, 307, 309
liquid crystal polymers (LCPs)
  applications of, 278–79
  phase separation and PDLC, 291–94
  phases of, 279–82
  photoabsorption and optical storage, 298–302
  photoisomerization of, 294–98
  photopolymerization of, 282, 284–91
local light intensity, 207
lophine, 90
lophyl radicals. **See** triarylimidazolyl radicals
lucirin TPO (BASF), 40

macromolecular photoinitiators, 260–61
mercaptobenzoxazole, 204
methacrylate monomer system, 315–16
methacrylic acid (MA), 394
methylene blue dichromated gelatin (MBDCG) films, 347–48
methyl methacrylate (MMA) monomer, 322–25, 394
methyl orange (MO)/PVA, 352–53
methyl red (MR)/PMMA system, 352
microcircuits, electronic, 26, 31
microlithography, 31. **See also** photoresists
mobile initiator, 243, 246, 248
monoacrylates, **285**
monomers
  fluorescent in high optical density photopolymer film, 190–200
  in situ polymerization of oriented LC, 284–86
  radiation curing chemistry, 46–52
Monte Carlo analysis, 188
multicomponent monomer system, 327–28

Nicéphore Niépce, Joseph, 3–4, 4–31
NMR imaging, 217
nonlinear optical (NLO) applications, 288–90
novolak resin, 374–78
NPG/TXD system, 130–42, **143**

off-axis holographic technique, 307, 309
oligomer media, 83–87, 264–66
oilgourethaneacrylate (OUA), 313
omnidex photopolymer system, 318–19
onium salts
  eosin/amine/onium salt system, 77–83
  ketocoumarin/amine/onium salt system, 75–77
  photochemistry of, 149, 381–88
  resist chemistry using, 388–94
optical storage, 294, 297, 298–302
organic salts, cationic photopolymerization, 63
organoborates, 63, 65–66
organometallic cationic photoinitiators, 151
oriented polymer LC films, 286–91
oxygen
  photopolymerization kinetics in films, 218–27
  polymer spatial distribution, 227–33
  sensitivity of HABI to, 205–207
  singlet sensitization, 143–46

patterned exposure, of photopolymer films, 223–34
pattern photobleaching method, 191–94

p-benzoquinone (PBQ), 344–45
peresters, 65, 72–73
peroxide/electron donating dyes, 123–29
phase hologram, 309
photoacid generators, 148–55
photocrosslinking polymers, 322–49, **350**
photocurable formulations, 270, **272**
photoengraving, 10–11, 26, 28, 31
photography
  first experiments in, 4
  first negative, 4–5
  history of, 3–4, 5–31
photoinitiated cationic polymerization, 36
photoinitiators
  chemically amplified resists and interaction with polymer, 394–400
  curing speed and choice of, 59–60
  differential through cure, 43
  kinetic model of anisotropic photopolymerization, 203–204
  metal salts as, 63
  photolysis of intiator, 43, 45–46
  reactivity of systems in bulk monomer/oligomer media in air, 83–87
  UV curing chemistry, 38–46
photolithography, 10, 26
photopolymerizable systems, holographic, 311–16
photoreactive liquid crystals, examples of, **285**
photoresists
  development of, 368–70
  diazoketones, 370–81
  interaction between polymer and photoinitiators in chemically amplified, 394–400
  onium salts and, 381–94
photosensitive microgels, 158–62, **163**, 164
photosensitizers (PS), laser-induced polymerization reactions, 61–64
plasticized polymer matrix
  basic principles, 185–90
fluorescent monomer in high optical density film, 190–200
  kinetic models of in films, 200–10
  modeling and experiments in photopolymerization anisotropy, 210–53
p-nitrobenzyl-9,10-dimethoxyanthracene-2-sulfonate, 152
polarization holographic recording, 349, 351
polyacrylates, 282
polyester host systems, 325–27
poly(4-hydroxstyrene) (poly-HOST), 380
polymer dispersed liquid crystal (PDLC), 291–94
polymer induced phase separation (PIPS), 291–94
polymer matrix, diffusion of
  basic principles, 185–90
  fluorescent monomer in high optical density film, 190–200
  kinetic models of in films, 200–10
  modeling and experiments in photopolymerization anisotropy, 210–53
polymethacrylates, 282
poly (methyl methacrylate) (PMMA), 322–25, 344–45
poly-(N-vinyl carbazole) (PVCz) system, 343–44
poly(phthalaldehyde), 391
polysiloxanes, 282, 298, 300
polyurethane-acrylate coating, **44**
poly(vinyl acetate) (PVA), 222, 227
poly(vinyl butyrate) (PVB), 222
prehardening, holographic films, 346–47
prepolymers, oligomer photoinitiators bound to, 264–66
printed circuit boards (PCBs), 392, 394
printing engineering, 162, 187. **See also** photoresists
propoxy-substituted thioxanthones, 41
propylene carbonate, 51–52
PTMG-bound BAE photoinitiator (PTMG-BAE), 269
PTMG-urethane diacrylate (PTMG-UA), **268**, 271
pyrogallol tris-(methanesulfonate) (PTM), 398–400

Quantacure CPTX, 41
quinones, as polymerization initiators, 63

radiation curing
  applications of, 34–35, 52–54
  chemistry of, 36–38
  development of, 35
  UV curing chemistry, 38–52
Rapicure DVE-3, 50–51
Rapicure PEPC, 51–52
reagents, spatial distribution of in photopolymer films, 216–18

real-time holographic recording material, 339
real-time UV spectroscopy, 45–46
reflection holograms, 338

safety, of radiation curing technology, 38
self-focusing, monomer diffusion and, 239–43
sensitizer molecules (JAW), 170–83
side chain (LCPs) (SCLCPs), 282, **283**
silanol systems, 155–58
silicone-acrylates, 47–49
silver, positive half-tone images on, 28–30
silver halide photographic emulsion (SHPE), 307
silylether photoinitiation system, 158
singlet oxygen sensitization, 143–46
sol-gel system, 357, 359
spectral hole burning, 356–57
spin-echo technique, 216
spiropyran LCPs, 294, **296**
squarilium dyes, 299
stationary-state hypothesis, 206–207
stepwise crosslinking, 112–13, **114–15**
stereolithography, 38
stilbene isomerisation, **296**, 297
succinimidoyl triflates (SIT), 398–400
sulfonic esters, 152–53
sulfonium salts, 149
supertwisted nematic (STN) liquid crystal
    display, 287–88
swelling, in photopolymer films, 234–35,
    236–39, **245**
switchable windows, 292

t-butyl peroxide (tBP), 341–42
tert-butyl and **tert**-butoxycarbonyloxy (TBOC)
    groups, 389, **390**, 391
tert-butylmethacrylate (TBMA), 394
tetrabenzoporphyrins (TBP), 145–46
tetra-functional BAE photoinitiator bound to
    acrylated epoxy prepolymer (TGAM-A-
    BAE), 270, 271, **272**, 273, 274–75, 276
tetrahydropyranyl (THP), 388–89
thick (or volume) hologram, 309
thin (or plane) hologram, 309
thiopyrylium salt/perester, 72–73, **74**
thioxanthene dye, 73, 83, 123–29, **130**
3-substituted coumarin dyes, 123–29
time-dependent fluorescence experiments, 172
time-modulated illumination, 189
time-resolved fluorescence experiments, 172
tin plate, bitumen image on, 23–25, **26**
titanocene complexes, 42–43, 66, 154–55, **156**
tobacco mosaic viruses, 281
transient absorption experiments, 172–73
trialkylstannanes, 64
triarylimidazolyl radicals
  electron transfer between hexaaylbiimidazole
    and visible sensitizer, 170–83
  low mobility of, 204
  oxygen reactivity of, 205–207
  stability of, 248, 250–53
triarylalkyl borates, 146–48
triarylsulfonium salts, 148, 150
triethanolamine (TEA), 314
triethylene glycol divinyl ether, 50–51
trimethylolpropane triacrylate (TMPTA), 48
triphenylsulfonium salts, 149, 381–88
triplet sensitized photolysis, 384–85
tris(4-tolyl)sulfonium salt (TTS), 383
two-photon holographic recording, 314
TXD/NPG system, 130–42, **143**

ultraviolet (UV) curing, chemistry of, 38–52
ultraviolet (UV) light
  coatings cured by, 37
  photocurable formulations and, 270
ultraviolet (UV) photoinitiators, 39–41

vacuum chamber, 220
varnish, photosensitive, 18–21
vinyl ether, 50–52, **53**
visible light photoinitiation
  acid generation by, 148–55
  aluminum complex-silanol systems, 155–58
  development of, 111–12
  examples of, **117–22**
  methods of increasing sensitivity, 112–13,
    116, 162, 164
  photoradical generation by intermolecular
    processes, 123–46
  photosensitive microgels, 158–62
  radiation curing chemistry, 41–43
  radical generation by intramolecular electron
    transfer, 146–48
visible sensitizer, electron transfer between
    hexaaylbiimidazole and, 170–83

Wedgwood, Thomas, 4
write/read/erase (WRE) cycles, 349

xerographic imaging, 298